Lecture Notes in Computer Science   13896

Founding Editors

Gerhard Goos
Juris Hartmanis

The series Lecture Notes in Computer Science (LNCS), including its subseries Lecture Notes in Artificial Intelligence (LNAI) and Lecture Notes in Bioinformatics (LNBI), has established itself as a medium for the publication of new developments in computer science and information technology research, teaching, and education.

LNCS enjoys close cooperation with the computer science R & D community, the series counts many renowned academics among its volume editors and paper authors, and collaborates with prestigious societies. Its mission is to serve this international community by providing an invaluable service, mainly focused on the publication of conference and workshop proceedings and postproceedings. LNCS commenced publication in 1973.

Roland Glück · Luigi Santocanale ·
Michael Winter

Editors

Relational and Algebraic Methods in Computer Science

20th International Conference, RAMiCS 2023
Augsburg, Germany, April 3–6, 2023
Proceedings

 Springer

Editors
Roland Glück (ID)
Deutsches Zentrum für Luft- und Raumfahrt
Augsburg, Germany

Luigi Santocanale (ID)
Aix-Marseille University
Marseille, France

Michael Winter
Brock University
St Catharines, ON, Canada

ISSN 0302-9743   ISSN 1611-3349 (electronic)
Lecture Notes in Computer Science
ISBN 978-3-031-28082-5   ISBN 978-3-031-28083-2 (eBook)
https://doi.org/10.1007/978-3-031-28083-2

This Springer imprint is published by the registered company Springer Nature Switzerland AG
The registered company address is: Gewerbestrasse 11, 6330 Cham, Switzerland

Preface

This volume contains the proceedings of the 20th International Conference on Relational and Algebraic Methods in Computer Science (RAMiCS 2023), which was held at the Technologiezentrum Augsburg in Augsburg, Germany, on April 3–6, 2023.

The RAMiCS conferences series aims to bring together a community of researchers to advance the development and dissemination of relation algebras, Kleene algebras, and similar algebraic formalisms. Topics covered range from mathematical foundations to applications as conceptual and methodological tools in computer science and beyond. More than 30 years after its foundation in 1991 in Warsaw, Poland, RAMiCS, initially named "Relational Methods in Computer Science", remains a main venue in this field. The series merged with the workshops on Applications of Kleene Algebra in 2003 and adopted its current name in 2009. Previous events were organized in Dagstuhl, Germany (1994), Paraty, Brazil (1995), Hammamet, Tunisia (1997), Warsaw, Poland (1998), Québec, Canada (2000), Oisterwijk, The Netherlands (2001), Malente, Germany (2003), St. Catharines, Canada (2005), Manchester, UK (2006), Frauenwörth, Germany (2008), Doha, Qatar (2009), Rotterdam, The Netherlands (2011), Cambridge, UK (2012), Marienstatt, Germany (2014), Braga, Portugal (2015), Lyon, France (2017), Groningen, The Netherlands (2018), Palaiseau, France (2020, online), and Marseille, France (2021).

RAMiCS 2023 attracted 26 submissions, of which 17 were selected for presentation by the Program Committee. Each submission was evaluated according to high academic standards by at least three independent reviewers, and scrutinized further during two weeks of intense electronic discussion. The organizers are very grateful to all Program Committee members for this hard work, including the lively and constructive debates, and to the external reviewers for their generous help and expert judgments. Without this dedication we could not have assembled such a high-quality program; we hope that all authors benefitted from these efforts.

Apart from the submitted articles, this volume features the abstracts of the presentations of the three invited speakers, Alexander Knapp, John Stell, and Valeria Vignudelli. We are delighted that all three invited speakers accepted our invitation to present their work at the conference.

Last, but not least, we would like to thank the members of the RAMiCS Steering Committee for their support and advice. We gratefully acknowledge financial and administrative support by the Zentrum für Leichtbauproduktions technologie of the Deutsches Zentrum für Luft- und Raumfahrt, the Deutsche Forschungsgemeinschaft and the Technologiezentrum Augsburg.

We also appreciate the excellent facilities offered by the EasyChair conference administration system, and Springer's help in publishing this volume. Finally, we are indebted to all authors and participants for supporting this conference.

January 2023

Roland Glück
Luigi Santocanale
Michael Winter

Organization

Program Committee

Adriana Balan	University Politehnica of Bucharest, Romania
Manuel Bodirsky	TU Dresden, Germany
Miguel Couceiro	University of Lorraine, CNRS, LORIA, France
Manfred Droste	Leipzig University, Germany
Uli Fahrenberg	LRE & EPITA Rennes, France
Hitoshi Furusawa	Kagoshima University, Japan
Wesley Fussner	University of Bern, Switzerland
Silvio Ghilardi	Università degli Studi di Milano, Italy
Roland Glück (Co-chair)	German Aerospace Center, Germany
Sam van Gool	Université Paris Cité, France
Walter Guttmann	University of Canterbury, New Zealand
Robin Hirsch	University College London, UK
Peter Höfner	Australian National University, Australia
Marcel Jackson	La Trobe University, Australia
Ali Jaoua	Oryx Universal College, Qatar (with LJMU, UK)
Peter Jipsen	Chapman University, USA
Sebastiaan Joosten	University of Minnesota, USA
Barbara König	University of Duisburg-Essen, Germany
Wendy MacCaull	St. Francis Xavier University, Canada
Roger Maddux	Iowa State University, USA
Nelma Moreira	University of Porto, Portugal
Martin Mueller	University of Applied Science Bonn-Rhein-Sieg, Germany
Damien Pous	CNRS - ENS Lyon, France
Luigi Santocanale (Co-chair)	LIS, Aix-Marseille University, France
Ana Sokolova	University of Salzburg, Austria
Sara Ugolini	Spanish National Research Council CSIC, Spain
Michael Winter (Co-chair)	Brock University, Canada

Additional Reviewers

Bonzio, Stefano
Calk, Cameron
Cirstea, Corina
de Groot, Jim
de Lacroix, Cédric
Gaubert, Stephane
Kahl, Wolfram
Kappé, Tobias
Kniazev, Roman
Knäuer, Simon

Marcos, Miguel Andrés
Marx, Maarten
Nishizawa, Koki
Santschi, Simon
Semanišinová, Žaneta
Starke, Florian
Tiwari, Ashish
Tsumagari, Norihiro
Youssef, Youssef Mahmoud

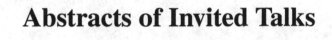
Abstracts of Invited Talks

Specifying Event/Data-based Systems

Alexander Knapp[1]

Universität Augsburg, Germany
`knapp@informatik.uni-augsburg.de`

Abstract. Event/data-based systems are controlled by events, their local data state may change in reaction to events. Numerous methods and notations for specifying such reactive systems have been designed, though with varying focus on the different development steps and their refinement relations. We first briefly review some of such methods, like temporal/modal logic, TLA, UML state machines, symbolic transition systems, CSP, synchronous languages, and Event-B with their support for parallel composition and refinement. We then present $\mathcal{E}^{\downarrow}$-logic for covering a broad range of abstraction levels of event/data-based systems from abstract requirements to constructive specifications in a uniform foundation. $\mathcal{E}^{\downarrow}$-logic uses diamond and box modalities over structured events adopted from dynamic logic, for recursive process specifications it offers (control) state variables and binders from hybrid logic. The semantic interpretation relies on event/data transition systems; specification refinement is defined by model class inclusion. Constructive operational specifications given by state transition graphs can be characterised by a single $\mathcal{E}^{\downarrow}$-sentence. Also a variety of implementation constructors is available in $\mathcal{E}^{\downarrow}$-logic to support, among others, event refinement and parallel composition. Thus the whole development process can rely on $\mathcal{E}^{\downarrow}$-logic and its semantics as a common basis.

---

[1] Joint work with Rolf Hennicker, LMU München, Germany, and Alexandre Madeira, CIDMA, U. Aveiro, Portugal

Algebra and Logic in Granularity

John G. Stell

University of Leeds, Leeds LS2 9JT, U.K
j.g.stell@leeds.ac.uk

Conceptually, a relation on a set can be understood as a lens through which subsets are seen in different ways. For example, at different levels of detail. This idea underlies some basic operations in mathematical morphology as used in image processing. In a more abstract setting, the idea also fits one view of quantale modules [1], where a quantale acting on a complete lattice models the operation of observing elements of the lattice by means of the elements of quantale. This talk will consider the role of relations, and their generalizations, in the development of a theory of granularity, or level of detail. Starting from basic mathematical morphology, algebraic and logical structures motivated by extensions of this will be discussed.

To explain the starting point, mathematical morphology is a body of techniques used in image processing since the 1960s. The algebraic basis of these techniques is already well-known through the work of Serra, Heijmans [3], Ronse, and others. A black and white image can be seen as a subset of a set of pixels. Operations mapping subsets to subsets provide ways of modifying images. Such operations arise from relations on the underlying set. Algebraically, a binary relation induces an adjunction yielding the dilation (left, or lower, adjoint) and erosion (right, or upper, adjoint) operations used in image processing. From these are derived openings and closings. Families of openings and closings are used to build "alternating sequential filters", "granulometries" and further practically useful operations. Mathematical morphology was extended from sets of pixels to graphs and subgraphs [2]. The appropriate notion of relation to deal with this was identified [5], and these can be seen as quantale-enriched distributors. A quite separate application of quantales to mathematical morphology was developed by Russo [4]. The talk will discuss connections between these accounts.

References

1. Abramsky, S., Vickers, S.: Quantales, observational logic and process semantics. Math. Struct. Comput. Sci. **3**, 161–227 (1993)
2. Cousty, J., Najman L., Dias, F., Serra, J.: Morphological filtering on graphs. Comput. Vision Image Underst. **117**, 370–385 (2013)
3. Heijmans, Henk J. A. M.: Morphological Image Operators. Academic Press Inc., San Diego

4. Russo, C.: Quantale Modules and their Operators, with Applications. J. Logic Comput. **20**(4), 917–946 (2008)
5. Stell, J. G.: Relations on Hypergraphs. In: Kahl, W., Griffin, T. G. (eds) Relational and Algebraic Methods in Computer Science. RAMICS 2012. Lecture Notes in Computer Science, vol 7560. Springer, Heidelberg. (2012). https://doi.org/10.1007/978-3-642-33314-9_22

Equational Theories and Distances for Computational Effects

Valeria Vignudelli

CNRS/ENS Lyon, France

Computational effects such as nondeterminism and probabilities can be abstractly modelled in category theory as monads, and can be syntactically described in universal algebra via equational theories.

Equational theories allow us to reason on the equivalence of programs with computational effects. In recent years, much work has been devoted to the development of techniques to reason not only on program equivalence, but also on program distances. The correspondence between monads and equational theories has been extended to capture such notions of distances, via the framework of quantitative equational theories.

In this talk, I will present such correspondence between monads and equational theories, as well as recent results applying the framework of quantitative equational theories to computational effects and distances on them. I will show several examples of axiomatizations, including more involved cases such as the combination of nondeterminism, probabilities and termination.

Contents

Amalgamation Property for Some Varieties of BL-Algebras Generated by One Finite Set of BL-Chains with Finitely Many Components

Stefano Aguzzoli[1] and Matteo Bianchi[2]([⊠])

[1] Department of Computer Science, Università degli Studi di Milano,
Via Celoria 18, 20133 Milan, Italy
aguzzoli@di.unimi.it
[2] Milan, Italy
matteob@gmail.com

Abstract. BL-algebras are the algebraic semantics of Basic logic BL, the logic of all continuous t-norms and their residua. In a previous work, we provided the classification of the amalgamation property (AP) for the varieties of BL-algebras generated by one BL-chain with finitely many components. As an open problem, we left the analysis of the AP for varieties of BL-algebras generated by one finite set of BL-chains with finitely many components. In this paper we provide a partial solution to this problem. We provide a classification of the AP for the varieties of BL-algebras generated by one finite set of BL-chains with finitely many components, which are either cancellative hoops or finite Wajsberg hoops. We also discuss the difficulties to generalize this approach to the more general case.

Keywords: BL-algebras · Hoops · Amalgamation property · Ordinal sums · Lattices of varieties

1 Introduction

Basic Logic BL was introduced by Petr Hájek in 1998 [15], and in [10] it was shown that BL is the logic of all continuous t-norms and their residua. BL is algebraizable in the sense of Blok and Pigozzi [8], and its equivalent algebraic semantics is given by a variety of residuated lattices, called BL-algebras. The lattice of varieties of BL-algebras $\mathcal{L}(\mathbb{BL})$ is mostly unknown, and a lot of research has been devoted to study its structure, and the properties of subvarieties of BL-algebras. Among them, the amalgamation property (AP) is an important one, since it is well known that a variety of BL-algebras has the AP if and only if the corresponding logic has the deductive interpolation property. The AP for

M. Bianchi—Independent Researcher.

BL and some of its subvarieties was shown in [17], and in [12] the analysis was further extended. In [16] the case of GBL-algebras (a far-reaching generalization of BL-algebras) was also tackled, by providing a partial classification. However, in contrast to the case of MV-algebras, the analysis of the AP for the varieties of BL-algebras is far from being complete. In our previous work [5] we provided a full classification of the AP for varieties generated by one BL-chain with finitely many components. As an open problem, we left the classification of the AP for all the varieties generated by one finite set of BL-chains with finitely many components. In this paper we provide a partial answer, by using some new results on the AP presented in [14] namely, the connection with a generally weaker form, called one-sided amalgamation property, 1AP. Our main result is the classification of the AP for all the varieties generated by one finite set of BL-chains with finitely many components which are either cancellative hoops or finite Wajsberg hoops. As a corollary, we obtain the classification of the AP for all the varieties of BL-algebras which are generated by one finite set of finite BL-chains. In the last section we discuss open problems.

2 Preliminaries

2.1 BL-Algebras

Definition 1 ([15]). *A BL-algebra is an algebra* $(A, *, \Rightarrow, \wedge, \vee, 0, 1)$ *such that:*

i) $(A, \wedge, \vee, 0, 1)$ *is a bounded lattice with minimum* 0 *and maximum* 1.
ii) $(A, *, 1)$ *is a commutative monoid.*
iii) $(*, \Rightarrow)$ *forms a* residuated pair*:* $z * x \leq y$ *iff* $z \leq x \Rightarrow y$ *for all* $x, y, z \in A$.
iv) *The following identities hold, for all* $x, y \in A$*:*

$$(x \Rightarrow y) \vee (y \Rightarrow x) = 1. \quad \text{(Prelinearity)}$$
$$x \wedge y = x * (x \Rightarrow y). \quad \text{(Divisibility)}$$

A totally ordered BL-algebra is called a BL-chain.

Every algebra $([0, 1], *, \Rightarrow, \min, \max, 0, 1)$, where $*$ is a continuous t-norm, and $\Rightarrow$ is its residuum, is a BL-algebra [11], called a standard BL-algebra. Two well-known examples are the standard MV-algebra $[0, 1]_{\text{Ł}}$ and the standard Gödel-algebra $[0, 1]_{\text{G}}$. In $[0, 1]_{\text{Ł}}$ we have $x * y = \max\{0, x + y - 1\}$, and $x \Rightarrow y = \min\{1, 1 - x + y\}$. In $[0, 1]_{\text{G}}$ it holds that $x * y = \min\{x, y\}$, whilst $x \Rightarrow y = 1$ if $x \leq y$, and $x \Rightarrow y = y$ if $x > y$. We define $\neg x \overset{\text{def}}{=} x \Rightarrow 0$.

2.2 BL-Algebras and Ordinal Sums

Every BL-chain can be decomposed as an *ordinal sum* of hoops. Before stating the result, we need some preparation.

Definition 2 ([13]). *A hoop is an algebra* $\mathcal{A} = (A, *, \Rightarrow, 1)$ *of type* $(2, 2, 0)$ *such that:*

(i) $(A, *, 1)$ *is a commutative monoid,*

(ii) $\Rightarrow$ *is a binary operation satisfying the following properties:*

- $x \Rightarrow x = 1$,
- $x * (x \Rightarrow y) = y * (y \Rightarrow x)$,
- $x \Rightarrow (y \Rightarrow z) = (x * y) \Rightarrow z$.

A bounded hoop is an algebra $\mathcal{A} = (A, *, \Rightarrow, 0, 1)$ such that $(A, *, \Rightarrow, 1)$ is a hoop, and $0 \leq x$ for all $x \in A$. The binary relation $\leq$ on $\mathcal{A}$ is defined as $x \leq y$ if and only if $x \Rightarrow y = 1$. It follows from the hoop axioms that this binary relation is indeed a partial order. An unbounded hoop is a hoop without minimum.

A *Wajsberg hoop* is a hoop $\mathcal{A}$ satisfying

$$(x \Rightarrow y) \Rightarrow y = (y \Rightarrow x) \Rightarrow x.$$

A *cancellative hoop* is a hoop satisfying

$$x \Rightarrow (x * y) = y.$$

It is well known that bounded Wajsberg hoops are term-equivalent to MV-algebras (see [1], and [9] for MV-algebras). We also recall that the variety of Wajsberg hoops $\mathbb{WH}$ contains all cancellative hoops. In particular, the class of totally ordered cancellative hoops coincides with the class of totally ordered unbounded Wajsberg hoops. The class of all cancellative hoops forms a variety, called $\mathbb{CH}$. Of course $\mathbb{CH} \subsetneq \mathbb{WH}$. BL-chains can be obtained by means of the *ordinal sum construction.*

Definition 3. *Let* $(I, \leq)$ *be a totally ordered set with minimum* 0. *For all* $i \in I$, *let* $\mathcal{A}_i = (A_i, *_i, \Rightarrow_i, 1)$ *be a hoop such that for* $i \neq j$, $A_i \cap A_j = \{1\}$. *Then* $\bigoplus_{i \in I} \mathcal{A}_i$ *is called the* ordinal sum *of the family* $\{\mathcal{A}_i\}_{i \in I}$, *whose universe is given by* $\bigcup_{i \in I} A_i$, *and whose operations[1]* $\Rightarrow, *$ *are given by:*

$$x \Rightarrow y \stackrel{\text{def}}{=} \begin{cases} x \Rightarrow_i y & \text{if } x, y \in A_i, \\ y & \text{if } j < i, \ x \in A_i, \ y \in A_j, \\ 1 & \text{if } i < j, \ 1 \neq x \in A_i, \ y \in A_j. \end{cases}$$

$$x * y \stackrel{\text{def}}{=} \begin{cases} x *_i y & \text{if } x, y \in A_i, \\ y & \text{if } j < i, \ x \in A_i, \ 1 \neq y \in A_j, \\ x & \text{if } i < j, \ 1 \neq x \in A_i, \ y \in A_j. \end{cases}$$

The hoops $\mathcal{A}_i$ *are called* components. *When* I *is finite, for example* $I = \{0, \ldots, k\}$, *we sometimes use the notation* $\mathcal{A}_0 \oplus \cdots \oplus \mathcal{A}_k$, *in place of* $\bigoplus_{i \in I} \mathcal{A}_i$.

As shown in [2] every BL-chain is canonically representable as an ordinal sum of hoops.

---

[1] The relation $x \leq y$ iff $x \Rightarrow y = 1$ equips $\bigoplus_{i \in I} \mathcal{A}_i$ with a total order. For every $x \in A_i, y \in A_j$, $x \leq y$ iff $x < 1$ and $i < j$ or $i = j$ and $x \leq_i y$.

Theorem 1 ([2]). *For every BL-chain $\mathcal{A}$ there are a unique (up to order-isomorphisms) totally ordered set $(I, \leq)$ with minimum 0 and a unique (up to isomorphisms) family $\{\mathcal{A}_i \mid i \in I\}$ of non-trivial totally ordered Wajsberg hoops whose first component $\mathcal{A}_0$ is bounded, such that $\mathcal{A} \cong \bigoplus_{i \in I} \mathcal{A}_i$.*

Observe that the idempotent elements in any ordinal sum of Wajsberg hoops are exactly 1 and the bottoms of every component with minimum. Let $\mathcal{A}$ be a BL-chain. With $\#\mathcal{A}$ we denote the number of the components of $\mathcal{A}$, i.e., $\#\mathcal{A} = |I|$, in the decomposition of $\mathcal{A}$ described in Theorem 1. With $(\mathcal{A})_i$ we denote the i-th component of $\mathcal{A}$. Clearly $(\mathcal{A})_0$ is an MV-chain, and $(\mathcal{A})_0 \hookrightarrow \mathcal{A}$.

Remark 1. – By slight abuse of terminology we shall often consider ordinal sums $\bigoplus_{i \in I} \mathcal{A}_i$, where there are some $\mathcal{A}_i$ (with $i \neq \min I$) being MV-chains, with the obvious meaning that we are actually considering the 0-free reduct of each such $\mathcal{A}_i$.

– By slight abuse of notation we shall sometimes consider ordinal sums $\bigoplus_{i \in I} \mathcal{A}_i$ where two or more components have elements in common distinct from 1 (for example, $\mathcal{A}_i = \mathcal{A}_j$ for some $i \neq j \in I$). In such cases we tacitly mean to consider an ordinal sum $\bigoplus_{i \in I} \mathcal{B}_i$, with $\mathcal{B}_i \simeq \mathcal{A}_i$ for every $i \in I$ and $\mathcal{B}_i \cap \mathcal{B}_j = \{1\}$ for $i \neq j$.

– Unless stated otherwise, from now on we assume that all the ordinal sums of Wajsberg hoops that we consider have non-trivial components.

With **2** we denote the two element boolean algebra. The hoop $\mathcal{C}_\infty = (C_\infty, *, \Rightarrow, 1)$ is defined as follows, for $x, y \in C_\infty = \{x \in \mathbb{Z} : x \leq 0\}$:

- $1^{\mathcal{C}_\infty} = 0$,
- $x *^{\mathcal{C}_\infty} y = x + y$,
- $x \Rightarrow^{\mathcal{C}_\infty} y = \begin{cases} 0 & \text{if } x \leq^{\mathbb{Z}} y, \\ y - x & \text{otherwise.} \end{cases}$

For $k \geq 2$, we define $\mathcal{L}_k$ as the subalgebra of $[0, 1]_{\text{L}}$ with carrier $\{0, \frac{1}{k-1}, \ldots, \frac{k-1}{k-1}\}$. For $k \geq 1$ we define $\mathcal{P}_k$ as $\mathbf{2} \oplus \underbrace{\mathcal{C}_\infty \oplus \cdots \oplus \mathcal{C}_\infty}_{k \text{ times}}$. For $k \geq 1$, $\mathcal{P}_k$ generates a variety, called $\mathbb{P}_k$, where $\mathbb{P}_1$ is the variety of product algebras. We refer the reader to [4] for further details. For $k \geq 2$ we define $\mathcal{G}_k$ as $\underbrace{\mathbf{2} \oplus \cdots \oplus \mathbf{2}}_{k-1 \text{ times}}$. $\mathcal{G}_k$ is a Gödel-chain (G-chain, for short) with k elements: see [11] for further details. With $\mathbb{G}_k$ we denote $\mathbf{V}(\mathcal{G}_k)$. With $\mathbb{B} = \mathbb{G}_2$ we denote the variety of boolean algebras.

The radical of a totally ordered Wajsberg hoop (MV-chain) $\mathcal{A}$, is the intersection of all the maximal filters of $\mathcal{A}$, and will be denoted by $Rad(\mathcal{A})$. Let $\mathcal{A}$ be an MV-chain. We say that $\mathcal{A}$ has a finite rank if $\mathcal{A}/Rad(\mathcal{A}) \simeq \mathcal{L}_k$, for some k (in this case $rank(\mathcal{A}) = k$), whilst $\mathcal{A}$ has infinite rank if $\mathcal{A}/Rad(\mathcal{A})$ is an infinite simple MV-chain. We can define *mutatis mutandis* the same notion for a totally ordered Wasjberg hoop[2]: the only difference is that $\mathcal{A}/Rad(\mathcal{A})$ would be

---

[2] Note that every non-trivial totally ordered cancellative hoop $\mathcal{A}$ does not have rank, since $\mathcal{A}/Rad(\mathcal{A})$ is an infinite cancellative hoop.

the 0-free reduct of a simple MV-chain (finite or infinite). Let $\mathcal{A}$ be a BL-chain or a totally ordered Wajsberg hoop. With $Ch(\mathcal{A})$ we denote the class of all the non-trivial chains of $\mathbf{V}(\mathcal{A})$.

Lemma 1 ([2])

- Let $\bigoplus_{i \in I} \mathcal{A}_i$ be a non-trivial BL-chain. Then $\mathbf{ISP}_u(\bigoplus_{i \in I} \mathcal{A}_i) = \mathbf{I}(\bigoplus_{i \in I} \mathbf{SP}_u(\mathcal{A}_i))$, where $\bigoplus_{i \in I} \mathbf{SP}_u(\mathcal{A}_i) = \{\bigoplus_{i \in I} \mathcal{B}_i : \mathcal{B}_i \in \mathbf{SP}_u(\mathcal{A}_i)\}$.
- If $\mathcal{A}$ is an infinite totally ordered cancellative hoop, then $\mathbf{ISP}_u(\mathcal{A}) = Ch(\mathbb{CH})$.
- If $\mathcal{A}$ is a totally ordered Wajsberg hoop with infinite rank, and for every $n \geq 2$, $\mathcal{L}_n \hookrightarrow \mathcal{A}$, then $\mathbf{ISP}_u(\mathcal{A}) = Ch(\mathcal{A})$.
- If $\mathcal{A}$ is a totally ordered Wajsberg hoop with $rank(\mathcal{A}) = n$, and $\mathcal{L}_n \hookrightarrow \mathcal{A}$, then $\mathbf{ISP}_u(\mathcal{A}) = Ch(\mathcal{A})$. If in addition $\mathcal{A}$ is also finite, then[3] $\mathbf{ISP}_u(\mathcal{A}) = \mathbf{IS}(\mathcal{A}) = Ch(\mathcal{A})$.

Lemma 2. Let S be a finite set of BL-chains such that, for every $\mathcal{A} \in S$.

- $\mathcal{A}$ has finitely many components.
- Each $(\mathcal{A})_i$ is either cancellative or it is a Wajsberg hoop with finite rank such that $(\mathcal{A})_i / Rad((\mathcal{A})_i) \hookrightarrow (\mathcal{A})_i$.

Let $\mathbb{L} = \mathbf{V}(S)$. Then the following hold.

1. $Ch(\mathbb{L}) = \mathbf{ISP_u}(S) = \bigcup_{\mathcal{T} \in S} \mathbf{ISP_u}(\mathcal{T})$.
2. In particular, for every $\mathcal{A} = \bigoplus_{i=0}^{k} \mathcal{A}_i \in S$ such that $(\mathcal{A})_0$ is finite, $Ch(\mathcal{A}) = \mathbf{I}(\mathbf{S}(\mathcal{A}_0) \oplus \bigoplus_{i=1}^{k} \mathbf{SP_u}(\mathcal{A}_i))$.
3. If every $\mathcal{A} \in S$ is finite, then $Ch(\mathbb{L}) = \mathbf{IS}(S) = \bigcup_{\mathcal{T} \in S} \mathbf{IS}(\mathcal{T})$.

Proof. Let S be a finite set of BL-chains with finitely many components which are cancellative or finite, and let $\mathbb{L} = \mathbf{V}(S)$.

1. The first equality follows by [2, Theorem 7.6], and the first part of its proof. The second equality follows by [6, Lemma 2].
2. This follows by Lemma 1, and [6, Lemma 2].
3. This follows by 1, and Lemma 1. □

We recall that a BL-algebra $\mathcal{A}$ is *subdirectly irreducible* (SI) whenever the trivial congruence Δ is strictly meet irreducible in the lattice of congruences $Con(\mathcal{A})$. We say that $\mathcal{A}$ is *finitely subdirectly irreducible* (FSI) whenever the trivial congruence Δ is meet irreducible in the lattice of congruences $Con(\mathcal{A})$. Subdirectly irreducible and finitely subdirectly irreducible BL-algebras are related to BL-chains, as the following theorem show.

Theorem 2 ([11])

- Every subdirectly irreducible BL-algebra is totally ordered.

---

[3] The assumption that $Ch(\mathcal{A})$ does not contain trivial chains is essential. Indeed, if $\mathcal{A}$ is non-trivial, then $\mathbf{ISP}_u(\mathcal{A})$ does not contain trivial algebras.

- *A BL-algebra is finitely subdirectly irreducible if and only if it is totally ordered.*

We conclude the section with some observations on the homomorphisms among BL-algebras.

Lemma 3. *Let $\mathcal{A}, \mathcal{B}$ be MV-chains, where $\mathcal{A}$ is simple, and $\mathcal{B}$ is non-trivial. If there is a homomorphism k from $\mathcal{A}$ to $\mathcal{B}$, then $\mathcal{A} \overset{k}{\hookrightarrow} \mathcal{B}$.*

Proof. Let $\mathcal{A}, \mathcal{B}$ as above, and suppose that there is a homomorphism k between $\mathcal{A}$ and $\mathcal{B}$. Then $ker(k)$ is a congruence on $\mathcal{A}$, which is a simple MV-chain. This means that $ker(k) \in \{\Delta, \nabla\}$. If $ker(k) = \nabla$, this implies that $k(a) = 1^{\mathcal{B}}$, for every $a \in A$. Therefore we have that $k(0^{\mathcal{A}}) = 1^{\mathcal{B}}$. However this is impossible, as by the definition of homomorphism $k(0^{\mathcal{A}}) = 0^{\mathcal{B}}$, and since $\mathcal{B}$ is non-trivial we must have $0^{\mathcal{B}} \neq 1^{\mathcal{B}}$. Therefore $ker(k) = \Delta$, and this implies $\mathcal{A} \overset{k}{\hookrightarrow} \mathcal{B}$. The proof is settled.   □

Note that Lemma 3 does not hold for Wajsberg hoops, due to the lack of the constant 0 in the signature.

Lemma 4. *Let $\mathcal{A}, \mathcal{B}$ be two non-trivial BL-chains, where $(\mathcal{A})_0$ simple. If there is a homomorphism k from $\mathcal{A}$ to $\mathcal{B}$, then $(\mathcal{A})_0 \xrightarrow{k \restriction (A_0)} (\mathcal{B})_0$ and $(\mathcal{A})_0 \overset{k}{\hookrightarrow} \mathcal{B}$.*

Proof. Immediate by Lemma 3.   □

3   Amalgamation Property for Varieties Generated by a Finite Set of BL-Chains with Finitely Many Components

We start with some definitions.

Definition 4. *Let $\mathbb{L}$ be a class of algebras. A V-formation is a tuple $(\mathcal{A}, \mathcal{B}, \mathcal{C}, i, j)$ such that $\mathcal{A}, \mathcal{B}, \mathcal{C} \in \mathbb{L}$, $\mathcal{A} \overset{i}{\hookrightarrow} \mathcal{B}$, and $\mathcal{A} \overset{j}{\hookrightarrow} \mathcal{C}$.*

- *We say that $\mathbb{L}$ has the* one-sided amalgamation property *(1AP), whenever for every V-formation $(\mathcal{A}, \mathcal{B}, \mathcal{C}, i, j)$ there is a tuple $(\mathcal{D}, h, k)$, called 1-amalgam, such that $\mathcal{D} \in \mathbb{L}$, $\mathcal{B} \overset{h}{\hookrightarrow} \mathcal{D}$, k is a homomorphism from $\mathcal{C}$ to $\mathcal{D}$, and $h \circ i = k \circ j$.*
- *We say that $\mathbb{L}$ has the* amalgamation property *(AP), whenever for every V-formation $(\mathcal{A}, \mathcal{B}, \mathcal{C}, i, j)$ there is a tuple $(\mathcal{D}, h, k)$, called an amalgam, such that $\mathcal{D} \in \mathbb{L}$, $\mathcal{B} \overset{h}{\hookrightarrow} \mathcal{D}$, $\mathcal{C} \overset{k}{\hookrightarrow} \mathcal{D}$, and $h \circ i = k \circ j$.*

Clearly the AP implies the 1AP. Interestingly, also the converse holds, if we assume that the class of algebras $\mathbb{L}$ satisfies some properties. By $\mathbb{L}_{\mathrm{FSI}}$ we denote the class of FSI algebras of $\mathbb{L}$.

Theorem 3 ([14]). *Let $\mathbb{L}$ be a variety with the congruence extension property such that $\mathbb{L}_{FSI}$ is closed under subalgebras. The following are equivalent:*

- $\mathbb{L}$ *has the amalgamation property.*
- $\mathbb{L}$ *has the one-sided amalgamation property.*
- $\mathbb{L}_{FSI}$ *has the one-sided amalgamation property.*
- *Every V-formation of finitely generated algebras from $\mathbb{L}_{FSI}$ has an amalgam in $\mathbb{L}_{FSI} \times \mathbb{L}_{FSI} = \{\mathcal{A} \times \mathcal{B} : \mathcal{A}, \mathcal{B} \in \mathbb{L}_{FSI}\}$.*
- *Every V-formation of finitely generated algebras from $\mathbb{L}_{FSI}$ has an amalgam in $\mathbb{L}$.*

Applying the previous theorem to the case of BL-algebras we obtain the following result.

Theorem 4. *A variety $\mathbb{L}$ of BL-algebras has the AP if and only if $Ch(\mathbb{L})$ has the 1AP.*

Proof. Let $\mathbb{L}$ be a variety of BL-algebras. By Theorem 2 we have that $\mathbb{L}_{FSI}$ coincides with the class of all chains in $\mathbb{L}$, and hence $\mathbb{L}_{FSI}$ is closed under subalgebras. Moreover, as shown in [18, p. 42] every variety of BL-algebras has the congruence extension property. Then the theorem's claim follows by Theorem 3. □

Proposition 1. *Let $\mathbb{L}$ be a variety of BL-algebras. If $\mathbb{L}$ has the AP, then $\mathbb{L} \cap MV$ is single-chain generated.*

Proof. Let $\mathbb{L}$ be a variety of BL-algebras having the AP. Suppose by contradiction that $\mathbb{L} \cap MV$ is not single-chain generated. Then, by [19] we have that $\mathbb{L} \cap MV$ does not have the AP. Pick a V-formation $(\mathcal{A}, \mathcal{B}, \mathcal{C}, i, j)$ with $\mathcal{A}, \mathcal{B}, \mathcal{C} \in Ch(\mathbb{L} \cap MV)$. Since $\mathbb{L}$ has the AP, by Theorem 4 $Ch(\mathbb{L})$ has the 1AP. Then there exists an 1-amalgam $(\mathcal{D}, h, k)$, with $\mathcal{D} \in Ch(\mathbb{L})$, for $(\mathcal{A}, \mathcal{B}, \mathcal{C}, i, j)$. Let $\mathcal{E}$ be the subalgebra of $\mathcal{D}$ generated by $h(\mathcal{B}) \cup k(\mathcal{C})$. Since $h(\mathcal{B})$ and $k(\mathcal{C})$ are both MV-chains, then $h(\mathcal{B}) \cup k(\mathcal{C}) \subseteq (\mathcal{D})_0$. Then $\mathcal{E}$ is an MV-chain, and hence $\mathcal{E} \in \mathbb{L} \cap MV$. Then by Theorem 4, $\mathbb{L} \cap MV$ has the AP, which is a contradiction. Therefore we must conclude that if $\mathbb{L}$ has the AP, then $\mathbb{L} \cap MV$ is single-chain generated. The proof is settled. □

Remark 2. The converse of Proposition 1 does not hold. For example $\mathbb{G}_4 \cap MV = \mathbb{B}$ is single-chain generated, but $\mathbb{G}_4$ does not have the AP.

Theorem 5 ([5]). *Let $\mathbb{L}$ be a variety of BL-algebras such that every chain has finitely many components. If $\mathbb{L}$ contains $\mathcal{G}_4$ or $\mathcal{P}_2$, then $\mathbb{L}$ does not have the AP.*

Corollary 1. *Let $\mathbb{L}$ be a variety of BL-algebras such that every chain has finitely many components. If $\mathbb{L}$ has the AP, then every $\mathcal{A} \in Ch(\mathbb{L})$ satisfies the following properties.*

- $\#\mathcal{A} \leq 3$.
- *If $\mathcal{A}$ is finite, then $\#\mathcal{A} \leq 2$.*
- *If $\#\mathcal{A} = 3$, then one between $(\mathcal{A})_1, (\mathcal{A})_2$ is cancellative, and the other one is finite.*

Theorem 6 ([5]). *Let $\mathbb{L}$ be a variety of BL-algebras generated by one chain with finitely many components. Then the following are equivalent:*

(i) *$\mathbb{L}$ has the AP.*

(ii) *Every BL-chain $\mathcal{A} = \bigoplus_{i \in I} \mathcal{A}_i$ such that $\mathbf{V}(\mathcal{A}) = \mathbb{L}$ satisfies the following conditions.*

- *$|I| \leq 3$.*
- *There is at most one $i \in I \setminus \{0\}$ such that $\mathcal{A}_i$ is infinite, and there is at most one $j \in I \setminus \{0\}$ such that $\mathcal{A}_j$ is bounded.*
- *If $|I| \geq 2$ then the following hold.*
  - *If $\mathcal{A}_0$ has infinite rank, then $\mathcal{L}_k \hookrightarrow \mathcal{A}_0$, for every $k \geq 2$.*
  - *If $\mathcal{A}_0$ is infinite and $rank(\mathcal{A}_0) = k$, then $\mathcal{L}_k \hookrightarrow \mathcal{A}_0$.*

3.1 Minimal Set of Generators and Related Properties

Definition 5. *A* minimal set of generators *(m-set) is a non-empty finite set of non-trivial BL-chains with finitely many components, say S, such that $\mathbf{V}(T) \subsetneq \mathbf{V}(S)$, for every $T \subsetneq S$.*

The proof of the following result is straightforward.

Lemma 5. *Let S be a non-empty finite set of non-trivial BL-chains with finitely many components. Then there exists an m-set (not necessarily unique) $T \subseteq S$ such that $\mathbf{V}(T) = \mathbf{V}(S)$.*

The m-sets have some interesting properties.

Lemma 6. *Let $\mathbb{L}$ be a variety generated by one finite set of BL-chains with finitely many components. Then,*

- *There exists an m-set S such that $\mathbf{V}(S) = \mathbb{L}$.*
- *If $\mathbb{L}$ is single-chain generated, then $|S| = 1$, for every m-set S such that $\mathbf{V}(S) = \mathbb{L}$.*
- *If every chain in $\mathbb{L}$ is finite, then there is an m-set S containing only finite BL-chains such that $\mathbf{V}(S) = \mathbb{L}$.*
- *Let S be an m-set such that $\mathbf{V}(S) = \mathbb{L}$. For every $\mathcal{A} \in S$, if $\mathcal{A} \hookrightarrow \mathcal{B}$, with $\mathcal{B} \in Ch(\mathbb{L})$, then $\mathcal{B} \in Ch(\mathcal{A})$.*

Proof. Let $\mathbb{L}$ be a variety generated by one finite set of BL-chains with finitely many components.

- Immediate by Lemma 5.
- Let S be an m-set such that $\mathbf{V}(S) = \mathbb{L}$. Note that this set, which is finite, necessarily exists, since $\mathbb{L}$ is generated by one BL-chain with finitely many components. Suppose by contradiction that $|S| \geq 2$. This means that $\mathbb{L} = \bigvee_{\mathcal{A} \in S} \mathbf{V}(\mathcal{A})$, and since S is an m-set we have that $\mathbf{V}(\mathcal{A}) \subsetneq \mathbb{L}$, for every $\mathcal{A} \in S$. But then this would imply that $\mathbb{L}$ is not join irreducible in $\mathcal{L}(\mathbb{L})$, and by [3, Theorem 5.1] this is a contradiction, since $\mathbb{L}$ is single-chain generated.

- If every chain in $\mathbb{L}$ is finite, then $\mathbb{L}$ is generated by one finite set of finite BL-chains, and the claim follows by Lemma 5.
- Let S be an m-set such that $\mathbf{V}(S) = \mathbb{L}$, and assume that $\mathcal{A} \hookrightarrow \mathcal{B}$, for some $\mathcal{B} \in Ch(\mathbb{L})$. We have that $Ch(\mathbb{L}) = Ch(S) = \bigcup_{\mathcal{C} \in S} Ch(\mathcal{C})$, where the last equality follows by [6, Lemma 2], as S is finite. Since S is an m-set, we have that $\mathcal{A} \notin Ch(\mathcal{C})$, for every $\mathcal{C} \in S \setminus \{\mathcal{A}\}$. Then we must conclude that $\mathcal{B} \in Ch(\mathcal{A})$. □

Proposition 2. *Let $\mathbb{L}$ be a variety of BL-algebras generated by an m-set S. If $\mathbb{L}$ has the AP, then:*

i) *For every $\mathcal{A} \in S$, $\#\mathcal{A} \le 3$.*
ii) *There exists $\mathcal{A} \in S$ such that $(\mathcal{B})_0 \in \mathbf{V}((\mathcal{A})_0)$, for every $\mathcal{B} \in S$. Moreover, $\mathbf{V}((\mathcal{A})_0) = \mathbb{L} \cap MV$.*
iii) *If S contains an MV-chain $\mathcal{A}$, then $\mathbf{V}(\mathcal{A}) = \mathbb{L} \cap MV$, and every chain in $S \setminus \{\mathcal{A}\}$ has at least two components.*
iv) *Suppose that the first component of every chain in S is finite. Then there exists $\mathcal{A} \in S$ such that $(\mathcal{B})_0 \hookrightarrow \mathcal{A}$, for every $\mathcal{B} \in S$.*
v) *Suppose that the first component of every chain in S has finite rank. Then there exists $\mathcal{A} \in S$ such that $rank((\mathcal{B})_0) - 1$ divides $rank((\mathcal{A})_0) - 1$, for every $\mathcal{B} \in S$.*

Proof. Let $\mathbb{L}$ be a variety of BL-algebras generated by an m-set S. Assume that $\mathbb{L}$ has the AP.

i) Immediate by Corollary 1.
ii) By Proposition 1 we know that there exists $\mathcal{C} \in Ch(\mathbb{L})$ such that $\mathbf{V}(\mathcal{C}) = \mathbb{L} \cap MV$. By [6, Lemma 2] and [2, Theorem 7.9] we have that $Ch(MV \cap \mathbb{L}) = \bigcup_{\mathcal{D} \in T} Ch(\mathcal{D})$, where $T = \{(\mathcal{D})_0 : \mathcal{D} \in S\}$. This means that $\mathcal{C} \in Ch((\mathcal{A})_0)$, for some $\mathcal{A} \in S$. Since the first component of every BL-chain is an MV-chain, we clearly have that $T \subset MV \cap \mathbb{L} = \mathbf{V}((\mathcal{A})_0)$. Then $(\mathcal{B})_0 \in \mathbf{V}((\mathcal{A})_0)$, for every $\mathcal{B} \in S$. Therefore ii) holds true.
iii) Suppose that S contains an MV-chain $\mathcal{A}$. Since S is an m-set, $\mathcal{A} \notin \mathbf{V}(\mathcal{B})$, for every $\mathcal{B} \in S \setminus \{\mathcal{A}\}$. By ii) we must have $(\mathcal{B})_0 \in \mathbf{V}(\mathcal{A})$, for every $\mathcal{B} \in S$. By [6, Lemma 2] and [2, Theorem 7.9] we have that $Ch(MV \cap \mathbb{L}) = \bigcup_{\mathcal{D} \in T} Ch(\mathcal{D})$, where $T = \{(\mathcal{D})_0 : \mathcal{D} \in S\}$. Then an easy check shows that $\mathbf{V}(\mathcal{A}) = \mathbb{L} \cap MV$. Suppose that $S \setminus \{\mathcal{A}\}$ contains an MV-chain $\mathcal{C}$. Then $\mathcal{C} = (\mathcal{C})_0 \in \mathbf{V}(\mathcal{A})$, but this is a contradiction, since S is an m-set. Therefore every chain in $S \setminus \{\mathcal{A}\}$ has at least two components.
iv) Suppose now that the first component of every chain in S is finite. By ii) we have that there exists $\mathcal{A} \in S$ such that $(\mathcal{B})_0 \in \mathbf{V}((\mathcal{A})_0)$, for every $\mathcal{B} \in S$. Since the first component of every chain in S is finite, we have that $(\mathcal{B})_0 \hookrightarrow \mathcal{A}$, for every $\mathcal{B} \in S$. Then iv) is true.
v) Suppose that the first component of every chain in S has finite rank. By ii) there exists $\mathcal{A} \in S$ such that $(\mathcal{B})_0 \in \mathbf{V}((\mathcal{A})_0)$, for every $\mathcal{B} \in S$. Since $(\mathcal{A})_0$ has finite rank, by known results on MV-algebras and the previous observation we have that $rank((\mathcal{B})_0) - 1$ divides $rank((\mathcal{A})_0) - 1$, for every $\mathcal{B} \in S$. Therefore v) holds true, and the proof is settled. □

Lemma 7. *Let* $\mathbb{L}$ *be a variety of BL-algebras generated by an m-set* S. *Assume that* $\mathbb{L}$ *has the AP. If* S *contains an MV-chain* $\mathcal{A}$ *which is either finite or with infinite rank, then* $S = \{\mathcal{A}\}$.

Proof. Let $\mathbb{L}$ be a variety of BL-algebras generated by an m-set S. Assume that $\mathbb{L}$ has the AP, and that S contains an MV-chain $\mathcal{A}$ which is either finite or with infinite rank. Suppose by contradiction that $|S| \geq 2$, and pick $\mathcal{B} \in S \setminus \{\mathcal{A}\}$. We have two cases.

Assume first that $\mathcal{A}$ is finite. Pick the V-formation $(\mathbf{2}, \mathcal{B}, \mathcal{A}, i, j)$, where i, j are defined in the unique and obvious way. Since by hypothesis $\mathbb{L}$ has the AP, by Theorem 4, $Ch(S)$ has the 1AP. Therefore there exists a 1-amalgam $(\mathcal{C}, h, k)$ for $(\mathbf{2}, \mathcal{B}, \mathcal{A}, i, j)$, with $\mathcal{C} \in Ch(S)$. Since $\mathcal{B} \overset{h}{\hookrightarrow} \mathcal{C}$, by Lemma 6 we have $\mathcal{C} \in Ch(\mathcal{B})$. Since k is a homomorphism from $\mathcal{A}$ to $\mathcal{C}$, and $\mathcal{A}$ is simple, by Lemma 4 k must be an embedding. But this would imply $\mathcal{A} \in \mathbf{V}(\mathcal{C}) \subseteq \mathbf{V}(\mathcal{B})$, which is impossible as S is an m-set. Therefore the 1AP for $Ch(S)$ fails, and by Theorem 4 the AP for $\mathbf{V}(S) = \mathbb{L}$ does not hold.

The last case is when $\mathcal{A}$ has infinite rank. Then $\mathbf{V}(\mathcal{A}) = \mathbb{MV}$ (see [9]), and hence $[0,1]_\mathbb{L} \in Ch(S)$. Clearly $\mathbf{V}(\mathcal{A}) = \mathbf{V}([0,1]_\mathbb{L})$. Pick the V-formation $(\mathbf{2}, \mathcal{B}, [0,1]_\mathbb{L}, i, j)$, where i, j are defined in the unique and obvious way. Since $\mathbb{L}$ has the AP by Theorem 4, $Ch(S)$ has the 1AP. Therefore there exists a 1-amalgam $(\mathcal{C}, h, k)$ for $(\mathbf{2}, \mathcal{B}, [0,1]_\mathbb{L}, i, j)$, with $\mathcal{C} \in Ch(S)$. Since $\mathcal{B} \overset{h}{\hookrightarrow} \mathcal{C}$, by Lemma 6 we have $\mathcal{C} \in Ch(\mathcal{B})$. Since k is a homomorphism from $[0,1]_\mathbb{L}$ to $\mathcal{C}$, and $[0,1]_\mathbb{L}$ is simple, by Lemma 4 k must be an embedding. But this would imply $[0,1]_\mathbb{L}, \mathcal{A} \in \mathbf{V}(\mathcal{C}) \subseteq \mathbf{V}(\mathcal{B})$, which is impossible as S is an m-set. Therefore the 1AP for $Ch(S)$ fails, and by Theorem 4 the AP for $\mathbf{V}(S)$ does not hold. Summing up, if $|S| \geq 2$, the AP for $\mathbf{V}(S)$ fails to hold. On the other side, in [19] it is shown that every variety generated by one MV-chain has the AP. Therefore we must conclude that $S = \{\mathcal{A}\}$, and the proof is settled. $\square$

Remark 3. One could ask if Lemma 7 could be extended to the general case, where $\mathcal{A}$ is any (non-trivial) MV-chain. The answer is negative. Pick the m-set $S = \{\mathcal{K}_2, \mathcal{G}_3\}$, where $\mathcal{K}_2 = \Gamma(\mathbb{Z}_G \times_{\text{lex}} \mathbb{Z}_G, (1,0))$ is Chang's MV-algebra: here $\mathbb{Z}_G$ denotes the additive group of integers, and Γ is the gamma functor (see [9] for details). By [6, Lemma 2], Lemma 1, Lemma 2 we have that the non-trivial chains in $\mathbf{V}(S)$, up to isomorphisms, are $\{\mathcal{K}_2, \mathbf{2}, \mathcal{G}_3\}$. We show that $Ch(S)$ has the 1AP. Let $(\mathcal{A}, \mathcal{B}, \mathcal{C}, i, j)$ be a V-formation. If $\mathcal{B}, \mathcal{C} \in Ch(\mathcal{G}_3)$ or $\mathcal{B}, \mathcal{C} \in Ch(\mathcal{K}_2)$, then since $Ch(\mathcal{G}_3)$ and $Ch(\mathcal{K}_2)$ have the 1AP, we can find a 1-amalgam for $(\mathcal{D}, h, k)$ for $(\mathcal{A}, \mathcal{B}, \mathcal{C}, i, j)$, with $\mathcal{D} \in Ch(S)$. The remaining cases are when $\mathcal{B} \neq \mathcal{C}$ and $\mathcal{B}, \mathcal{C} \in \{\mathcal{K}_2, \mathcal{G}_3\}$. Suppose first that $\mathcal{B} = \mathcal{K}_2$ and $\mathcal{C} = \mathcal{G}_3$. Consider the triple $(\mathcal{K}_2, h, k)$, where $h = id_{\mathcal{K}_2}$, and k is defined as follows. $k(0) = 0$, and $k(c) = 1$ for every $0 < c \in \mathcal{C}$. It is easy to see that h is an embedding from $\mathcal{B}$ to $\mathcal{K}_2$, and that k is a homomorphism from $\mathcal{C}$ to $\mathcal{K}_2$. Finally, an easy check shows that $\mathcal{A} \simeq \mathbf{2}$, and that $h(i(a)) = k(j(a))$, for every $a \in \mathcal{A}$. The last case is when $\mathcal{B} = \mathcal{G}_3$ and $\mathcal{C} = \mathcal{K}_2$. Consider the triple $(\mathcal{G}_3, h, k)$, where $h = id_{\mathcal{G}_3}$, and k is defined as follows. $k(0) = 0$, and $k(c) = 1$ for every $0 < c \in \mathcal{C}$. It is easy to see that h is an embedding from $\mathcal{B}$ to $\mathcal{G}_3$, and that k is a homomorphism from $\mathcal{C}$ to $\mathcal{G}_3$. Finally, an easy check shows that $\mathcal{A} \simeq 2$, and that $h(i(a)) = k(j(a))$, for every $a \in \mathcal{A}$. Therefore $Ch(S)$ has the 1AP, and by Theorem 4, $\mathbf{V}(S)$ has the AP.

3.2 A Classification of the AP, for Varieties Generated by M-Sets with Either Cancellative or Finite Components

Lemma 8. *Let S be an m-set in which every chain has either cancellative or finite components. Suppose that at least one of the following conditions holds:*

- *There are $\mathcal{A}, \mathcal{B} \in S$ such that $\#\mathcal{A} = \#\mathcal{B} = k$, and $\mathcal{A} \neq \mathcal{B}$, with $k \geq 1$, $k \neq 2$.*
- *There are $\mathcal{A}, \mathcal{B} \in S$ such that $\#\mathcal{A} = \#\mathcal{B} = 2$, and $\mathcal{A} \neq \mathcal{B}$, where either $\mathcal{A}, \mathcal{B}$ are both finite or $(\mathcal{A})_1, (\mathcal{B})_1$ are both cancellative.*
- *There are $\mathcal{A}, \mathcal{B} \in S$ such that $\#\mathcal{A} = 2$ and $\#\mathcal{B} = 3$.*

Then $\mathbf{V}(S)$ does not have the AP.

Proof. Let S be an m-set in which every chain has either cancellative or finite components.

- Assume that there are $\mathcal{A}, \mathcal{B} \in S$ such that $\#\mathcal{A} = \#\mathcal{B}$, and $\mathcal{A} \neq \mathcal{B}$.
  If $\#\mathcal{A} = \#\mathcal{B} = 1$, by Lemma 7, $\mathbf{V}(S)$ does not have the AP.
  The next case is $\#\mathcal{A} = \#\mathcal{B} = 3$. Note that if either $\mathcal{A}$ or $\mathcal{B}$ has no cancellative components or more than one, then by Corollary 1, $\mathbf{V}(S)$ does not have the AP. So, let us assume that both $\mathcal{A}, \mathcal{B}$ have exactly one cancellative component (by hypothesis the other components are finite). Pick the V-formation $(\mathcal{G}_3, \mathcal{A}, \mathcal{B}, i, j)$, where i, j are defined in the unique and obvious way. Suppose by contradiction that $\mathbf{V}(S)$ has the AP. By Theorem 4, $Ch(S)$ has the 1AP, and there exists a 1-amalgam $(\mathcal{C}, h, k)$, with $\mathcal{C} \in Ch(S)$. Since $\mathcal{A} \xrightarrow{h} \mathcal{C}$, then by Lemma 6 we have $\mathcal{C} \in \mathbf{V}(\mathcal{A})$. On the other side $\mathcal{B} \nrightarrow \mathcal{C} \in \mathbf{V}(\mathcal{A})$, since S is an m-set. Since k is a homomorphism between $\mathcal{B}$ and $\mathcal{C}$, then by Lemma 4, $(\mathcal{B})_0$ must embed into $(\mathcal{C})_0$, as $(\mathcal{B})_0$ is a finite MV-chain. Let $(\mathcal{B})_j$, with $j \in \{1, 2\}$ be the other finite component of $\mathcal{B}$ different from the first one (indeed, by hypothesis $\mathcal{B}$ has exactly one cancellative component). Note that if $(\mathcal{B})_j$ embeds into $(\mathcal{C})_j$, then by Theorem 2, $\mathcal{B} \in \mathbf{V}(\mathcal{A})$, and this is impossible as S is an m-set. This means that $(\mathcal{B})_j \nrightarrow (\mathcal{C})_j$. However, this implies that $k(b) = 1$, for every $b \in (\mathcal{B})_j$. Let m be the bottom of the second component of $\mathcal{G}_3$. Then $k(j(m)) = 1$, whilst since i, h are both embeddings, $h(i(m)) < 1$. But this means that $(\mathcal{C}, h, k)$ is not a 1-amalgam for $(\mathcal{G}_3, \mathcal{A}, \mathcal{B}, i, j)$, a contradiction. By Theorem 4, $\mathbf{V}(S)$ does not have the AP. Finally, if $\#\mathcal{A} = \mathcal{B} \geq 4$, by Corollary 1, $\mathbf{V}(S)$ does not have the AP.
- Suppose now that $\#\mathcal{A} = \#\mathcal{B} = 2$. If $\mathcal{A}, \mathcal{B}$ are both finite, then pick the V-formation $(\mathcal{G}_3, \mathcal{A}, \mathcal{B}, i, j)$, where i, j are defined in the unique and obvious way. Suppose by contradiction that $\mathbf{V}(S)$ has the AP. By Theorem 4, $Ch(S)$ has the 1AP, and there exists a 1-amalgam $(\mathcal{C}, h, k)$, with $\mathcal{C} \in Ch(S)$. Since $\mathcal{A} \xrightarrow{h} \mathcal{C}$, by Lemma 6 we have $\mathcal{C} \in \mathbf{V}(\mathcal{A})$. On the other side $\mathcal{B} \nrightarrow \mathcal{C} \in \mathbf{V}(\mathcal{A})$, since S is an m-set. Since k is a homomorphism between $\mathcal{B}$ and $\mathcal{C}$,

then $ker(k) \in Con(\mathcal{B})$. Since $\mathcal{B}$ is a finite BL-chain with two components, it has three congruences: Δ, ∇, θ, where θ is the congruence corresponding to the filter $Rad(\mathcal{A})$ (which is the carrier of $(\mathcal{B})_1$). Now, since $\mathcal{B} \not\hookrightarrow \mathcal{C}$, and $\mathcal{C}$ is non-trivial we have that $ker(k) \notin \{\Delta, \nabla\}$. Therefore $ker(k) = \theta$. This implies that $k(b) = 1$, for every $b \in (\mathcal{B})_1$. Let m be the bottom element of the second component of $\mathcal{G}_3$. We have that $k(j(m)) = 1$, however since i, h are both embeddings we have $h(i(m)) < 1$. However this is a contradiction, as $(\mathcal{C}, h, k)$ should be a 1-amalgam. By Theorem 4, $\mathbf{V}(S)$ does not have the AP. Assume now that $(\mathcal{A})_1, (\mathcal{B})_1$ are both cancellative. Pick the V-formation $(\mathbf{2}, \mathcal{A}, \mathcal{B}, i, j)$, where i, j are defined in the unique and obvious way. Suppose by contradiction that $\mathbf{V}(S)$ has the AP. By Theorem 4, $Ch(S)$ has the 1AP, and there exists a 1-amalgam $(\mathcal{C}, h, k)$, with $\mathcal{C} \in Ch(S)$. Since $\mathcal{A} \overset{h}{\hookrightarrow} \mathcal{C}$, then by Lemma 6 we have $\mathcal{C} \in \mathbf{V}(\mathcal{A})$. By Lemma 2 $(\mathcal{C})_1$ must be cancellative. Since k is a homomorphism between $\mathcal{B}$ and $\mathcal{C}$, and $(\mathcal{B})_0$ is a finite MV-chain, by Lemma 4, $(\mathcal{B})_0 \xrightarrow{k_{\restriction(\mathcal{B}_0)}} (\mathcal{C})_0$. However, since $(\mathcal{B})_1, (\mathcal{C})_1$ are both cancellative, by Lemma 2 we would have $\mathcal{B} \in \mathbf{V}(\mathcal{C})$, and hence $\mathcal{B} \in \mathbf{V}(\mathcal{A})$. However this is impossible, as S is an m-set. Therefore $Ch(S)$ does not have the 1AP, and by Theorem 4, $\mathbf{V}(S)$ does not have the AP.

– Assume that there are $\mathcal{A}, \mathcal{B} \in S$ such that $\#\mathcal{A} = 2$ and $\#\mathcal{B} = 3$. If $\mathcal{B}$ does not have exactly one cancellative component, then by Theorem 5, $\mathbf{V}(S)$ does not have the AP, and we are done. So, suppose that $\mathcal{B}$ has exactly one cancellative component. Assume first that $(\mathcal{A})_1$ is cancellative. Consider the V-formation $(\mathbf{2}, \mathcal{B}, \mathcal{A}, i, j)$, where i, j are defined in the unique and obvious way. Suppose by contradiction that $\mathbf{V}(S)$ has the AP. By Theorem 4, $Ch(S)$ has the 1AP, and there exists a 1-amalgam $(\mathcal{C}, h, k)$ with $\mathcal{C} \in Ch(S)$. In particular h is an embedding from $\mathcal{B}$ to $\mathcal{C}$, and $\mathcal{C} \in Ch(\mathcal{B})$ by Lemma 6. By Lemma 2 $\mathcal{C}$ has one cancellative component. Since k is a homomorphism between $\mathcal{A}$ and $\mathcal{C}$, and $(\mathcal{A})_0$ is a finite MV-chain, by Lemma 4, $(\mathcal{A})_0 \xrightarrow{k_{\restriction(\mathcal{A}_0)}} (\mathcal{C})_0$. By Lemma 2 a direct inspection shows that $\mathcal{A} \in \mathbf{V}(\mathcal{C})$, and hence $\mathcal{A} \in \mathbf{V}(\mathcal{B})$. However this is impossible, as S is an m-set. Therefore $Ch(S)$ does not have the 1AP, and by Theorem 4, $\mathbf{V}(S)$ does not have the AP. The last case is when $(\mathcal{A})_1$ is a finite Wajsberg hoop. This means that $\mathcal{A}$ is a finite BL-chain. Pick the V-formation $(\mathcal{G}_3, \mathcal{B}, \mathcal{A}, i, j)$, where i, j are defined in the unique and obvious way (as $\mathcal{B}$ has exactly one cancellative component, whilst the others are finite). Suppose by contradiction that $\mathbf{V}(S)$ has the AP. By Theorem 4, $Ch(S)$ has the 1AP, and there exists a 1-amalgam $(\mathcal{C}, h, k)$ with $\mathcal{C} \in Ch(S)$. In particular h is an embedding from $\mathcal{B}$ to $\mathcal{C}$, and hence $\#\mathcal{C} = 3$, and $\mathcal{C} \in Ch(\mathcal{B})$ by Lemma 6. Since S is an m-set then $\mathcal{A} \notin \mathbf{V}(\mathcal{B})$, and hence $\mathcal{A} \not\hookrightarrow \mathcal{C}$. So, as k is a homomorphism from $\mathcal{A}$ to $\mathcal{C}$, then $ker(k)$ is a congruence on $\mathcal{A}$. Since $\mathcal{A} = (\mathcal{A})_0 \oplus (\mathcal{A})_1$, and both the components are finite Wajsberg hoops, then $\mathcal{A}$ has exactly three filters: $\{1\}, A, Rad(\mathcal{A})$, where $Rad(\mathcal{A})$ is the carrier of $(\mathcal{A})_1$. This means that $\mathcal{A}$ has exactly three congruences: Δ, ∇, θ, where θ is the congruence corresponding to $Rad(\mathcal{A})$. Now, since $\mathcal{A} \not\hookrightarrow \mathcal{C}$, then $ker(k) \neq \Delta$. Moreover, since $\mathcal{C}$ is non-trivial we also have $ker(k) \neq \nabla$.

So the only possibility is that $ker(k) = \theta$. However this implies that $k(a) = 1$, for every $a \in (\mathcal{A})_1$. Let m be the bottom element of the second component of $\mathcal{G}_3$. Then we have that $k(j(m)) = 1$, but since h, i are both embeddings, we have that $h(i(m)) < 1$. However this is a contradiction, since $(\mathcal{C}, h, k)$ is a 1-amalgam for $(\mathcal{G}_3, \mathcal{B}, \mathcal{A}, i, j)$. By Theorem 4 we conclude that $\mathbf{V}(S)$ does not have the AP also in this case, and the proof is settled.   □

Theorem 7. *Let* $\mathbb{L}$ *be a variety generated by one finite set of BL-chains with finitely many components, that are either finite or cancellative. Then* $\mathbb{L}$ *has the AP if and only if one of the following two cases holds.*

- $\mathbb{L} = \mathbf{V}(\mathcal{A})$, *where* $\mathcal{A} \in Ch(\mathbb{L})$, *and satisfies one of the following conditions:*
  a) $\#\mathcal{A} \leq 2$, *there is at most one cancellative component, and the others are finite (including the first-one).*
  b) $\#\mathcal{A} = 3$, *two components (including the first-one) are finite, and the other one is cancellative.*
- $\mathbb{L} = \mathbf{V}(\{\mathcal{B}, \mathcal{C}\})$, *where:*
  c) $\mathcal{B}, \mathcal{C} \in Ch(\mathbb{L})$, $\#\mathcal{B} = \#\mathcal{C} = 2$.
  d) $(\mathcal{B})_1$ *is finite,* $(\mathcal{C})_1$ *is cancellative, and* $(\mathcal{B})_0 \simeq (\mathcal{C})_0$.

Proof. Let $\mathbb{L}$ be a variety generated by one finite set of BL-chains with finitely many components, that are either finite or cancellative. Then there exists an m-set S such that $\mathbf{V}(S) = \mathbb{L}$. If $S = \{\mathcal{A}\}$, with $\mathcal{A} \in Ch(\mathbb{L})$, and $\mathcal{A}$ satisfies the theorem's hypothesis, then by Theorem 6, $\mathbb{L}$ has the AP.

Suppose now that $S = \{\mathcal{B}, \mathcal{C}\}$, and c), d) are satisfied. To show that $\mathbf{V}(S)$ has the AP, we prove that $Ch(S)$ has the 1AP. Pick a V-formation $(\mathcal{D}, \mathcal{E}, \mathcal{F}, i, j)$, where $\mathcal{D}, \mathcal{E}, \mathcal{F} \in Ch(S)$. By [6, Lemma 2] $Ch(S) = Ch(\mathcal{B}) \cup Ch(\mathcal{C}) = \mathbf{IS}(\mathcal{B}) \cup \mathbf{IS}((\mathcal{C})_0) \oplus \mathbf{ISP}_u((\mathcal{C})_1)$. This implies that every chain with one cancellative component in $Ch(S)$ belongs to $Ch(\mathcal{C})$. Moreover, since $(\mathcal{B})_0 \simeq (\mathcal{C})_0$ we have that every finite chain in $Ch(S)$ belongs to $Ch(\mathcal{B})$.

We distinguish three cases.

- If $\mathcal{E}, \mathcal{F}$ have both a cancellative component (note that every chain in $\mathbb{L}$ has at most two components), then $\mathcal{D}, \mathcal{E}, \mathcal{F} \in Ch(\mathcal{C})$. Since by Theorem 6, $\mathbf{V}(\mathcal{C})$ has the AP, then $Ch(\mathcal{C})$ has the 1AP. This means that there exists a 1-amalgam $(\mathcal{G}, h, k)$ for $(\mathcal{D}, \mathcal{E}, \mathcal{F}, i, j)$, with $\mathcal{G} \in Ch(\mathcal{C})$.
- If $\mathcal{E}, \mathcal{F}$ are both finite, then $\mathcal{D}, \mathcal{E}, \mathcal{F} \in Ch(\mathcal{B})$. Since by Theorem 6, $\mathbf{V}(\mathcal{B})$ has the AP, then $Ch(\mathcal{B})$ has the 1AP. This means that there exists a 1-amalgam $(\mathcal{G}, h, k)$ for $(\mathcal{D}, \mathcal{E}, \mathcal{F}, i, j)$, with $\mathcal{G} \in Ch(\mathcal{B})$.
- The last case is when one chain between $\mathcal{E}, \mathcal{F}$, has one cancellative component, and the other chain is finite.
  Suppose first that $(\mathcal{E})_1$ is cancellative, and $\mathcal{F}$ is finite. Consider the triple $(\mathcal{G}, h, k)$, where $\mathcal{G} = (\mathcal{C})_0 \oplus (\mathcal{E})_1$, and h, k will be defined later. By Lemma 2 $\mathcal{G} \in Ch(\mathcal{C})$, and by d) an easy check shows that $(\mathcal{E})_0$ embeds into $(\mathcal{G})_0$. Since $(\mathcal{E})_0$ is a finite MV-chain, this embedding is unique: let us call it r. Let h be a map from $\mathcal{E}$ to $\mathcal{G}$ defined as follows: $h(e) = r(e)$ if $e \in (\mathcal{E})_0$, and $h(e) = e$ if $e \in (\mathcal{E})_1$. An easy check shows that h is an embedding from $\mathcal{E}$ to $\mathcal{G}$.

Now, by Lemma 2, [6, Lemma 2] and d) we have that the first component of every chain in $\mathbb{L}$ can be embedded in $(\mathcal{C})_0$. This means that $(\mathcal{F})_0$ embeds into $(\mathcal{C})_0$, and since they are both finite MV-chains, this embedding is unique: let us call it s. Define now a map k from $\mathcal{F}$ to $\mathcal{G}$ as follows: $k(f) = s(f)$ if $f \in (\mathcal{F})_0$, and $k(f) = 1$ if $f \in (\mathcal{F})_1$ (this last case holds true only if $\mathcal{F}$ has two components, otherwise $\mathcal{F} = (\mathcal{F})_0$). An easy check shows that k is a homomorphism from $\mathcal{F}$ to $\mathcal{G}$. It remains to show that, for every $d \in \mathcal{D}$, $h(i(d)) = k(j(d))$. Note that $\#\mathcal{D} = 1$. Indeed, if $\#\mathcal{D} = 2$ then either $\mathcal{D} \not\hookrightarrow \mathcal{E}$ or $\mathcal{D} \not\hookrightarrow \mathcal{F}$, as $(\mathcal{E})_1$ is cancellative, $\mathcal{F}$ is finite, and $(\mathcal{D})_1$ must be either cancellative or finite. Note also that $i(\mathcal{D})$, $h(i(\mathcal{D}))$, $j(\mathcal{D})$, $k(j(\mathcal{D}))$ are subalgebras of, respectively, $(\mathcal{E})_0$, $(\mathcal{G})_0$, $(\mathcal{F})_0$, $(\mathcal{G})_0$. Now, $(\mathcal{G})_0 = (\mathcal{C})_0$, and by Lemma 2, [6, Lemma 2] and d) the first component of every chain in $\mathbb{L}$ embeds into $(\mathcal{C})_0$. Since $Ch((\mathcal{C})_0)$ has the AP, a direct inspection shows that $h(i(d)) = k(j(d))$, for every $d \in \mathcal{D}$. Therefore $(\mathcal{G}, h, k)$ is a 1-amalgam for $(\mathcal{D}, \mathcal{E}, \mathcal{F}, i, j)$.

Finally, let us assume that $\mathcal{E}$ is finite, and $(\mathcal{F})_1$ is cancellative. Consider the triple $(\mathcal{B}, h, k)$. By d), [6, Lemma 2], and Lemma 2, an easy check shows that $\mathcal{E} \in Ch(\mathcal{B}) = \mathbf{IS}(\mathcal{B})$. Since the components of $\mathcal{E}$ (or the component, if $\#\mathcal{E} = 1$) are finite Wajsberg hoops, there is only one way to embed $\mathcal{E}$ into $\mathcal{B}$: let us call h such embedding. By d), [6, Lemma 2], and Lemma 2, an easy check shows that $(\mathcal{F})_0$ embeds into $(\mathcal{B})_0$, and since $(\mathcal{F})_0$ is finite, this embedding is unique: let us call it t. Let k be a map from $\mathcal{F}$ to $\mathcal{B}$ defined as follows: $k(f) = t(f)$ if $f \in (\mathcal{F})_0$, and $k(f) = 1$ if $f \in (\mathcal{F})_1$. An easy check shows that k is a homomorphism from $\mathcal{F}$ to $\mathcal{B}$. Finally, with a proof similar to the previous case, we can show that $\#\mathcal{D} = 1$, and that $h(i(d)) = k(j(d))$, for every $d \in \mathcal{D}$. Therefore $(\mathcal{G}, h, k)$ is a 1-amalgam for $(\mathcal{D}, \mathcal{E}, \mathcal{F}, i, j)$.

This shows that $Ch(S)$ has the 1AP. Therefore, by Theorem 4, $\mathbf{V}(S)$ has the AP.

Conversely, assume that $\mathbb{L}$ has the AP. Note that by Proposition 2 every algebra in S must have at most three components, as otherwise the AP would fail. By Lemma 7 we have that if S contains a chain $\mathcal{A}$ with $\#\mathcal{A} = 1$, then $S = \{\mathcal{A}\}$. Then, by Lemma 8 an easy check shows that there are only two cases: either $|S| = 1$ or $S = \{\mathcal{B}, \mathcal{C}\}$, where $\#\mathcal{B} = \#\mathcal{C} = 2$, and one of $(\mathcal{B})_1, (\mathcal{C})_1$ is cancellative, whilst the other is finite.

– Suppose first that $S = \{\mathcal{B}, \mathcal{C}\}$, where $\#\mathcal{B} = \#\mathcal{C} = 2$ and one of $(\mathcal{B})_1, (\mathcal{C})_1$ is cancellative, whilst the other is finite. Note that c) is already satisfied, so it remains to check that d) holds true. Since $\mathbb{L}$ has the AP, by Theorem 4, $Ch(S)$ has the 1AP. Pick the V-formation $(\mathbf{2}, \mathcal{C}, \mathcal{B}, i, j)$, where i, j are defined in the obvious way. Then there exists a 1-amalgam $(\mathcal{D}, h, k)$ for $(\mathbf{2}, \mathcal{C}, \mathcal{B}, i, j)$, with $\mathcal{D} \in Ch(S)$. Since $\mathcal{C} \overset{h}{\hookrightarrow} \mathcal{D}$, by Lemma 6, $\mathcal{D} \in \mathbf{V}(\mathcal{C})$. Since k is a homomorphism between $\mathcal{B}$ and $\mathcal{D}$, by Lemma 4 we have that $(\mathcal{B})_0 \hookrightarrow (\mathcal{D})_0 \in \mathbf{V}((\mathcal{C}_0))$.

Pick now the V-formation $(\mathbf{2}, \mathcal{B}, \mathcal{C}, i, j)$, where i, j are defined in the obvious way. Since $\mathbb{L}$ has the AP, by Theorem 4, $Ch(S)$ has the 1AP. Then there exists a 1-amalgam $(\mathcal{E}, h, k)$ for $(\mathbf{2}, \mathcal{B}, \mathcal{C}, i, j)$, with $\mathcal{E} \in Ch(S)$. Since $\mathcal{B} \xhookrightarrow{h} \mathcal{E}$, by Lemma 6, $\mathcal{E} \in \mathbf{V}(\mathcal{B})$. Since k is a homomorphism between $\mathcal{C}$ and $\mathcal{E}$, by Lemma 4 we have that $(\mathcal{C})_0 \hookrightarrow (\mathcal{E})_0 \in \mathbf{V}((\mathcal{B})_0)$.

Then we have that $(\mathcal{C})_0 \in \mathbf{V}((\mathcal{B})_0)$ and $(\mathcal{B})_0 \in \mathbf{V}((\mathcal{C})_0)$. Since $(\mathcal{B})_0$ and $(\mathcal{C})_0$ are both finite MV-chains, we must have that $(\mathcal{B})_0 \simeq (\mathcal{C})_0$. Therefore d) is satisfied.

- The last case is when $|S| = 1$. By Theorem 6 the only member of S, say $\mathcal{A}$, must satisfy a) and b). The proof is settled. $\qquad\square$

By Theorem 7 we have the following result.

Corollary 2. *Let S be a finite set of finite BL-chains. Then $\mathbf{V}(S)$ has the AP if and only $S = \{\mathcal{A}\}$, and $\#\mathcal{A} \leq 2$.*

4   Discussion and Future Works

The classification of the AP for varieties of BL-algebras generated by a finite set of BL-chains with finitely many components remains an open problem. However, we can state the following result.

Lemma 9. *Let S be an m-set containing a BL-chain $\mathcal{A}$ such that $\#\mathcal{A} \geq 2$, and one of the following holds:*

- $(\mathcal{A})_0$ *has infinite rank and $\mathcal{L}_n \nrightarrow (\mathcal{A})_0$, for some $n \in \mathbb{N}$.*
- $(\mathcal{A})_0$ *is infinite, has rank n, and $\mathcal{L}_n \nrightarrow (\mathcal{A})_0$.*

Then $\mathbf{V}(S)$ does not have the AP.

Proof. Let S be an m-set containing a BL-chain $\mathcal{A}$ as above. By hypothesis $(\mathcal{A})_0$ is infinite, and either has rank n or it has infinite rank. By [2, Theorem 7.9] we have that $\mathcal{L}_n \in Ch((\mathcal{A})_0) \subsetneq Ch(S)$. Pick the V-formation $(\mathbf{2}, \mathcal{A}, \mathcal{L}_n, i, j)$, where i, j are defined in the unique and obvious way. Suppose by contradiction that $\mathbf{V}(S)$ has the AP: then $Ch(S)$ has the 1AP, and there exists a 1-amalgam $(\mathcal{B}, h, k)$, with $\mathcal{B} \in Ch(\mathcal{A})$, for $(\mathbf{2}, \mathcal{A}, \mathcal{L}_n, i, j)$. Since $\mathcal{A} \xhookrightarrow{h} \mathcal{B}$, by Lemma 6, $\mathcal{B} \in Ch(\mathcal{A})$. Now, since k is a homomorphism from $\mathcal{L}_n$ to $\mathcal{B}$, by Lemma 4, $\mathcal{L}_n \xhookrightarrow{k} \mathcal{B}$. This implies that $\mathcal{L}_n \in \mathbf{V}(\mathcal{A})$, but by [2, Lemma 4.6] this is a contradiction, as $\mathcal{L}_n \nrightarrow (\mathcal{A})_0$. Therefore $\mathbf{V}(S)$ does not have the AP, and the proof is settled. $\quad\square$

One of the main issues with this general case concerns the following problem:

Problem 1. Let $\mathcal{A}, \mathcal{B}$ be two non-trivial MV-chains. When is it possible to define a homomorphism from $\mathcal{A}$ to $\mathcal{B}$?

If $\mathcal{A}$ is simple, by Lemma 3 every homomorphism is an embedding. In general, however, Problem 1 is non-trivial. This makes difficult to prove or disprove the 1AP for $Ch(S)$, when S is an m-set containing chains with non-cancellative and non-simple components. Future works will be devoted to classify the AP for this case. Another interesting case is the one of locally finite varieties of BL-algebras (see [7]), since their chains are ordinal sums of finite Wajsberg hoops. We will tackle the analysis of the AP for those varieties in a future work.

Acknowledgements. This work was partially supported by Istituto Nazionale di Alta Matematica (Indam).

References

1. Aglianò, P., Ferreirim, I., Montagna, F.: Basic hoops: an algebraic study of continuous t-norms. Stud. Logica. **87**, 73–98 (2007)
2. Aglianò, P., Montagna, F.: Varieties of BL-algebras I: general properties. J. Pure Appl. Algebra **181**(2–3), 105–129 (2003)
3. Aguzzoli, S., Bianchi, M.: Single chain completeness and some related properties. Fuzzy Sets Syst. **301**, 51–63 (2016)
4. Aguzzoli, S., Bianchi, M.: On linear varieties of MTL-algebras. Soft. Comput. **23**, 2129–2146 (2019)
5. Aguzzoli, S., Bianchi, M.: Amalgamation property for varieties of BL-algebras generated by one chain with finitely many components. In: Fahrenberg, U., Gehrke, M., Santocanale, L., Winter, M. (eds.) RAMiCS 2021. LNCS, vol. 13027, pp. 1–18. Springer, Cham (2021). https://doi.org/10.1007/978-3-030-88701-8_1
6. Aguzzoli, S., Bianchi, M.: Strictly join irreducible varieties of BL-algebras: the missing pieces. Fuzzy Sets Syst. **418**, 84–100 (2021)
7. Bianchi, M., Montagna, F.: n-contractive BL-logics. Arch. Math. Log. **50**(3–4), 257–285 (2011)
8. Blok, W., Pigozzi, D.: Algebraizable Logics. Memoirs of the American Mathematical Society, vol. 77. American Mathematical Society (1989)
9. Cignoli, R., D'Ottaviano, I., Mundici, D.: Algebraic Foundations of Many-Valued Reasoning. Trends in Logic, vol. 7. Kluwer Academic Publishers (1999)
10. Cignoli, R., Esteva, F., Godo, L., Torrens, A.: Basic fuzzy logic is the logic of continuous t-norms and their residua. Soft. Comput. **4**(2), 106–112 (2000)
11. Cintula, P., Hájek, P., Noguera, C.: Handbook of Mathematical Fuzzy Logic, vol. 1 and 2. College Publications (2011)
12. Cortonesi, T., Marchioni, E., Montagna, F.: Quantifier elimination and other model-theoretic properties of BL-algebras. Notre Dame J. Formal Logic **52**, 339–379 (2011)
13. Esteva, F., Godo, L., Hájek, P., Montagna, F.: Hoops and fuzzy logic. J. Log. Comput. **13**(4), 532–555 (2003)
14. Fussner, W., Metcalfe, G.: Transfer theorems for finitely subdirectly irreducible algebras (2022). Preprint https://doi.org/10.48550/ARXIV.2205.05148
15. Hájek, P.: Metamathematics of Fuzzy Logic. Trends in Logic, vol. 4, Paperback edn. Kluwer Academic Publishers (1998)
16. Metcalfe, G., Montagna, F., Tsinakis, C.: Amalgamation and interpolation in ordered algebras. J. Alg. **402**, 21–82 (2014)
17. Montagna, F.: Interpolation and Beth's property in propositional many-valued logics: a semantic investigation. Ann. Pure. Appl. Log. **141**(1–2), 148–179 (2006)
18. Noguera, C.: Algebraic study of axiomatic extensions of triangular norm based fuzzy logics. Ph.D. thesis, IIIA-CSIC (2006)
19. Nola, A.D., Lettieri, A.: One chain generated varieties of MV-algebras. J. Alg. **225**(2), 667–697 (2000)

Comer Schemes, Relation Algebras, and the Flexible Atom Conjecture

Jeremy F. Alm[1], David Andrews[2], and Michael Levet[3(✉)]

[1] Department of Mathematics, Lamar University, Beaumont, TX 77707, USA
[2] Department of Mathematics, University of Dallas, Dallas, TX 75080, USA
andrews@udallas.edu
[3] Department of Computer Science, University of Colorado Boulder,
Boulder, CO 80309, USA
michael.levet@colorado.edu

Abstract. In this paper, we consider relational structures arising from Comer's finite field construction, where the cosets need not be sum free. These Comer schemes generalize the notion of a Ramsey scheme and may be of independent interest. As an application, we give the first finite representation of 34_{65}. We complement our upper bounds with some lower bounds. Using a SAT solver, we establish that neither 33_{65} nor 34_{65} are representable on fewer than 24 points.

Keywords: Flexible Atom Conjecture · Comer schemes · Representations

1 Introduction

Given a class of finite algebraic structures, it is natural to ask which members can be *instantiated* or *represented* over a finite set S, where there exist natural operations on S corresponding to the operations of the algebraic structure. In the setting of finite groups, the representation question is answered by Cayley's theorem: every finite group can be instantiated as a finite permutation group. In this paper, we consider the class of finite relation algebras, which are Boolean lattices that satisfy certain equational axioms that capture the notion of relational composition (see Sect. 2 for a more precise formulation). There exist finite relation algebras that do not admit representations even over infinite sets- see for instance [10] and the citations therein. It is essentially folklore that there are relation algebras that admit representations over infinite sets, but are not finitely representable. The so called *point algebra* is one such example[1]. On the other

---

[1] A proof will be included in the full version.

ML was partially supported by J. A. Grochow's NSF award CISE-2047756 and a Summer Research Fellowship through the Department of Computer Science at the University of Colorado Boulder. We wish to thank the anonymous referees for helpful feedback.

R. Glück et al. (Eds.): RAMiCS 2023, LNCS 13896, pp. 17–33, 2023.
https://doi.org/10.1007/978-3-031-28083-2_2

hand, Comer [11, Theorem 5.3] showed that every finite integral relation algebra with a flexible atom (i.e., an atom that does not participate in any forbidden diversity cycles) is representable over a countable set.

It is natural to ask whether Comer's result can be strengthened to hold in the setting of finite sets. This is precisely the *Flexible Atom Conjecture*, which states that every finite integral relation algebra with a flexible atom is representable over a finite set. Jipsen, Maddux, & Tuza showed that the finite symmetric integral relation algebras in which every diversity atom is flexible (denoted $\mathfrak{E}_{n+1}(1,2,3)$), are finitely representable. In particular, the algebra with n flexible atoms is representable over a set of size $(2+o(1))n^2$ [15]. We note that if all cycles are present, then all diversity atoms are flexible. Hence, the case considered in [15] is intuitively the *big end* of the Flexible Atom Conjecture.

The other extreme is when just enough cycles are present for one atom to be flexible. This case was handled by Alm, Maddux, & Manske [1], who exhibited a representation of the algebra A_n obtained from splitting the non-flexible diversity atom of 6_7 into n symmetric atoms. In particular, this construction yielded a representation of $A_2 = 32_{65}$ over a set of size $416,714,805,91$ (here, we use Maddux's [17] numbering for relation algebras such as 6_7 and 32_{65}). Dodd & Hirsch [13] subsequently improved the bound of the minimum representation size of 32_{65} to $63,432,274,896$. This was subsequently improved to 8192 by J.F. Alm & D. Sexton (unpublished), and later 3432 by [5]. Finally, in [2], the authors exhibited a representation over a set of size 1024 for 32_{65}, as well as the first polynomial upper bound on $\min(\mathrm{Spec}(A_n))$.

In the quest for finite representations, it is desirable to constrain the search space. This simultaneously motivates the study of small representations, the study of highly symmetric representations such as over groups, as well as lower bounds. There are few lower bounds in the literature. Jipsen, Maddux, & Tuza exhibited a lower bound of n^2+n+1 for the relation algebra $\mathfrak{E}_{n+1}(1,2,3)$ [15]. In [2], the authors showed that any representation of A_n requires at least $2n^2+4n+1$ points, which is asymptotically double the trivial lower bound of $n^2 + 2n + 3$. The key technique involved analyzing the combinatorial substructure induced by the flexible atom. In the special case of $A_2 = 32_{65}$, the authors used a SAT solver to further improve the lower bound. Namely, they showed that 32_{65} is not representable on a set of fewer than 26 points.

Alm & Levet [6] subsequently improved the lower bound of A_n to $2n^2 + \Omega(n\sqrt{\ln(n)})$ by incorporating the Ramsey number $R(3,k)$. Furthermore, Alm & Levet showed that 31_{37} is not representable using fewer than 16 points and 35_{37} is not representable using fewer than 14 points. In the process, Alm & Levet produced two generalizations of $31_{37}, 33_{37}$, and 35_{37}, obtaining lower bounds against these generalizations.

The notion of a Ramsey scheme was first introduced (but not so named) by Comer [12], where he used them to obtain finite representations of relation algebras. Kowalski [16] later introduced the term *Ramsey relation algebra* to refer to the (abstract) relation algebras obtained from embeddings into Ramsey schemes.

In [19, Problem 2.7], Maddux raised the question as to whether Ramsey schemes exist for all number of colors. Comer communicated this problem to Trotter; and in the mid-80's, Erdös, Szemerédi, & Trotter gave a purported proof on the existence of Ramsey schemes for sufficiently many colors. Trotter communicated this proof to Comer via email, who in turn communicated it to Maddux [18]. Unfortunately, the construction provided by Erdös, Szemerédi, & Trotter was not correct, as it did not satisfy the mandatory cycle condition. It remains open as to whether there even exist Ramsey schemes for infinitely many values of n. Similarly, it remains open as to whether all of the *small* (those with four or fewer atoms) integral symmetric relation algebras with at least one flexible atom outlined in [17] admit finite representations. As of 2009 [20], $33_{65}, 34_{65}$, and 59_{65} were the only symmetric 4-atom integral relation algebras with a flexible atom not known to admit finite representations. As of 2013 [22], 33_{37} and 35_{37} were the only non-symmetric 4-atom integral relation algebras with a flexible atom not known to admit finite representations. In 2017, Alm & Maddux [3] gave a finite representation for 59_{65}. In [6], Alm & Levet gave finite representations for 33_{37} and 35_{37}. In this paper, we give a finite representation for 34_{65}, which leaves 33_{65} as the only remaining 4-atom integral relation algebra with a flexible atom from [17] that is not known to admit a finite representation.

There has been considerable progress in constructing Ramsey schemes for a finite number of colors using a technique due to Comer [12]. Intuitively, Comer's method takes a positive integer m and considers finite fields $\mathbb{F}_q$, where $q \equiv 1$ (mod $2m$). We next consider the unique multiplicative subgroup $H \leq \mathbb{F}_q^\times$ of order $(q - 1)/m$, and check whether the cosets of $\mathbb{F}_q^\times/H$ yield a representation (namely, taking the cosets to be the atoms of the relation algebra). With the sole exception of an alternate construction of the 3-color algebra using $(\mathbb{Z}/4\mathbb{Z})^2$ [23], all known constructions have been due to the guess-and-check finite field method of Comer [12].

The m-color Ramsey number

$$R_m(3) = R(\underbrace{3, \ldots, 3}_{m})$$

provides an upper bound on q, restricting the search space for the finite fields to be considered.

Using this method, Comer [12] produced Ramsey schemes for $m = 2, 3, 4, 5$ colors. Using a computer, Maddux [21] extended this work for $m = 6, 7$. Independently, Kowalski [16] constructed Ramsey schemes over prime fields for $2 \leq m \leq 120$ colors (with the exception of $m = 8, 13$), and Alm & Manske [7] constructed Ramsey schemes over prime fields for $2 \leq m \leq 400$ colors, (again with the exception of $m = 8, 13$). Kowalski [16] also considered non-prime fields. Alm [4] subsequently produced Ramsey schemes for $2 \leq m \leq 2000$ colors (excluding $m = 8, 13$) using the fast algorithm of [8]. He also substantially improved the upper bound on p with respect to m, to $p < m^4 + 5$, finally showing that no construction over prime fields exists for $m = 13$ [4]. Alm & Levet [6] further improved this bound to $p < m^4 - (2 - o(1))m^3 + 5$.

Alm & Levet [6] generalized the notion of a Ramsey scheme to the directed (antisymmetric) setting. As a consequence, they gave finite representations for the relation algebras $33_{37}, 35_{37}, 77_{83}, 78_{83}, 80_{83}, 82_{83}, 83_{83}, 1310_{1316}, 1313_{1316}, 1315_{1316}$, and 1316_{1316}. Only 83_{83} and 1316_{1316} were previously known to be finitely representable, by a slight generalization of [15].

Given the success of Comer's method [12], we ask the following inverse question: what relation algebras admit finite constructions via Comer's method? We investigate this question here.

Main Results. In this paper, we extend the notion of a Ramsey scheme [12] by relaxing the notion that the cosets need to be sum-free. We refer to such relational structures as *Comer schemes*. A Comer scheme naturally generates an abstract integral symmetric relation algebra. A priori, the cycle structure of these algebras is not clear. We investigate this question here.

As a first application, we provide the first finite representation for the relation algebra 34_{65}. This relation algebra has four symmetric atoms $1', a, b$, and c, with forbidden cycles bbc and ccb. The atom a is flexible. Hence, by [11, Theorem 5.3], 34_{65} admits a representation over a countable set.

Theorem 1. *The relation algebra 34_{65} admits a representation on $p = 3697$ points.*

We obtain our finite representation of 34_{65} by embedding it into the integral symmetric relation algebra with 24 diversity atoms $a_0, \ldots, a_{23}$. The forbidden cycles are of the form $\{a_i a_i a_{i+12} : 0 \le i \le 23\}$, where the indices $i, i, i+12$ are all taken modulo 24. This relation algebra admits a finite representation via a Comer scheme. See Sect. 5 for full details.

We next turn to investigating the cycle structure of these algebras. We are in fact able to produce a large number of these objects for different forbidden cycle configurations.

Theorem 2. *We have the following.*

(a) *For $n \in \{1, \ldots, 2000\} \setminus \{8, 13\}$, the integral symmetric relation algebra on n atoms with forbidden cycles $\{a_i a_i a_i : 0 \le i < n\}$ admits a finite representation over a prime field Comer scheme.*

(b) *For $n \in \{5, \ldots, 14\} \cup \{16, \ldots, 33\} \cup \{35, \ldots, 500\}$, the integral symmetric relation algebra on n atoms with forbidden cycles $\{a_i a_i a_{i+1} : 0 \le i < n\}$, where the indices $i, i, i+1$ are all taken modulo n, admits a finite representation over a prime field Comer scheme.*

(c) *For $n \in \{5, \ldots, 500\} \setminus \{12, 14, 18, 28, 36\}$, the integral symmetric relation algebra on n atoms with forbidden cycles $\{a_i a_i a_{i+2} : 0 \le i < n\}$, where the indices $i, i, i+2$ are all taken modulo n, admits a finite representation over a prime field Comer scheme.*

Remark 1. We note that when n is odd, forbidding the cycles $\{a_i a_i a_{i+1} : 0 \le i < n\}$ (where all indices are taken modulo n) yields the same relation algebra as when we forbid the cycles $\{a_i a_i a_{i+2} : 0 \le i < n\}$ (where all indices are again taken modulo n). See Lemma 3.

Remark 2. We note that Ramsey schemes also have close connections with other combinatorial structures such as association schemes, coherent configurations, and permutation groups [11,12]. Thus, our Comer scheme constructions may be of independent interest.

We complement our upper bounds with some lower bounds. Here, we consider not only 34_{65}, but also 33_{65}, which is the integral symmetric relation algebra with atoms $1', a, b, c$, with a flexible and forbidden cycles ccc, bcc, cbb. By analyzing the combinatorial structure of the relation algebras in tandem with a SAT solver, we obtain the following.

Theorem 3. *We have the following.*

(a) Any square representation of 33_{65} requires at least 24 points.
(b) Any square representation of 34_{65} requires at least 24 points.

2 Preliminaries

Definition 1. *A relation algebra is an algebra $\langle A, \wedge, \vee, \neg, 0, 1, \circ, \breve{}, 1' \rangle$ that satisfies the following.*

- $\langle A, \wedge, \vee, \neg, 0, 1 \rangle$ *is a Boolean algebra, with $\neg$ our unary negation operator, 0 the identity for $\vee$, and 1 the identity for $\wedge$,*
- $\langle A, \circ, 1' \rangle$ *is a monoid, with $1'$ our nullary identity. That is, relational composition is associative, and there is an identity relation $1'$.*
- $\breve{}$ *is our unary converse operation and is an involution with respect to composition. Namely, $\breve{\breve{a}} = a$ for all $a \in A$, and $(a \circ b)\breve{} = \breve{b} \circ \breve{a}$ for all $a, b \in A$,*
- *Converse and composition both distribute over disjunction. Precisely, for all $a, b, c \in A$, we have that:*

$$(a \vee b)\breve{} = \breve{a} \vee \breve{b}, \text{ and}$$
$$(a \vee b) \circ c = (a \circ c) \vee (b \circ c).$$

- *(Triangle Symmetry) For all $a, b \in A$, we have that:*

$$(a \circ b) \wedge c = 0 \iff (\breve{c} \circ a) \wedge \breve{b} = 0 \iff (b \circ \breve{c}) \wedge \breve{a} = 0.$$

Note that the Triangle Symmetry axiom defines an equivalence relation on triples of diversity atoms that corresponds to the symmetries of the triangle.

When the relation algebra is understood, we simply write A rather than $\langle A, \wedge, \vee, \neg, 0, 1, \circ, \breve{}, 1' \rangle$.

Definition 2. *Let A be a relation algebra. We say that $a \in A$ is an atom if $a \neq 0$ and $b < a \implies b = 0$. Furthermore, we say that a is a diversity atom if a also satisfies $a \wedge 1' = 0$.*

Definition 3. *For diversity atoms* a, b, c, *the triple* (a, b, c)—*usually denoted* abc—*is called a* diversity cycle. *We say that* abc *is* forbidden *if* $(a \circ b) \wedge c = 0$ *and* mandatory *if* $a \circ b \geq c$. *We say that a diversity atom is* symmetric *if* $a = \breve{a}$.

Remark 3. We note that for integral relation algebras, the composition operation $\circ$ is determined by the mandatory diversity cycles.

Definition 4. *Let* f *be a symmetric diversity atom. We say that* f *is* flexible *if for all diversity atoms* a, b, *we have that* abf *is mandatory.*

Definition 5. *We say that a relation algebra* A *is* representable *if there exists a set* U *and an equivalence relation* $E \subseteq U \times U$ *such that* A *embeds into*

$$\langle 2^E, \cup, {}^c, \circ, {}^{-1}, Id_E \rangle.$$

Here, ^{c} *is set complementation, and* ${}^{-1}$ *is the relational inverse.*

In this paper, we will only be concerned with *simple* relation algebras, where there exists a set U such that $E = U \times U$. We call such a representation *square*.

Definition 6. *Let* A *be a finite relation algebra. Define*

$$Spec(A) := \{\alpha \leq \omega : A \text{ has a square representation over a set of cardinality } \alpha\}.$$

Remark 4. A relation algebra admits a square representation on m points precisely if we can color a complete graph on m vertices to satisfy prescribed constraints.

2.1   Ramsey Schemes

Definition 7. *Let* U *be a set, and* $m \in \mathbb{Z}^+$. *A* Ramsey scheme *in* m *colors is a partition of* $U \times U$ *into* $m + 1$ *binary relations* $Id, R_0, \ldots, R_{m-1}$ *such that the following conditions hold:*

(A) $R_i^{-1} = R_i$,
(B) $R_i \circ R_i = R_i^c$, *and*
(C) For all pairs of distinct $i, j \in [m-1]$, $R_i \circ R_j = Id^c$.

Here, $Id = \{(u, u) : u \in U\}$ *is the identity relation over* U.

The usual method of constructing the relations $R_0, \ldots, R_{m-1}$ is a *guess-and-check* approach due to Comer [12], which works as follows. Fix $m \in \mathbb{Z}^+$. We search over primes $p \equiv 1 \pmod{2m}$, where a desirable p satisfies the following. Let $X_0 := H \leq \mathbb{F}_p^\times$ be the unique subgroup of order $(p-1)/m$. Now let $X_1, \ldots, X_{m-1}$ be the cosets of $\mathbb{F}_p^\times / X_0$. In particular, as $\mathbb{F}_p^\times$ is cyclic, we may write $X_i = g^i X_0 = \{g^{am+i} : a \in \mathbb{Z}^+\}$, where g is a generator of $\mathbb{F}_p^\times$. Suppose that $X_0, \ldots, X_{m-1}$ satisfy the following conditions:

(a) $X_i = -X_i$, for all $0 \le i \le m - 1$,
(b) $X_i + X_i = \mathbb{F}_p \setminus X_i$, for all $0 \le i \le m - 1$, and
(c) For all distinct $0 \le i, j \le m - 1$, $X_i + X_j = \mathbb{F}_p \setminus \{0\}$.

For each $0 \le i \le m - 1$, define $R_i := \{(x, y) \in \mathbb{F}_p \times \mathbb{F}_p : x - y \in X_i\}$. Here, the sets $R_0, \ldots, R_{m-1}$ together with $\mathrm{Id} = \{(u, u) : u \in \mathbb{F}_p\}$ are the atoms in our relation algebra. It is easy to check that conditions (a)–(c) on the sets $X_0, \ldots, X_{m-1}$ imply that conditions (A)–(C) from the definition of a Ramsey scheme are satisfied for the relations $R_0, \ldots, R_{m-1}$.

We note that condition (b), that $X_i + X_i = \mathbb{F}_p \setminus X_i$, indicates that each X_i is sum-free. The fact that $p \equiv 1 \pmod{2m}$ implies that X_0 has even order. It follows that X_0 is symmetric; i.e., $X_0 = -X_0$. In [4], Ramsey schemes were constructed for all $m \le 2000$ except for $m = 8, 13$, and it was shown that if $p > m^4 + 5$, then X_0 contains a solution to $x + y = z$. In such cases, Comer's construction fails to yield an m-color Ramsey scheme.

3   Integral Symmetric Relation Algebras with Forbidden Cycle Configurations

In this section, we generalize the notion of a Ramsey scheme to the setting where the cosets need not be sum-free. We then examine the relation algebras generated by these schemes.

Definition 8. *Let p be a prime, $U = \mathbb{F}_p^\times$, and m a divisor of $(p - 1)/2$. Let $H \le \mathbb{F}_p^\times$ be the unique subgroup of index m, and let $X_0, \ldots, X_{m-1}$ be the cosets of $\mathbb{F}_p^\times/H$. A Comer scheme in m colors is a partition of $U \times U$ into $m + 1$ binary relations $\mathrm{Id}, R_0, \ldots, R_{m-1}$ as follows. Let $\mathrm{Id} = \{(x, x) : x \in \mathbb{F}_p\}$ and $R_i = \{(x, y) \in \mathbb{F}_p \times \mathbb{F}_p : x - y \in X_i\}$.*

Remark 5. We will be particularly interested in Comer schemes that satisfy the following additional condition: for all pairs of distinct $i, j \in [m-1]$, $R_i \circ R_j = \mathrm{Id}^c$. Here, $\mathrm{Id} = \{(u, u) : u \in U\}$ is the identity relation over U.

We construct Comer schemes using Comer's method [12]. Fix $m \in \mathbb{Z}^+$, and let $p \equiv 1 \pmod{2m}$ be a prime. Let $X_0 := H \le \mathbb{F}_p^\times$ be a subgroup of order $(p - 1)/m$. Now let $X_1, \ldots, X_{m-1}$ be the cosets of $\mathbb{F}_p^\times/X_0$. In particular, as $\mathbb{F}_p^\times$ is cyclic, we may write $X_i = g^i X_0 = \{g^{am+i} : a \in \mathbb{Z}^+\}$, where g is a generator of $\mathbb{F}_p^\times$. Suppose that $X_0, \ldots, X_{m-1}$ satisfy the following conditions:

(a) $X_i = -X_i$, for all $0 \le i \le m - 1$,
(b) For all distinct $0 \le i, j \le m - 1$, $X_i + X_j = \mathbb{F}_p \setminus \{0\}$.

For each $0 \le i \le m - 1$, define $R_i := \{(x, y) \in \mathbb{F}_p \times \mathbb{F}_p : x - y \in X_i\}$. Here, the sets $R_0, \ldots, R_{m-1}$ are the atoms in our relation algebra. It is easy to check that conditions (a) and (b) on the sets $X_0, \ldots, X_{m-1}$ imply that conditions (A) and (B) from the definition of a Comer scheme are satisfied for the relations $R_0, \ldots, R_{m-1}$.

The fact that $p \equiv 1 \pmod{2m}$ implies that X_0 has even order. It follows that X_0 is symmetric; i.e., $X_0 = -X_0$.

We collect some preliminary observations about these Comer schemes. First, we observe that Comer's construction yields relation algebras that have *rotational symmetry*.

Lemma 1. *Let $n \in \mathbb{Z}^+$ and let $p = nk + 1$ be a prime number and g a primitive root modulo p.*

For $i \in \{0, 1, \ldots, n-1\}$, define

$$X_i = \left\{ g^i, g^{n+i}, g^{2n+i}, \ldots, g^{(k-1)n+i} \right\} \subseteq \mathbb{Z}/p\mathbb{Z}.$$

Then $(X_0 + X_j) \cap X_k = \emptyset$ if and only if $(X_i + X_{i+j}) \cap X_{i+k} = \emptyset$.

Proof. Multiply through by g^i.

We now formalize the notion of an automorphism that respects this rotational symmetry.

Definition 9. *Let $p = nk + 1$ be prime with k even, and let the X_i's as be as in Lemma 1. Let $C(p, n)$ denote the proper relation algebra generated by the sets R_i as in the definition of a Comer scheme. Then automorphisms of $C(p, n)$ can be seen as permutations of the atoms, as follows.*

$$\mathrm{Aut}(C(p, n)) = \{ \pi \in S_n : A_i \circ A_j \supseteq A_k \Leftrightarrow A_{\pi(i)} \circ A_{\pi(j)} \supseteq A_{\pi(k)} \}.$$

Of course, the condition on the relations R_i is equivalent to the following condition on the X_i's:

$$X_i + X_j \supseteq X_k \Leftrightarrow X_{\pi(i)} + X_{\pi(j)} \supseteq X_{\pi(k)}.$$

Remark 6. Lemma 1 implies that for any Comer algebra $C(p, n)$ in n colors, we have that $\mathrm{Aut}(C)$ contains a subgroup isomorphic to $\mathbb{Z}/n\mathbb{Z}$.

It is natural to ask about relation algebras with forbidden cycles. In particular, given a forbidden cycle scheme and a prescribed number of atoms n, can we realize the corresponding relation algebra with a Comer scheme? This motivates the following definition.

Definition 10. *Let $\mathcal{A}_n([i, i+j, i+\ell])$ denote the integral RA with n symmetric diversity atoms $a_0, \ldots, a_{n-1}$ whose forbidden cycles are those of the form $\{a_i a_{i+j} a_{i+\ell} : 0 \le i < n\}$, with indices considered modulo n.*

Remark 7. We note that for any m-color Comer scheme, the automorphism group has a copy of $\mathbb{Z}/m\mathbb{Z}$. Thus, if we forbid one cycle—say $a_0 a_{0+j} a_{0+\ell}$—then we forbid $a_i a_{i+j} a_{i+\ell}$ for all i.

We will be particularly interested in $\mathcal{A}_n([i, i, i+j])$, as these forbid a rotational class of 2-cycles (bichromatic triangles). Furthermore, setting $j = 0$ yields Ramsey schemes.

Lemma 2. $\mathrm{Aut}(\mathcal{A}_n([i,i,i+1])) \cong \mathbb{Z}/n\mathbb{Z}$.

Proof. Suppose $\pi \in \mathrm{Aut}(\mathcal{A}_n([i,i,i+1]))$ and $\pi(0) = x$. Since $a_0 a_0 a_1$ is forbidden, $a_{\pi(0)} a_{\pi(0)} a_{\pi(1)}$ is forbidden as well, so $a_x a_x a_{\pi(1)}$ is forbidden. But this forces $\pi(1) = x + 1 \pmod{n}$, since if xxy is forbidden, y must be $x + 1$. Similarly, $\pi(2)$ must be $x + 2 \pmod{n}$, and so on. So π must take the form $\pi(s) = x + s \pmod{n}$. All such permutations are clearly in $\mathrm{Aut}(\mathcal{A}_n([i,i,i+1]))$, and we have shown they are the only ones. So $\mathrm{Aut}(\mathcal{A}_n([i,i,i+1])) \cong \mathbb{Z}/n\mathbb{Z}$. $\square$

Lemma 3. *If* $\gcd(j,n) = 1$, *then* $\mathcal{A}_n([i,i,i+j]) \cong \mathcal{A}_n([i,i,i+1])$.

Proof. Let $\rho : \mathrm{At}(\mathcal{A}_n([i,i,i+1])) \rightarrow \mathrm{At}(\mathcal{A}_n([i,i,i+j]))$ be given by $a_i \mapsto a_{j \cdot i \pmod{n}}$. Since $\gcd(j,n) = 1$, ρ is a bijection. It is easy to check that ρ preserves the forbidden cycles. $\square$

The next lemma tells us that for Comer algebras, the isomorphism in Lemma 3 arises in a particularly nice way.

Lemma 4. *If* $C(p,n)$ *has forbidden cycles* $[X_i, X_i, X_{i+j}]$ *and* $\gcd(j,n) = 1$, *then* X_j *contains a primitive root* g, *and reindexing using* g *as a generator will give forbidden cycles* $[X_i, X_i, X_{i+1}]$.

Proof. Let g be the primitive root that gives the indexing with forbidden cycles $[X_i, X_i, X_{i+j}]$. Now g^ℓ is also a primitive root modulo p if $\gcd(\ell, p-1) = 1$. We want $g^\ell \in X_j$, so we want to find an integer a with $\gcd(an+j, p-1) = 1$. Since $\gcd(j,n) = 1$, Dirichlet's theorem on primes in arithmetic progressions gives some prime $p' = an + j$, and clearly $\gcd(p', p-1) = 1$. Then $g^{p'}$ is a primitive root and is in X_j. $\square$

Lemma 5. *If* $\gcd(j,n) > 1$, *then* $\mathrm{Aut}(\mathcal{A}_n([i,i,i+j]))$ *contains a non-identity permutation* π *that has fixed points. Hence* $\mathcal{A}_n([i,i,i+j]) \not\cong \mathcal{A}_n([i,i,i+1])$.

Proof. Let $x = \gcd(j,n) > 1$. Consider the permutation $\pi = (0 \; x \; 2x \; 3x \ldots)$, written in cycle notation. We claim that $\pi \in \mathrm{Aut}(\mathcal{A}_n([i,i,i+j]))$. Consider the forbidden cycle $a_0 a_0 a_j$. Write $j = bx$ for some positive integer b. Under π, this cycle $a_0 a_0 a_{bx}$ gets mapped to $a_x a_x a_{(b+1)x}$, and since $(b+1)x = x + j$, the cycle $a_x a_x a_{(b+1)x}$ is forbidden. In fact, π just permutes the forbidden cycles $a_{\ell x} a_{\ell x} a_{\ell x + j}$ and leaves the other forbidden cycles fixed. $\square$

Example 1. Consider $\mathcal{A}_6([i,i,i+2])$. Then $j = x = 2$. The permutation $(0\;2\;4)$ permutes the forbidden cycles $a_0 a_0 a_2$, $a_2 a_2 a_4$, and $a_4 a_4 a_0$. See Fig. 1.

The following lemma from Alon and Bourgain gives us just what we need to show that if p is large relative to n, then $C(p,n)$ has only flexible diversity atoms.

Lemma 6 ([9], **Proposition 1.4).** *Let* q *be a prime power and let* A *be a multiplicative subgroup of the finite field* $\mathbb{F}_q$ *of size* $|A| = d \geq q^{1/2}$. *Then, for any two subsets* $B, C \subset \mathbb{F}_q$ *satisfying* $|B||C| \geq q^3/d^2$, *there are* $x \in B$ *and* $y \in C$ *so that* $x + y \in A$.

Lemma 7. *If $p > n^4 + 5$, then every diversity atom of $C(p,n)$ is flexible.*

Proof. We need to show $(X_i + X_j) \cap X_0 \neq \varnothing$ for arbitrary i and j. Set $q = p$, $A = X_0$, $B = X_i$, and $C = X_j$ in Lemma 6. Then $|A| = |B| = |C| = (p-1)/n$. Then we need $|B||C| \geq q^3/d^2$, which translates to $(p-1)^4 \geq n^4 p^3$, which is satisfied when $p > n^4 + 5$. Then all diversity cycles are mandatory by Lemma 1.

4 Constructing Comer Schemes

In this section, we document our constructions for Theorem 2. Some data are summarized in Table 1. While for some small n, there is no construction of a Comer RA representation for $\mathcal{A}_n([i,i,i+j])$ for $j = 0, 1, 2$, it would seem for large enough n there is always some modulus p that works.

Representations of $\mathcal{A}_n([i,i,i+1])$ exist for all $35 \leq n \leq 500$. In Fig. 2, we compare the smallest modulus p for representations over $C(p,n)$ for $\mathcal{A}_n([i,i,i])$ vs $\mathcal{A}_n([i,i,i+1])$. The growth is a bit slower for the latter.

5 A Cyclic Group Representation of Relation Algebra 34_{65}

As an application, we give the first known finite representation of 34_{65}. Relation algebra 34_{65} has four symmetric atoms $1'$, a, b, and c, with forbidden cycles bbc and ccb. The atom a is flexible, hence 34_{65} is representable over a countable set.

We noticed that it would be sufficient to find a prime $p = nk + 1$, k and n both even, such that $C(p,n)$ has $[i,i,i+n/2]$ as its only forbidden class. Then we could map b to X_0, map c to $X_{n/2}$, and map a to the union of all the other X_i's; in other words, 34_{65} embeds in $\mathcal{A}_n([i,i,i+n/2])$ for all even $n > 4$. There's no limit to how big p can be, since n can also be as large as necessary; we just throw "everything else" into the image of a. A computer search using the fast algorithm from [8] quickly found a hit: for $p = 3697$ and $n = 24$, $[i,i,i+12]$ alone is forbidden. (This indexing is for the primitive root $g = 5$ in $\mathbb{F}_{3697}$.)

6 Lower Bounds

6.1 $\mathcal{A}_n([i, i+j, i+\ell])$

Proposition 1. *Let $i,j,\ell \in \{0,\ldots,n-1\}$. Any finite representation of $\mathcal{A}_n([i,i+j,i+\ell])$ requires at least n^2 points. If furthermore $j = \ell$, then at least $n^2 + 1$ points are required.*

Proof. We phrase the proof in the language of graph theory (see Remark 4). The atoms constitute the edge colors of a complete graph. Fix $i \in [n]$, and let uv be an edge colored a_i (where here, a_i is a diversity atom of $\mathcal{A}_n([i,i+j,i+\ell])$).

Any non-forbidden triangle must be included in our graph. Let $c_1, c_2, c_3 \in \mathcal{A}_n([i,i+j,i+\ell])$ be diversity atoms (colors on the graph). Suppose that $c_1 c_2 c_3$

is a mandatory triangle. If xy is an edge colored c_1 and w is a vertex such that xw is colored c_2 and yw is colored c_3, then we say that w witnesses the (c_2, c_3) need for xy.

If $j \neq \ell$, then $(a_{i+j}, a_{i+\ell})$ and $(a_{i+\ell}, a_{i+j})$ are forbidden. So uv has $n^2 - 2$ needs. Together with u, v, this yields n^2 points in total.

If instead $j = \ell$, we only forbid (a_{i+j}, a_{i+j}). Thus, uv has $n^2 - 1$ needs in this case. Together with u, v, this yields $n^2 + 1$ points in total.

Remark 8. In light of Proposition 1, we obtain that the representations we found for $\mathcal{A}_n([i, i, i])$ with $n = 1, 2$ were indeed minimal representations. Proposition 1 only provides a lower bound of 10 points for $\mathcal{A}_3([i, i, i])$. However, using a SAT solver (see here), we were able to show that there is no representation on fewer than 13 points. Thus, the representation we found was indeed minimal. (This result seems to be folklore).

The algebra $\mathcal{A}_3([i, i, i])$, is in fact the 3-color Ramsey algebra; that is, the relation algebra with diversity atoms $1', a, b, c$ and forbidden cycles aaa, bbb, ccc. It is folklore amongst relation algebra specialists that $\mathrm{Spec}(\mathcal{A}_3([i, i, i])) = \{13, 16\}$, though a proof appears not to have been written down. Our result verifies that the minimum element in $\mathrm{Spec}(\mathcal{A}_3([i, i, i]))$ is indeed 13.

In the case of $n = 4$, we used a SAT solver (see here) to establish a lower bound of 21 points required to represent $\mathcal{A}_4([i, i, i])$. In particular, it follows that if there exists a prime field Comer scheme representation, then at least 23 points are required. Details regarding the SAT Solver will appear in the full version.

6.2 33_{65}

In this section, we consider the relation algebra 33_{65}, which has atoms $1', a, b, c$, with a flexible and the following cycles forbidden: ccc, bcc, cbb. We use the language of graph theory to discuss the relation algebra (see Remark 4). The atoms constitute the edge colors of a complete graph. We use the color red to correspond to the flexible atom, green to correspond to the atom b, and blue to correspond to the atom c. Thus, the forbidden triangles are precisely the ones containing all-blue edges, or only blue and green edges with at least one blue edge.

Lemma 8. *If 33_{65} is finitely representable, then any finite representation must have at least 15 points.*

Proof. We refer to Fig. 3. Let $u_0 u_1$ be colored according to the flexible atom, for which we use the color red. The needs of $u_0 u_1$ are precisely $\{r, b, g\} \times \{r, b, g\}$, yielding a total of 11 points. Now fix $i = 0, 1$. Suppose that $u_i x_j, u_i x_k$ incident to u_i where at least one such edge is blue, and the other is either blue or green. As the all blue, blue-blue-green, and blue-green-green triangles are forbidden, $x_j x_k$ is necessarily red. In particular, this implies that the following edges are red:

- $x_1 : x_1x_2, x_1x_3, x_1x_4, x_1x_5, x_1x_6, x_1x_7, x_1x_8$ (7 edges)
- $x_2 : x_2x_3, x_2x_5, x_2x_6, x_2x_7$ (4 edges)
- $x_3 : x_3x_4, x_3x_5, x_3x_6, x_3x_7$ (4 edges)
- $x_4 : x_4x_5, x_4x_6, x_4x_8$ (3 edges)
- $x_5 : x_5x_6, x_5x_8$ (2 edges)

Consider the red edge x_1x_5, which has 9 needs. The b-g need is met by u_1u_2. Now the r-r need is met by x_2, x_3, x_4, x_6, x_8. We can meet the r-g need with x_7 and an arbitrary need with x_9. This leaves 4 unsatisfied needs, necessitating 4 additional points. So we require at least 15 points.

We now use a SAT solver to improve our lower bound. For $n \geq 16$, we build a Boolean formula $\Phi^{(n)}$ whose satisfiability is a necessary condition for 33_{65} to be representable on n points. So if $\Phi^{(n)}$ is not satisfiable, then 33_{65} does not admit a representation on n points. Details regarding the SAT solver will appear in the full version.

Lemma 9. *If* $16 \leq n \leq 23$, *then* $n \notin \mathrm{Spec}(33_{65})$.

Remark 9. Our SAT solver code can be found here.

6.3 34_{65}

We again phrase our results in the language of graph theory (see Remark 4). We use the color red to correspond to the flexible atom, green to correspond to the atom b, and blue to correspond to the atom c. Thus, the forbidden triangles are the blue-blue-green and blue-green-green ones.

Lemma 10. *Any finite representation of* 34_{65} *must have at least* 13 *points.*

Proof. The proof is similar to that of Lemma 8, referring instead to Fig. 4. Details will appear in the full version.

We use a SAT solver almost similarly as in the case of 33_{65} to improve the lower bound. Details regarding the SAT solver will appear in the full version.

Theorem 4. *Let* $13 \leq n \leq 23$. *Then* $n \notin \mathrm{Spec}(34_{65})$.

Remark 10. Our SAT solver code can be found here.

7   Conclusion

We extended the notion of a Ramsey scheme by relaxing the condition that the cosets need to be sum-free. Using this combinatorial construction, we investigated the integral symmetric relation algebras $\mathcal{A}_n([i, i+j, i+\ell])$ that forbid the cycle configurations $\{a_i a_{i+j} a_{i+\ell} : 0 \leq i < n\}$. As an application, we showed that 34_{65} admits a finite representation over the Comer scheme $C(24, 3697)$.

We also established several lower bounds. We showed that the minimum element in $\mathrm{Spec}(\mathcal{A}_3([i,i,i]))$ is at least 13 (folklore- see Remark 8), and the minimum element in $\mathrm{Spec}(\mathcal{A}_4([i,i,i]))$ is at least 21. Furthermore, we showed that the minimum element in both $\mathrm{Spec}(33_{65})$ and $\mathrm{Spec}(34_{65})$ is at least 24.

We conclude with several open problems.

Conjecture 1 (Strong Flexible Atom Conjecture). Every finite integral RA with a flexible atom is representable over a Comer scheme.

This paper contains some evidence for this conjecture, as does [6], which contains many new finite representations over Comer schemes. The infinite families of Directed Anti-Ramsey algebras from [6], and the infinite family $\{\mathcal{A}_n([i,i,i+1]) : n > 4\}$, both appear to be "mostly" representable, based on the evidence. Except for some exceptions for small n, Comer schemes seem to yield representations. The quasirandom nature of the sum-product interaction in finite fields (see [14] for example) seems to suggest a heuristic similar to one used in number theory: *any potential structure in the primes not ruled out by obvious considerations can probably be found.* Case in point: we looked for an n and a p such that $\mathcal{A}_n([i,i,i+n/2])$ was representable over $\mathbb{F}_p$. And we found one. There are likely infinitely many such n and p, and 34_{65} will embed in all of them.

There are (at least) two ingredients needed to complete the proof: (i) prove that Comer schemes are actually quasirandom in a relevant sense; and (ii) prove that for every RA with a flexible atom, there is an algebra (for instance, $\mathcal{A}_n([i,i+j,i+\ell])$, but presumably more general) into which it embeds that admits a representation over a Comer scheme.

Problem 1. Formulate a suitable notion of quasirandomness for sequences of relation algebra atom structures or Comer schemes.

The relation algebra 33_{65} is not known to admit a finite representation. In fact, it is the last remaining algebra in the family N_{65} that has a flexible atom and for which no finite representation is known.

Problem 2. Find a forbidden scheme that would admit a representation of 33_{65}.

Problem 3. The third author conjectures that there may exist a group representation of 33_{65} over a symmetric group. As the minimum element of $\mathrm{Spec}(33_{65})$ is at least 24, we ask whether 33_{65} admits a group representation over S_4. Similarly, we ask whether 33_{65} admits a group representation over S_5.

Problem 4. Algebra $\mathcal{A}_3([i,i,i+1])$ is 42_{65} and is not representable. Algebra $\mathcal{A}_4([i,i,i+1])$ is not representable via Comer's method. Is it representable by some other method?

Problem 5. Algebra $\mathcal{A}_6([i,i,i+2])$ is not representable via Comer's method. Is it representable by some other method?

Problem 6. Algebra $\mathcal{A}_5([i,i,i+1])$ is representable over $\mathbb{F}_{61}$. Is it representable over an infinite set? Over $\mathbb{Z}$?

A   Tables and Figures

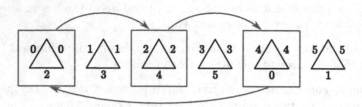

Fig. 1. Depiction of the action of the permutation (0 2 4) on the forbidden cycles of $\mathcal{A}_6([i, i, i+2])$.

Table 1. Smallest modulus for a representation over a $C(p, n)$, or x if none exists. The "–" indicates that an entry is redundant (in light of Lemma 3).

n	$\mathcal{A}_n([i, i, i]))$	$\mathcal{A}_n([i, i, i+1]))$	$\mathcal{A}_n([i, i, i+2]))$
1	2	x	x
2	5	x	x
3	13	x	x
4	41	x	x
5	71	61	–
6	97	109	x
7	491	127	–
8	x	257	x
9	523	307	–
10	1181	641	421
11	947	331	–
12	769	673	x
13	x	667	–
14	1709	953	x
15	1291	x	x
16	1217	2593	1697

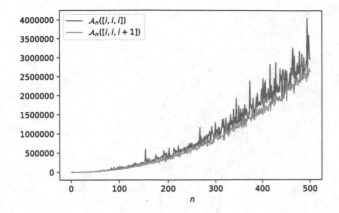

Fig. 2. Smallest modulus p over which $\mathcal{A}_n([i,i,i])$ and $\mathcal{A}_n([i,i,i+1])$ are representable as a $C(p,n)$

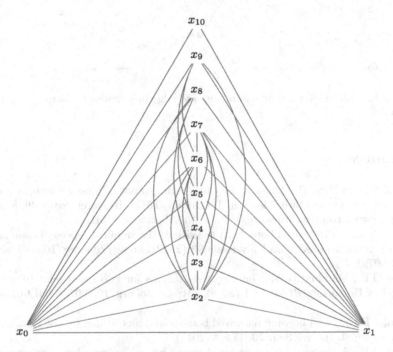

Fig. 3. Subgraph which must appear off any red edge in a representation of 33_{65}. (Color figure online)

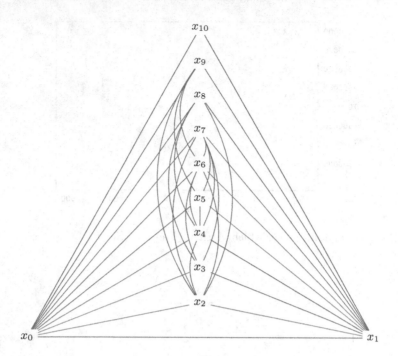

Fig. 4. Subgraph which must appear off any red edge in a representation of 34_{65}. (Color figure online)

References

1. Alm, J., Maddux, R., Manske, J.: Chromatic graphs, Ramsey numbers and the flexible atom conjecture. Electron. J. Combin. **15**(1), Research paper 49, 8 (2008). http://www.combinatorics.org/Volume_15/Abstracts/v15i1r49.html
2. Alm, J., et al.: Improved bounds on the size of the smallest representation of relation algebra 32_{65}. Algebra Universalis **83** (2022). https://doi.org/10.1007/s00012-022-00791-4
3. Alm, J.F., Maddux, R.D.: Finite representations for two small relation algebras. Algebra Universalis **79**(4), 1–4 (2018). https://doi.org/10.1007/s00012-018-0570-4
4. Alm, J.F.: 401 and beyond: improved bounds and algorithms for the Ramsey algebra search. J. Integer Seq. **20**, 17.8.4 (2017)
5. Alm, J.F., Andrews, D.A.: A reduced upper bound for an edge-coloring problem from relation algebra. Algebra Universalis **80**(2), 1–11 (2019). https://doi.org/10.1007/s00012-019-0592-6
6. Alm, J.F., Levet, M.: Directed Ramsey and anti-Ramsey schemes and the flexible atom conjecture (2022). https://doi.org/10.48550/ARXIV.1901.06781
7. Alm, J.F., Manske, J.: Sum-free cyclic multi-bases and constructions of Ramsey algebras. Discret. Appl. Math. **180**, 204–212 (2015). https://doi.org/10.1016/j.dam.2014.08.002

8. Alm, J.F., Ylvisaker, A.: A fast coset-translation algorithm for computing the cycle structure of Comer relation algebras over Z/pZ. Theor. Comput. Sci. **791**, 127–131 (2019). https://doi.org/10.1016/j.tcs.2019.05.019
9. Alon, N., Bourgain, J.: Additive patterns in multiplicative subgroups. Geom. Funct. Anal. **24**(3), 721–739 (2014). https://doi.org/10.1007/s00039-014-0270-y
10. Andrèka, H., Németi, I., Givant, S.: Nonrepresentable relation algebras from groups. Rev. Symb. Logic **13**(4), 861–881 (2020). https://doi.org/10.1017/S1755020319000224
11. Comer, S.: Combinatorial aspects of relations. Algebra Universalis **18**, 77–94 (1984). https://doi.org/10.1007/BF01182249
12. Comer, S.D.: Color schemes forbidding monochrome triangles. In: Proceedings of the Fourteenth Southeastern Conference on Combinatorics, Graph Theory and Computing (Boca Raton, Fla., 1983), vol. 39, pp. 231–2366 (1983)
13. Dodd, L., Hirsch, R.: Improved lower bounds on the size of the smallest solution to a graph colouring problem, with an application to relation algebra. J. Relational Methods Comput. Sci. **2**, 18–26 (2013)
14. Green, B.: Sum-product phenomena in $\mathbb{F}_p$: a brief introduction. arXiv:0904.2075 [math.NT] (2009)
15. Jipsen, P., Maddux, R.D., Tuza, Z.: Small representations of the relation algebra $\mathcal{E}_{n+1}(1,2,3)$. Algebra Universalis **33**(1), 136–139 (1995)
16. Kowalski, T.: Representability of Ramsey relation algebras. Algebra Universalis **74** (2015). https://doi.org/10.1007/s00012-015-0353-0
17. Maddux, R.: Relation Algebras. Studies in Logic and the Foundations of Mathematics, vol. 150. Elsevier B.V., Amsterdam (2006)
18. Maddux, R.: Personal communication
19. Maddux, R.: Some varieties containing relation algebras. Trans. Am. Math. Soc. **272**, 501 (1982). https://doi.org/10.2307/1998710
20. Maddux, R.: Edge-coloring problems (2009)
21. Maddux, R.: Do all the Ramsey algebras exist? (2011). Talk presented at the AMS sectional meeting in Iowa City, 18 March 2011
22. Maddux, R.: Personal communication (2013)
23. Whitehead, E.G.: Difference sets and sum-free sets in groups of order 16. Discret. Math. **13**(4), 399–407 (1975). https://doi.org/10.1016/0012-365X(75)90060-6

A General Method for Representing Sets of Relations by Vectors

Rudolf Berghammer[1] and Michael Winter[2(✉)]

[1] Institut für Informatik, Universität Kiel, 24098 Kiel, Germany
[2] Department of Computer Science, Brock University, St. Catharines, ON, Canada
mwinter@brocku.ca

Abstract. We present a general method for computing vector representations $\mathfrak{r}$ of sets of relations. This method is used for obtaining $\mathfrak{r}$ from an inclusion $\mathfrak{R} \subseteq \mathfrak{S}$, where $\mathfrak{R}$ and $\mathfrak{S}$ are relation-algebraic expressions over a relation rather than its vector representation. The core of the method is a theorem that shows how $\mathfrak{r}$ can be obtained from $\mathfrak{R} \subseteq \mathfrak{S}$ in one step. As applications we consider some problems concerning kernels of relations.

1 Introduction

Reduced ordered binary decision diagrams (ROBDDs) are a very efficient data structure for representing sets and relations. This has been shown by numerous applications of RelView, a ROBDD-based tool for the manipulation and visualization of relations and relational programming (see [1]). The use of ROBDDs often leads to an amazing computational power, in particular, if a hard problem is solved by computing a very large set of 'interesting objects' or 'potential solutions' and subsequently selecting an (optimal) solution among them.

In many applications the set of potential solutions is a subset $\mathcal{R}$ of the power set 2^X for some set X. A method for solving such a task is to start with a logical formula $\varphi(Y)$ that characterizes whether a set $Y \in 2^X$ belongs to $\mathcal{R}$ or not. This formula is then transformed (using correspondences between logical and relation-algebraic constructions) into the form $\mathfrak{r}_Y$, where $\mathfrak{r}$ is a relation-algebraic expression that evaluates to a relational vector in the sense of [10]. Finally, $\mathfrak{r}$ is translated into the programming language of RelView for execution. For some typical examples, see [2,3,5]. When Y is represented by a vector v, then very often $\varphi(Y)$ is equivalent to an inclusion $\mathfrak{v} \subseteq \mathfrak{w}$, where $\mathfrak{v}$ and $\mathfrak{w}$ are relation-algebraic expressions over v with a close or simple relationship between the inclusion and the expression $\mathfrak{r}$. Based on this observation, in [4] a general method was developed that allows obtaining the expression $\mathfrak{r}$ directly from the inclusion $\mathfrak{v} \subseteq \mathfrak{w}$ if $\mathfrak{v}$ and $\mathfrak{w}$ are of a specific syntactic form, called column-wise extendible vector expressions. Normally this leads to a substantial simplification of the development of $\mathfrak{r}$.

M. Winter—The author gratefully acknowledges support from the Natural Sciences and Engineering Research Council of Canada (283267).

R. Glück et al. (Eds.): RAMiCS 2023, LNCS 13896, pp. 34–51, 2023.
https://doi.org/10.1007/978-3-031-28083-2_3

In certain situations X is a Cartesian product $X_1 \times X_2$ which means that $\mathcal{R}$ is a subset of the set of relations with source X_1 and target X_2. Examples can be found in [5,7], where the original logic-based development method is applied to get $\mathfrak{r}$, and in [4,6], where the method of [4] is used instead. In particular the examples of [6] demonstrate the superiority of the method of [4] with regard to the logic-based method. However, the method of [4] has an obvious disadvantage. It requires that both sides of the inclusion $\mathfrak{v} \subseteq \mathfrak{w}$ describing the property $R \in \mathcal{R}$ are relation-algebraic expressions over a vector representation of R and not over R itself, which is usually given and much simpler.

In this paper we present a new and very general method for computing vector representations $\mathfrak{r}$ of sets $\mathcal{R}$ of relations. This overcomes the just mentioned problem, i.e., it can be used for computing $\mathfrak{r}$ from an inclusion $\mathfrak{R} \subseteq \mathfrak{S}$, where both sides are relation-algebraic expressions over a relation R and not over a vector representation of R. Decisive for this method is an equation that reduces the vector representation of a composition $R; S$ of relations to those of R and S. Together with known results concerning the remaining relation-algebraic operations, this allows the computation of vector representations via a recursive function ν_r and the proof of a theorem that shows how $\mathfrak{r}$ can be obtained from $\mathfrak{R} \subseteq \mathfrak{S}$ directly. As applications we treat some of the problems of [7] with the new method.

2   Relational Preliminaries

In this section we want to recall some basic facts about (binary, set-theoretic) relations and their operations that are used throughout this paper. For more details on relation algebras, see [10,11] for example.

Given sets X and Y, the power set $2^{X \times Y}$ of $X \times Y$ is the set of relations with *source* X and *target* Y, which we will denote by $[X \leftrightarrow Y]$. We will write $R : X \leftrightarrow Y$ instead of $R \in [X \leftrightarrow Y]$ and call $X \leftrightarrow Y$ the *type* of R. If X and Y are finite, then we may represent any relation $R : X \leftrightarrow Y$ by a Boolean matrix, in which an entry 'true' (or 1) in the row corresponding to $x \in X$ and in the column corresponding to $y \in Y$ indicates that $(x, y) \in R$. Analogously, the entry 'false' (or 0) indicates that the elements are not in relation R, i.e., $(x, y) \notin R$. This matrix interpretation is also used by the RELVIEW system to visualize relations. In addition, we will often use Boolean matrix notation and terminology in the remainder of this paper. In particular, we write $R_{x,y}$ instead of $(x, y) \in R$. Then we speak of a point-wise notation.

The following basic operations on relations will be used: $\overline{R}$ (*complement*), $R \cup S$ (*union*), $R \cap S$ (*intersection*), R^{T} (*transposition* or *converse*), and $R; S$ (*composition*). We assume that transposition and complementation bind stronger than composition and composition binds stronger than union and intersection. In addition, we have the constants O (*empty relation*), L (*universal relation*) and I (*identity relation*). Note that these constants are polymorphic, e.g., there is an identity relation of type $X \leftrightarrow X$ for all X. We also will use $R \subseteq S$ to indicate that R is included in S. We assume that the reader is familiar with these concepts.

As derived operation we will use $syq(R, S) := \overline{R^\mathsf{T}; \overline{S}} \cap \overline{\overline{R}^\mathsf{T}; S}$, the *symmetric quotient* of $R : X \leftrightarrow Y$ and $S : X \leftrightarrow Z$. It easily can be shown that $syq(R, S)$ has type $Y \leftrightarrow Z$ and for all $y \in Y$ and $z \in Z$ it holds $syq(R, S)_{y,z}$ iff for all $x \in X$ it holds $R_{x,y}$ iff $S_{x,z}$.

Besides the well-known lattice theoretic properties and basic properties of composition and transposition, i.e., $Q; (R \cup S) = Q; R \cup Q; S$, $Q; (R \cap S) \subseteq Q; R \cap Q; S$, $(R \cup S)^\mathsf{T} = R^\mathsf{T} \cup S^\mathsf{T}$, $(R \cap S)^\mathsf{T} = R^\mathsf{T} \cap S^\mathsf{T}$, $(R; S)^\mathsf{T} = S^\mathsf{T}; R^\mathsf{T}$, $(R^\mathsf{T})^\mathsf{T} = R$ and $\overline{R^\mathsf{T}} = \overline{R}^\mathsf{T}$, relations also satisfy the so-called *modular inclusion* $Q; R \cap S \subseteq Q; (R \cap Q^\mathsf{T}; S)$, and the so-called *Schröder equivalences* $Q^\mathsf{T}; \overline{S} \subseteq \overline{R}$ iff $Q; R \subseteq S$ iff $\overline{S}; R^\mathsf{T} \subseteq \overline{Q}$. In the remainder of the paper we will use these properties without mentioning. Some additional properties are summarized in the following lemma. A proof can be found in [10, 11].

Lemma 2.1. *Let be* $Q : X \leftrightarrow Y$, $R : X \leftrightarrow Z$, $S : Y \leftrightarrow Z$, *and* $T : X \leftrightarrow Z$. *Then we have*

(1) $(Q \cap R; \mathsf{L}); S = Q; S \cap R; \mathsf{L}$,
(2) $(Q; S \cap T); \mathsf{L} = (Q \cap T; S^\mathsf{T}); \mathsf{L}$.

An important class of relations are given by maps (or functions). We call a relation $Q : X \leftrightarrow Y$ *univalent* (or a partial function) iff $Q^\mathsf{T}; Q \subseteq \mathsf{I}$, *total* iff $\mathsf{I} \subseteq Q; Q^\mathsf{T}$, *injective* iff Q^T is univalent, *surjective* iff Q^T is total, and a *map* iff Q is total and univalent. The following lemma collects some important properties of univalent relations.

Lemma 2.2. *Assume* $f : X \leftrightarrow Y$ *to be univalent. Furthermore, let be* $Q, R : Y \leftrightarrow Z$, $S : W \leftrightarrow X$, $T : W \leftrightarrow Y$, *and* $U : X \leftrightarrow Y$. *Then we have*

(1) $f; (Q \cap R) = f; Q \cap f; R$,
(2) $(S \cap T; f^\mathsf{T}); f = S; f \cap T$,
(3) $f \cap (f \cap U); \mathsf{L} = f \cap U$.

Another important concept is the notion of pairs and the *projection relations* $\pi : X \times Y \leftrightarrow X$ and $\rho : X \times Y \leftrightarrow Y$. The projection relations have the Cartesian product $X \times Y$ as source and X resp. Y as target and are defined by $\pi_{(u_1,u_2),x}$ iff $u_1 = x$ and $\rho_{(u_1,u_2),y}$ iff $u_2 = y$, for all $(u_1, u_2) \in X \times Y$, $x \in X$ and $y \in Y$. These relations and the corresponding object $X \times Y$ can also be defined abstractly (up to isomorphism) by the formulas $\pi^\mathsf{T}; \pi \subseteq \mathsf{I}$, $\rho^\mathsf{T}; \rho \subseteq \mathsf{I}$, $\pi; \pi^\mathsf{T} \cap \rho; \rho^\mathsf{T} = \mathsf{I}$ and $\pi^\mathsf{T}; \rho = \mathsf{L}$, see [13][1]. To enhance presentation, in the remainder of this paper we will overload the projection relations, i.e., consider them as polymorphic. In all such cases it is easy to determine their types from the context using the typing rules of the operations of relation algebra.

---

[1] In [10, 11] instead of inclusions the first two axioms are equations. This leads to difficulties if precisely one of the sets of $X \times Y$ is empty. If e.g., $X \neq \emptyset$, then $\pi : X \times Y \leftrightarrow X$ is not surjective, whereas $\pi^\mathsf{T}; \pi = \mathsf{I}$ implies surjectivity. The weaker notion used here already implies the uniqueness of the construction up to isomorphism.

Based upon the projection relations we can define the *left pairing* of two relations $R : X \leftrightarrow Z$ and $S : Y \leftrightarrow Z$ by $[\![R, S]\!] := \pi; R \cap \rho; S$ of type $X \times Y \leftrightarrow Z$. When using a point-wise notation we have $[\![R, S]\!]_{(u_1, u_2), z}$ iff $R_{u_1, z}$ and $S_{u_2, z}$, for all $(u_1, u_2) \in X \times Y$ and $z \in Z$. Similar to the left pairing we can define the *right pairing* of two relations $R : Z \leftrightarrow X$ and $S : Z \leftrightarrow Y$ by $[R, S] := R; \pi^\mathsf{T} \cap S; \rho^\mathsf{T}$. Here we obtain $Z \leftrightarrow X \times Y$ as type and that $[R, S]_{z, (u_1, u_2)}$ iff R_{z, u_1} and S_{z, u_2}, for all $(u_1, u_2) \in X \times Y$ and $z \in Z$. Finally, the *parallel composition* (or *product*) $R \parallel S : X \times X' \leftrightarrow Y \times Y'$ of two relations $R : X \leftrightarrow Y$ and $S : X' \leftrightarrow Y'$ is defined by $R \parallel S := \pi; R; \pi^\mathsf{T} \cap \rho; S; \rho^\mathsf{T}$, i.e., we have $(R \parallel S)_{(u_1, u_2), (v_1, v_2)}$ iff R_{u_1, v_1} and S_{u_2, v_2}, for all $(u_1, u_2) \in X \times X'$ and $(v_1, v_2) \in Y \times Y'$. As a consequence we have $R \parallel S = [\![\pi; R, \rho; S]\!] = [R; \pi^\mathsf{T}, S; \rho^\mathsf{T}]$, where the right pairing is formed w.r.t. the projection relations of $Y \times Y'$ and the left pairing is formed w.r.t. the projection relations of $X \times X'$.

From Lemma 2.2(2) we obtain

$$[\![Q, R]\!]; \pi = (Q; \pi^\mathsf{T} \cap R; \rho^\mathsf{T}); \pi = Q \cap R; \rho^\mathsf{T}; \pi = Q \cap R; \mathsf{L},$$

so that $[\![Q, R]\!]; \pi = Q$ follows if R is total. Analogously, we get $[\![Q, R]\!]; \rho = R$ if Q is total and similar results for the left pairing and the parallel composition.

The sharpness property of relational products is the question whether the following equation

$$[\![Q, R]\!]; [S, T] = Q; S \cap R; T$$

holds for all relations Q, R, S and T of suitable types. Note that the equation only involves one Cartesian product, and is easy to verify for set-theoretic relations. However, the equation does not follow from the axioms of a relation algebra and of the corresponding projection relations alone. But if we require sufficient additional structure, i.e., the existence of at least one additional Cartesian product, we are able to show sharpness. The lemma below generalizes the equation above slightly. If instantiated with $S_1 = \pi$ and $S_2 = \rho$ it verifies sharpness under the assumption of the existence of the additional Cartesian product $W \times X$. The lemma itself is an immediate consequence of the approach developed in [8].

Lemma 2.3. *Let be* $Q_1 : W \leftrightarrow X$, $Q_2 : W \leftrightarrow Y$, $R_1 : X \leftrightarrow Z$, $R_2 : Y \leftrightarrow Z$, $S_1 : V \leftrightarrow X$, *and* $S_2 : V \leftrightarrow Y$ *with* S_1 *and* S_2 *univalent and*

$$Q_1^\mathsf{T}; Q_2 \cap R_1^\mathsf{T}; R_2 \subseteq S_1^\mathsf{T}; S_2.$$

Furthermore, assume that the Cartesian product $W \times X$ *exists. Then we have*

$$(Q_1; S_1^\mathsf{T} \cap Q_2; S_2^\mathsf{T}); (S_1; R_1^\mathsf{T} \cap S_2; R_2^\mathsf{T}) = Q_1; R_1^\mathsf{T} \cap Q_2; R_2^\mathsf{T}.$$

In the remainder of this paper we will assume that the Cartesian product for every pair of sets (objects) exists, together with the corresponding projection relations, such that we will always have sharpness.

The relation $\mathsf{S}^{X, X'} : X \times X' \leftrightarrow X' \times X$ is defined by $\mathsf{S}^{X, X'} := [\![\rho, \pi]\!] = [\rho^\mathsf{T}, \pi^\mathsf{T}]$. As usual we will drop the sets X and X' and write simply S instead of $\mathsf{S}^{X, X'}$ if the sets of the Cartesian products are clear from the context.

The relation exchanges the two components of a pair, i.e., we get $S_{(u_1,u_2),(v_1,v_2)}$ iff $u_1 = v_2$ and $u_2 = v_1$, for all $(u_1, u_2) \in X \times X'$ and $(v_1, v_2) \in X' \times X$. Furthermore, we have $S^\mathsf{T} = S$ and $S; S = I$, where the two occurrences of S in both equations are different versions of the polymorphic relation.

The partial identity $I \cap \pi; \rho; \pi^\mathsf{T}; \rho^\mathsf{T}$ has source and target $(X \times Y) \times (Y \times Z)$. From this type information we can infer that the first occurrence of π denotes the first projection relation of the Cartesian product $(X \times Y) \times (Y \times Z)$, whereas the second occurrence of π denotes the first projection relation of the Cartesian product $Y \times Z$, and the first occurrence of ρ denotes the second projection relation of $X \times Y$, whereas the second occurrence of ρ denotes the second projection relation of $(X \times Y) \times (Y \times Z)$. The relation $I \cap \pi; \rho; \pi^\mathsf{T}; \rho^\mathsf{T}$ acts as a filter when composing it with a suitable relation. Using point-wise notation, a quadruple $((u_1, u_2), (v_1, v_2)) \in (X \times Y) \times (Y \times Z)$ is related to itself by the relation $I \cap \pi; \rho; \pi^\mathsf{T}; \rho^\mathsf{T}$ iff $u_2 = v_1$, representing the condition under with the pair $(u_1, v_2) \in X \times Z$ would be in the composition of two relations, where the first relation contains the pair $(u_1, u_2) \in X \times Y$ and the second relation contains the pair $(v_1, v_2) \in Y \times Z$. Note that we have

$$I \cap \pi; \rho; \pi^\mathsf{T}; \rho^\mathsf{T} = \left(I \cap \pi; \rho; \pi^\mathsf{T}; \rho^\mathsf{T} \right)^\mathsf{T} = I \cap \rho; \pi; \rho^\mathsf{T}; \pi^\mathsf{T}.$$

By means of the partial identity $I \cap \pi; \rho; \pi^\mathsf{T}; \rho^\mathsf{T}$ we now define the relation $C^{X,Y,Z}$ of type $(X \times Y) \times (Y \times Z) \leftrightarrow X \times Z$ by $C^{X,Y,Z} := (I \cap \pi; \rho; \pi^\mathsf{T}; \rho^\mathsf{T}); (\pi \parallel \rho)$. This relation removes the common intermediate element. Again, we will usually drop the three sets X, Y and Z of $C^{X,Y,Z}$ and overload the relation.

In the next sections we also will use the relation-level equivalents of the set-theoretic symbol '$\in$' as basic relations. These are the (again polymorphic) *membership relations* $M : X \leftrightarrow 2^X$, which are point-wisely described by $M_{x,Y}$ iff $x \in Y$, for all elements $x \in X$ and sets $Y \in 2^X$. There exists a relation-algebraic axiomatization of membership relations which specifies these up to isomorphism. See [11], for example. But for the applications of the present paper the above point-wise description suffices.

3   Vector Representation of Relations

Relational vectors are relations $v : X \leftrightarrow Y$ with $v; L = v$. Such a v can be interpreted as a subset of X in the following sense: If represented by a Boolean matrix, the relation v is a matrix in which every row consists completely either of 1's (or 'true'-entries) or of 0's (or 'false'-entries) indicating that the element corresponding to that row either belongs to the subset of X or not. Since Y is irrelevant in this representation we will always consider vectors with target $\mathbf{1}$, where $\mathbf{1} = \{\perp\}$ is a specific singleton set. In this case all relations in the set $[X \leftrightarrow \mathbf{1}]$ of relations are vectors since for a singleton set we have $L = I$. Note that a singleton set can also be defined abstractly (up to isomorphism) as a so-called unit. A unit is an object $\mathbf{1}$ such that $L = I$ for $L : \mathbf{1} \leftrightarrow \mathbf{1}$ and $L : X \leftrightarrow \mathbf{1}$ is total for every X.

Relations are specific sets. In the following we concentrate on their representations by means of vectors.

The first step in providing a vector representation of arbitrary relations is to establish a Boolean lattice isomorphism between the set of relations $R : X \leftrightarrow Y$ and the set of vectors $v : X \times Y \leftrightarrow \mathbf{1}$. Given a relation $R : X \leftrightarrow Y$ and a vector $v : X \times Y \leftrightarrow \mathbf{1}$ we define $vec(R) : X \times Y \leftrightarrow \mathbf{1}$ and $Rel(v) : X \leftrightarrow Y$ as shown in the following figure (Fig. 1).

$$vec(R) := [\![R, \mathsf{I}]\!]; \mathsf{L},$$

$$Rel(v) := \pi^{\mathsf{T}}; (v; \mathsf{L} \cap \rho).$$

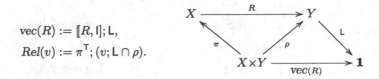

Fig. 1. Definition of functions vec and Rel

The functions $vec : [X \leftrightarrow Y] \to [X \times Y \leftrightarrow \mathbf{1}]$ and $Rel : [X \times Y \leftrightarrow \mathbf{1}] \to [X \leftrightarrow Y]$ are inverse to each other, i.e., we have $Rel(vec(R)) = R$ and $vec(Rel(v)) = v$, for all $R : X \leftrightarrow Y$ and $v : X \times Y \leftrightarrow \mathbf{1}$. Furthermore, vec and Rel are monotone w.r.t. inclusion, i.e., $R \subseteq S$ implies $vec(R) \subseteq vec(S)$, for all $R, S : X \leftrightarrow Y$, and $v \subseteq w$ implies $Rel(v) \subseteq Rel(w)$, for all $v, w : X \times Y \leftrightarrow \mathbf{1}$. We even have $R \subseteq S$ iff $vec(R) \subseteq vec(S)$, for all $R, S : X \leftrightarrow Y$, and $v \subseteq w$ iff $Rel(v) \subseteq Rel(w)$, for all $v, w : X \times Y \leftrightarrow \mathbf{1}$. Proofs of these facts can be found in [10], for example.

The next lemma shows how the five basic operations of relation algebra can be performed directly on the corresponding vectors. Together with the bijectivity properties (1), (3) and (4) of Lemma 3.1 imply that the function vec is an isomorphism from the Boolean lattice $[X \leftrightarrow Y]$ of all relations between X and Y to the Boolean lattice $[X \times Y \leftrightarrow \mathbf{1}]$ of all vectors with source $X \times Y$, with $vec^{-1} = Rel$, such that $vec(\mathsf{O}) = \mathsf{O}$, $vec(\mathsf{L}) = \mathsf{L}$, $Rel(\mathsf{O}) = \mathsf{O}$ and $Rel(\mathsf{L}) = \mathsf{L}$ immediately follows.

Lemma 3.1. *Let be* $Q, R : X \leftrightarrow Y$ *and* $S : Y \leftrightarrow Z$. *Then we have*

(1) $vec(\overline{Q}) = \overline{vec(Q)}$,
(2) $vec(Q^{\mathsf{T}}) = \mathsf{S}; vec(Q)$,
(3) $vec(Q \cup R) = vec(Q) \cup vec(R)$,
(4) $vec(Q \cap R) = vec(Q) \cap vec(R)$,
(5) $vec(R; S) = \mathsf{C}^{\mathsf{T}}; [\![vec(R), vec(S)]\!]$.

Proof. The four properties (1) to (4) were already shown in [10]. In order to prove property (5), we would like to show that the composition $\mathsf{C}^{\mathsf{T}}; [\![vec(R), vec(S)]\!]$ can basically be written as the left-hand side of the equation in Lemma 2.3 for suitable S_1 and S_2, and then apply that lemma. Therefore, we will use the following abbreviations

$$i := \mathsf{I} \cap \pi; \rho; \pi^{\mathsf{T}}; \rho^{\mathsf{T}}, \qquad \tilde{\pi} := \pi \cap \rho; \pi; \rho^{\mathsf{T}},$$

$$f := \pi; \rho \cap \rho; \pi, \qquad \overset{\smile}{\rho} := \rho \cap \pi; \rho; \pi^{\mathsf{T}}.$$

Obviously, all four relations are univalent and we have $i; \pi = \tilde{\pi}$, $i; \rho = \tilde{\rho}$, $\tilde{\pi}; \rho = f$ and $\tilde{\rho}; \pi = f$, where the last four equations follow immediately from Lemma 2.2(2). Furthermore, we have $\tilde{\pi}; [\![R, \mathsf{I}]\!] \subseteq \tilde{\pi}; \rho = f$ and

$$
\begin{aligned}
\tilde{\rho}; [\![S, \mathsf{I}]\!]; \mathsf{L} \cap f &= \tilde{\rho}; (\pi; S \cap \rho); \mathsf{L} \cap f \\
&= \tilde{\rho}; (\pi \cap \rho; S^{\mathsf{T}}); \mathsf{L} \cap f && \text{Lemma 2.1(2)} \\
&= (\tilde{\rho}; \pi \cap \tilde{\rho}; \rho; S^{\mathsf{T}}); \mathsf{L} \cap f && \text{Lemma 2.2(1)} \\
&= (f \cap \tilde{\rho}; \rho; S^{\mathsf{T}}); \mathsf{L} \cap f \\
&= f \cap \tilde{\rho}; \rho; S^{\mathsf{T}} && \text{Lemma 2.2(3)} \\
&= \tilde{\rho}; \pi \cap \tilde{\rho}; \rho; S^{\mathsf{T}} \\
&= \tilde{\rho}; (\pi \cap \rho; S^{\mathsf{T}}) && \text{Lemma 2.2(1)} \\
&= \tilde{\rho}; [\![\mathsf{I}, S^{\mathsf{T}}]\!].
\end{aligned}
$$

These two properties imply

$$
\begin{aligned}
i; [\![vec(R), vec(S)]\!] &= i; \pi; [\![R, \mathsf{I}]\!]; \mathsf{L} \cap i; \rho; [\![S, \mathsf{I}]\!]; \mathsf{L} && \text{Lemma 2.2(1)} \\
&= \tilde{\pi}; [\![R, \mathsf{I}]\!]; \mathsf{L} \cap \tilde{\rho}; [\![S, \mathsf{I}]\!]; \mathsf{L} \\
&= (\tilde{\pi}; [\![R, \mathsf{I}]\!] \cap \tilde{\rho}; [\![S, \mathsf{I}]\!]; \mathsf{L}); \mathsf{L} && \text{Lemma 2.1(1)} \\
&= (\tilde{\pi}; [\![R, \mathsf{I}]\!] \cap f \cap \tilde{\rho}; [\![S, \mathsf{I}]\!]; \mathsf{L}); \mathsf{L} && \text{see above} \\
&= (\tilde{\pi}; [\![R, \mathsf{I}]\!] \cap \tilde{\rho}; [\![\mathsf{I}, S^{\mathsf{T}}]\!]); \mathsf{L} && \text{see above.}
\end{aligned}
$$

On the other hand, we have

$$
\begin{aligned}
\mathsf{C} &= i; (\pi \,\|\, \rho) \\
&= i; (\pi; \pi; \pi^{\mathsf{T}} \cap \rho; \rho; \rho^{\mathsf{T}}) \\
&= i; \pi; \pi; \pi^{\mathsf{T}} \cap i; \rho; \rho; \rho^{\mathsf{T}} && \text{Lemma 2.2(1)} \\
&= \tilde{\pi}; \pi; \pi^{\mathsf{T}} \cap \tilde{\rho}; \rho; \rho^{\mathsf{T}}
\end{aligned}
$$

so that the composition $\mathsf{C}^{\mathsf{T}}; [\![vec(R), vec(S)]\!]$ in fact can be written as the left-hand side of Lemma 2.3 with $S_1 = \tilde{\pi}$ and $S_2 = \tilde{\rho}$.

The following calculation now shows that the additional assumption of Lemma 2.3 is satisfied:

$$[\![R, \mathsf{I}]\!]; [\![\mathsf{I}, S^\mathsf{T}]\!]^\mathsf{T} \cap (\pi; \pi^\mathsf{T})^\mathsf{T}; \rho; \rho^\mathsf{T}$$

$$= [\![R, \mathsf{I}]\!]; [\![\mathsf{I}, S^\mathsf{T}]\!]^\mathsf{T} \cap \pi; \pi^\mathsf{T}; \rho; \rho^\mathsf{T}$$

$$= [\![R, \mathsf{I}]\!]; [\![\mathsf{I}, S^\mathsf{T}]\!]^\mathsf{T} \cap \pi; \mathsf{L}; \rho^\mathsf{T}$$

$$= [\![R, \mathsf{I}]\!]; [\![\mathsf{I}, S^\mathsf{T}]\!]^\mathsf{T} \qquad\qquad\qquad \pi \text{ and } \rho \text{ total}$$

$$\subseteq \rho; \pi^\mathsf{T}$$

$$= \rho; \pi^\mathsf{T} \cap \rho; \pi^\mathsf{T}$$

$$= [\![\mathsf{I}, \rho; \pi^\mathsf{T}]\!]; [\![\rho; \pi^\mathsf{T}, \mathsf{I}]\!] \qquad\qquad \text{sharpness}$$

$$= (\pi^\mathsf{T} \cap \rho; \pi^\mathsf{T}; \rho^\mathsf{T}); (\rho \cap \pi; \rho; \pi^\mathsf{T})$$

$$= \tilde{\pi}^\mathsf{T}; \tilde{\rho},$$

As a consequence we conclude

$$(\pi; \pi^\mathsf{T}; \tilde{\pi}^\mathsf{T} \cap \rho; \rho^\mathsf{T}; \tilde{\rho}^\mathsf{T}); (\tilde{\pi}; [\![R, \mathsf{I}]\!] \cap \tilde{\rho}; [\![\mathsf{I}, S^\mathsf{T}]\!]) = \pi; \pi^\mathsf{T}; [\![R, \mathsf{I}]\!] \cap \rho; \rho^\mathsf{T}; [\![\mathsf{I}, S^\mathsf{T}]\!].$$

We finally obtain

$$vec(R; S)$$

$$= [\![R; S, \mathsf{I}]\!]; \mathsf{L}$$

$$= (\pi; R; S \cap \rho); \mathsf{L}$$

$$= (\pi; R \cap \rho; S^\mathsf{T}); \mathsf{L} \qquad\qquad\qquad\qquad \text{Lemma 2.1(2)}$$

$$= (\pi; \pi^\mathsf{T}; [\![R, \mathsf{I}]\!] \cap \rho; \rho^\mathsf{T}; [\![\mathsf{I}, S^\mathsf{T}]\!]); \mathsf{L}$$

$$= (\pi; \pi^\mathsf{T}; \tilde{\pi}^\mathsf{T} \cap \rho; \rho^\mathsf{T}; \tilde{\rho}^\mathsf{T}); (\tilde{\pi}; [\![R, \mathsf{I}]\!] \cap \tilde{\rho}; [\![\mathsf{I}, S^\mathsf{T}]\!]); \mathsf{L} \quad \text{see above}$$

$$= \mathsf{C}^\mathsf{T}; i; [\![vec(R), vec(S)]\!]$$

$$= \mathsf{C}^\mathsf{T}; [\![vec(R), vec(S)]\!], \qquad\qquad\qquad i^\mathsf{T}; i = i; i = i = i^\mathsf{T}$$

i.e., the desired property (5).   □

The relation C allows us to express a composition of two relations as an operation on their corresponding vectors. Sometimes, when computing all relations that satisfy a certain property, it is sufficient to convert only one relation in a series of compositions into a vector. Quite often this even leads to a more efficient implementation (e.g., in RELVIEW) of testing the property. The following lemma was shown in [9] and provides exactly such a translation.

Lemma 3.2. *Let* $Q : W \leftrightarrow X$, $R : X \leftrightarrow Y$, *and* $S : Y \leftrightarrow Z$. *Then we have*

$$vec(Q; R; S) = (Q \,\|\, S^\mathsf{T}); vec(R).$$

As specific cases we get $vec(Q; R) = (Q \,\|\, \mathsf{I}); vec(R)$ by taking S as identity relation and $vec(R; S) = (\mathsf{I} \,\|\, S^\mathsf{T}); vec(R)$ by taking Q as identity relation.

4   Vector Representation of Sets of Relations

In [9] so-called vector predicates are introduced for the relational treatment of evolutionary algorithms. They are functions in the usual mathematical sense on relations and model those sets of relations which are built from vectors using as operations only complementation, union, intersection and a restricted version of composition. This approach is continued and refined in [4] in view of the specification of vectors $v : 2^X \leftrightarrow \mathbf{1}$ which represent subsets $\mathcal{R}$ of given power sets 2^X. The aim is to avoid lengthy and complex logical calculations and to work rather with relation-algebraic specifications of the elements of $\mathcal{R}$ via inclusions $\mathfrak{v} \subseteq \mathfrak{w}$ such that v can be obtained from $\mathfrak{v} \subseteq \mathfrak{w}$ in one step using a general procedure. Decisive for this is that both sides of the inclusions are so-called column-wise extendible vector expressions. These are specific relation-algebraic expressions which can be seen as syntactical counterpart of vector predicates. Formally, they are defined as follows:

Definition 4.1. *Given a variable v of type $X \leftrightarrow \mathbf{1}$, the set $\mathbf{VE}_v$ of typed column-wise extendible vector expressions over v is inductively defined as follows:*

(1) We have $v \in \mathbf{VE}_v$ and its type is $X \leftrightarrow \mathbf{1}$.
(2) If $w : Y \leftrightarrow \mathbf{1}$, then $w \in \mathbf{VE}_v$ and its type is $Y \leftrightarrow \mathbf{1}$.
(3) If $\mathfrak{v} \in \mathbf{VE}_v$ is of type $Y \leftrightarrow \mathbf{1}$, then $\overline{\mathfrak{v}} \in \mathbf{VE}_v$ and its type is $Y \leftrightarrow \mathbf{1}$.
(4) If $\mathfrak{v}, \mathfrak{w} \in \mathbf{VE}_v$ are of type $Y \leftrightarrow \mathbf{1}$, then $\mathfrak{v} \cup \mathfrak{w} \in \mathbf{VE}_v$ and $\mathfrak{v} \cap \mathfrak{w} \in \mathbf{VE}_v$ and their types are $Y \leftrightarrow \mathbf{1}$.
(5) If $\mathfrak{v} \in \mathbf{VE}_v$ is of type $Y \leftrightarrow \mathbf{1}$ and $\mathfrak{R}$ is a relation-algebraic expression of type $Z \leftrightarrow Y$ in which v does not occur, then $\mathfrak{R}; \mathfrak{v} \in \mathbf{VE}_s$ and its type is $Z \leftrightarrow \mathbf{1}$.

So, the vector expressions from $\mathbf{VE}_v$ are built from the variable v using vectors and as operations only complementation, union, intersection and left-composition with a relation-algebraic expression in which v does not occur. Note that v is the only variable in such an expression. In the following we also allow the use of derived operations like symmetric quotients and pairings, but these are only seen as abbreviations. For example, $R; [\![v, S; v]\!]$ is considered as a column-wise extendible vector expression over the variable v since unfolding the definition of the left pairing yields $R; (\pi; v \cap \rho; S; v)$.

In a column-wise extendible vector expression $\mathfrak{v}$ over v the variable v can be replaced by a relation R with the same source as v. The result is denoted as $\mathfrak{v}[R/v]$ and is inductively defined as follows.

Definition 4.2. *Given a variable v of type $X \leftrightarrow \mathbf{1}$, $\mathfrak{v} \in \mathbf{VE}_v$ and $R : X \leftrightarrow Z$, we define $\mathfrak{v}[R/v]$ as follows, using induction on the structure of $\mathfrak{v}$:*

(1) $v[R/v] = R$.
(2) $w[R/v] = w; \mathsf{L}$, with $\mathsf{L} : \mathbf{1} \leftrightarrow Z$.
(3) $\overline{\mathfrak{w}}[R/v] = \overline{\mathfrak{w}[R/v]}$.
(4) $(\mathfrak{w} \cup \mathfrak{u})[R/v] = \mathfrak{w}[R/v] \cup \mathfrak{u}[R/v]$ and $(\mathfrak{w} \cap \mathfrak{u})[R/v] = \mathfrak{w}[R/v] \cap \mathfrak{u}[R/v]$.
(5) $(\mathfrak{R}; \mathfrak{w})[R/v] = \mathfrak{R}; (\mathfrak{w}[R/v])$.

For example, for the variable v of type $X \leftrightarrow \mathbf{1}$, $S : Y \leftrightarrow X$ and $w : Y \leftrightarrow \mathbf{1}$ we have $S; v \cap w \in \mathbf{VE}_v$. A replacement of v in $S; v \cap w$ by the membership relation $\mathsf{M} : X \leftrightarrow 2^X$ then yields $S; \mathsf{M} \cap w; \mathsf{L}$, which has type $Y \leftrightarrow 2^X$.

The general procedure to obtain the vectors $v : 2^X \leftrightarrow \mathbf{1}$ from the inclusions $\mathfrak{v} \subseteq \mathfrak{w}$, that we have mentioned above, is shown in [4] as Theorem 1. If we instantiate this theorem in such a way that instead of a general power set 2^X a power set of a Cartesian product $X \times Y$ is taken, i.e., a vector $v : [X \leftrightarrow Y] \leftrightarrow \mathbf{1}$ representing a subset $\mathcal{R}$ of the set $[X \leftrightarrow Y]$ of relations is to be computed, then we get the following result.

Theorem 4.1. *Let the subset $\mathcal{R}$ of the set $[X \leftrightarrow Y]$ of relations be specified as $\mathcal{R} = \{Rel(r) \mid r \in [X \times Y \leftrightarrow \mathbf{1}] \wedge \mathfrak{v} \subseteq \mathfrak{w}\}$, where $\mathfrak{v}, \mathfrak{w} \in \mathbf{VE}_r$. Using the membership relation $\mathsf{M} : X \times Y \leftrightarrow [X \leftrightarrow Y]$ the set $\mathcal{R}$ is represented by the vector*

$$v := \mathsf{L}; \overline{(\mathfrak{v}[\mathsf{M}/r] \cap \overline{\mathfrak{w}[\mathsf{M}/r]})}^{\mathsf{T}} : [X \leftrightarrow Y] \leftrightarrow \mathbf{1}.$$

Using this result, in [6] a lot of vectors are obtained which represent important classes of relations. A certain disadvantage of Theorem 4.1 is that it works not directly with relations but with their vector representations. If, e.g., $\mathcal{R}$ is the set of all transitive relations on a set X, then one would like to specify that R is a member of $\mathcal{R}$ by $R; R \subseteq R$ instead of by $\mathsf{C}^{\mathsf{T}}; [\![r, r]\!] \subseteq r$ as in [6], where r is the vector representation of R. Using the properties of the two functions vec and Rel given in Sect. 3, it is not hard to calculate the inclusion $\mathsf{C}^{\mathsf{T}}; [\![r, r]\!] \subseteq r$ (between column-wise extendible vector expressions over r) from the common specification $Rel(r); Rel(r) \subseteq Rel(r)$ of $Rel(r)$ to be transitive.

In the following we generalize the example of transitive relations and consider arbitrary inclusions $\mathfrak{R} \subseteq \mathfrak{S}$, where $\mathfrak{R}$ and $\mathfrak{S}$ are relation-algebraic expressions that are constructed from a variable R of type $X \leftrightarrow Y$ using certain relations (e.g., membership relations and projection relations) and the constants and operations (including again also derived ones) of relation algebra. We denote the set of all these relation-algebraic expressions as $\mathbf{RE}_R$. Note that R is the only variable in such an expression. Our aim is to get from the specification $\mathcal{R} = \{R \in [X \leftrightarrow Y] \mid \mathfrak{R} \subseteq \mathfrak{S}\}$ a vector representation of this set in one step similar to Theorem 4.1. Decisive is the following function that transforms expressions from $\mathbf{RE}_R$ into expressions from $\mathbf{VE}_r$, where $r : X \times Y \leftrightarrow \mathbf{1}$.

Definition 4.3. *Given a variable r of type $X \times Y \leftrightarrow \mathbf{1}$ and a variable R of type $X \leftrightarrow Y$, we define the function $\nu_r : \mathbf{RE}_R \to \mathbf{VE}_r$ as follows, using induction on the structure of the argument:*

(1) $\nu_r(R) = r$.
(2) $\nu_r(S) = vec(S)$ for all relations S.
(3) $\nu_r(\overline{\mathfrak{R}}) = \overline{\nu_r(\mathfrak{R})}$.
(4) $\nu_r(\mathfrak{R}^{\mathsf{T}}) = \mathsf{S}; \nu_r(\mathfrak{R})$.
(5) $\nu_r(\mathfrak{R} \cup \mathfrak{S}) = \nu_r(\mathfrak{R}) \cup \nu_r(\mathfrak{S})$.
(6) $\nu_r(\mathfrak{R} \cap \mathfrak{S}) = \nu_r(\mathfrak{R}) \cap \nu_r(\mathfrak{S})$.
(7) $\nu_r(\mathfrak{R}; \mathfrak{S}) = \mathsf{C}^{\mathsf{T}}; [\![\nu_r(\mathfrak{R}), \nu_r(\mathfrak{S})]\!]$.

In the next lemma we verify that ν_r in fact yields a column-wise extendible vector expression over the variable r. Furthermore, we show that if r is instantiated as $vec(R)$, then ν_r equals the function vec.

Lemma 4.1. *Let be a variable r of type $X \times Y \leftrightarrow \mathbf{1}$ and a variable R of type $X \leftrightarrow Y$. For all $\mathfrak{R} \in \mathbf{RE}_R$ we then have $\nu_r(\mathfrak{R}) \in \mathbf{VE}_r$ and, provided r is instantiated as $vec(R)$, also $\nu_r(\mathfrak{R}) = vec(\mathfrak{R})$.*

Proof. We use induction on the structure of the expression $\mathfrak{R}$.

The induction base is that $\mathfrak{R}$ is the variable R or a relation S. In the first case we have $\nu_r(R) = r \in \mathbf{VE}_r$ and $\nu_r(R) = r$. So, if r is instantiated as $vec(R)$, then $\nu_r(\mathfrak{R}) = vec(\mathfrak{R})$. The case that $\mathfrak{R}$ is a relation S is trivial.

In the first case of the induction step, assume $\mathfrak{R}$ to be of the form $\overline{\mathfrak{S}}$ and that the induction hypothesis holds for the expression $\mathfrak{S}$. Then $\nu_r(\mathfrak{S}) \in \mathbf{VE}_r$ and Definition 4.3(3) yield $\nu_r(\overline{\mathfrak{S}}) \in \mathbf{VE}_r$. If r is instantiated as $vec(R)$, we have $\nu_r(\mathfrak{S}) = vec(\mathfrak{S})$ and we obtain

$$\nu_r(\overline{\mathfrak{S}}) = \overline{\nu_r(\mathfrak{S})} = \overline{vec(\mathfrak{S})} = vec(\overline{\mathfrak{S}})$$

using Definition 4.3(3) and Lemma 3.1(1). Next, let $\mathfrak{R}$ be of the form $\mathfrak{S}^\mathsf{T}$ and the induction hypothesis hold for the expression $\mathfrak{S}$. Here $\nu_r(\mathfrak{S}) \in \mathbf{VE}_r$ and Definition 4.3(4) show $\nu_r(\mathfrak{S}^\mathsf{T}) \in \mathbf{VE}_r$. As an instantiation of r as $vec(R)$ yields $\nu_r(\mathfrak{S}) = vec(\mathfrak{S})$, we get

$$\nu_r(\mathfrak{S}^\mathsf{T}) = \mathsf{S}; \nu_r(\mathfrak{S}) = \mathsf{S}; vec(\mathfrak{S}) = vec(\mathfrak{S}^\mathsf{T}),$$

using Definition 4.3(4) and Lemma 3.1(2). In the same way the cases can be treated where $\mathfrak{R}$ is a union, an intersection and a composition of expressions. When $\mathfrak{R}$ is a union, Definition 4.3(5) and Lemma 3.1(3) apply, when it is an intersection, Definition 4.3(6) and Lemma 3.1(4) apply, and when it is a composition, Definition 4.3(7) and Lemma 3.1(5) apply. □

Now, we are able to prove the following variant of Theorem 4.1. The main difference in this modified version is that the elements of the set of relations are specified by an inclusion of relation-algebraic expressions over the variable R rather than as an inclusion between column-wise extendible vector expressions over a variable r that stands for the vector representation of R.

Theorem 4.2. *Let the subset $\mathcal{R}$ of the set $[X \leftrightarrow Y]$ of relations be specified as $\mathcal{R} = \{R \in [X \leftrightarrow Y] \mid \mathfrak{R} \subseteq \mathfrak{S}\}$, where $\mathfrak{R}, \mathfrak{S} \in \mathbf{RE}_R$. Taking a variable r of type $X \times Y \leftrightarrow \mathbf{1}$ and using the membership relation $\mathsf{M} : X \times Y \leftrightarrow [X \leftrightarrow Y]$, the set $\mathcal{R}$ is represented by the vector*

$$v := \overline{\mathsf{L}; (\overline{\nu_r(\mathfrak{R})[\mathsf{M}/r]} \cap \overline{\nu_r(\mathfrak{S})[\mathsf{M}/r]})}^{\mathsf{T}} : [X \leftrightarrow Y] \leftrightarrow \mathbf{1}.$$

Proof. First, we prove that the original specification of the set $\mathcal{R}$ is equivalent to the specification

$$\mathcal{R} = \{Rel(r) \mid r \in [X \times Y \leftrightarrow \mathbf{1}] \wedge \nu_r(\mathfrak{R}) \subseteq \nu_r(\mathfrak{S})\}.$$

Let $R : X \leftrightarrow Y$ be given. We have to verify that $\mathfrak{R} \subseteq \mathfrak{S}$ holds iff there exists $r : X{\times}Y \leftrightarrow \mathbf{1}$ such that $\nu_r(\mathfrak{R}) \subseteq \nu_r(\mathfrak{S})$ and $R = Rel(r)$. Here is the proof, where r ranges over $[X{\times}Y \leftrightarrow \mathbf{1}]$:

$$\mathfrak{R} \subseteq \mathfrak{S} \iff vec(\mathfrak{R}) \subseteq vec(\mathfrak{S}) \qquad\qquad \text{see Sect. 3}$$
$$\iff \nu_{vec(R)}(\mathfrak{R}) \subseteq \nu_{vec(R)}(\mathfrak{S}) \qquad \text{by Lemma 4.1}$$
$$\iff \exists r : r = vec(R) \wedge \nu_r(\mathfrak{R}) \subseteq \nu_r(\mathfrak{S})$$
$$\iff \exists r : R = Rel(r) \wedge \nu_r(\mathfrak{R}) \subseteq \nu_r(\mathfrak{S}) \qquad \text{see Sect. 3}$$

From Lemma 4.1 we also get that both sides of the inclusion of the second specification of $\mathcal{R}$ are column-wise extendible vector expressions over r. Hence, Theorem 4.1 is applicable and yields the desired result. $\qquad\qquad\qquad\square$

As an example for applying the previous theorem, we consider again transitive relations, i.e., we use the specification

$$\mathcal{R} = \{R \in [X \leftrightarrow X] \mid R; R \subseteq R\}.$$

For arbitrary $R : X \leftrightarrow X$ and $r : X{\times}X \leftrightarrow \mathbf{1}$ we get for the left-hand side of the inclusion $R; R \subseteq R$ that $\nu_r(R; R) = \mathsf{C}^\mathsf{T}; [\![\nu_r(R), \nu_r(R)]\!] = \mathsf{C}^\mathsf{T}; [\![r, r]\!]$, such that $\nu_r(R; R)[\mathsf{M}/r] = \mathsf{C}^\mathsf{T}; [\![\mathsf{M}, \mathsf{M}]\!]$, with $\mathsf{M} : X{\times}X \leftrightarrow [X \leftrightarrow X]$ as membership relation. In case of the right-hand side of $R; R \subseteq R$ we have $\nu_r(R) = r$, and this yields $\nu_r(R)[\mathsf{M}/r] = \mathsf{M}$. So, Theorem 4.2 implies that the vector

$$trans := \overline{\mathsf{L}; (\mathsf{C}^\mathsf{T}; [\![\mathsf{M}, \mathsf{M}]\!] \cap \overline{\mathsf{M}})}^\mathsf{T}$$

of type $[X \leftrightarrow X] \leftrightarrow \mathbf{1}$ represents $\mathcal{R}$, i.e., the set of transitive relations on X.

5   Applications: Kernels and Richardson's Criterion

A directed simple graph, i.e., a directed graph without multiple edges between the same pair of vertices, on a set of vertices X can be represented by a relation $R : X \leftrightarrow X$. A subset K of X is a *kernel* of R if for all $x \in X$ it holds $x \notin K$ iff there exists $y \in K$ such that $R_{x,y}$. Kernels have been introduced in [12] as a generalization of a solution of a cooperative game. Not every relation has a kernel and it is known that determining whether a relation has a kernel is a NP-complete problem.

There exist a series of sufficient criteria for the existence of kernels which can be tested efficiently. In [7] it was investigated how well these criteria characterize the class of relations that have kernels. This was done by computing the number of relations with a kernel for all sets X up to 7 elements using RELVIEW. Then the four most popular criteria of the above mentioned series were considered and for each of them the number of relations satisfying the criteria was computed – again for all X with $|X| \leq 7$ using RELVIEW. The numerical data of [7] show that even in case of the most general of the criteria, the absence of odd cycles

(Richardson's criterion), only a very small portion of the relations with kernels satisfy this property. So, the criteria are very far away from characterizing the class of relations with kernels. The data also led to the following conjecture for finite sets X: The probability that a relation selected uniformly at random from $[X \leftrightarrow X]$ has a kernel tends to zero if $|X|$ tends to infinity.

The key for getting the data of [7] are relation-algebraic specifications of five vectors of type $[X \leftrightarrow X] \leftrightarrow \mathbf{1}$ which represent the set of relations on X having kernels and the set of relations on X satisfying one of the criteria. The REL-VIEW-programs then are nothing else than translations of the specifications into the programming language of the system. Each vector is obtained from a logical description of the relations in question and its transformation into an equivalent relation-algebraic expression. Especially in case of Richardson's criterion the development is rather technical and complex. In the following we solve two of the problems of [7] using the new method. A comparison with [7] shows that the new solutions are much more simple and many steps are very straight-forward.

We start with the characterization of relations having kernels. Assume $R : X \leftrightarrow X$ and $K \subseteq X$. Using the definition of a kernel and the point-wise description of symmetric quotients, we have that K is a kernel of R iff $syq(\overline{\mathsf{M}}, R; \mathsf{M})_{K,K}$, where $\mathsf{M} : X \leftrightarrow 2^X$ is a membership relation. As a consequence R has a kernel iff $syq(\overline{\mathsf{M}}, R; \mathsf{M}) \cap \mathsf{I} \neq \mathsf{O}$. Thus, R has no kernel iff $syq(\overline{\mathsf{M}}, R; \mathsf{M}) \subseteq \overline{\mathsf{I}}$. In order to obtain a vector $kernel : [X \leftrightarrow X] \leftrightarrow \mathbf{1}$ that represents the set of relations on X with a kernel, we use our method to transform the inclusion $syq(\overline{\mathsf{M}}, R; \mathsf{M}) \subseteq \overline{\mathsf{I}}$ into a vector $nokernel : [X \leftrightarrow X] \leftrightarrow \mathbf{1}$ and define $kernel$ as complement of $nokernel$.

For the development of the vector $nokernel$, assume an arbitrary $r : X \times X \leftrightarrow \mathbf{1}$ to be given. Using the definition of the symmetric quotient, the definition of the function ν_r and Lemma 3.1 in combination with the abbreviations $m := vec(\mathsf{M})$ and $mt := vec(M^{\mathsf{T}})$, we get

$$\nu_r(syq(\overline{\mathsf{M}}, R; \mathsf{M}))$$

$$= \nu_r(\overline{\overline{\mathsf{M}}^{\mathsf{T}}; \overline{R; \mathsf{M}}} \cap \overline{\overline{\overline{\mathsf{M}}}^{\mathsf{T}}; R; \mathsf{M}})$$

$$= \nu_r(\overline{\overline{\mathsf{M}}^{\mathsf{T}}; \overline{R; \mathsf{M}}}) \cap \nu_r(\overline{\mathsf{M}^{\mathsf{T}}; R; \mathsf{M}})$$

$$= \overline{\nu_r(\overline{\mathsf{M}}^{\mathsf{T}}; \overline{R; \mathsf{M}})} \cap \overline{\nu_r(\mathsf{M}^{\mathsf{T}}; R; \mathsf{M})}$$

$$= \overline{\mathsf{C}^{\mathsf{T}}; [\nu_r(\overline{\mathsf{M}}^{\mathsf{T}}), \nu_r(\overline{R; \mathsf{M}})]} \cap \overline{\mathsf{C}^{\mathsf{T}}; [\nu_r(\mathsf{M}^{\mathsf{T}}), \nu_r(R; \mathsf{M})]}$$

$$= \overline{\mathsf{C}^{\mathsf{T}}; [vec(\overline{\mathsf{M}}^{\mathsf{T}}), \overline{\nu_r(R; \mathsf{M})}]} \cap \overline{\mathsf{C}^{\mathsf{T}}; [vec(\mathsf{M}^{\mathsf{T}}), \nu_r(R; \mathsf{M})]}$$

$$= \overline{\mathsf{C}^{\mathsf{T}}; [vec(\overline{\mathsf{M}^{\mathsf{T}}}), \overline{\mathsf{C}^{\mathsf{T}}; [\nu_r(R), \nu_r(\mathsf{M})]}]} \cap \overline{\mathsf{C}^{\mathsf{T}}; [vec(\mathsf{M}^{\mathsf{T}}), \mathsf{C}^{\mathsf{T}}; [\nu_r(R), \nu_r(\mathsf{M})]]}$$

$$= \overline{\mathsf{C}^{\mathsf{T}}; [\overline{vec(\mathsf{M}^{\mathsf{T}})}, \overline{\mathsf{C}^{\mathsf{T}}; [r, vec(\mathsf{M})]}]} \cap \overline{\mathsf{C}^{\mathsf{T}}; [vec(\mathsf{M}^{\mathsf{T}}), \mathsf{C}^{\mathsf{T}}; [r, vec(\mathsf{M})]]}$$

$$= \overline{\mathsf{C}^{\mathsf{T}}; [\overline{mt}, \overline{\mathsf{C}^{\mathsf{T}}; [r, m]}]} \cap \overline{\mathsf{C}^{\mathsf{T}}; [mt, \mathsf{C}^{\mathsf{T}}; [r, m]]}$$

for the left-hand side of the inclusion $syq(\overline{M}, R; M) \subseteq \overline{I}$, hence

$$\nu_r(syq(\overline{M}, R; M))[M/r] = \overline{C^T; [\![mt; L, C^T; [\![M, m; L]\!]}\!] \cap \overline{C^T; [\![mt; L, C^T; [\![M, m; L]\!]}\!]}$$

for the replacement of r by the membership relation $M : X \times X \leftrightarrow [X \leftrightarrow X]$. In case of the right-hand side of the inclusion $syq(\overline{M}, R; M) \subseteq \overline{I}$ we have $\nu_r(\overline{I}) = vec(\overline{I}) = \overline{vec(I)}$ and with the abbreviation $i := vec(I)$ we get

$$\nu_r(\overline{I})[M/r] = \overline{i}; L = \overline{i; L}$$

for the replacement of r by M. Now, Theorem 4.2 immediately yields

$$nokernel := \overline{L; (\overline{C^T; [\![mt; L, \overline{H}]\!]} \cap \overline{C^T; [\![mt; L, H]\!]} \cap i; L)}^{-T}, \text{ where } H := C^T; [\![M, m; L]\!].$$

Next, we consider Richardson's criterion and assume, as in [7], a finite set X with $|X| \leq 7$. To compute a vector that represents the set of relations on X without odd cycles, in [7] four vectors $\mathfrak{cyc1}$, $\mathfrak{cyc3}$, $\mathfrak{cyc5}$ and $\mathfrak{cyc7}$ of type $[X \leftrightarrow X] \leftrightarrow \mathbf{1}$ were developed which represent the set of relations on X having a cycle of length 1, 3, 5 and 7, respectively. Then the intersection of $\overline{\mathfrak{cyc1}}$, $\overline{\mathfrak{cyc3}}$, $\overline{\mathfrak{cyc5}}$ and $\overline{\mathfrak{cyc7}}$ yielded the desired vector. The developments of $\mathfrak{cyc5}$ and especially of $\mathfrak{cyc7}$ are very technical and complex. For example, in case of $\mathfrak{cyc7}$ the relation-algebraic expression describes that for a given relation $R : X \leftrightarrow X$ there exist three pairs $(u_1, u_2) \in X \times X$, $(v_1, v_2) \in X \times X$ and $(w_1, w_2) \in X \times X$ and an element $x \in X$ such that the tuples (u_2, u_1, x, v_2, v_1) and (v_1, w_1, w_2, u_2) are paths in R and, therefore, the tuple $(u_2, u_1, x, v_2, v_1, w_1, w_2, u_2)$ is a cycle of length 7 in R.

A relation $R : X \leftrightarrow X$ has no cycle of length 1 if $R \subseteq \overline{I}$, no cycle of length 3 if $R^3 \subseteq \overline{I}$, no cycle of length 5 if $R^5 \subseteq \overline{I}$ and no cycle of length 7 if $R^7 \subseteq \overline{I}$, where powers are defined as usual. The method of this paper allows using these specifications directly leading to four vectors $nocyc1$, $nocyc3$, $nocyc5$ and $nocyc7$ of type $[X \leftrightarrow X] \leftrightarrow \mathbf{1}$ which represent the set of relations on X having no cycle of length 1, 3, 5 and 7, respectively. Then the intersection of $nocyc1$, $nocyc3$, $nocyc5$ and $nocyc7$ yields the vector we are looking for.

For the following, let an arbitrary $r : X \times X \leftrightarrow \mathbf{1}$ be given. With regard to the inclusion $R \subseteq \overline{I}$ we have $\nu_r(R) = r$ and $\nu_r(\overline{I}) = vec(\overline{I}) = \overline{vec(I)}$, hence $\nu_r(R)[M/r] = M$ and $\nu_r(\overline{I})[M/r] = \overline{vec(I)}; L = \overline{vec(I); L}$ for the replacement of r by the membership relation $M : X \times X \leftrightarrow [X \leftrightarrow X]$. So, Theorem 4.2 yields

$$nocyc1 := \overline{L; (M \cap \overline{vec(I); L})}^T.$$

In case of the inclusions $R^3 \subseteq \overline{I}$, $R^5 \subseteq \overline{I}$ and $R^7 \subseteq \overline{I}$ we work with the equivalent inclusions $(R^2)^T \subseteq \overline{R}$, $(R^3)^T \subseteq \overline{R^2}$ and $(R^4)^T \subseteq \overline{R^3}$ since these lead to more efficient RELVIEW-programs than the original inclusions. For $(R^2)^T \subseteq \overline{R}$ we get

$$\nu_r((R^2)^T) = S; \nu_r(R; R) = S; C^T; [\nu_r(R), \nu_r(R)] = S; C^T; [r, r]$$

and, hence, $\nu_r((R^2)^T)[M/r] = S; C^T; [M, M]$ for the left-hand-side, and

$$\nu_r(\overline{R}) = \overline{\nu_r(R)} = \overline{r}$$

and, hence, $\nu_r(\overline{R})[M/r] = \overline{M}$ for the right-hand-side. If we apply Theorem 4.2 to these results, we obtain the vector

$$nocyc3 := \overline{\mathsf{L}; (\mathsf{S}; \mathsf{C}^\mathsf{T}; [\![M, M]\!] \cap M)}^\mathsf{T}.$$

The remaining inclusions can be treated analogously. From $(R^3)^\mathsf{T} \subseteq \overline{R^2}$ we get

$$nocyc5 := \overline{\mathsf{L}; (\mathsf{S}; \mathsf{C}^\mathsf{T}; [\![M, H]\!] \cap H)}^\mathsf{T}, \text{ where } H := \mathsf{C}^\mathsf{T}; [\![M, M]\!],$$

and inclusion $(R^4)^\mathsf{T} \subseteq \overline{R^3}$ leads to

$$nocyc7 := \overline{\mathsf{L}; (\mathsf{S}; \mathsf{C}^\mathsf{T}; [\![H, H]\!] \cap \mathsf{C}^\mathsf{T}; [\![M, H]\!])}^\mathsf{T}, \text{ where } H := \mathsf{C}^\mathsf{T}; [\![M, M]\!].$$

We have implemented the relation-algebraic specifications developed in this section in RELVIEW and have compared the running times with those given in [7]. Doing so, we have used the same environment as mentioned in [7], i.e., version 8.2 of RELVIEW[2] on a PC with 2 CPUs of type Intel® Xeon® E5-2698, each with 20 cores and 3.60 GHz base frequency, 512 GByte RAM and running Arch Linux 5.2.0. We only present the data for $|X| = 7$. Using the RELVIEW-programs resulting from [7], it takes 138.67 s to compute the vector that represents the set of 188 553 949 010 868 relations on X which have a kernel and 32 220.55 s to compute the vector that represents the set of 16 230 843 049 relations on X which satisfy Richardson's criterion. The RELVIEW-programs we have obtained from the specifications of this paper need 201.31 s for computing the first vector and 18 843.34 s for computing the second one.

From Lemma 3.2 we get $\nu_r(\mathfrak{R}; R; \mathfrak{S}) = (\mathfrak{R} \,\|\, \mathfrak{S}^\mathsf{T}); \nu_r(R)$ for all $R : X \leftrightarrow Y$ and $\mathfrak{R}, \mathfrak{S} \in \mathbf{RE}_R$, where $r := vec(R)$. In Sect. 3 we have mentioned that this often leads to more efficient implementations. If we proceed the above presented calculation for $\nu_r(syq(\overline{M}, R; M))$ with $\overline{\nu_r(R; M)} = \overline{(\mathsf{I} \,\|\, \mathsf{M}^\mathsf{T}); \nu_r(R)}$ and $\nu_r(\mathsf{M}^\mathsf{T}; R; M) = (\mathsf{M}^\mathsf{T} \,\|\, \mathsf{M}^\mathsf{T}); \nu_r(R)$ after the third step, we obtain the variant

$$nokernel := \overline{\mathsf{L}; (\mathsf{C}^\mathsf{T}; [\![\overline{mt; \mathsf{L}}, \overline{(\mathsf{I} \,\|\, \mathsf{M}^\mathsf{T}); M}]\!] \cap \overline{(\mathsf{M}^\mathsf{T} \,\|\, \mathsf{M}^\mathsf{T}); M} \cap i; \mathsf{L})}^\mathsf{T}$$

the RELVIEW-implementation of which allows to compute the vector representation of the set of relations on X having kernels for $|X| = 7$ in 189.89 s.

6 Further Applications

We have applied our method to many other classes of specific relations. These include the remaining three criteria for the existence of kernels treated in [7], i.e., bipartite relations, progressively finite relations and symmetric and irreflexive relations. Also many of the vectors presented in [6] have been redeveloped using

---

[2] This is the newest version of the tool. It is described at the Web-site [14] and the source code is available from Github via [15] and from Zenodo via [16].

the new method. We also have applied the new method to classes of relations not treated so far, viz. lattices, bounded partial orders, finite directed acyclic graphs and arborescences, tournaments, rectangular relations, difunctional relations, general Ferrers relations, strongly connected relations and maps having fixpoints.

Sometimes the specification of a property $P(R)$ of a relation R leads to an inequation $\mathfrak{R} \neq O$, with a relation-algebraic expression $\mathfrak{R} \in \mathbf{RE}_R$. An example is the specification of R to have a kernel by $syq(\overline{M}, R; M) \cap I \neq O$ in Sect. 5. Also the specification of R to be a bounded partial order and of R to be a map with a fixed point leads to such inequations, viz. to $\overline{\overline{R}; L} \neq O$ (existence of a least element) and $\overline{\overline{R^T}; L} \neq O$ (existence of a greatest element) in the first case and to $R \cap I \neq O$ (existence of a reflexive element) in the second case.

To cope with the inequation $syq(\overline{M}, R; M) \cap I \neq O$, in Sect. 5 we consider the equation $syq(\overline{M}, R; M) \cap I = O$ instead, which specifies R to have no kernel. Then we transform it into the equivalent inclusion $syq(\overline{M}, R; M) \subseteq \overline{I}$, apply our method to the latter and, finally, form the complement of the result to obtain the vector we are actually interested in. Since the carrier set of R is non-empty, $syq(\overline{M}, R; M) \cap I \neq O$ holds iff $L \subseteq L; (syq(\overline{M}, R; M) \cap I); L$. We have applied our method also to that inclusion. But the corresponding RELVIEW-program proved to be less efficient than that we have obtained from the approach of Sect. 5.

That the inequation $\mathfrak{R} \neq O$ is equivalent to the inclusion $L \subseteq L; \mathfrak{R}; L$ for relations on non-empty carrier sets we also have used to get vector representations of the set of bounded partial orders and of maps with fixed points. In each case the specific form of the expression $\mathfrak{R}$ allows to apply Lemma 3.2 for computing $\nu_r(L; \mathfrak{R}; L)$. E.g., for the inclusion $L \subseteq L; (R \cap I); L$ we get $\nu_r(L) = L$ for the left-hand side and $\nu_r(L; (R \cap I); L) = (L \| L^T); \nu_r(R \cap I) = L; (r \cap vec(I))$ for the right-hand side. In combination with Theorem 4.2 this yields $L; \overline{\overline{L; (M \cap vec(I); L)}}^T$ as vector representation of the set of relations with a reflexive element, which still can be simplified to $(L; (M \cap vec(I); L))^T$. With the RELVIEW-program obtained from the simplified specification and the vectors of [6] for the classes of univalent and total relations, respectively, we have been able to compute the vector representation of the set of maps on X having fixed points up to $|X| = 240$. For 240 elements 5.55 s are needed to get the result and to store it in a ROBDD with 229 439 vertices. Our RELVIEW-experiments show that the percentage of maps having fixed points decreases from 100% if $|X| = 1$ to 63.284 if $|X| = 240$. This is in accordance with the well-known result that, if P_n denotes the probability that a map on an n-element set selected uniformly at random has a fixed point, then P_n tends to $1 - \frac{1}{e}$ if n tends to infinity.

7   Concluding Remarks

The computational power obtained by the use of ROBDDs and RELVIEW becomes clear if we compare the running times mentioned in Sect. 5 and 6 with the times needed in case of a "classical" brute-force approach. E.g., if we assume that some algorithm could generate every map on a given finite set X and test the existence of a fixed point in, say, 10^{-6} seconds, it would take $2.82 \cdot 10^{11}$ seconds (i.e., more than 9000 years) for this task already if $|X| = 15$, since in this case there are $282\,325\,794\,823\,047\,151$ maps on X having a fixed point. RELVIEW only needs $0.019\,\mathrm{s}$ to compute a vector representing this set.

Membership relations $\mathsf{M} : X{\times}Y \leftrightarrow [X \leftrightarrow Y]$ play a central role in our approach. The variable ordering used in RELVIEW allows to implement M by a ROBDD the number of vertices is linear in the size of $X{\times}Y$. Besides the very efficient ROBDD-implementation of the relational operations this specific feature of the tool seems to be the main reason for the amazing computational power in case of problems that deal with the computation of a subset of a powerset.

In addition to applying the theory to further examples and applications, there are at least two theoretical topics related to the material of this paper we would like to investigate in the future. The first topic is the question whether it is possible to find for any property $P(R)$ of relations $R : X \leftrightarrow Y$ expressed in second order logic an equivalent finite set of inclusions $\mathfrak{R}_i \subseteq \mathfrak{S}_i$ with $\mathfrak{R}_i, \mathfrak{S}_i \in \mathbf{RE}_R$ for all i. If this is the case, then the method of this paper could be applied immediately to the inclusions to get a vector that represents the set of all these relations. If this is not the case, it would be interesting to characterize the second order properties for which an equivalent description by a finite set of inclusions $\{\mathfrak{R}_1 \subseteq \mathfrak{S}_1, \ldots, \mathfrak{R}_n \subseteq \mathfrak{S}_n\}$ is possible, i.e., for which our method is applicable. The second topic is the above mentioned conjecture about the probability of a relation selected uniformly at random from $[X \leftrightarrow X]$ to have a kernel. The conjecture states that this probability tends to zero if $|X|$ tends to infinity. So far we only have been able to prove that, given any (but fixed) $k \in \mathbb{N}$, the probability of a relation selected uniformly at random from $[X \leftrightarrow X]$ to have a kernel with at most k elements tends to zero if $|X|$ tends to infinity.

Acknowledgement. We thank the referees for their very valuable remarks.

References

1. Berghammer, R., Neumann, F.: RELVIEW – an OBDD-based computer algebra system for relations. In: Ganzha, V.G., Mayr, E.W., Vorozhtsov, E.V. (eds.) CASC 2005. LNCS, vol. 3718, pp. 40–51. Springer, Heidelberg (2005). https://doi.org/10.1007/11555964_4
2. Berghammer, R.: Applying relation algebra and REL VIEW to solve problems on orders and lattices. Acta Infor. **45**, 211–236 (2008)
3. Berghammer, R.: Relation-algebraic modeling and solution of chessboard independence and domination problems. J. Logic and Alg. Progr. **81**, 625–642 (2012)
4. Berghammer, R.: Column-wise extendible vector expressions and the relational computation of sets of sets. In: Hinze, R., Voigtländer, J. (eds.) MPC 2015. LNCS, vol. 9129, pp. 238–256. Springer, Cham (2015). https://doi.org/10.1007/978-3-319-19797-5_12
5. Berghammer, R.: Tool-based relational investigation of closure-interior relatives for finite topological spaces. In: Höfner, P., Pous, D., Struth, G. (eds.) RAMICS 2017. LNCS, vol. 10226, pp. 60–76. Springer, Cham (2017). https://doi.org/10.1007/978-3-319-57418-9_4

6. Berghammer, R.: Relational computation of sets of relations. In: Fahrenberg, U., Gehrke, M., Santocanale, L., Winter, M. (eds.) RAMiCS 2021. LNCS, vol. 13027, pp. 54–71. Springer, Cham (2021). https://doi.org/10.1007/978-3-030-88701-8_4
7. Berghammer, R., Kulczynski, M.: Experimental investigation of sufficient criteria for relations to have kernels. In: Fahrenberg, U., Gehrke, M., Santocanale, L., Winter, M. (eds.) RAMiCS 2021. LNCS, vol. 13027, pp. 72–89. Springer, Cham (2021). https://doi.org/10.1007/978-3-030-88701-8_5
8. Desharnais, J.: Monomorphic characterization of n-ary direct products. Inf. Sci. **119**(3–4), 275–288 (1999)
9. Kehden, B.: Evaluating sets of search points using relational algebra. In: Schmidt, R.A. (ed.) RelMiCS 2006. LNCS, vol. 4136, pp. 266–280. Springer, Heidelberg (2006). https://doi.org/10.1007/11828563_18
10. Schmidt, G., Ströhlein, T.: Relations and Graphs. Springer, Heidelberg (1993)
11. Schmidt, G.: Relational Mathematics. Cambridge University Press, Cambridge (2010)
12. von Neumann, J., Morgenstern, O.: Theory of Games and Economic Behaviour. Princton University Press, Princeton (1944)
13. Winter, M.: Goguen Categories - A Categorical Approach to L-Fuzzy Relations. Trends in Logic, vol. 25. Springer, Heidelberg (2007). https://doi.org/10.1007/978-1-4020-6164-6
14. https://www.rpe.informatik.uni-kiel.de/en/research/relview
15. https://github.com/relview
16. https://zenodo.org/record/4708085#.YICAmS0RppR

Contextuality in Distributed Systems

Nasos Evangelou-Oost$^{(\boxtimes)}$ ⓘ, Callum Bannister ⓘ, and Ian J. Hayes ⓘ

The University of Queensland, St Lucia, Australia
a.evangelouoost@uq.edu.au

Abstract. We present a lattice of distributed program specifications, whose ordering represents implementability/refinement. Specifications are modelled by families of subsets of *relative* execution traces, which encode the local orderings of state transitions, rather than their absolute timing according to a global clock. This is to overcome fundamental physical difficulties with synchronisation. The lattice of specifications is assembled and analysed with several established mathematical tools. Sets of nondegenerate cells of a simplicial set are used to model relative traces, presheaves model the parametrisation of these traces by a topological space of variables, and information algebras reveal novel constraints on program correctness. The latter aspect brings the enterprise of program specification under the widening umbrella of *contextual semantics* introduced by Abramsky et al. In this model of program specifications, *contextuality* manifests as a failure of a consistency criterion comparable to Lamport's definition of *sequential consistency*. The theory of information algebras also suggests efficient local computation algorithms for the verification of this criterion. The novel constructions in this paper have been verified in the proof assistant *Isabelle/HOL*.

Keywords: Information algebras · Presheaves · Refinement lattices

1 Introduction

Lattices of sets of traces have been successful in the field of formal methods as algebraic models for program specification and verification. For concurrent programs, two important examples are trace models of Concurrent Kleene Algebra [11] and Concurrent Refinement Algebra [10]. A core advantage of these models is that they facilitate *compositional reasoning*, which mitigates the inherent difficulties in analysing the exponential proliferation of program behaviours that occur when programs run in parallel.

These models do not explicitly account for the topological structure inherent to a distributed system, which can make reasoning about local behaviour, e.g. local variable blocks, cumbersome. Moreover, trace models for concurrency often implicitly assume a "global clock" with which all traces progress in lockstep—and

Supported through an Australian Government Research Training Program Scholarship and Discovery Grant DP190102142 from the Australian Research Council (ARC).

R. Glück et al. (Eds.): RAMiCS 2023, LNCS 13896, pp. 52–68, 2023.
https://doi.org/10.1007/978-3-031-28083-2_4

this assumption limits their applicability to systems distributed over significant distances in space, due to physical (e.g. from relativistic physics) constraints on synchronisation.

In this article we describe a lattice of program specifications, that encodes the possible behaviours of a distributed system as subsets of *relative* traces, as well as its configuration into independent parallel processors.

Inspirations for this work are previous uses of presheaves in concurrency [8,12], the aforementioned refinement algebras [10,11], topological models of concurrency such as [6,17], and the diverse applications of valuation and information algebras [18], their relationship to sheaves, and their associated notion of *contextuality* [1,4].

In Sect. 2 we introduce the notion of a *relative trace*, which is a chain in a proset (preordered set) of states. Such relative traces are assigned to each variable $v \in V$ in a distributed system, and are modelled as a subset of the cartesian product $\prod_{v \in V} \Omega_v$ of the prosets Ω_v of states for each variable. The variables $v \in V$ are topologised in a space $(V, \mathcal{D})$ representing the physical configuration of the variables, and the inclusion ordering of open sets in this space induces a restriction action on the traces. These traces are called *relative* because this restriction action does not preserve the absolute timing of their states, but only the ordering of state transitions.

In Sect. 3, we introduce the notion of a *specification*, which is a pair $(\mathbf{A}, \mathcal{U})$ where $\mathbf{A}$ is a presheaf on $\mathcal{D}$ whose values are subsets of possible relative traces, and $\mathcal{U}$ is a maximal cover of $(V, \mathcal{D})$, representing the distribution of the specification into independent asynchronous components. Moreover, we explain that such specifications form a lattice $(\mathbf{\Lambda}, \preceq)$ whose ordering represents the refinement relation between specifications.

In Sect. 4, we define an *information algebra* and some associated constructs. We show how the lattice of specifications $\mathbf{\Lambda}$ corresponds to a particular ordered, adjoint information algebra. This associated information algebra permits the definition of *local* and *global consistency* for specifications, which we introduce in Sect. 5.

In Sect. 6, we show how these consistency criteria can arise in a classical scenario in distributed systems, namely the *dining philosophers*.

In the following we assume familiarity with the basic definitions of order theory and category theory, e.g. of a proset, a lattice, a category, a functor, a natural transformation, etc. Possibly less familiar structures—topological spaces, presheaves, and (augmented) simplicial sets—are briefly reviewed.

Proofs of the results in this paper are available in the appendix of the version published on arXiv [5]. Several of the constructions and proofs in this paper have been formalised in the *Isabelle/HOL* proof assistant.[1]

---

[1] https://github.com/onomatic/ramics23-proofs.

2   Relative Traces

A *topological space* is an abstract model of a geometric space, but without built-in notions of angle, distance, curvature, etc. It is formalised set-theoretically as a set X of *points*, and a set $\mathcal{T} \subseteq \mathbf{P}X$ of subsets of X, called *open sets*, which are closed under unions and finite intersections[2]. Roughly speaking, the open sets measure proximity, where nearby points occupy many open sets in common.

Let V be a finite[3] set of *variables* of a distributed computer system and $\mathcal{D}$ a topology on V. We call the open sets of $\mathcal{D}$ *domains*. We consider that a single computer is also a distributed system (as in [14]); then V could be the set of memory locations over its CPU caches, RAM modules, hard disks, etc., and the topology $\mathcal{D}$ encodes the connectivity between these parts. Or, V could be the set of memory locations over a distributed database comprised of many individual computers, and $\mathcal{D}$ represents the network topology of this distributed system.

The Frame Functor Ω. To each variable $v \in V$ we associate a *nonempty* proset Ω_v of *states* whose order represents reachability or causality, and we extend this assignment to open sets $U \in \mathcal{D}$ by setting

$$\Omega U := \prod_{v \in U} \Omega_v \tag{1}$$

where the right-hand side is a cartesian product of prosets (for which the ordering is given componentwise). Moreover, each inclusion of open sets $U \subseteq U'$ induces a function $\Omega U' \to \Omega U$ by projection of tuples (i.e. function restriction). Such restriction maps $\Omega U' \to \Omega U$ have the effect of discarding information involving variables outside U.

We evidently have that the function induced by $U \subseteq U$ is the identity, and if $U'' \subseteq U' \subseteq U$ then the functions $\Omega U \to \Omega U''$ and the composite of $\Omega U' \to \Omega U''$ and $\Omega U \to \Omega U'$ are equal—it does not matter whether we restrict tuples immediately to U'', or first restrict to U' and then restrict to U''. These assignments and properties are summarised in saying that $\Omega : \mathcal{D}^{\mathrm{op}} \to \mathscr{P}\!\mathit{ro}$ is a contravariant functor from the posetal category $\mathcal{D}$ to the category of prosets, dubbed the *frame functor*.

The Augmented Simplicial Nerve Functor N. A *presheaf* is a contravariant functor valued in sets. An *augmented simplicial set*[4] is a presheaf $S : \Delta_+{}^{\mathrm{op}} \to \mathcal{S}et$ whose domain is the *augmented simplex category* Δ_+. This is the category with

---

[2] Consequently, $\mathcal{T}$ contains at least $\emptyset$ and X, being the union and intersection respectively of an empty family of open sets.

[3] Finiteness is not crucial, but it simplifies our presentation, and in real-world examples finiteness is a realistic assumption.

[4] An ordinary *simplicial set* is a presheaf on the full subcategory $\Delta \subset \Delta_+$ consisting of only *nonempty* linearly ordered posets.

- objects as linearly ordered posets $\langle n \rangle := \{ i \in \mathbb{N} \mid 0 \leq i \leq n \}$ for integers $n \geq$ 0, as well as $\langle -1 \rangle := \emptyset$,
- morphisms as weakly monotone functions $f : \langle n \rangle \to \langle m \rangle$, i.e. those satisfying $i \leq j \implies fi \leq fj$.

Augmented simplicial sets are the objects of a category, denoted by Set_{Δ_+}, with natural transformations as morphisms. It is conventional to write the application $S\langle n \rangle$ of an augmented simplicial set S on $\langle n \rangle \in \Delta_+$ as S_n, and the application Sf of S on a morphism $f \in \Delta_+$ as f^*.

For an augmented simplicial set $S \in Set_{\Delta_+}$, an element $x \in S_n$ is called an *n-cell of* S, or a *cell of* S *(of degree n)*. The cell x is *degenerate* if there exists a non-injective function $f \in \Delta_+$ and a cell y of S with $x = f^*y$. Note that any cell of degree -1 is nondegenerate.

To any proset $\mathcal{P}$ we can produce an augmented simplicial set $\mathbf{N}\mathcal{P}$, called the *augmented simplicial nerve of* $\mathcal{P}$, whose action on objects $\langle n \rangle$ for $n \in \mathbb{N} \cup \{-1\}$ is given by

$$(\mathbf{N}\mathcal{P})_n := \mathcal{P}ro(\langle n \rangle, \mathcal{P}) \tag{2}$$

where $\mathcal{P}ro(\langle n \rangle, \mathcal{P})$ is the set of monotone functions from $\langle n \rangle$ to $\mathcal{P}$, or equivalently the set of all *chains of length n* in $\mathcal{P}$, including the empty chain, denoted $[\,]$, which is said to have length -1. Given a morphism $f : \mathcal{P} \to \mathcal{P}'$ in $\mathcal{P}ro$, we define $\mathbf{N}f : \mathbf{N}\mathcal{P} \to \mathbf{N}\mathcal{P}'$ as postcomposition with f, i.e.

$$(x : \langle n \rangle \to \mathcal{P}) \mapsto (f \circ x : \langle n \rangle \to \mathcal{P}') \tag{3}$$

This evidently defines a functor $\mathbf{N} : \mathcal{P}ro \to Set_{\Delta_+}$. Note that for $\mathcal{P} \in \mathcal{P}ro$ a proset, a nondegenerate n-cell of $\mathbf{N}\mathcal{P}$ is a chain of length n in $\mathcal{P}$ *that contains no repeated adjacent elements*. In particular, the empty chain $[\,]$ is nondegenerate.

For each domain $U \in \mathcal{D}$, $(\mathbf{N} \circ \mathbf{\Omega})U$ is an augmented simplicial set, such that for each $n \in \mathbb{N} \cup \{-1\}$, $((\mathbf{N} \circ \mathbf{\Omega})U)_n$ is the set of all possible sequences of states in $\mathbf{\Omega}U$, and the functions $((\mathbf{N} \circ \mathbf{\Omega})U)_n \to ((\mathbf{N} \circ \mathbf{\Omega})U)_m$ are generated by *mumbling* and *stuttering* maps on traces [3] (i.e. maps that omit or repeat elements of a sequence, respectively).

The nondegenerate cells functor D. A basic result in the theory of simplicial sets is the following:

Lemma 1 (Eilenberg-Zilber [7, II.3.1, pp. 26–27][5]). *For each cell x of an augmented simplicial set S there exists a unique nondegenerate cell x' such that there exists a unique surjection $f_x \in \Delta_+$ with $x = f_x^* x'$.*

For any augmented simplicial set S we can produce a plain set $\mathbf{D}S$ consisting of only the nondegenerate cells of S (as in Lemma 1). Moreover, we can extend this assignment to morphisms of Set_{Δ_+},

$$\mathbf{D} : Set_{\Delta_+} \to Set \tag{4}$$

$$S \mapsto \text{set of nondegenerate cells of } S \tag{5}$$

$$\alpha \mapsto (x \mapsto (\alpha x)') \tag{6}$$

---

[5] The cited result is stated for ordinary simplicial sets, but it clearly also applies to augmented simplicial sets as there are no degeneracies of a cell of degree -1.

56   N. Evangelou-Oost et al.

In other words, the action of **D** on morphisms of augmented simplicial sets $\alpha : S \to S'$ gives a function $\mathbf{D}\alpha : \mathbf{D}S \to \mathbf{D}S'$ that sends a nondegenerate cell x of $\mathbf{D}S$ to the nondegenerate cell $(\alpha x)'$ of S' that generates αx, whose existence and uniqueness are assured by Lemma 1. These assignments assemble to a functor.

Lemma 2. **D** *is a functor.*

Postcomposing $\mathbf{N} \circ \Omega$ by **D**, we obtain the presheaf

$$\Theta := \mathbf{D} \circ \mathbf{N} \circ \Omega : \mathcal{D}^{\mathrm{op}} \to \mathcal{S}et \tag{7}$$

that we call **chaos**.

Definition 1 (relative trace). *For $U \in \mathcal{D}$, an element $t \in \Theta U$ is a U-relative trace, or for short, a U-trace or simply a trace.*

Notes 1. We use the standard shorthand for restriction maps of a presheaf $F : \mathcal{T}^{\mathrm{op}} \to \mathcal{S}et$ (e.g. $F = \Theta$) on a topological space $(X, \mathcal{T})$; $t|_U := (Fi)t$ for $t \in FU'$ and $U, U' \in \mathcal{D}$ with $U' \subseteq U$ and i the unique morphism $i : U \to U'$.

Notes 2. By fixing a linear ordering of V, a U-trace $t \in \Theta U$ for some $U \in \mathcal{D}$ can be represented by a unique matrix with rows labelled in increasing order by the $v \in U$, and columns indexed by "time", with the property that adjacent columns of the matrix are always distinct (the empty trace $[] \in \Theta U$ is therefore represented by the unique matrix with $|U|$ rows and zero columns.). This property could be restated as saying that the traces $t \in \Theta U$ do not contain *stutterings* [3]. The qualifier "relative" applied to "traces" emphasises the latter property, which entails that a trace only records the relative ordering of its states, rather than their absolute timing according to an implied "global clock". This is illustrated in the example below.

Example 1. Let $(V, \mathbf{P}V)$ be the discrete space on a set $V := \{a, b\}$ of variables, and $\Omega_a := \{a_0, a_1\}$ and $\Omega_b := \{b_0, b_1\}$ their corresponding prosets of states, both with the *total ordering*, i.e. for which all pairs of elements are related. The trace

$$t = \begin{bmatrix} a_0 & a_0 & a_1 \\ b_0 & b_1 & b_1 \end{bmatrix} \in \Theta V \tag{8}$$

informally corresponds to an ordered set of observations of the system, where the state transition $a_0 \rightsquigarrow a_1$ is observed after the transition $b_0 \rightsquigarrow b_1$. On the other hand, the trace

$$t' = \begin{bmatrix} a_0 & a_1 \\ b_0 & b_1 \end{bmatrix} \in \Theta V \tag{9}$$

corresponds to a discretely ordered set of observations where neither $a_0 \rightsquigarrow a_1$ is observed before $b_0 \rightsquigarrow b_1$ or vice versa (although we refrain from saying they are simultaneous/synchronous). Moreover, we have

$$t|_{\{a\}} = t'|_{\{a\}} = \begin{bmatrix} a_0 & a_1 \end{bmatrix} \in \Theta\{a\} \tag{10}$$

It is for this reason we refer to such traces as *relative*, because only the relative ordering of states is preserved under restriction maps: the transition $a_0 \rightsquigarrow a_1$ in t was at "time = 2" but in $t|_{\{a\}}$ it is at "time = 1".

3   Specifications

A *cover* of a topological space $(X, \mathcal{T})$ is a family $\mathcal{U} = \{U_i\}_i$ of open sets whose union $\bigcup_i U_i$ equals X. A *maximal cover* is a cover that is also an *antichain*, meaning $U_i \subseteq U_j$ if and only if $i = j$. We call a maximal cover of $(V, \mathcal{D})$ a *context*, where each $U_i \in \mathcal{U}$ is a *domain* (of an independent process of the distributed system).

A *subpresheaf* $\mathbf{G}$ of a presheaf $\mathbf{F} : \mathcal{C}^{\mathrm{op}} \to \mathit{Set}$ on a category $\mathcal{C}$, written $\mathbf{G} \subseteq \mathbf{F}$, is a family of subsets $\mathbf{G}X \subseteq \mathbf{F}X$ for each object $X \in \mathcal{C}$ that assemble to a presheaf, where $\mathbf{G}$ inherits the action of $\mathbf{F}$ on morphisms $X' \to X$ (i.e. by function restriction).

Definition 2 (specification). *A pair* $(\mathbf{A}, \mathcal{U})$, *where* $\mathbf{A} \subseteq \Theta$ *is a subpresheaf of* Θ, *and* $\mathcal{U} = \{U_i\}_i$ *is a context, is called a* **specification**.

The first factor of a specification records the possible relative execution traces of a distributed system, and the second defines the domains of the independent asynchronous processes that make up the system. Specifications are partially ordered, where

$$(\mathbf{A}, \mathcal{U}) \preceq (\mathbf{B}, \mathcal{W}) \tag{11}$$

if and only if both $\mathbf{A} \subseteq \mathbf{B}$ and $\mathcal{U}$ *refines* $\mathcal{W}$, meaning every open $U \in \mathcal{U}$ is contained in some $W \in \mathcal{W}$. This ordering represents *implementation* (or *refinement*) of specifications: $(\mathbf{A}, \mathcal{U}) \preceq (\mathbf{B}, \mathcal{W})$ means the left-hand side *implements* (or *refines*) the right-hand side.

Example 2. Let the set of variables $V := \{a, b\}$ equipped with the discrete topology, and let the subpresheaf $\mathbf{A}$ be defined by $\mathbf{A}\emptyset := \{[]\}$, $\mathbf{A}\{a\} := \{[a_0\, a_1]\}$, and $\mathbf{A}\{b\} := \mathbf{A}V := \emptyset$, then it holds that $\mathbf{A}$ with the context $\{\{a\}, \{b\}\}$ refines chaos with the *trivial context* $\{V\}$, i.e. $(\mathbf{A}, \{\{a\}, \{b\}\}) \preceq (\Theta, \{V\})$.

Note that refinement in the first factor represents *reduction of nondeterminism*, whereas in the second factor it is *increase of parallelism*.

The subpresheaves of Θ form a complete distributive lattice[6] Sub Θ [16, §III.8 Prop. 1][7] with meet and join given by pointwise intersection and union, i.e.

$$(\mathbf{A} \cap \mathbf{B})U = \mathbf{A}U \cap \mathbf{B}U, \qquad (\mathbf{A} \cup \mathbf{B})U = \mathbf{A}U \cup \mathbf{B}U \tag{12}$$

Theorem 1. *The set of maximal covers* $\mathsf{Cov}_{\max}\mathcal{T}$ *of a space* $(X, \mathcal{T})$ *with the refinement ordering* $\leq$ *described above, forms a complete distributive lattice, with meet and join given for all* $\mathcal{U}, \mathcal{W} \in \mathsf{Cov}_{\max}\mathcal{T}$ *respectively by*

$$\mathcal{U} \wedge \mathcal{W} = \{U \in L_{\mathcal{U}} \cap L_{\mathcal{W}} \mid \nexists V \in L_{\mathcal{U}} \cap L_{\mathcal{W}}.\ U \subset V\} \tag{13}$$

$$\mathcal{U} \vee \mathcal{W} = \{U \in \mathcal{U} \cup \mathcal{W} \mid \nexists V \in \mathcal{U} \cup \mathcal{W}.\ U \subset V\} \tag{14}$$

where $L_{\mathcal{U}} := \{U \in \mathcal{T} \mid \exists V \in \mathcal{U}.\ U \subseteq V\}$.

---

[6] Actually, a *complete bi-Heyting algebra* [19, Cor. 9.1.13].

[7] The cited result is stated more generally for the lattice of sub*sheaves* of a given *sheaf* over a *site*. Here we take the *trivial site*, over which sheaves are equivalent to presheaves.

The set of specifications $\mathbf{\Lambda}$ is then defined as the cartesian product of the distributive lattices,

$$\mathbf{\Lambda} := \mathsf{Sub}\,\Theta \times \mathsf{Cov}_{\mathsf{max}}\mathcal{D} \tag{15}$$

where $\mathsf{Cov}_{\mathsf{max}}\mathcal{D}$ is the lattice of maximal covers of $(V, \mathcal{D})$, and this is again a complete distributive lattice [2, p. 12], with meet and joint defined pointwise, i.e.

$$(\mathbf{A}, \mathcal{U}) \wedge (\mathbf{B}, \mathcal{W}) = (\mathbf{A} \cap \mathbf{B}, \mathcal{U} \wedge \mathcal{W}) \tag{16}$$

$$(\mathbf{A}, \mathcal{U}) \vee (\mathbf{B}, \mathcal{W}) = (\mathbf{A} \cup \mathbf{B}, \mathcal{U} \vee \mathcal{W}) \tag{17}$$

for $(\mathbf{A}, \mathcal{U}), (\mathbf{B}, \mathcal{W}) \in \mathbf{\Lambda}$. Informally, the meet of two specifications in $(\mathbf{\Lambda}, \preceq)$ is the specification that contains all behaviours common to both while increasing parallelism to the minimal extent, whereas the join of two specifications is the specification containing the union of their behaviours while decreasing parallelism to the maximal extent.

The lattice $(\mathbf{\Lambda}, \preceq)$ has as top element $\top = (\Theta, \{V\})$, and as bottom element $\bot = (\emptyset, \bigwedge \mathsf{Cov}_{\mathsf{max}}\mathcal{D})$, where $\emptyset$ is the constant functor with $\emptyset U = \emptyset$ and with $\emptyset i = 1_\emptyset$ the identity function on the empty set, for all open sets $U \in \mathcal{D}$ and inclusions $i : U \subseteq U'$. Note that the meet $\bigwedge \mathsf{Cov}_{\mathsf{max}}\mathcal{D}$ over all contexts in $\mathsf{Cov}_{\mathsf{max}}\mathcal{D}$ exists because $\mathcal{D}$ is finite; this is the *finest* context of $\mathcal{D}$.

Our goal in the next subsection is to show that the specifications of $(\mathbf{\Lambda}, \preceq)$ can be profitably analysed through a structure known as an *information algebra*.

4 Information Algebras

An *information algebra* is an algebraic structure modelling information parameterised over a lattice of domains, together with *combination* and *projection* operators. These specialise the *valuation algebras* introduced by Shenoy [21]. Our use of information algebras is motivated by the theory of *contextual semantics* developed in [1,4]. However, [1,4] assume a discrete topology, whereas we prefer to allow arbitrary finite topological spaces to make a closer connection between the mathematical model of a distributed system and its physical topological configuration. Therefore, in the following we mildly generalise the definitions and results of [1,4] to arbitrary finite topological spaces.

Definition 3 (information algebra). *Let $(X, \mathcal{T})$ be a topological space over a finite set of variables X. An **information algebra** over $\mathcal{T}$ is a quintuple $(\Phi, \mathcal{T}, \mathrm{d}, \downarrow, \otimes)$, where Φ is a set, d a function, $\otimes$ a binary operation, and $\downarrow$ a partially defined operation,*

1. Labelling: $\mathrm{d} : \Phi \to \mathcal{T}, \phi \mapsto \mathrm{d}\phi$,
2. Projection: $\downarrow : \Phi \times \mathcal{T} \to \Phi, (\phi, U) \mapsto \phi^{\downarrow U}$, defined for all $U \subseteq \mathrm{d}\phi$,
3. Combination: $- \otimes - : \Phi \times \Phi \to \Phi, (\phi, \psi) \mapsto \phi \otimes \psi$,

such that the following properties (explained below) hold, where for $U \in \mathcal{T}$, $\Phi_U := \{\phi \in \Phi \mid \mathrm{d}\phi = U\}$, and where $\phi, \psi \in \Phi$:

(I1) Commutative semigroup: $(\Phi, \otimes)$ *is associative and commutative.*

(I2) Projection: *given* $U \subseteq \mathrm{d}\phi$, $\mathrm{d}(\phi^{\downarrow U}) = U$.

(I3) Transitivity: *given* $W \subseteq U \subseteq \mathrm{d}\phi$, $(\phi^{\downarrow U})^{\downarrow W} = \phi^{\downarrow W}$.

(I4) Domain: $\phi^{\downarrow \mathrm{d}\phi} = \phi$.

(I5) Labelling: $\mathrm{d}(\phi \otimes \psi) = \mathrm{d}\phi \cup \mathrm{d}\psi$.

(I6) Combination: *for* $U := \mathrm{d}\phi$, $W := \mathrm{d}\psi$ *and* $Q \in \mathcal{T}$ *such that* $U \subseteq Q \subseteq U \cup W$, *we have* $(\phi \otimes \psi)^{\downarrow Q} = \phi \otimes \psi^{\downarrow Q \cap W}$.

(I7) Neutrality: *for each* $U \in \mathcal{T}$, *there exists a* neutral element $1_U \in \Phi_U$ *such that* $\phi \otimes 1_U = 1_U \otimes \phi = \phi$ *for all* $\phi \in \Phi_U$. *Moreover, these neutral elements satisfy* $1_U \otimes 1_W = 1_{U \cup W}$ *for all* $U, W \in \mathcal{T}$.

(I8) Nullity: *for each* $U \in \mathcal{T}$, *there exists a* null element $0_U \in \Phi_U$ *such that* $\phi \otimes 0_U = 0_U \otimes \phi = 0_U$. *Moreover, for all* $U, W \in \mathcal{T}$ *with* $W \subseteq U$ *and* $\phi \in \Phi_U$, *these null elements satisfy* $\phi^{\downarrow W} = 0_W \iff \phi = 0_U$.

(I9) Idempotence: *For all* $U \subseteq \mathrm{d}\phi$, *it holds that* $\phi \otimes \phi^{\downarrow U} = \phi$.

The elements $\phi \in \Phi$ of an information algebra $(\Phi, \mathcal{T}, \mathrm{d}, \downarrow, \otimes)$ are called **valuations**. An element $U \in \mathcal{D}$ is called a **domain**. The **domain of a valuation** ϕ is the set $\mathrm{d}\phi \in \mathcal{D}$.

Some explanation for these axioms may be helpful. Axiom (I1) says the order in which information is combined is irrelevant. Axioms (I2)–(I4) essentially say that the triple $(\Phi, \mathrm{d}, \downarrow)$ defines the structure of a presheaf (see Note 3 below). (I5) is clear. (I6) is the subtlest of the axioms; it says that to add a new piece of information, we can first strip its irrelevant parts. This turns out to be crucial in developing efficient computational algorithms [13,18]. (I7) posits neutral elements, that contain "irrelevant" information, in the sense that combining with them adds nothing new, whereas (I8) posits null elements of "destructive" or "contradictory" information, that "corrupt" any other information combined with them. (I9) distinguishes information algebras from their more general cousins, *valuation algebras*, and is "the signature axiom of qualitative or logical, rather than quantitative, e.g. probabilistic, information. It says that counting how many times we have a piece of information is irrelevant" [1].

Notes 3. Any information algebra $(\Phi, \mathcal{T}, \mathrm{d}, \downarrow, \otimes)$ determines a presheaf $\Phi :$ $\mathcal{T}^{\mathrm{op}} \to \mathcal{S}et$, defined on objects $U \in \mathcal{T}$ by $\Phi U := \Phi_U$, and such that if $W \subseteq U$ we have an action of restriction defined by projection, i.e. $\phi|_W := \phi^{\downarrow W}$, for $\phi \in \Phi U$. This presheaf is called the **prealgebra associated to the information algebra** Φ. We sometimes use this without mention.

Information algebras may often be enriched with a partial ordering on valuations, enabling the relative quantification of their information content [9].

Definition 4 (ordered information algebra). *Let* Φ *be an information algebra on a space of variables* $(X, \mathcal{T})$. *Then* $(\Phi, \mathcal{T}, \mathrm{d}, \downarrow, \otimes, \leq)$ *is an* **ordered information algebra** *if and only if* $\leq$ *is a partial order on* Φ *such that the following axioms hold:*

(O1) Partial order: *for all* $\psi, \psi \in \Phi$, $\phi \leq \psi$ *implies* $\mathrm{d}\phi = \mathrm{d}\psi$. *Moreover, for every* $U \in \mathcal{T}$ *and* $\Psi \subseteq \Phi_U$, *the infimum* Inf Ψ *exists.*

(O2) Null element: *for all $U \in \mathcal{T}$, we have* $\operatorname{Inf} \Phi_U = 0_U$.
(O3) Monotonicity of combination: *for all $\phi_1, \phi_2, \psi_1, \psi_2 \in \Psi$ such that $\phi_1 \leq \phi_2$ and $\psi_1 \leq \psi_2$ we have $\phi_1 \otimes \psi_1 \leq \phi_2 \otimes \psi_2$.*
(O4) Monotonicity of projection: *for all $\phi, \psi \in \Phi$, if $\phi \leq \psi$, then $\phi^{\downarrow U} \leq \psi^{\downarrow U}$, for all $U \subseteq \mathrm{d}\phi = \mathrm{d}\psi$.*

Generically, we can interpret $\phi \leq \psi$ for $\phi, \psi \in \Phi$ as meaning ψ is *less*[8] informative than ϕ. Null elements 0_U for $U \in \mathcal{T}$ represent *over-constrained*, or *contradictory* information involving the variables U.

Tuple System Structure. Our goal now is to show that the lattice $(\Lambda, \preceq)$ of specifications introduced in Sect. 3 is naturally associated to a particular ordered information algebra. To this end, we first introduce an auxiliary construction known as a *tuple system*, which generalises the idea of a parameterised set of cartesian (i.e. ordinary) tuples.

Definition 5 (tuple system). *A **tuple system** over a lattice $\mathcal{L}$ is a quadruple $(\mathbf{T}, \mathcal{L}, \mathrm{d}, \downarrow)$, where $\mathbf{T}$ is a set, $\mathrm{d} : \mathbf{T} \to \mathcal{L}$ a function, and $\downarrow : \mathbf{T} \times \mathcal{L} \to \mathbf{T}$ a partially defined operation, such that $x_{\downarrow U}$ is defined only when $U \leq \mathrm{d}x$, and which satisfy the following axioms: for $x, y \in \mathbf{T}$ and $U, W \in \mathcal{L}$,*

(T1) if $U \leq \mathrm{d}x$ then $\mathrm{d}(x_{\downarrow U}) = U$,
(T2) if $W \subseteq U \subseteq \mathrm{d}x$ then $(x_{\downarrow U})_{\downarrow W} = x_{\downarrow W}$,
(T3) if $\mathrm{d}x = U$ then $x_{\downarrow U} = x$,
(T4) for $U := \mathrm{d}x$, $W := \mathrm{d}y$, if $x_{\downarrow U \wedge W} = y_{\downarrow U \wedge W}$, then there exists $z \in \mathbf{T}$ such that $\mathrm{d}z = U \vee W$, $z_{\downarrow U} = x$ and $z_{\downarrow W} = y$,
(T5) for $\mathrm{d}x = U$ and $U \leq W$, there exists $y \in \mathbf{T}$ such that $\mathrm{d}y = W$ and $y_{\downarrow U} = x$.

Notes 4. Similar to Note 3, axioms (T1)–(T3) imply that $\mathbf{T}$ is associated to a presheaf $\mathbf{T} : \mathcal{L}^{\mathrm{op}} \to Set$ in an evident way. Also, any information algebra defines a tuple system, with the same domain and projection operations [13, Lemma 6.11, p. 170].

Theorem 2. *The set $\Theta := \coprod_{U \in \mathcal{D}} \Theta U$ with $\mathrm{d} := \pi_1 : \Theta \to \mathcal{D}$ the first projection from the disjoint union Θ, i.e. $(U, \phi) \mapsto U$, and $\downarrow$ defined by restriction relative to the presheaf Θ, i.e. $x_{\downarrow U} := x|_U := (\Theta i)x$, where i is the inclusion $i : U \hookrightarrow \mathrm{d}x$, defines a tuple system over the space $\mathcal{D}$ of domains.*

For a tuple system $(\mathbf{T}, \mathcal{L})$ and $U \in \mathcal{L}$, a subset $A \subseteq \mathbf{T}_U := \{x \in \mathbf{T} \mid \mathrm{d}x = U\}$ is called a *relation*[9]. From any tuple system, we can generate an ordered information algebra of relations in a canonical way.

---

[8] Note that the ordering is in the "wrong" sense; this is so it corresponds to subset inclusion in Theorem 3 below. A different, *canonical ordering*, is used in [13], defined $\phi \leq_{\mathrm{can}} \psi \iff \phi \otimes \psi = \psi$ for all $\phi, \psi \in \Phi$. Actually, we have $\phi \leq \psi \implies \psi \leq_{\mathrm{can}} \phi$.
[9] In [4], a relation is instead called an *information set*.

Theorem 3 ([13, **Theorem 6.10**]). *Let* $(\mathbf{T}, \mathcal{L})$ *be a tuple system. Define a* **relation over** $U \in \mathcal{L}$ *to be a subset* $R \subseteq \mathbf{T}$ *such that* $\mathrm{d}x = U$ *for all* $x \in R$, *and define the* **domain of** R *as* $\mathrm{d}R := U$. *For* $U \leq \mathrm{d}R$, *the projection of* R *onto* U *is defined*

$$R^{\downarrow U} := \{x_{\downarrow U} \in \mathbf{T} \mid x \in R\} \tag{18}$$

For relations R, S *define the* **join of** R *and* S *as*

$$R \otimes S := \{x \in \mathbf{T} \mid \mathrm{d}x = \mathrm{d}R \vee \mathrm{d}S, x_{\downarrow \mathrm{d}R} \in R, x_{\downarrow \mathrm{d}S} \in S\} \tag{19}$$

For each $U \in \mathcal{L}$, *define* $0_U := \emptyset$, *called the* **empty relation on** U, *and* $1_U := \mathbf{T}_U$, *called the* **universal relation on** U.

Then the set $\mathcal{R}_{\mathbf{T}} := \bigsqcup_{U \in \mathcal{L}} \mathbf{P}(\mathbf{T}_U)$ *of all relations, where* $\mathbf{P}$ *is the (covariant) powerset functor, is an ordered information algebra, with ordering given by subset inclusion* $\subseteq$, *with null elements* 0_U *and neutral elements* 1_U, *for all* $U \in \mathcal{L}$.

We associate to the lattice $(\Lambda, \preceq)$ of specifications, the ordered information algebra $(\mathcal{R}_\Theta, \mathcal{D}, \mathrm{d}, \downarrow, \otimes, \subseteq)$, whose valuations represent *nondeterministic computations*; the nondeterminism corresponding to the multiplicity of traces in its relations. On each domain $U \in \mathcal{D}$, the ordering $\subseteq$ on $(\mathcal{R}_\Theta)U$ encodes *implementability* (or *refinement*) via reduction of nondeterminism, i.e. $R \subseteq S$ if and only if every trace of R is also a trace of S. The top element 1_U consists of all possible traces on U, whereas the bottom element 0_U is an empty set of traces.

Often, the combination operation of an information algebra has a canonical description via an adjunction [1]. It is convenient to note that this holds for $(\mathcal{R}_\Theta, \mathcal{D}, \mathrm{d}, \downarrow, \otimes, \subseteq)$.

Adjoint Structure. The following definition is adapted from [1] to an arbitrary finite base space. Let $(\Phi, \mathcal{T}, \mathrm{d}, \downarrow, \otimes, \leq)$ be an ordered information algebra. Due to the universal property of products in the category $\mathcal{S}et$, we have, for all opens $U, W \in \mathcal{T}$, the following commutative diagram,

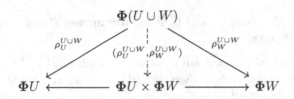

where Φ is viewed as a prealgebra, and where $\rho_W^U : \Phi U \to \Phi W$ are the restriction maps $x \mapsto x|_W$ for all $U, W \in \mathcal{T}$ with $W \subseteq U$.

Definition 6 (adjoint information algebra). *An* **adjoint information algebra** *is an ordered information algebra* $(\Phi, \mathcal{T}, \mathrm{d}, \downarrow, \otimes, \leq)$ *such that each restriction of its combination operation* $- \otimes - : \Phi U \times \Phi W \to \Phi(U \cup W)$ *is the right adjoint of the map* $(\rho_U^{U \cup W}, \rho_W^{U \cup W})$, *defined in the diagram above. Hence,* $\otimes$

is the unique map such that both,

$$1_{\Phi(U \cup W)} \leq \otimes \circ (\rho_U^{U \cup W}, \rho_W^{U \cup W}) \tag{20}$$

$$(\rho_U^{U \cup W}, \rho_W^{U \cup W}) \circ \otimes \leq 1_{\Phi U \times \Phi W} \tag{21}$$

where $\leq$ is the pointwise order induced from the partial order of the algebra, and $1_{\Phi U} : \Phi U \to \Phi U$ is the identity function on ΦU for each $U \in \mathcal{J}$.

In other words, in an adjoint information algebra Φ with $U, W \in \mathcal{J}$, Definition 6 says for all $\phi \in \Phi_{U \cup W}$, it holds

$$\phi \leq \phi^{\downarrow U} \otimes \phi^{\downarrow W} \tag{22}$$

and Definition 6 says for all $\phi \in \Phi_U$ and $\psi \in \Phi_W$, both the following inequalities hold

$$(\phi \otimes \psi)^{\downarrow U} \leq \phi, \qquad (\phi \otimes \psi)^{\downarrow W} \leq \psi \tag{23}$$

Theorem 4. *An information algebra of relations $\mathcal{R}_\mathbf{T}$ over a tuple system $\mathbf{T}$ is adjoint.*

Corollary 1. *The information algebra $\mathcal{R}_\Theta$ is adjoint.*

5   Local and Global Consistency

In this subsection, we introduce two[10] concepts of *agreement* that have an interesting interpretation for specifications in $(\Lambda, \preceq)$.

Let $(\Phi, \mathcal{J}, \mathrm{d}, \downarrow, \otimes)$ be an information algebra over a space $(X, \mathcal{J})$. A finite set of valuations $K := \{\phi_1, \ldots, \phi_n\} \subseteq \Phi$ is called a ***knowledgebase (on Φ)***. We are often interested in the case where $\bigcup_{\phi \in K} \mathrm{d}\phi = X$.

Definition 7 (local agreement). *Two valuations $\phi, \psi \in \Phi$* **locally agree** *if and only if*

$$\phi^{\downarrow \mathrm{d}\phi \cap \mathrm{d}\psi} = \psi^{\downarrow \mathrm{d}\phi \cap \mathrm{d}\psi} \tag{24}$$

A knowledgebase $K = \{\phi_1, \ldots, \phi_n\} \subseteq \Phi$ **locally agrees** *if and only if every pair ϕ_i, ϕ_j in K locally agrees.*

Definition 8 (global agreement). *A knowledgebase $K = \{\phi_1, \ldots, \phi_n\} \subseteq \Phi$* **globally agrees** *if and only if there exists[11] a valuation $\gamma \in \Phi_U$, where $U = \bigvee_{i=1}^n \mathrm{d}\phi_i$ for which, for all $1 \leq i \leq n$,*

$$\gamma^{\downarrow \mathrm{d}\phi_i} = \phi_i \tag{25}$$

---

[10] A third notion of *complete disagreement* is introduced in [1,4], but we do not make use of it here.

[11] Unlike in the definition of a *sheaf*, which is a presheaf on a topological space satisfying a certain continuity condition, there is no requirement that the amalgamation of local data (here γ) should be unique. Actually, it is common in physical applications that global sections are not unique; see for example [20] for applications of sheaf theory to the field of signal processing, where this is generally the case.

The γ of Definition 8 is called a ***truth valuation for*** K.

Notes 5. Global agreement implies local agreement: for any pair ϕ, ψ in a globally agreeing knowledgebase K, we have

$$\phi^{\downarrow d\phi \cap d\psi} = (\gamma^{\downarrow d\phi})^{\downarrow d\phi \cap d\psi} \overset{(13)}{=} \gamma^{\downarrow d\phi \cap d\psi} \overset{(13)}{=} (\gamma^{\downarrow d\psi})^{\downarrow d\phi \cap d\psi} = \psi^{\downarrow d\phi \cap d\psi} \qquad (26)$$

The converse is generally false, as we see in Example 3 below.

To any specification $(\mathbf{A}, \mathcal{U})$ we associate a knowledgebase on $\mathcal{R}_\Theta$,

$$K_{(\mathbf{A}, \mathcal{U})} := \{\mathbf{A}U\}_{U \in \mathcal{U}} \qquad (27)$$

where $\mathrm{d}(\mathbf{A}U) = U$ for each $U \in \mathcal{U}$.

Definition 9 (local/global consistency). *The specification* $(\mathbf{A}, \mathcal{U})$ *is* **locally consistent** *if and only if the associated knowledgebase* $K_{(\mathbf{A}, \mathcal{U})}$ *locally agrees. The specification* $(\mathbf{A}, \mathcal{U})$ *is* **globally consistent** *if and only if* $K_{(\mathbf{A}, \mathcal{U})}$ *globally agrees, and the corresponding truth valuation* $\gamma \in \mathcal{R}_\Theta$ *is a section of* $\mathbf{A}$.

Local consistency of a specification is a basic prerequisite for correctness. Global consistency is a subtler correctness criterion, and is related to Lamport's definition of *sequential consistency* for concurrent programs [15]:

"...the result of any execution is the same as if the operations of all the processors were executed in some sequential order, and the operations of each individual processor appear in this sequence in the order specified by its program."

Indeed, a globally consistent specification is one that can be represented by a subset of execution traces on the union of the domains of all the valuations, each one encoding a sequential ordering of states, such that when restricted to an individual domain in the context, the states occur in the same order as specified by the valuation on that domain.

The following characterises local consistency of a specification in terms of a property of the associated subpresheaf of chaos, and suggests a convenient approach to its verification.

Theorem 5. *A specification* $(\mathbf{A}, \mathcal{U})$ *is locally consistent if* $\mathbf{A}$ *is* **flasque beneath the cover** $\mathcal{U}$, *i.e. if every restriction map* $\mathbf{A}W' \to \mathbf{A}W$ *is surjective, whenever* $W \subseteq W' \subseteq U$ *for some* $U \in \mathcal{U}$.

The next result shows that in the case of an adjoint information algebra, a global valuation must take on the particular form of a solution to a so-called *inference problem* [13,18], and thereby suggests a method to determine global consistency for specifications in $\mathbf{\Lambda}$.

Theorem 6. *Let* $\mathbf{\Phi}$ *be an adjoint information algebra, let* $K = \{\phi_1, \ldots, \phi_n\} \subseteq \mathbf{\Phi}$ *be a knowledgebase, and let* $\gamma = \bigotimes_{i=1}^{n} \phi_i$. *Then* K *agrees globally if and only if* $\gamma^{\downarrow d\phi_i} = \phi_i$ *for all* $1 \leq i \leq n$. *In this case,* γ *is the greatest truth valuation for* K.

Determining if a knowledgebase is locally consistent is computationally straightforward. Global consistency, however, is computationally intensive to verify. To give an indication of the computational cost, assume for simplicity that for each variable $v \in V$, $\mathbf{\Omega}_v = \omega$ is a constant value. Let $K = \{\phi_1, \ldots, \phi_n\}$ be a knowledgebase. Then to determine if K is globally consistent, according to Theorem 6 we must compute $(\otimes K)^{\downarrow \mathrm{d}\phi_j}$ for each $1 \leq j \leq n$. To compute the join $\otimes K$ involves "filtering" from the valuations on the cartesian product of the state spaces $\mathbf{\Omega}_{\mathrm{d}\phi_i} \cong \omega^{\mathrm{d}\phi_i}$, i.e. the proset

$$\prod_i \omega^{\mathrm{d}\phi_i} \cong \omega^{\amalg_i \, \mathrm{d}\phi_i} \tag{28}$$

whose underlying set has cardinality exponential in the number of variables, and is generally intractable to compute in practice.

Fortunately, by applying the combination axiom (I6) of Definition 3 inductively, we can avoid computing the join $\otimes K$ directly, and instead compute for each j,

$$\phi_j \otimes \left(\bigotimes_{i \neq j} \phi_i^{\downarrow \mathrm{d}\phi_i \cap \mathrm{d}\phi_j} \right) \tag{29}$$

which is still exponential in the variables, but the number of variables in the exponent has been reduced, often significantly.

This is the starting point for *local computation* algorithms, such as the *fusion* and *collect* algorithms, which are generic algorithms for computing global agreement in information algebras, which in some applications are best-in-class [4, 13].

6   Example: The Dining Philosophers

In [1], a knowledgebase that locally agrees but globally disagrees is called *contextual*. We next give an example of this phenomenon—a locally consistent but globally inconsistent specification—in a classical scenario in concurrency, the "dining philosophers".

Example 3. This example models a group of philosophers sat at a circular table wanting to eat a meal, with one chopstick on the table between each adjacent pair of philosophers. A philosopher can either think or eat. To eat, a philosopher must hold both their adjacent chopsticks. Our presentation here is based on the one in [8].

Let $n \geq 2$, let $\{p_0, \ldots, p_{n-1}\}$ be variables corresponding to the philosophers, and let $\{c_0, \ldots, c_{n-1}\}$ be variables corresponding to the chopsticks. For each $0 \leq i < n$, let

$$\mathbf{\Omega}_{p_i} := \{t, e\}, \qquad \mathbf{\Omega}_{c_i} := \{i - 1, *, i\} \tag{30}$$

where t and e stand respectively for "thinking" and "eating", and $i - 1, i$ refer respectively to the philosophers p_{i-1}, p_i who may hold chopstick c_i, and $*$ to the neutral state of the chopstick on the table (all indices taken mod n).

Define a context $\mathcal{U} = \{U_i\}_{i=0}^{n-1}$ where $U_i := \{c_i, p_i, c_{i+1}\}$ represents the frame of reference of the philosopher i as an independent asynchronous process in the distributed system.

For example, if $n = 3$, we have

$$\mathcal{U} = \{\{c_0, p_0, c_1\}, \{c_1, p_1, c_2\}, \{c_2, p_2, c_0\}\} \tag{31}$$

Let $(V := \cup\mathcal{U}, \mathcal{D})$ be the topological space generated by the *subbasis* $\mathcal{U}$; i.e. $\mathcal{D}$ consists of all unions of intersections of elements of $\mathcal{U}$.

(A visual representation of the situation is given by the *Čech nerve* of the context $\mathcal{U}$; this is a simplicial complex whose n-cells are nonempty n-fold intersections of the U_i with distinct indices. In the case that $n = 3$, the Čech nerve of $\mathcal{U}$ is (the boundary of) a triangle, with only 0-cells and 1-cells (see Fig. 1).)

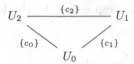

Fig. 1. Čech nerve of the context $\mathcal{U}$ for $n = 3$.

Informally, let $(\mathbf{A}, \mathcal{U})$ be the specification containing all traces according to the following protocol: the legal state transitions on the $\{c_i\}$ for $x \in \Omega_{c_i}$ are:

$$* \mapsto x \tag{33}$$
$$x \mapsto * \tag{34}$$
$$x \mapsto x \tag{35}$$

meaning, a chopstick may either be picked up or put down, or remain in its current state. The legal state transitions on the U_i are:

$$\rightsquigarrow (*, t, *) \qquad \text{(rule 1, initial state)}$$
$$(l, x, r) \rightsquigarrow (l', x, r') \qquad\qquad l, l', r, r' \neq i \qquad \text{(rule 2)}$$
$$(l, t, r) \rightsquigarrow (l', e, r') \qquad\qquad l, l', r, r' \neq i \qquad \text{(rule 3)}$$
$$(l, e, *) \rightsquigarrow (l', e, i) \qquad\qquad l, l' \neq i \qquad \text{(rule 4)}$$
$$(l, e, i) \rightsquigarrow (l', e, i) \qquad\qquad l, l' \neq i \qquad \text{(rule 5)}$$
$$(*, e, i) \rightsquigarrow (i, e, i) \qquad\qquad \text{(rule 6)}$$
$$(i, e, i) \rightsquigarrow (*, t, *) \qquad\qquad \text{(rule 7)}$$

The first rule says philosophers begin thinking, without chopsticks. The second rule says if a philosopher has no chopsticks, they may continue in their present state, without constraining the actions of their two neighbours. The third rule says if they have no chopsticks and are thinking, then they may become hungry, without constraining their neighbours. The fourth rule says that if they have no chopsticks, they may pick up their right one, if it is available, without constraining their left neighbour. The fifth rule says they can remain in the state of having just a right chopstick, without constraining their left neighbour. The sixth rule says if they have a right chopstick, they may pick up the left one if it is available. The last rule says that if they have both chopsticks and are eating, they can put them both down and think.

Consider a sub-specification $(\mathbf{B}, \mathcal{U}) \preceq (\mathbf{A}, \mathcal{U})$ whose corresponding knowledgebase is $K_{(\mathbf{B}, \mathcal{U})} := (\{\phi_0, \phi_1, \phi_2\}, \mathcal{U})$, where for $0 \leq i \leq 2$, each ϕ_i is a singleton

$$\phi_i = \left\{ \begin{bmatrix} * * * & (i-1) & * & i & * & * \\ t & e & e & e & e & e & t & t \\ * & * & i & i & i & i & * & (i+1) \end{bmatrix} \right\} \tag{36}$$

using the matrix representation of Note 2, where the first row corresponds to the variable c_i, the second to p_i, and the third to c_{i+1}. For example, the command ϕ_1 contains the single trace corresponding to the following linear sequence of events:

1. p_1 becomes hungry (rule 3);
2. p_1 picks up the right chopstick (rule 4);
3. p_0 picks up the left chopstick (rule 5);
4. p_0 puts down the left chopstick (rule 5);
5. p_1 picks up the left chopstick (rule 6);
6. p_1 eats and puts down both chopsticks (rule 7);
7. p_2 picks up the right chopstick (rule 2).

Clearly, the specification is legal according to the protocol described above, and enables each philosopher to eat their meal. Moreover, the specification is locally consistent; we have for each i,

$$\phi_i{}^{\downarrow \{c_i\}} = \begin{bmatrix} * & (i-1) & * & i \end{bmatrix} = \phi_{i-1}{}^{\downarrow \{c_i\}} \tag{37}$$

On the other hand, it is intuitively clear that the specification cannot be globally consistent, because

1. ϕ_0 says p_0 picks up c_1 before p_2 picks up c_0,
2. ϕ_2 says p_2 picks up c_0 before p_1 picks up c_2,
3. ϕ_1 says p_1 picks up c_2 before p_0 picks up c_1,

and together these events form a causal loop, which is physically impossible, and moreover, not representable as a trace on $V = U_0 \cup U_1 \cup U_2$. This can be calculated formally using Eq. (29), but we omit the details for reasons of space.

This example illustrates that global consistency of a specification is an important criterion for correctness.

7   Conclusion

We have presented a refinement lattice of specifications to model distributed programs, using mathematical structures that emphasise the intrinsic topological structure of the distributed system. The specifications in our lattice consist of subpresheaves of *relative* traces, for which the absolute timing of events is not preserved under restriction maps, but only their relative ordering. This aspect was emphasised to reflect fundamental physical constraints on synchronisation— at high speeds, such as those of modern computer technology, Einstein has taught us that the idea of synchronous events loses its meaning. This structure of relative traces then revealed an interesting correctness criterion for specifications, related to Lamport's definition of *sequential consistency*.

Acknowledgements. Nasos is grateful to his PhD advisor Larissa Meinicke for helpful feedback; to the Category Theory Zulip community (https://categorytheory.zulipchat.com) for many helpful conversations, and especially to Amar Hadzihasanovic who suggested the proof of Lemma 3, and also for the support of the Australian Government Research Training Program Scholarship. This research was supported by Discovery Grant DP190102142 from the Australian Research Council (ARC). We thank the anonymous reviewers for their helpful comments and suggestions.

References

1. Abramsky, S., Carù, G.: Non-locality, contextuality and valuation algebras: a general theory of disagreement. Philos. Trans. Roy. Soc. A **377**(2157), 20190036, 22 (2019). https://doi.org/10.1098/rsta.2019.0036
2. Birkhoff, G.: Lattice Theory. American Mathematical Society Colloquium Publications, vol. 25, 3rd edn. American Mathematical Society, Providence (1979)
3. Brookes, S.D.: Full abstraction for a shared-variable parallel language. Inf. Comput. **127**(2), 145–163 (1996). https://doi.org/10.1006/inco.1996.0056
4. Carù, G.: Logical and topological contextuality in quantum mechanics and beyond. Ph.D. thesis, University of Oxford, UK (2019)
5. Evangelou-Oost, N., Bannister, C., Hayes, I.J.: Contextuality in distributed systems (2022). https://doi.org/10.48550/ARXIV.2210.09476
6. Fajstrup, L., Goubault, E., Haucourt, E., Mimram, S., Raussen, M.: Directed Algebraic Topology and Concurrency. Springer, Cham (2016). https://doi.org/10.1007/978-3-319-15398-8
7. Gabriel, P., Zisman, M.: Calculus of fractions and homotopy theory. Ergebnisse der Mathematik und ihrer Grenzgebiete, Band 35, Springer, New York (1967)
8. Goguen, J.A.: Sheaf semantics for concurrent interacting objects. Math. Struct. Comput. Sci. **2**(2), 159–191 (1992). https://doi.org/10.1017/S0960129500001420
9. Haenni, R.: Ordered valuation algebras: a generic framework for approximating inference. Int. J. Approx. Reason. **37**(1), 1–41 (2004). https://doi.org/10.1016/j.ijar.2003.10.009
10. Hayes, I.J., Colvin, R.J., Meinicke, L.A., Winter, K., Velykis, A.: An algebra of synchronous atomic steps. In: Fitzgerald, J., Heitmeyer, C., Gnesi, S., Philippou, A. (eds.) FM 2016. LNCS, vol. 9995, pp. 352–369. Springer, Cham (2016). https://doi.org/10.1007/978-3-319-48989-6_22
11. Hoare, T., van Staden, S., Möller, B., Struth, G., Zhu, H.: Developments in concurrent Kleene algebra. J. Log. Algebraic Methods Program. **85**(4), 617–636 (2016). https://doi.org/10.1016/j.jlamp.2015.09.012
12. Joyal, A., Nielsen, M., Winskel, G.: Bisimulation from open maps. Inf. Comput. **127**(2), 164–185 (1996). https://doi.org/10.1006/inco.1996.0057

13. Kohlas, J.: Information Algebras - Generic Structures for Inference. DISCMATH, Springer, Heidelberg (2003). https://doi.org/10.1007/978-1-4471-0009-6
14. Lamport, L.: Time, clocks, and the ordering of events in a distributed system. Commun. ACM **21**(7), 558–565 (1978). https://doi.org/10.1145/359545.359563
15. Lamport, L.: How to make a multiprocessor computer that correctly executes multiprocess programs. IEEE Trans. Comput. **28**(9), 690–691 (1979). https://doi.org/10.1109/TC.1979.1675439
16. Mac Lane, S., Moerdijk, I.: Sheaves in Geometry and Logic: A First Introduction to Topos Theory. UTX, Springer, New York (1994). https://doi.org/10.1007/978-1-4612-0927-0. Corrected reprint of the 1992 edition
17. Porter, T.: Enriched categories and models for spaces of dipaths. In: Kopperman, R., Panangaden, P., Smyth, M.B., Spreen, D. (eds.) Computational Structures for Modelling Space, Time and Causality. Dagstuhl Seminar Proceedings (DagSemProc), vol. 6341. Schloss Dagstuhl - Leibniz-Zentrum für Informatik, Dagstuhl, Germany (2007). https://doi.org/10.4230/DagSemProc.06341.5. https://drops.dagstuhl.de/opus/volltexte/2007/898
18. Pouly, M., Kohlas, J.: Generic Inference: A Unifying Theory for Automated Reasoning. Wiley, Hoboken (2012)
19. Reyes, M.L.P., Reyes, G.E., Zolfaghari, H.: Generic figures and their glueings: a constructive approach to functor categories. Polimetrica (2004)
20. Robinson, M.: Topological Signal Processing. ME, Springer, Heidelberg (2014). https://doi.org/10.1007/978-3-642-36104-3
21. Shenoy, P.P.: A valuation-based language for expert systems. Int. J. Approx. Reason. **3**(5), 383–411 (1989). https://doi.org/10.1016/0888-613X(89)90009-1

The Structure of Locally Integral Involutive Po-monoids and Semirings

José Gil-Férez$^{(\boxtimes)}$, Peter Jipsen, and Siddhartha Lodhia

Chapman University, Orange, CA 92866, USA
{gilferez,jipsen,lodhia}@chapman.edu

Abstract. We show that every locally integral involutive partially ordered monoid (ipo-monoid) $\mathbf{A} = (A, \leqslant, \cdot, 1, \sim, -)$, and in particular every locally integral involutive semiring, decomposes in a unique way into a family $\{\mathbf{A}_p : p \in A^+\}$ of integral ipo-monoids, which we call its *integral components*. In the semiring case, the integral components are semirings. Moreover, we show that there is a family of monoid homomorphisms $\Phi = \{\varphi_{pq} \colon \mathbf{A}_p \to \mathbf{A}_q : p \leqslant q\}$, indexed on the positive cone $(A^+, \leqslant)$, so that the structure of $\mathbf{A}$ can be recovered as a glueing $\int_\Phi \mathbf{A}_p$ of its integral components along Φ. Reciprocally, we give necessary and sufficient conditions so that the Płonka sum of any family of integral ipo-monoids $\{\mathbf{A}_p : p \in D\}$, indexed on a lower-bounded join-semilattice $(D, \vee, 1)$, along a family of monoid homomorphisms Φ is an ipo-monoid.

Keywords: Residuated lattices · Involutive partially ordered monoids · Semirings · Płonka sums · Frobenius quantales

1 Introduction

Idempotent semirings are algebras of the form $(A, \vee, \cdot, 1)$ where $(A, \vee)$ is a semilattice (with order $x \leqslant y \iff x \vee y = y$), $(A, \cdot, 1)$ is a monoid, and the monoid operation distributes over the join. They play an important role in mathematics, logic, and theoretical computer science, since they generalize distributive lattices and expand to Kleene algebras and residuated lattices. An *involutive semiring* is an idempotent semiring with operations $\sim$ and $-$ satisfying:

$$x \leqslant y \iff x \cdot \sim y \leqslant -1 \iff -y \cdot x \leqslant -1.$$

These algebras are term-equivalent to involutive residuated lattices and, in the case that the lattice is complete, to Frobenius quantales (see [5] and [4]). Furthermore, algebras of binary relations are involutive semirings under the operations of union, composition, and complement-converse. The structural characterization obtained in this paper is valid for more general partially ordered structures called *involutive po-monoids* where the semilattice $(A, \vee)$ is replaced by a poset $(A, \leqslant)$.

An ipo-monoid is *integral* when the monoid identity 1 is also the top element of the order, that is, the inequality $x \leqslant 1$ holds. This is a very important property

© The Author(s), under exclusive license to Springer Nature Switzerland AG 2023
R. Glück et al. (Eds.): RAMiCS 2023, LNCS 13896, pp. 69–86, 2023.
https://doi.org/10.1007/978-3-031-28083-2_5

for residuated lattices, since it is equivalent to the proof-theoretical rule called weakening. In this work we identify a much larger class, namely the class of locally integral ipo-monoids.

The main result in this paper is that every locally integral ipo-monoid $\mathbf{A}$ decomposes in a unique way into a family of integral ipo-monoids $\{\mathbf{A}_p : 1 \leqslant p\}$, which we call its *integral components*. Two locally integral ipo-monoids can have the same integral components, but may differ in the way these components are *glued* together. We find in the literature similar situations in which a number of structures are glued together to form a new one: for instance, in [6] it is described how chains can be attached to an odd Sugihara monoid in order to form a commutative idempotent residuated chain, and in [8] how Boolean algebras can be glued together to form commutative idempotent involutive residuated lattices.

In our case, we associate to every locally integral ipo-monoid $\mathbf{A}$ a join-semilattice indexed family of monoid homomorphisms $\Phi = \{\varphi_{pq}: \mathbf{A}_p \to \mathbf{A}_q : 1 \leqslant p \leqslant q\}$ between its integral components so that the structure of $\mathbf{A}$ can be completely recovered as an aggregate or *glueing* $\int_\Phi \mathbf{A}_p$ of these integral components along Φ in two stages: first, the monoid part of $\mathbf{A}$ turns out to be the Płonka sum of the family Φ, and the involutive negations can be defined componentwise. Then, we recover the order of $\mathbf{A}$ using the product, the negations, and the local identities.

As an application of our results, we can combine certain semantics for fuzzy logics with semantics for relevance logic using, for example, the well-understood MV-algebras as building blocks of a glueing.

We exploit this decomposition in order to prove that several properties of locally integral ipo-monoids are *local*, in that a locally integral ipo-monoid satisfies them if and only if all its integral components satisfy them. One of the most significant local properties established here is local finiteness.

Previous research into the structure of doubly-idempotent semirings can be found in [1,2]. The structure of all finite commutative idempotent involutive residuated lattices is completely described in [8] in a step-by-step decomposition. In the current paper, this is significantly generalized to all locally integral ipo-monoids, without any restrictions regarding finiteness, commutativity, or full idempotence. A similar use of Płonka sums can be found in [7], where the structure of even and odd involutive commutative residuated chains is studied.

We set the terminology and notation in Sect. 2, and describe the fundamental properties of ipo-monoids needed in the rest of the paper. In Sect. 3, we introduce the class of locally integral ipo-monoids and show that every locally integral ipo-monoid is the glueing of its integral components. Finally, in Sect. 4, we solve the reverse problem, that is, we provide necessary and sufficient conditions so that the glueing of a system of integral ipo-monoids is an ipo-monoid.

2   Involutive Partially Ordered Monoids and Semirings

An *involutive partially ordered monoid*, or *ipo-monoid* for short, is a structure of the form $(A, \leqslant, \cdot, 1, \sim, -)$ such that $(A, \leqslant)$ is a poset (i.e., $\leqslant$ is a reflexive, antisymmetric, and transitive binary relation on A), $(A, \cdot, 1)$ is a monoid (i.e., $\cdot$ is an associative binary operation on A and 1 is its identity element) satisfying:

$$x \leqslant y \iff x \cdot {\sim}y \leqslant 0 \iff -y \cdot x \leqslant 0, \tag{ineg}$$

where, by definition, $0 = -1$.[1] The unary operations $\sim$ and $-$ are called *involutive negations*. If there is no danger of confusion, we will write xy instead of $x \cdot y$. Given an ipo-monoid $\mathbf{A}$, we say that $\mathbf{A}$ is *cyclic* if it satisfies ${\sim}x = -x$. An element x of $\mathbf{A}$ is *central* if $x \cdot y = y \cdot x$ for any other $y \in A$, and $\mathbf{A}$ is *commutative* if all its elements are central. An element x of $\mathbf{A}$ is *idempotent* if $x \cdot x = x$, and $\mathbf{A}$ is *idempotent* if all its elements are idempotent. We will be specially interested in ipo-monoids with a lattice order. These can be then presented as algebraic structures $(A, \wedge, \vee, \cdot, 1, \sim, -)$ called *iℓ-monoids* or *involutive semirings*.[2]

Lemma 1. *Every ipo-monoid satisfies the following properties:*

1. *double negation:* ${\sim}{-}x = x = {-}{\sim}x$ (dn)

2. *rotation:* $x \cdot y \leqslant z \iff y \cdot {\sim}z \leqslant {\sim}x \iff -z \cdot x \leqslant -y$ (rot)

3. *antitonicity:* $x \leqslant y \iff {\sim}y \leqslant {\sim}x \iff -y \leqslant -x$ (ant)

4. *residuation:* $xy \leqslant z \iff x \leqslant -(y \cdot {\sim}z) \iff y \leqslant {\sim}(-z \cdot x)$ (res)

5. *constants:* $0 = {\sim}1, \ {\sim}0 = 1, \ and \ -0 = 1.$ (ct)

    The properties of the previous lemma will often be used without mentioning them explicitly. Notice also that the multiplication is residuated, with *left* and *right residuals* $z/y = -(y \cdot {\sim}z)$ and $x \backslash z = {\sim}(-z \cdot x)$, respectively, as (res) can be rewritten as:

$$x \cdot y \leqslant z \iff x \leqslant z/y \iff y \leqslant x \backslash z.$$

The fact that $\cdot$ preserves arbitrary existing joins, and therefore is order-preserving, in both arguments follows easily from these observations. It can be also readily checked that the involutive negations can be expressed in terms of the residuals as follows: ${\sim}x = x \backslash 0$ and $-x = 0/x$. Since in any commutative ipo-monoid the equality $y/x = x \backslash y$ holds, every commutative ipo-monoid is cyclic.

Lemma 2. *Every ipo-monoid satisfies the following properties:*

1. $-({\sim}x \cdot {\sim}y) = {\sim}(-x \cdot -y)$,
2. ${\sim}x$ *is idempotent if and only if* $-x$ *is idempotent.*

Proof. 1. Using (res), (rot), (dn), and (res) again, we obtain $z \leqslant -({\sim}x \cdot {\sim}y) \iff z \cdot {\sim}x \leqslant y \iff -y \cdot z \leqslant -{\sim}x \iff -y \cdot z \leqslant x \iff z \leqslant {\sim}(-x \cdot -y)$. Since z is arbitrary, we deduce that $-({\sim}x \cdot {\sim}y) = {\sim}(-x \cdot -y)$.

---

[1] Notice that the symmetry of all the properties of Lemma 1, and specially (ct), suggests that we would obtain the same results had we defined $0 = {\sim}1$.

[2] This terminology is based on the observation that $(A, \vee, \cdot, 1)$ is an idempotent unital semiring since the residuation property of Lemma 1 implies that $x(y \vee z) = xy \vee xz$ and $(x \vee y)z = xz \vee yz$, and $\wedge$ is term definable by the De Morgan laws.

2. Assume that $\sim x$ is idempotent. Then, by (dn) and the previous part, $-x \cdot -x = -\sim(-x \cdot -x) = --(\sim x \cdot \sim x) = --\sim x = -x$. The rest is analogous.   □

Lemma 3. *For every ipo-monoid* **A**, *the following conditions are equivalent:*

1. *The identity* $-x \cdot x = x \cdot \sim x$ *holds in* **A**,
2. *The identity* $\sim(-x \cdot x) = -(x \cdot \sim x)$, *that is,* $x \backslash x = x / x$, *holds in* **A**.

Proof. Suppose that the equation $-x \cdot x = x \cdot \sim x$ holds in **A**. In particular, we have that $-\sim x \cdot \sim x = \sim x \cdot \sim\sim x$, that is, $x \cdot \sim x = \sim x \cdot \sim\sim x$. Hence,

$$\sim(-x \cdot x) = \sim(-x \cdot -\sim x) = -(\sim x \cdot \sim\sim x) = -(x \cdot \sim x),$$

where the middle equality follows from Lemma 2(1). In order to prove the other implication, suppose that the equation $\sim(-x \cdot x) = -(x \cdot \sim x)$ holds in **A**. In particular, we have that $\sim(x \cdot \sim x) = \sim(-\sim x \cdot \sim x) = -(\sim x \cdot \sim\sim x) = \sim(-x \cdot -\sim x) = \sim(-x \cdot x)$, where again the last but one equality follows from Lemma 2(1). Using (dn), we deduce that $-x \cdot x = x \cdot \sim x$.   □

Given an ipo-monoid **A**, we call $A^+ = \{x \in A : 1 \leqslant x\}$ the *positive cone* of **A**, and its elements the *positive* elements of **A**, and $\downarrow 0 = \{x \in A : x \leqslant 0\}$ the principal order-ideal generated by 0. We say that an ipo-monoid **A** is $\downarrow 0$-*idempotent* if all the elements in $\downarrow 0$ are idempotent. Thus, an involutive semiring is $\downarrow 0$-idempotent if and only if the quasiequation $x \wedge 0 = x \implies x^2 = x$ holds in **A**. Furthermore, this property can be expressed by the identity $(x \wedge 0)^2 = x \wedge 0$. Our next result characterizes $\downarrow 0$-idempotence in ipo-monoids.

Lemma 4. *An ipo-monoid is* $\downarrow 0$-*idempotent* $\iff$ *for all* $x, y \leqslant 0$, $x \cdot y = x \wedge y$.

Proof. If **A** is $\downarrow 0$-idempotent, then $0 \cdot 0 \leqslant 0$, and applying (rot) we obtain $0 = 0 \cdot 1 = 0 \cdot \sim 0 \leqslant \sim 0 = 1$. Thus, if $x, y \leqslant 0$, in particular $x, y \leqslant 1$, and therefore $x \cdot y \leqslant x$ and $x \cdot y \leqslant y$. Also, if $z \leqslant x$ and $z \leqslant y$, then in particular $z \leqslant 0$ and so it is idempotent. Thus, $z = z \cdot z \leqslant x \cdot y$ and hence $x \cdot y = x \wedge y$. Conversely, if **A** satisfies that for all $x, y \leqslant 0$, $x \cdot y = x \wedge y$, then for any $x \leqslant 0$, $x \cdot x = x \wedge x = x$.   □

The next result shows that $\downarrow 0$-idempotence implies that all the elements in the positive cone are idempotent. The converse is not always true.

Theorem 5. *If* **A** *is an ipo-monoid so that* $0 \leqslant 1$ *and* $x \in \downarrow 0$ *is idempotent, then both* $\sim x$ *and* $-x$ *are idempotent. Thus, all positive elements of a* $\downarrow 0$-*idempotent ipo-monoid are idempotent.*

Proof. Suppose that **A** is an ipo-monoid so that $0 \leqslant 1$ and $x \leqslant 0$ is idempotent. By (ant), $1 \leqslant -x$ and so, $-x = -x \cdot 1 \leqslant -x \cdot -x$. Also, $x \leqslant x$ implies $-xx \leqslant 0 \leqslant 1$, and therefore, $-x \cdot x = -x \cdot x \cdot x \leqslant 1x = x$, and by (rot), $-x \cdot -x \leqslant -x$. Thus, $-x \cdot -x = -x$. By Lemma 2(2), $\sim x$ is also idempotent. Finally, if **A** is $\downarrow 0$-idempotent, in particular $0 \cdot 0 = 0$, which implies that $0 \leqslant 1$ by (rot), and for every $1 \leqslant x$, we have that $\sim x \leqslant 0$ is idempotent and therefore so is $-\sim x = x$.   □

3   Locally Integral IPO-Monoids and Involutive Semirings

An ipo-monoid is *integral* if it satisfies the inequality $x \leqslant 1$. Thus, integral ipo-monoids form a po-subvariety of the po-variety of ipo-monoids, in the sense of [9]. Notice that, since the inequality $x \leqslant 1$ can be expressed as $x \vee 1 = 1$ in the language of involutive semirings, the integral involutive semirings form a subvariety of the involutive semirings. We will introduce in what follows another po-subvariety of ipo-monoids and the corresponding subvariety of the variety of involutive semirings. We say that an ipo-monoid is *locally integral*[3] if

1. it satisfies the identity $-x \cdot x = x \cdot {\sim}x$,
2. multiplication is square-decreasing, that is, $x^2 \leqslant x$,
3. it is $\downarrow 0$-idempotent.

The main goal of this section is a decomposition theorem stating that every locally integral ipo-monoid (involutive semiring, respectively) can be decomposed in a very particular way into integral involutive ipo-monoids (involutive semirings, respectively). Let's start by proving that integrality implies local integrality.

Proposition 6. *Every integral ipo-monoid is locally integral.*

Proof. Suppose that **A** is an integral ipo-monoid. The inequality $1 \cdot x \leqslant x$ implies that $1 \leqslant x \backslash x$, and therefore $x \backslash x = 1$, by the integrality of **A**. Analogously, $x / x = 1$, and hence $x \backslash x = x / x$, which by Lemma 3 is equivalent to $-x \cdot x = x \cdot {\sim}x$.

The square decreasing property follows immediately from the monotonicity of multiplication, since $x \leqslant 1$ implies that $xx \leqslant 1x = x$.

Finally, ${\sim}x \leqslant 1$ implies that $0 \leqslant x$, for all x in **A**, and in particular $\downarrow 0 = \{0\}$. Furthermore, $1 \cdot 0 \leqslant 1$ implies that $0 \cdot 0 = 0 \cdot {\sim}1 \leqslant {\sim}1 = 0$, whence we deduce that $0 \cdot 0 = 0$, proving that **A** is $\downarrow 0$-idempotent.   □

Given a locally integral ipo-monoid **A**, we define for every x in A the elements $0_x = x \cdot {\sim}x$ and $1_x = -0_x$. Local integrality implies that $0_x = -x \cdot x$ and $1_x = {\sim}0_x$, by Lemma 3, and hence ${\sim}1_x = -1_x = 0_x$. Notice also that $1_x = x \backslash x = x / x$, and hence $0_x \leqslant 0$ and $1 \leqslant 1_x$. Thus, both 0_x and 1_x are idempotent. We will use the interval notation $[0_x, 1_x] = \{y \in A : 0_x \leqslant y \leqslant 1_x\}$. The equivalence relation $x \equiv y$ if and only if $1_x = 1_y$ partitions every locally integral ipo-monoid in its equivalence classes $A_x = \{y \in A : 1_x = 1_y\}$ and, obviously, $x \in A_x$. The next lemma offers a very useful description of A_x.

Lemma 7. *For any locally integral ipo-monoid* **A** *and all* x *and* y *in* A:

1. $0_{{\sim}x} = 0_{-x} = 0_x$ *and* $1_{{\sim}x} = 1_{-x} = 1_x$,
2. $x \in [0_x, 1_x]$, *and therefore* $0_x \leqslant 1_x$,
3. $1_x \cdot y = y \iff 1_x \leqslant 1_y$,

---

[3] This class forms a po-quasivariety, by definition. It is not known whether it is a po-variety or a proper po-quasivariety.

4. $y \in [0_x, 1_x] \iff [0_y, 1_y] \subseteq [0_x, 1_x]$,
5. $y \in A_x \iff y \in [0_x, 1_x]$ and $1_x \cdot y = y$.

Proof. 1. A simple computation shows that $0_{\sim x} = -\sim x \cdot \sim x = x \cdot \sim x = 0_x$, and therefore $1_{\sim x} = -0_{\sim x} = -0_x = 1_x$. The proof that $0_{-x} = 0_x$ and $1_{-x} = 1_x$ is analogous.

2. The square-decreasing property, namely, $x \cdot x \leqslant x$, can also be expressed as $x \leqslant x \backslash x = 1_x$, by residuation. Thus, using part (1), we have that $0_x = 0_{\sim x} = -1_{\sim x} \leqslant -\sim x = x$. That is, $x \in [0_x, 1_x]$.

3. $1_x \leqslant 1_y = y/y$ is equivalent to $1_x \cdot y \leqslant y$, by residuation. And since $1 \leqslant 1_x$, we also have that $y \leqslant 1_x \cdot y$. Hence, $1_x \cdot y \leqslant y$ is equivalent to $1_x \cdot y = y$.

4. For the left-to-right implication, notice that $0_x \leqslant y \leqslant 1_x$ implies that $0_x \leqslant \sim y \leqslant 1_x$, by (ant), and then $0_x = 0_x \cdot 0_x \leqslant y \cdot \sim y = 0_y$. By (ant) again, we obtain that $1_y \leqslant 1_x$. The reverse implication is a consequence of part (2).

5. If $y \in A_x$, then $1_y = 1_x$, and thus $y \in [0_y, 1_y] = [0_x, 1_x]$, by part (2). Moreover, $1_x \cdot y = 1_y \cdot y = y$, by part (3). For the reverse implication, notice that if $y \in [0_x, 1_x]$ and $1_x \cdot y = y$, then $1_y \leqslant 1_x$, by part (4), and $1_x \leqslant 1_y$, by part (3).   □

Next, we will use the description of A_x of the previous lemma in order to show that the sets A_x are closed under several operations of **A**.

Lemma 8. *Let* **A** *be a locally integral ipo-monoid. For every x in A:*

1. *A_x is closed under the involutive negations,*
2. *A_x is closed under multiplication,*
3. *A_x is closed under all existing nonempty joins and nonempty meets.*

Proof. 1. By Lemma 7(1), if $y \in A_x$ then $1_{\sim y} = 1_y = 1_x$, and hence $\sim y \in A_x$.

2. If $y, z \in A_x$ then $y, z \in [0_x, 1_x]$, by Lemma 7(5). Hence, $0_x = 0_x \cdot 0_x \leqslant y \cdot z \leqslant 1_x \cdot 1_x = 1_x$. Also by Lemma 7(5), we have $1_x \cdot (y \cdot z) = (1_x \cdot y) \cdot z = y \cdot z$, since $y \in A_x$. Thus, again by Lemma 7(5), $y \cdot z \in A_x$.

3. Suppose that $\emptyset \neq Y \subseteq A_x$ and the join $\bigvee Y$ exists in A. Since for every y in Y, $y \in A_x \subseteq [0_x, 1_x]$, we obtain that also $\bigvee Y \in [0_x, 1_x]$. And since multiplication distributes with respect to all existing joins, we have that $1_x \cdot \bigvee Y = \bigvee_{y \in Y} 1_x \cdot y = \bigvee_{y \in Y} y = \bigvee Y$. Thus, by Lemma 7(5), $\bigvee Y \in A_x$. The closure under all existing nonempty meets can be obtained from the fact that A_x is also closed under negations and $\bigwedge Y = -\bigvee_{y \in Y} \sim y$.   □

Our next goal is to find a canonical representative for each equivalence class A_x. But first, we will provide useful characterizations of A^+ and $\downarrow 0$.

Lemma 9. *Let* **A** *be a locally integral ipo-monoid. For all p and a in A, we have that $p \in A^+$ if and only if $p = 1_p$ and $a \in \downarrow 0$ if and only if $a = 0_a$. In particular, both involutive negations coincide for positive elements and for elements in $\downarrow 0$.*

Proof. We already know that $p \leqslant 1_p$ is valid for all p in A. If moreover $1 \leqslant p$, then $1_p = p/p \leqslant p/1 = p$. The other implication is trivial, since we know that $1 \leqslant 1_p$ is true for all p. The second part follows from the following equivalences:

$$a \in {\downarrow}0 \iff {\sim}a \in A^+ \iff 1_a = 1_{{\sim}a} = {\sim}a \iff a = -1_a = 0_a.$$

The last part is true, since for every $p \in A^+$, ${\sim}p = {\sim}1_p = -1_p = -p$, and analogously for the elements of ${\downarrow}0$. $\qquad\square$

Lemma 10. *Let* $\mathbf{A}$ *be a locally integral ipo-monoid. For every* x *in* A, 1_x *is the only positive element of* A_x *and* 0_x *is the only element of* A_x *below* 0.

Proof. Obviously, $1_x \in [0_x, 1_x]$ and also $1_x \cdot 1_x = 1_x$, since $1_x \in A^+$. Thus, $1_x \in A_x$, by Lemma 7. Also, as we mentioned before, $1 \leqslant 1_x$. For any positive $p \in A_x$, we would have that $p = 1_p = 1_x$, by Lemma 9. The second part follows from (ant) and the fact that A_x is closed under the involutive negations. $\qquad\square$

Remark 11. Notice that the previous lemma tells us that for every x in A, there is only one positive element p so that $A_x = A_p$. This means that the family $\{A_x : x \in A\}$ is actually indexed by A^+ and that for all $p, q \in A^+$, we have $A_p = A_q$ if and only if $p = q$. Furthermore, from Lemma 9 and the previous comment, ${\downarrow}0 = \{0_x : x \in A\} = \{0_p : p \in A^+\}$.

We can now show that the relation and operations of a locally integral ipo-monoid furnish each equivalence class A_x with the structure of an integral ipo-monoid, with a suitable identity.

Proposition 12. *If* $\mathbf{A}$ *is a locally integral ipo-monoid, then for every* p *in* A^+, *the structure* $\mathbf{A}_p = (A_p, \leqslant, \cdot, 1_p, {\sim}, -)$, *where the relation and the operations are the restrictions to* A_p *of the corresponding relation and operations of* $\mathbf{A}$, *is an integral ipo-monoid. If in addition* $\mathbf{A}$ *is a semiring, cyclic, or commutative, then* $\mathbf{A}_p$ *is also a semiring, cyclic, or commutative, respectively, for all* p *in* A^+.

Proof. By Lemma 8, every A_p is closed under multiplication and the involutive negations, and $1_p \in A_p$. Therefore, the structure $\mathbf{A}_p$ is well defined, $(A_p, \leqslant)$ is a poset, and since $1_p \cdot x = x$ for all $x \in A_p$ by Lemma 7, $(A_p, \cdot, 1_p)$ is a monoid. Moreover, since the only element of A_p below 0 is $0_p = -1_p$ by Lemma 10, we deduce from the property (ineg) of $\mathbf{A}$ that for all $x, y \in A_p$,

$$x \leqslant y \iff x \cdot {\sim}y \leqslant 0_p \iff -y \cdot x \leqslant 0_p,$$

which is precisely the property (ineg) for the structure $\mathbf{A}_p$. Finally, by Lemma 7 again, $A_p \subseteq [0_p, 1_p]$, and therefore $x \leqslant 1_p$ for all $x \in A_p$.

The proof for the locally integral involutive semirings follows from the fact that A_p is also closed under all binary joins and meets, by Lemma 8. $\qquad\square$

We call every $\mathbf{A}_p$ an *integral component* of $\mathbf{A}$. As we saw in Proposition 12, some properties of $\mathbf{A}$ are inherited by every of its integral components.

Sometimes the opposite is also true. We say that a property of ipo-monoids is *local* whenever an ipo-monoid has it if and only if all its integral components have it.

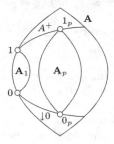

Fig. 1. Representation of the structure of a locally integral ipo-monoid

Given a locally integral ipo-monoid $\mathbf{A}$, the sets A^+ and $\downarrow 0$ are obviously partially ordered by the order of $\mathbf{A}$. The next proposition describes these two posets (Fig. 1).

Proposition 13. *Let $\mathbf{A}$ be a locally integral ipo-monoid. Then $(A^+, \cdot, 1)$ is a lower-bounded join-semilattice whose order coincides with the order of $\mathbf{A}$. Also, $(\downarrow 0, \cdot, 0)$ is an upper-bounded meet-semilattice, whose order coincides with the order of $\mathbf{A}$, and is dually isomorphic to $(A^+, \cdot, 1)$. If, in addition, A^+ is finite, then $(A^+, \leqslant)$ is a distributive lattice dual to $(\downarrow 0, \leqslant)$.*

Proof. For the first part, notice that A^+ is closed under products. For all $p, q \in A^+$, we have that $p = 1 \cdot p \leqslant pq$ and analogously $q \leqslant pq$. Furthermore, if $p \leqslant r$ and $q \leqslant r$, then $pq \leqslant r^2 \leqslant r$. This shows that $(A^+, \cdot, 1)$ is a join semilattice whose induced order is the restriction of $\leqslant$, and whose lower bound is 1.

As for the second part, the map $\eta \colon A^+ \to \downarrow 0$ given by $\eta(p) = {\sim}p = 0_p$ is bijective (Remark 11) and for any two elements $p, q \in A^+$, we have that $p \leqslant q$ if and only if $\eta(q) = 0_q = {\sim}q \leqslant {\sim}p = 0_p = \eta(p)$, by (ant), and $\eta(1) = {\sim}1 = 0$. Therefore, the restriction of $\leqslant$ to $\downarrow 0$ is a meet-semilattice ordering with upper bound 0. And by Lemma 4, given two elements $0_p, 0_q \in \downarrow 0$, we have $0_p \cdot 0_q = 0_p \wedge 0_q$.

Finally, if A^+ is finite, then also $\downarrow 0$ is finite and therefore a lattice with respect to the restricted order. Since meet and multiplication coincide in $\downarrow 0$, and multiplication distributes with respect to joins, $(\downarrow 0, \leqslant)$ is distributive, and therefore also $(A^+, \leqslant)$ is distributive. $\square$

Remark 14. Notice that the dual isomorphism $\eta \colon (A^+, \cdot, 1) \to (\downarrow 0, \cdot, 0)$ sends joins to meets, and therefore, for any two positive elements p and q, we have that

$$0_p \cdot 0_q = \eta(p) \cdot \eta(q) = \eta(p \cdot q) = 0_{pq}.$$

Also, since we showed in the previous proposition that the product of two positive elements is their join, then we deduce that multiplication of positive elements is commutative. We can actually improve on this result.

Proposition 15. *All positive elements of any locally integral ipo-monoid are central.*

Proof. Suppose that p is positive and let x be an arbitrary element. The equality $p \cdot 0_{px} = p(-(px) \cdot px) = p(px \cdot \sim(px)) = ppx \cdot \sim(px) = px \cdot \sim(px) = 0_{px}$ implies by (rot) that $-(px)x \leqslant -(px) \cdot px = 0_{px} \leqslant 0_{px} \cdot 1_{px} \leqslant \sim p = -p$, since $1 \leqslant p$ and $1 \leqslant 1_{px}$. Hence, $xp \leqslant px$, by (rot).

Now, applying (rot) to $xp \leqslant xp$, we obtain $-(xp)x \leqslant -p = \sim p$, and by (rot) again, $p \cdot -(xp) \leqslant -x$. Finally, since $xp \leqslant px$ is true for any x, in particular we have that $-(xp)p \leqslant p \cdot -(xp) \leqslant -x$, and by (rot) one last time, $px \leqslant xp$. □

As we saw in Lemma 8, every integral component of a locally integral ipo-monoid is closed under multiplication. But, what happens when we multiply elements from different components? The following lemma answers this question.

Lemma 16. *Given a locally integral ipo-monoid* **A**, *positive elements p and q, and elements $x \in A_p$ and $y \in A_q$, the product xy is in A_{pq}.*

Proof. The inequalities $1_p = p \leqslant pq = 1_{pq}$ and $1_q = q \leqslant pq = 1_{pq}$ imply that $A_p \cup A_q \subseteq [0_p, 1_p] \cup [0_q, 1_q] \subseteq [0_{pq}, 1_{pq}]$, and therefore $x, y \in [0_{pq}, 1_{pq}]$, whence we deduce that $xy \in [0_{pq}, 1_{pq}]$. Moreover, $1_{pq} \cdot (xy) = pqxy = pxqy = 1_x \cdot x \cdot 1_y \cdot y = xy$, by Lemma 9, Proposition 15, and Lemma 7. Hence, by Lemma 7, $xy \in A_{pq}$. □

All these results point toward the idea that locally integral ipo-monoids are built up from integral ones, or at least their monoid reducts are, by means of a Płonka sum. This construction was first introduced and studied in [10–12]; for more recent expositions see [13] and [3]. Given a compatible family of homomorphisms between algebras of the same type $\{\varphi_{ij} : \mathbf{A}_i \to \mathbf{A}_j : i \leqslant j\}$, indexed by the order of a lower-bounded join-semilattice $(I, \vee, \perp)$, its *Płonka sum* is the algebra **S** of the same type defined on the disjoint union of their universes $S = \biguplus_{i \in I} A_i$, so that for every constant symbol c, $c^{\mathbf{S}} = c^{\mathbf{A}_\perp}$, and for every n-ary operation symbol σ and elements $a_1 \in A_{i_1}, \ldots, a_n \in A_{i_n}$, $\sigma^{\mathbf{S}}(a_1, \ldots, a_n) = \sigma^{\mathbf{A}_j}(\varphi_{i_1 j}(a_1), \ldots, \varphi_{i_n j}(a_n))$, where $j = i_1 \vee \cdots \vee i_n$. The compatibility condition of the family of homomorphisms says that for every $i \in I$, φ_{ii} is the identity on $\mathbf{A}_i$, and that if $i \leqslant j \leqslant k$ then $\varphi_{jk} \circ \varphi_{ij} = \varphi_{ik}$. One can readily prove that the Płonka sum of a compatible family of homomorphisms is well defined and it satisfies all regular equations that hold in all the algebras of the family. Recall that a *regular* equation is an equation in which the variables that appear on the left-hand side are the same as the variables that appear on the right-hand side.

Given a locally integral ipo-monoid, we would like to find a compatible family Φ of monoid homomorphisms indexed on the order of A^+, so that the monoid reduct of **A** can be reconstructed as the Płonka sum of Φ. Consider, for every pair of positive elements $p \leqslant q$, the map $\varphi_{pq} : A_p \to A_q$ given by $\varphi_{pq}(x) = qx$.

Lemma 17. *Let* $\mathbf{A}$ *be a locally integral ipo-monoid and* $p \leqslant q$ *two positive elements. Then* $\varphi_{pq} \colon \mathbf{A}_p \to \mathbf{A}_q$ *is a well defined monoid homomorphism. Moreover, it respects arbitrary nonempty existing joins and therefore is monotone.*

Proof. For all positive elements p and q, and $x \in A_p$, we have that $qx \in A_{qp}$, by Lemma 16. Moreover, by Proposition 13, the inequality $p \leqslant q$ implies that $pq = q$. Hence, the map $\varphi_{pq} \colon A_p \to A_q$ is well defined. Furthermore, $\varphi_{pq}(1_p) = q1_p = qp = q = 1_q$ and for all $x, y \in A_p$,

$$\varphi_{pq}(x \cdot y) = qxy = qqxy = qxqy = \varphi_{pq}(x) \cdot \varphi_{pq}(y),$$

since q is positive and therefore idempotent and central, by Proposition 15. This shows that φ_{pq} is a monoid homomorphism. Finally, if $\emptyset \neq Y \subseteq A_p$ is such that $\bigvee Y$ exists, then $\varphi_{pq}(\bigvee Y) = q \cdot \bigvee Y = \bigvee_{y \in Y} qy = \bigvee_{y \in Y} \varphi_{pq}(y)$.   □

Proposition 18. *Let* $\mathbf{A}$ *be a locally integral ipo-monoid. Then, its associated family* $\Phi = \{\varphi_{pq} \colon \mathbf{A}_p \to \mathbf{A}_q\}$ *is compatible family of monoid homomorphisms indexed by the order of the join semilattice* $(A^+, \cdot, 1)$.

Proof. For every positive element p and $x \in A_p$, we have $\varphi_{pp}(x) = px = 1_p x = x$, by Lemma 7, since $x \in A_p$. That is, φ_{pp} is the identity homomorphism on $\mathbf{A}_p$. And if $p \leqslant q \leqslant r$ are positive elements, then $\varphi_{qr}(\varphi_{pq}(x)) = rqx = rx = \varphi_{pr}(x)$, since $rq = r$ by Proposition 13, because $q \leqslant r$.   □

As we will show in the next result, the monoid reduct of a locally integral ipo-monoid is the Płonka sum of the family above. Although this is not the case for the rest of the structure, still we can recover it from its integral components. Recall that a property is *local* if it is satisfied by an ipo-monoid if and only if it is satisfied by all its local components.

Theorem 19. *Let* $\mathbf{A}$ *be a locally integral ipo-monoid and* Φ *its associated family of monoid homomorphisms defined above. Then, its Płonka sum* $\mathbf{S} = (\biguplus \mathbf{A}_p, \cdot^{\mathbf{S}}, 1^{\mathbf{S}})$ *is the monoid reduct of* $\mathbf{A}$. *Moreover, if we define* $\sim^{\mathbf{S}} x = \sim^{\mathbf{A}_p} x$ *and* $-^{\mathbf{S}} x = -^{\mathbf{A}_p} x$, *for every* $x \in A_p$ *with* p *positive, and*

$$x \leqslant^{\mathbf{S}} y \iff x \cdot^{\mathbf{S}} \sim^{\mathbf{S}} y = 0_{pq}, \quad \text{for all } x \in A_p \text{ and } y \in A_q,$$

then $(\biguplus \mathbf{A}_p, \leqslant^{\mathbf{S}}, \cdot^{\mathbf{S}}, \sim^{\mathbf{S}}, -^{\mathbf{S}})$ *is* $\mathbf{A}$. *Furthermore, cyclicity and commutativity are local properties.*

Proof. By Remark 11, the set $\{A_p : p \in A^+\}$ is a partition of A, and therefore $\biguplus A_p = A$. The element $1^{\mathbf{S}} = 1^{\mathbf{A}_1} = 1$, and given two elements $x \in A_p$ and $y \in A_q$, for arbitrary positive elements p and q, and $r = pq$, we have that

$$x \cdot^{\mathbf{S}} y = \varphi_{pr}(x) \cdot^{\mathbf{A}_r} \varphi_{qr}(x) = rx \cdot ry = rrxy = rxy = 1_r \cdot (xy) = xy,$$

since r is positive, and therefore central and idempotent, and $xy \in A_r$ by Lemma 16. The involutive negations of every integral component $\mathbf{A}_p$ are the restrictions of the corresponding operations of $\mathbf{A}$, by Proposition 12, and there-fore $\sim^{\mathbf{S}}x = \sim^{\mathbf{A}_p}x = \sim x$ and $-^{\mathbf{S}}x = -^{\mathbf{A}_p}x = -x$.

Notice also that for every $x \in A_p$ and $y \in A_q$, for p and q positive, $x \leqslant y$ if and only if $x \cdot \sim y \leqslant 0$, by (ineg). Since $x \cdot \sim y \in A_{pq}$ and the only element below 0 in A_{pq} is 0_{pq} by Lemma 9, we have that

$$x \leqslant y \iff x \cdot \sim y \leqslant 0 \iff x \cdot \sim y = 0_{pq} \iff x \cdot^{\mathbf{S}} \sim^{\mathbf{S}} y = 0_{pq} \iff x \leqslant^{\mathbf{S}} y.$$

Finally, $\mathbf{A}$ is commutative if and only if all its integral components are commutative, since commutativity is expressible by the regular equation $x \cdot y = y \cdot x$. The same is true for cyclicity. $\square$

Corollary 20. *A locally integral ipo-monoid $\mathbf{A}$ is idempotent if and only if all its integral components are Boolean algebras. In particular, any idempotent ipo-monoid is commutative if and only if it satisfies $-x \cdot x = x \cdot \sim x$.*

Proof. An integral ipo-monoid is idempotent if and only if it is a Boolean algebra, because if $\mathbf{A}$ is idempotent then for all $x, y \in A$, $x \cdot y = x \wedge y$. Indeed, $x \cdot y \leqslant 1 \cdot y = y$ and analogously $x \cdot y \leqslant x$. And if $z \leqslant x$ and $z \leqslant y$, then $z = z \cdot z \leqslant x \cdot y$. Hence, the result follows from the fact that a locally integral ipo-monoid is idempotent if and only if all its integral components are idempotent. $\square$

The previous corollary covers the structural decomposition results in [8]. In this paper it is also shown that the variety of commutative idempotent involutive residuated lattices fails to be locally finite. Without the lattice operations, however, we have the following result.

Corollary 21. *Local finiteness is a local property of ipo-monoids.*

Proof. Suppose that the integral components of $\mathbf{A}$ are locally finite and let $X \subseteq A$ be a finite set and $J = \{1_x : x \in X\}$. Without loss of generality, we can assume that J is closed under binary joins (i.e., products), and that $J \subseteq X$. We will prove the proposition by induction on the cardinality of J. Let p be a minimal element in J and Y_p the closure of $X_p = X \cap A_p$ under products and involutive negations. Since $\mathbf{A}_p$ is locally finite, Y_p is also finite. Consider the finite set $X' = (X \smallsetminus X_p) \cup \{ry : y \in Y_p, \ p < r \in J\}$ and notice that $J' = \{1_x : x \in X'\} = J \smallsetminus \{p\}$, which is closed under binary joins, and $J' \subseteq X'$. By the inductive hypothesis, the subalgebra $\mathbf{B}$ generated by X' is finite. And since J' is closed under binary joins, $B \subseteq \bigcup_{q \in J'} A_q$. Now, for any $y \in Y_p$ and $x \in B$, $yx = (ry)x \in B$ and $xy = x(ry) \in B$, where $r = p \cdot 1_x \in J \smallsetminus \{p\}$. Since $1 \in B$ and both Y_p and B are closed under products and involutive negations, the universe of the subalgebra generated by X is $Y_p \cup B$, which is finite. The reciprocal is obvious. $\square$

4   Glueing Constructions

The last theorem of the previous section shows how every ipo-monoid is an aggregate of its integral components. Our next question is, what are the conditions that a family of integral ipo-monoids and a family of homomorphisms should satisfy so that the construction of Theorem 19 is a (locally integral) ipo-monoid?

To make this question precise, let's assume that $\mathbf{D} = (D, \vee, 1)$ is a lower-bounded join semi-lattice, $\mathcal{A} = \{\mathbf{A}_p : p \in D\}$ is family of integral ipo-monoids, and $\Phi = \{\varphi_{pq} \colon \mathbf{A}_p \to \mathbf{A}_q : p \leqslant^{\mathbf{D}} q\}$ is a compatible family of monoid homomorphisms. We call $(\mathbf{D}, \mathcal{A}, \Phi)$ a *semilattice direct system of integral ipo-monoids*. Letting $\mathbf{A}_p = (A_p, \leqslant_p, \cdot_p, 1_p, \sim_p, -_p)$, for all p in D, we define the structure

$$\textstyle\int_{\Phi} \mathbf{A}_p = \left(\biguplus_D A_p, \leqslant^{\mathbf{G}}, \cdot^{\mathbf{G}}, 1^{\mathbf{G}}, \sim^{\mathbf{G}}, -^{\mathbf{G}} \right),$$

where $\left(\biguplus_D A_p, \cdot^{\mathbf{G}}, 1^{\mathbf{G}} \right)$ is the Płonka sum of the family Φ, and therefore a monoid, and for all $p, q \in D$, $a \in A_p$, and $b \in A_q$, $\sim^{\mathbf{G}} a = \sim_p a$ and $-^{\mathbf{G}} a = -_p a$, and

$$a \leqslant^{\mathbf{G}} b \iff a \cdot^{\mathbf{G}} \sim^{\mathbf{G}} b = 0_{p \vee q}.$$

We call this structure $\int_{\Phi} \mathbf{A}_p$ the *glueing of $\mathcal{A}$ along the family* Φ (Fig. 2).

With this definition, one can restate Theorem 19 as saying that every locally integral ipo-monoid $\mathbf{A}$ is the glueing $\int_{\Phi} \mathbf{A}_p$ of its integral components along the family of homomorphisms $\Phi = \{\varphi_{pq} \colon \mathbf{A}_p \to \mathbf{A}_q\}$ determined by $\varphi_{pq}(x) = qx$. Our question is, given a system $(\mathbf{D}, \mathcal{A}, \Phi)$ of integral ipo-monoids, what are the conditions that Φ must satisfy in order to ensure that $\int_{\Phi} \mathbf{A}_p$ is an ipo-monoid?

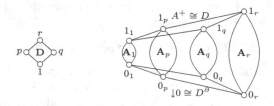

Fig. 2. Structure of a locally integral ipo-monoid

We start our analysis by identifying some relevant elements of $\int_{\Phi} \mathbf{A}_p$. But first, notice that since $\sim^{\mathbf{G}}$ and $-^{\mathbf{G}}$ are defined componentwise on the disjoint union $\biguplus_D A_p$, we can safely drop the superscripts and subscripts of these operations. Now, according to the definition of the Płonka sum, $1^{\mathbf{G}} = 1^{\mathbf{A}_1} = 1_1$. Let's set $0^{\mathbf{G}} = -1^{\mathbf{G}} = -1_1 = 0_1$.

The next lemma can be interpreted as saying that the glueing $\int_{\Phi} \mathbf{A}_p$ is indeed an "aggregate" of the integral ipo-monoids $\mathbf{A}_p$, although not necessarily an ipo-monoid itself, since the relation $\leqslant^{\mathbf{G}}$ could be not transitive. In the example of Fig. 3, $0_r \leqslant^{\mathbf{G}} 0_p \leqslant^{\mathbf{G}} 1_p \leqslant^{\mathbf{G}} 1_q$, but $0_r \not\leqslant^{\mathbf{G}} 1_q$.

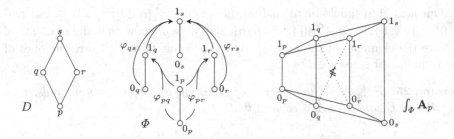

Fig. 3. A glueing of integral ipo-monoids that is not an ipo-monoid

Lemma 22. *If $\int_\Phi A_p$ is the glueing of a system of integral ipo-monoids $(\mathbf{D}, \mathcal{A}, \Phi)$, then the restrictions of $\leqslant^{\mathbf{G}}$, $\cdot^{\mathbf{G}}$, $\sim^{\mathbf{G}}$, and $-^{\mathbf{G}}$ to A_p are $\leqslant_p$, $\cdot_p$, $\sim_p$, and $-_p$, respectively. Moreover, for all $p \leqslant^{\mathbf{D}} q$ and $a \in A_p$, we have that $\varphi_{pq}(a) = 1_q \cdot^{\mathbf{G}} a$.*

Proof. The fact that $\sim^{\mathbf{G}}$ and $-^{\mathbf{G}}$ restricted to A_p are $\sim_p$ and $-_p$ is immediate, by the definitions. Now, if $a, b \in A_p$, then $p \vee p = p$, and by the definition of $\leqslant^{\mathbf{G}}$ we have that $a \leqslant^{\mathbf{G}} b \iff a \cdot^{\mathbf{G}} \sim b = 0_p \iff \varphi_{pp}(a) \cdot_p \varphi_{pp}(\sim b) = 0_p \iff a \cdot_p \sim b = 0_p \iff a \leqslant_p b$, since φ_{pp} is the identity on $\mathbf{A}_p$. For the same reason, $a \cdot^{\mathbf{G}} b = \varphi_{pp}(a) \cdot_p \varphi_{pp}(b) = a \cdot_p b$. Finally, if $p \leqslant^{\mathbf{D}} q$ and $a \in A_p$, then $\varphi_{pq}(a) = \varphi_{pq}(1_p \cdot_p a) = \varphi_{pq}(1_p) \cdot_q \varphi_{pq}(a) = 1_q \cdot_q \varphi_{pq}(a) = \varphi_{qq}(1_q) \cdot_q \varphi_{pq}(a) = 1_q \cdot^{\mathbf{G}} a$. □

Remark 23. An immediate consequence of this result is that $\leqslant^{\mathbf{G}}$ is a reflexive relation, since for every $p \in D$ and $a \in A_p$, we have that $a \leqslant^{\mathbf{G}} a$ if and only if $a \leqslant_p a$, which we know is true. This result also implies that $\sim^{\mathbf{G}}$ and $-^{\mathbf{G}}$ satisfy (dn), since $\sim^{\mathbf{G}} -^{\mathbf{G}} a = \sim_p -_p a = a = -_p \sim_p a = -^{\mathbf{G}} \sim^{\mathbf{G}} a$.

It seems obvious that, for $\int_\Phi A_p$ to be an ipo-monoid, the condition (ineg) has to be satisfied, what imposes on the family Φ the following *balance* condition:

$$\text{for all } p, q \in D, a \in A_p, b \in A_q, \quad a \cdot^{\mathbf{G}} \sim b = 0_{p \vee q} \iff -b \cdot^{\mathbf{G}} a = 0_{p \vee q}. \quad \text{(bal)}$$

One can readily check that the commutativity of $\int_\Phi A_p$ implies that Φ is balanced. We will prove next that when Φ is balanced, the operations $\sim^{\mathbf{G}}$ and $-^{\mathbf{G}}$ are involutive with respect to the relation $\leqslant^{\mathbf{G}}$.

Lemma 24. *If $\int_\Phi A_p$ is the glueing of a system of integral ipo-monoids $(\mathbf{D}, \mathcal{A}, \Phi)$ so that Φ satisfies (bal), then for all $p, q \in D$, $a \in A$, and $b \in B$,*

$$a \leqslant^{\mathbf{G}} b \iff -b \leqslant^{\mathbf{G}} -a \iff \sim b \leqslant^{\mathbf{G}} \sim a.$$

Proof. The first equivalence can be proven as follows: $a \leqslant^{\mathbf{G}} b \iff a \cdot^{\mathbf{G}} \sim b = 0_{p \vee q} \iff -b \cdot^{\mathbf{G}} a = 0_{p \vee q} \iff -b \cdot^{\mathbf{G}} (\sim -a) = 0_{p \vee q} \iff -b \leqslant^{\mathbf{G}} -a$. For the other equivalence, just notice that $\sim b \leqslant^{\mathbf{G}} \sim a \iff a = -\sim a \leqslant^{\mathbf{G}} -\sim b = b$. □

Our next step should be to analyze the sets $G^+ = \{a \in \biguplus A_p : 1_1 \leqslant^{\mathbf{G}} a\}$ and $\downarrow^{\mathbf{G}} 0_1 = \{a \in \biguplus A_p : a \leqslant^{\mathbf{G}} 0_1\}$.[4] In particular we will show that the elements of G^+ are the elements of the form 1_p, and the elements of $\downarrow^{\mathbf{G}} 0_1$ are the ones of the form 0_p, for some $p \in D$.

Lemma 25. *If $\int_\Phi \mathbf{A}_p$ is the glueing of a system of integral ipo-monoids, then for all p in D and a in A_p, we have that*

$$1_1 \leqslant^{\mathbf{G}} a \iff a = 1_p \qquad and \qquad a \leqslant^{\mathbf{G}} 0_1 \iff a = 0_p.$$

Proof. Since $1 \leqslant^{\mathbf{D}} p$, for all p, in particular $p = 1 \vee p$. Hence, $1_1 \leqslant^{\mathbf{G}} a \iff \varphi_{1p}(1_1) \cdot_p \varphi_{pp}(\sim_p a) = 0_p \iff 1_p \cdot_p \sim_p a = 0_p \iff \sim_p a = 0_p \iff a = -_p 0_p = 1_p$. The proof of the second equivalence is analogous. $\qquad \square$

Reflecting on Proposition 13, we would like to show that the relation $\leqslant^{\mathbf{G}}$ endows G^+ with a structure of join semilattice isomorphic to D, and $\downarrow^{\mathbf{G}} 0_1$ with a structure of meet semilattice dually isomorphic to D. In general, this will not be true. For this to hold, it will be necessary to assume an extra property of Φ. We will prove first that this property is valid for the family of monoid homomorphisms associated to a locally integral ipo-monoid.

Lemma 26. *Let $\mathbf{A}$ be a locally integral ipo-monoid and $p \leqslant q$ positive elements. Then, $\varphi_{pq}(0_p) = 0_q$ if and only if $p = q$.*

Proof. The implication from left to right is obvious, since $p = q$ implies that φ_{pq} is the identity map. As for the other implication, just notice that $p < q$ implies that $0_q < 0_p \leqslant q \cdot 0_p = \varphi_{pq}(0_p)$, and therefore $\varphi_{pq}(0_p) \neq 0_q$. $\qquad \square$

This suggests the following condition for Φ, which we call *zero avoidance*:

$$\text{for all } p \leqslant^{\mathbf{D}} q, \quad \varphi_{pq}(0_p) = 0_q \iff p = q. \qquad (za)$$

Lemma 27. *If $\int_\Phi \mathbf{A}_p$ is the glueing of a system of integral ipo-monoids $(\mathbf{D}, \mathcal{A}, \Phi)$ and Φ satisfies* (za)*, then for all p and q in D, $1_p \leqslant^{\mathbf{G}} 1_q \iff p \leqslant^{\mathbf{D}} q$ and $0_p \leqslant^{\mathbf{G}} 0_q \iff q \leqslant^{\mathbf{D}} p$.*

Proof. For all $p, q \in D$, with $r = p \vee q$, we have the following equivalences: $1_p \leqslant^{\mathbf{G}} 1_q \iff \varphi_{pr}(1_p) \cdot_r \varphi_{qr}(0_q) = 0_r \iff 1_r \cdot_r \varphi_{qr}(0_q) = 0_r \iff \varphi_{qr}(0_q) = 0_r \iff q = r = p \vee q \iff p \leqslant^{\mathbf{D}} q$. The proof of the second equivalence is analogous. $\qquad \square$

The previous lemma seems to capture the spirit of Proposition 13. Notice though that this proposition is more specific, as it says that the join of two positive elements, as well as the meet of two elements below 0, is their product.

---

[4] Notice that, even though we don't know whether $\leqslant^{\mathbf{G}}$ is a partial order (and actually, it will not be one in general), these definitions still make sense.

We can readily see that, in the glueing $\int_\Phi \mathbf{A}_p$ along a family Φ satisfying (za), for any two elements p and q in D, with $r = p \vee q$, we have

$$1_p \cdot^{\mathbf{G}} 1_q = \varphi_{pr}(1_p) \cdot_r \varphi_{qr}(1_q) = 1_r \cdot_r 1_r = 1_r = 1_{p \vee q} = 1_p \vee 1_q.$$

But, for the case of the elements of $\downarrow^{\mathbf{G}} 0_1$, this will not always be true: for instance, in the example of Fig. 3, $0_q \cdot 0_r = 1_s \neq 0_s = 0_{q \vee r}$. We will need to impose an extra condition on Φ:

$$\text{for all } p, q \in D, \quad 0_p \cdot^{\mathbf{G}} 0_q = 0_{p \vee q}. \tag{$*$}$$

Notice that condition $(*)$ is not spurious, as it is equivalent to the fact that for all $p, q \in D$, $0_p \leqslant^{\mathbf{G}} 1_q$, which is a desirable property, since we know that $0_p \leqslant^{\mathbf{G}} 0_1$, $0_1 \leqslant^{\mathbf{G}} 1_1$, and $1_1 \leqslant^{\mathbf{G}} 1_q$, and we want $\leqslant^{\mathbf{G}}$ to be a partial order, and in particular transitive. Thus, the condition $(*)$ will be a consequence of a much more general condition on Φ:

$$\text{for all } a, b, c \in \biguplus A_p, \quad \text{if } a \leqslant^{\mathbf{G}} b \text{ and } b \leqslant^{\mathbf{G}} c, \text{ then } a \leqslant^{\mathbf{G}} c. \tag{tr}$$

Our next result characterizes the condition (tr) in simpler terms.

Lemma 28. *If $\int_\Phi \mathbf{A}_p$ is the glueing of a system of integral ipo-monoids $(\mathbf{D}, \mathcal{A}, \Phi)$ and Φ satisfies* (bal), *then Φ satisfies* (tr) *if and only if it satisfies:*

1. *for all $p \leqslant^{\mathbf{D}} q$, and $a, b \in A_p$, $a \leqslant_p b \implies \varphi_{pq}(a) \leqslant_q \varphi_{pq}(b)$;* (mon)
2. *for all $p \leqslant^{\mathbf{D}} q$, $p \leqslant^{\mathbf{D}} r$, and $a \in A_p$, $\sim\varphi_{pq}(a) \leqslant^{\mathbf{G}} \varphi_{pr}(\sim a)$;* (lax)
3. *for all $p \vee r \leqslant^{\mathbf{D}} v$, $a \in A_p$, and $b \in A_r$,*

$$\varphi_{rv}(\sim b) \leqslant_v \sim\varphi_{pv}(a) \implies a \leqslant^{\mathbf{G}} b. \tag{$\sim$lax}$$

Proof. First, notice that for all $p \leqslant^{\mathbf{D}} q$ and $a \in A_p$, we have that $a \cdot^{\mathbf{G}} \sim\varphi_{pq}(a) = \varphi_{pq}(a) \cdot_q \varphi_{qq}(\sim\varphi_{pq}(a)) = \varphi_{pq}(a) \cdot_q \sim\varphi_{pq}(a) = 0_q$, what implies that $a \leqslant^{\mathbf{G}} \varphi_{pq}(a)$. We will use this property several times in what follows. Suppose now that Φ satisfies both (bal) and (tr).

(mon) Suppose that $a, b \in A_p$ are such that $a \leqslant_p b$, and let $p \leqslant^{\mathbf{D}} q$. Then, by the property above, $a \leqslant^{\mathbf{G}} b \leqslant^{\mathbf{G}} \varphi_{pq}(b)$, and by (tr), we obtain that $a \leqslant^{\mathbf{G}} \varphi_{pq}(b)$. Hence, $\varphi_{pq}(a) \cdot_q \sim\varphi_{pq}(b) = a \cdot^{\mathbf{G}} \sim\varphi_{pq}(b) = 0_q$, and therefore $\varphi_{pq}(a) \leqslant_q \varphi_{pq}(b)$.

(lax) By the property above, we have that $a \leqslant^{\mathbf{G}} \varphi_{pq}(a)$ and $\sim a \leqslant^{\mathbf{G}} \varphi_{pr}(\sim a)$, and by Lemma 24, $\sim\varphi_{pq}(a) \leqslant^{\mathbf{G}} \sim a$. We deduce by (tr) that $\sim\varphi_{pq}(a) \leqslant^{\mathbf{G}} \varphi_{pr}(\sim a)$.

($\sim$lax) By the property above, we have that $a \leqslant^{\mathbf{G}} \varphi_{pv}(a)$ and $\sim b \leqslant^{\mathbf{G}} \varphi_{rv}(\sim b)$, and by Lemma 24, $\sim\varphi_{pv}(a) \leqslant^{\mathbf{G}} \sim a$. If in addition we have $\varphi_{rv}(\sim b) \leqslant_v \sim\varphi_{pv}(a)$, then $\varphi_{rv}(\sim b) \leqslant^{\mathbf{G}} \sim\varphi_{pv}(a)$ and we deduce by (tr) that $\sim b \leqslant^{\mathbf{G}} \sim a$, and so $a \leqslant^{\mathbf{G}} b$.

In order to prove the reverse implication, suppose that Φ satisfies (bal) and the three above conditions, and $p, q, r \in D$, with $s = p \vee q$, $t = q \vee r$, $u = p \vee r$, $a \in A_p$, $b \in A_q$, and $c \in A_r$ are such that $a \leqslant^{\mathbf{G}} b$ and $b \leqslant^{\mathbf{G}} c$. Then, by definition of $\leqslant^{\mathbf{G}}$, we have that $\varphi_{ps}(a) \cdot_s \varphi_{qs}(\sim b) = 0_s$ and $\varphi_{qt}(b) \cdot_t \varphi_{rt}(\sim c) = 0_t$, whence we deduce that $\varphi_{ps}(a) \leqslant_s -\varphi_{qs}(\sim b)$ and $\varphi_{rt}(\sim c) \leqslant_t \sim\varphi_{qt}(b)$. Taking $v = s \vee t$, we deduce by (mon) that $\varphi_{pv}(a) \leqslant_v \varphi_{sv}(-\varphi_{qs}(\sim b))$ and $\varphi_{rv}(\sim c) \leqslant_v \varphi_{tv}(\sim\varphi_{qt}(b))$. Moreover, by (lax), we have that $\sim\varphi_{qt}(b) \leqslant^{\mathbf{G}} \varphi_{qs}(\sim b)$ and by Lemma 24, we deduce that $-\varphi_{qs}(\sim b) \leqslant^{\mathbf{G}} -\sim\varphi_{qt}(b) = \varphi_{qt}(b)$, and therefore

$$\varphi_{pv}(a) \cdot_v \varphi_{rv}(\sim c) \leqslant_v \varphi_{sv}(-\varphi_{qs}(\sim b)) \cdot_v \varphi_{tv}(\sim\varphi_{qt}(b)) = 0_v,$$

which implies that $\varphi_{pv}(a) \cdot_v \varphi_{rv}(\sim c) = 0_v$ and hence $\varphi_{rv}(\sim c) \leqslant_v \sim\varphi_{pv}(a)$, and applying $(\sim\text{lax})$, $a \leqslant^{\mathbf{G}} c$.   □

Remark 29. Notice that if a compatible family Φ satisfies (bal), then it satisfies (lax) if and only if for all $p \leqslant^{\mathbf{D}} q$, $p \leqslant^{\mathbf{D}} r$, and $a \in A_p$, $-\varphi_{pq}(a) \leqslant^{\mathbf{G}} \varphi_{pr}(-a)$.

Lemma 30. *If $\int_\Phi \mathbf{A}_p$ is the glueing of a system of integral ipo-monoids $(\mathbf{D}, \mathcal{A}, \Phi)$ and Φ satisfies (bal), (za), and (lax), then $\leqslant^{\mathbf{G}}$ is antisymmetric.*

Proof. Suppose that $p, q \in D$ with $r = p \vee q$, and $a \in A_p$ and $b \in A_q$ are such that $a \leqslant^{\mathbf{G}} b$ and $b \leqslant^{\mathbf{G}} a$. That is, $\varphi_{pr}(a) \cdot_r \varphi_{qr}(\sim b) = 0_r$ and $\varphi_{qr}(b) \cdot_r \varphi_{pr}(\sim a) = 0_r$, or equivalently $\varphi_{pr}(a) \leqslant_r -\varphi_{qr}(\sim b)$ and $\varphi_{qr}(b) \leqslant_r -\varphi_{pr}(\sim a)$. By (lax), we get

$$\varphi_{pr}(a) \leqslant_r -\varphi_{qr}(\sim b) \leqslant_r \varphi_{qr}(-\sim b) = \varphi_{qr}(b) \leqslant_r -\varphi_{pr}(\sim a).$$

Hence, we would have that $\varphi_{pr}(0_p) = \varphi_{pr}(a \cdot_p \sim_p a) = \varphi_{pr}(a) \cdot_r \varphi_{pr}(\sim_p a) = 0_r$. By (za), this only is possible if $p = r$. By a symmetric argument, we also obtain that $q = r$, and therefore $p = q$. Thus, by Lemma 22, we have that $a \leqslant_p b$ and $b \leqslant_p a$, and therefore $a = b$.   □

We are now in the position to prove our main result.

Theorem 31. *A structure $\mathbf{A}$ is a locally integral ipo-monoid if and only if there is a system $(\mathbf{D}, \mathcal{A}, \Phi)$ of integral ipo-monoids satisfying (bal), (za), and (tr) so that $\mathbf{A} = \int_\Phi \mathbf{A}_p$.*

Proof. As we showed in Theorem 19, if $\mathbf{A}$ is a locally integral ipo-monoid, then $(A^+, \cdot, 1)$ is a lower-bounded join-semilattice, its integral components form a family $\{\mathbf{A}_p : p \in A^+\}$ of integral ipo-monoids, and we have a compatible family of monoid homomorphisms $\Phi = \{\varphi_{pq} : \mathbf{A}_p \to \mathbf{A}_q : p \leqslant q\}$ given by $\varphi_{pq}(x) = qx$, so that $\mathbf{A} = \int_\Phi \mathbf{A}_p$. Moreover, Φ satisfies condition (bal) since $\mathbf{A}$ satisfies (ineg), condition (za) by Lemma 26, and condition (tr) since $\leqslant$ is a partial order.

Conversely, if $(D, \vee, 1)$ is a lower-bounded join-semilattice, $\{\mathbf{A}_p : p \in D\}$ is a family of integral ipo-monoids, and $\Phi = \{\varphi_{pq} : \mathbf{A}_p \to \mathbf{A}_q : p \leqslant^{\mathbf{D}} q\}$ is a compatible family of monoid homomorphisms satisfying (bal), (za), and (tr), then $\leqslant^{\mathbf{G}}$ is a reflexive binary relation on $\biguplus A_p$ by Remark 23, which is also transitive since it satisfies (tr), and antisymmetric by Lemma 30. That is, $(\biguplus A_p, \leqslant^{\mathbf{G}})$ is a poset. By construction, $(\biguplus A_p, \cdot^{\mathbf{G}}, 1^{\mathbf{G}})$ is a monoid. Furthermore, since Φ satisfies (bal) and the only element in $\downarrow^{\mathbf{G}} 0_1 \cap A_p$ is 0_p, for every $p \in D$, by Lemma 25, we deduce that $\int_\Phi \mathbf{A}_p$ satisfies (ineg) and therefore it is an ipo-monoid. It can be readily checked that for all $p \in D$ and $x \in A_p$,

$-^{\mathbf{G}}x \cdot^{\mathbf{G}} x = x \cdot^{\mathbf{G}} \sim^{\mathbf{G}}x$, since these involutive negations and products are computed inside $\mathbf{A}_p$, which is integral. For the same reasons, one can check that $x \cdot^{\mathbf{G}} x \leqslant^{\mathbf{G}} x$, since the product is computed inside $\mathbf{A}_p$ and the restriction of $\leqslant^{\mathbf{G}}$ to $\mathbf{A}_p$ is $\leqslant_p$, by Lemma 22. And since $\downarrow^{\mathbf{G}}0_1 = \{0_p : p \in D\}$ by Lemma 25 and $0_p \cdot^{\mathbf{G}} 0_p = 0_{p \vee p} = 0_p$ by (∗), which is a consequence of (tr), we also have that $\int_{\Phi} \mathbf{A}_p$ is ↓0-idempotent. In summary, $\int_{\Phi} \mathbf{A}_p$ is locally integral. Since for all $p \in D$ and $x \in A_p$, $1_x = \sim^{\mathbf{G}}(-^{\mathbf{G}}x \cdot^{\mathbf{G}} x) = \sim_p(-_px \cdot_p x) = 1_p$, we deduce that $\{\mathbf{A}_p : p \in D\}$ is the family of integral components of $\int_{\Phi} \mathbf{A}_p$. Also, by Lemma 22, we know that $\varphi_{pq}(x) = 1_q \cdot^{\mathbf{G}} x$, for all $p \leqslant^{\mathbf{D}} q$ and $x \in A_p$, that is, Φ is the family of homomorphisms of the decomposition of Theorem 19.   □

We illustrate the construction with two interesting examples (Fig. 4) and with a list of small integral components from which other examples can be constructed (Fig. 5).

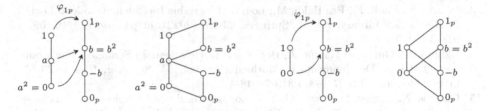

Fig. 4. Two glueings, one being a semiring, the other just an ipo-monoid

Corollary 32. *Given any nonempty family of nontrivial integral ipo-monoids (involutive semirings, respectively) there is a locally integral ipo-monoid (involutive semiring, respectively) whose integral components are the given ones.*

Proof. If $\{\mathbf{A}_p : p \in D\}$ is a nonempty family of nontrivial ipo-monoids, let's choose a lower-bounded linear order on D and let $\mathbf{D} = (\mathbf{D}, \vee, \mathbf{1})$ be the associated lower-bounded join-semilattice. Then, the set $\Phi = \{\varphi_{pq} \colon \mathbf{A}_p \to \mathbf{A}_q : p \leqslant^{\mathbf{D}} q\}$ of maps so that $\varphi_{pq}(x) = 1_q$ if $p < q$ and $\varphi_{pq}(x) = x$ if $p = q$ is a compatible family of monoid homomorphisms satisfying (bal), (za), (mon), (lax), and (∼lax). By

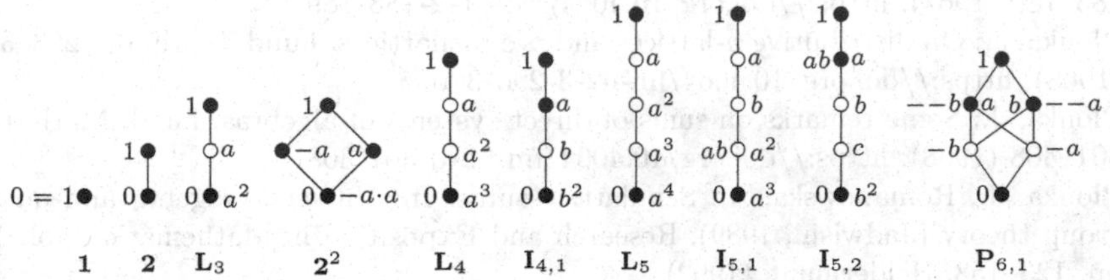

Fig. 5. All integral involutive semirings up to size 5 and an integral ipo-monoid of size 6, as components for constructing locally integral idempotent semirings and ipo-monoids.

Theorem 31, $\int_\Phi \mathbf{A}_p$ is a locally integral ipo-monoid whose integral components are $\{\mathbf{A}_p : p \in D\}$. If in addition all the integral components are involutive semirings, then $\int_\Phi \mathbf{A}_p$ is also an involutive semiring, since the join of $a \in A_p$ and $b \in \mathbf{A}_q$ in $\int_\Phi \mathbf{A}_p$ is either their join in $\mathbf{A}_p$, if $p = q$, or $1_{p \vee q}$, if $p \neq q$.   □

References

1. Alpay, N., Jipsen, P.: Commutative doubly-idempotent semirings determined by chains and by preorder forests. In: Fahrenberg, U., Jipsen, P., Winter, M. (eds.) RAMiCS 2020. LNCS, vol. 12062, pp. 1–14. Springer, Cham (2020). https://doi.org/10.1007/978-3-030-43520-2_1
2. Alpay, N., Jipsen, P., Sugimoto, M.: Unary-determined distributive ℓ-magmas and bunched implication algebras. In: Fahrenberg, U., Gehrke, M., Santocanale, L., Winter, M. (eds.) RAMiCS 2021. LNCS, vol. 13027, pp. 19–36. Springer, Cham (2021). https://doi.org/10.1007/978-3-030-88701-8_2
3. Bonzio, S., Paoli, F., Pra Baldi, M.: Logics of Variable Inclusion. Trends in Logic-Studia Logica Library, vol. 59. Springer, Cham (2022). https://doi.org/10.1007/978-3-031-04297-3
4. Eklund, P., Gutiérrez García, J., Höhle, U., Kortelainen, J.: Semigroups in Complete Lattices. Developments in Mathematics, vol. 54. Springer, Cham (2018). https://doi.org/10.1007/978-3-319-78948-4
5. Galatos, N., Jipsen, P., Kowalski, T., Ono, H.: Residuated Lattices: An Algebraic Glimpse at Substructural Logics. Studies in Logic and the Foundations of Mathematics, vol. 151. Elsevier B. V., Amsterdam (2007)
6. Gil-Férez, J., Jipsen, P., Metcalfe, G.: Structure theorems for idempotent residuated lattices. Algebra Univers. **81**(2), 1–25 (2020). https://doi.org/10.1007/s00012-020-00659-5
7. Jenei, S.: Group representation for even and odd involutive commutative residuated chains. Stud. Log. **110**(4), 881–922 (2022). https://doi.org/10.1007/s11225-021-09981-y
8. Jipsen, P., Tuyt, O., Valota, D.: The structure of finite commutative idempotent involutive residuated lattices. Algebra Univers. **82**(4), 1–23 (2021). https://doi.org/10.1007/s00012-021-00751-4
9. Pigozzi, D.: Partially ordered varieties and quasivarieties. Technical report, Iowa State University (2004). web.archive.org/web/20060902114300/, http://orion.math.iastate.edu/dpigozzi/notes/santiago_notes.pdf
10. Płonka, J.: On a method of construction of abstract algebras. Fund. Math. **61**, 183–189 (1967). https://doi.org/10.4064/fm-61-2-183-189
11. Płonka, J.: On distributive n-lattices and n-quasilattices. Fund. Math. **62**, 293–300 (1968). https://doi.org/10.4064/fm-62-3-293-300
12. Płonka, J.: Some remarks on sums of direct systems of algebras. Fund. Math. **62**, 301–308 (1968). https://doi.org/10.4064/fm-62-3-301-308
13. Płonka, J., Romanowska, A.: Semilattice sums. In: Universal algebra and quasigroup theory (Jadwisin, 1989). Research and Exposition in Mathematics, vol. 19, pp. 123–158. Heldermann (1992)

Compatibility of Refining and Controlling Plant Automata with Bisimulation Quotients

Roland Glück[(⊠)] [ID]

Deutsches Zentrum für Luft- und Raumfahrt, 86159 Augsburg, Germany
`roland.glueck@dlr.de`

Abstract. This paper is concerned with the refinement and control of a certain class of labelled transition systems, called plant automata, via bisimulation quotients. Refinement means that arbitrary transitions may be removed whereas control allows only removing edges with the same edge label. The goal is to ensure given LTL properties in the resulting plant automaton. We give a hardness result for refinement and control and investigate, in particular, the question whether refineability and controllability can be decided by looking at bisimulation quotients.

1 Introduction

Bisimulations are a well-known tool used in areas like automata theory (as in the classic papers [11,12]) or model checking (see the comprehensive survey [2]). The interest in bisimulations can be seen by work characterizing them in a relational (see [18]), relation algebraic (see [24]), semiring (see [7]) or lattice based (see [14]) setting. A series of works uses bisimulations as a tool in the context of model refinement. The problem there is to remove undesired transitions from a transition system in order to achieve desired properties like liveness (see [7]) or optimality with respect to criteria like shortest paths or maximum capacity paths (see [6]). The main idea is that under favorable circumstances, a solution can be found more efficiently if the problem is solved on the coarsest bisimulation quotient of the original system and this solution is propagated back to the original system. A general description of this idea is given in [8] and applies even to infinite systems, provided the coarsest bisimulation quotient is finite (a precursor of this idea appears already in [19]). The use of bisimulation quotients can serve two purposes: first, it can possibly lead to a speed up of solving the problem under consideration, and second, it can make infinite systems treatable if the coarsest quotient is finite.

In this paper we apply bisimulations to the problem of controlling automata in order to achieve certain properties expressible in linear temporal logic (LTL). First steps of controlling automata date back to the work in [16,21,22] which aims to control an automaton in a way that it satisfies the acceptance conditions of a Rabin automaton as introduced in [15]. More recent work in this area deals

© The Author(s), under exclusive license to Springer Nature Switzerland AG 2023
R. Glück et al. (Eds.): RAMiCS 2023, LNCS 13896, pp. 87–104, 2023.
https://doi.org/10.1007/978-3-031-28083-2_6

e.g. with energy problems as in [1], traffic routing as in [23] or safety-critical systems in general as in [17]. The LTL aspect of acceptance conditions of Rabin automata is used in [10] as a basic tool for controlling timed automata, and [9] (see [5] for an online preprint) shows how bisimulation quotients can be used to obtain a possible speed up for these basic procedures.

In the present work we continue this approach in several directions: first, we investigate a broader variety of LTL properties (the mentioned work only deals with four LTL formulae). Second, we consider other variants of control ([5,9,10] deal only with control in the sense of [16,21,22] where control takes place via enabling or disabling transitions with common edge labe; here we admit enabling and disabling of arbitrary sets of transitions). Third, we focus more on algorithms for refinement and investigate in which cases the detour via bisimulation quotients could be useful. To do so we introduce some basic definitions and properties in Sect. 2. In the following short Sect. 3 we give a hardness result and a useful canonical form of control. The main part of the paper is constituted by Sect. 4 where we give algorithms for refinement and controlling of plant automata with respect to selected LTL formulae and discuss the usefulness of bisimulation quotient in the respective cases. A short summary and outlook to future work in Sect. 5 finishes the paper.

2   Notations and Basic Properties

The main object of our investigations are a special kind of labelled transition systems. In the following, Σ and Π denote two disjoint alphabets. Letters from Σ will be denoted by $\alpha, \beta \ldots$ and variants thereof, symbols from Π by F, G, $\ldots$ and variants thereof. Γ^* denotes the set of all finite words over an alphabet Γ, likewise, Γ^ω denotes the set of all infinite words. The power set of a set A will be denoted by 2^A.

The following definition gives the general framework for our considerations:

Definition 2.1. A *set-labelled graph* is a tuple $((V, E), g)$ where (V, E) is a (possibly infinite) directed graph with node set V and edge set E and $g : E \to 2^\Sigma$ is the *edge labelling* function. A *model* is a tuple $M = (((V, E), g), a)$ where $((V, E), g)$ is a set-labelled graph and $a : V \to 2^\Pi$ is the *node labelling* function. A *plant automaton* is a model with a unique node v_0 such that $I \in a(v) \Leftrightarrow v = v_0$ holds for all $v \in V$ where $I \in \Pi$ is a special symbol.

The terms (set-)labelled graph and model are common and often used whereas the term plant automaton stems from [10] which stimulated also the present work. Note that an edge may be labelled by the empty set; however, in our framework (cf. the next Definition 2.2) such edges have no effect. In a plant automaton, the node v_0 will play the role of a starting state. In slight notational abuse, we will write $g(v_1, v_2)$ instead of the type theoretically correct $g((v_1, v_2))$ and $(v_1, \alpha, v_2) \in E$ as a more intuitive notation for $\alpha \in g(v_1, v_2)$. To avoid cumbersome notation, we use $g(v)$ as an abbreviation for the set

$\{\alpha \mid \exists w : (v, \alpha, w) \in E\}$. Also, we will omit set braces when dealing with single-ton sets, i.e., we simply write F instead of $\{\mathsf{F}\}$. Similarly, we use $M = (V, E, g, a)$ instead of the full form $M = (((V, E), g), a)$. A plant automaton can be regarded as a labeled transition system; the presentation above was chosen to ease compatibility with previous work as [5,6,9,10].

The next definition is a natural adaptation from model and automata theory:

Definition 2.2. A *run* r is a finite or infinite sequence from $V(\Sigma V)^* \cup V(\Sigma V)^\omega$ such that $(v_i, \alpha, v_{i+1}) \in E$ holds for all subsequences of r from $V\Sigma V$. A run is called a *trace* if it starts with v_0 (the unique node with $\mathsf{I} \in a(v_0)$).

In the next definition we adapt the concept of bisimulations to our framework:

Definition 2.3. Given two models $M = (V, E, g, a)$ and $\hat{M} = (\hat{V}, \hat{E}, \hat{g}, \hat{a})$ we call a relation $B \subseteq V \times \hat{V}$ a *bisimulation* between M and $\hat{M}$ if B is both left and right total and fulfills the following conditions:

- $(v, \hat{v}) \in B \Rightarrow a(v) = \hat{a}(\hat{v})$
- $(v, \alpha, w) \in E \land (v, \hat{v}) \in B \Rightarrow \exists \hat{w} \in \hat{V} : (w, \hat{w}) \in B \land (\hat{v}, \alpha, \hat{w}) \in \hat{E}$
- $(\hat{v}, \alpha, \hat{w}) \in \hat{E} \land (v, \hat{v}) \in B \Rightarrow \exists w \in V : (w, \hat{w}) \in B \land (v, \alpha, w) \in E$

A bisimulation between M and itself is called an *autobisimulation*. If an autobisimulation is also an equivalence it is called a *bisimulation equivalence*. Because the identity relation is an autobisimulation and autobisimulations over M are closed under union and composition there is a greatest (with respect to $\subseteq$) bisimulation equivalence which we call the *coarsest bisimulation equivalence* for M. The coarsest bisimulation quotient can be determined in $\mathcal{O}(|E| \cdot log(|V|))$ time by an algorithm given in [13]. The equivalence class of an element v under a bisimulation equivalence B will be denoted by v/B. Note that for a plant automaton the equivalence class v_0/B is the singleton set $\{v_0\}$ since v_0 is the unique node whose label set contains I.

Definition 2.4. Let B be a bisimulation equivalence for $M = (V, E, g, a)$. The *quotient* M/B is the model $(V/B, E/B, g/B, a/B)$, defined as follows:

- $V/B =_{def} \{v/B \mid v \in V\}$
- $(v/B, \alpha, w/B) \in E/B \Leftrightarrow_{def} \exists v' \in v/B, w' \in v/B : (v', \alpha, w') \in E$
- $(a/B)(v/B) =_{def} a(v)$

Note that this defines g/B implicitly since the second item determines the labels of an edge in M/B and that the three items are well-defined due to the properties of a bisimulation equivalence. A well-known fact is that a model and every of its quotients are bisimilar. In particular, the existence of the run $v_1 \alpha_1 v_2 \alpha_2 v_3/B \ldots$ in M implies the existence of the run $(v_1/B)\alpha_1(v_2/B)\alpha_2(v_3/B) \ldots$ in M/B. Vice versa, the existence of the run $(v_1/B)\alpha_1(v_2/B)\alpha_2(v_3/B) \ldots$ in M/B implies the existence of a run $w_1 \alpha_1 w_2 \alpha_2 w_3/B \ldots$ in M with $w_i \in v_i/B$.

A frequent task in control theory is to remove undesired transitions. This is captured by the next definition:

Definition 2.5. A model $M' = (V', E', g', a')$ is called a *refinement* of a model $M = (V, E, g, a)$ if the following conditions hold:

- $V' = V$
- $(v, \alpha, w) \in E' \Rightarrow (v, \alpha, w) \in E$
- $a'(v) = a(v)$

Intuitively, a refinement removes labels from edges or even entire edges but does not change the node set and the node labels. A stronger form of control is given in the following definition:

Definition 2.6. Given a model $M = (V, E, g, a)$ a *controller* of M is a mapping $c : V \to 2^{\Sigma}$ such that for all $v \in V$ the inclusion $c(v) \subseteq \{\alpha \mid \exists w : (v, \alpha, w) \in E\}$ holds. The model $M|c = (V|c, E|c, g|c, a|c)$, also called M *controlled by* c, is defined as follows:

- $V|c = V$
- $(v, \alpha, w) \in E|c \Leftrightarrow (v, \alpha, w) \in E \wedge \alpha \in c(v)$
- $(a|c)(v) = a(v)$

Analogously to refinement, a controller leaves nodes and node labels unchanged. In contrast to refinement, a controller can enable or disable transitions only via groups of edges bearing the same label. Formally, a controller can not remove edges but it can prevent transitions via an edge if a controller causes an edge to be labelled by the empty set.

A recurring idea in previous work is to refine or control a bisimulation quotient of a model and to play back this refinement or controller to the original model. This motivates the following definition:

Definition 2.7. Given a model $M = (V, E, g, a)$, a bisimulation equivalence B for M and a refinement $(M/B)' = ((V/B)', (E/B)', (g/B)', (a/B)')$ of M/B we define the *expansion* $(M/B)'\backslash B = ((V/B)'\backslash B, (E/B)'\backslash B, (g/B)'\backslash B, (a/B)'\backslash B)$ as follows:

- $(V/B)'\backslash B = V$
- $(v, \alpha, w) \in (E/B)'\backslash B \Leftrightarrow (v, \alpha, w) \in E \wedge (v/B, \alpha, w/B) \in (E/B)'$
- $((a/B)'\backslash B)(v) = a(v)$

In this case, $(M/B)'$ and $(M/B)'\backslash B$ are bisimilar, see [5,9]. Clearly, the expansion of a refinement yields a refinement again. Moreover, the expansion of a controlled plant automaton induces a controller on the original plant automaton by transferring the controller values of v/B to v (note that this is well-defined since v occurs in exactly one equivalence class of B).

Next we introduce some terminology from linear time logic. We denote LTL formulae over Π by φ or π and say that an infinite run $r = v_1 \alpha_1 v_2 \alpha_2 v_3 \ldots$ *fulfills* an LTL formula φ, denoted by $r \models \varphi$, if the following routinely defined conditions hold:

- $r \models F \Leftrightarrow F \in a(v_1)$

- $r \models \varphi \wedge \pi \Leftrightarrow r \models \varphi \wedge r \models \pi$
- $r \models \neg\varphi \Leftrightarrow \neg(r \models \varphi)$
- $r \models \bigcirc\varphi \Leftrightarrow v_2\alpha_2v_3\alpha_3v_4 \ldots \models \varphi$
- $r \models \Box\varphi \Leftrightarrow \forall i : v_i\alpha_iv_{i+1}\alpha_{i+1}v_{i+2} \ldots \models \varphi$
- $r \models \Diamond\varphi \Leftrightarrow \exists i : v_i\alpha_iv_{i+1}\alpha_{i+1}v_{i+2} \ldots \models \varphi$
- $r \models \varphi\mathcal{U}\pi \Leftrightarrow \exists i : v_i\alpha_iv_{i+1}\alpha_{i+1}v_{i+2} \ldots \models \pi \wedge \forall k : k < i : v_1\alpha_1v_2 \ldots v_k \models \varphi$

Other operators can be derived from this (not minimal) rule set; we restrict ourselves to the above set since these operators will be of importance in the sequel. Since a plant automaton may admit finite traces and the validity of LTL formulae is defined only over infinite traces we introduce the following definition:

Definition 2.8. A plant automaton M is called *live* if it has at least one trace and for every finite trace $v_0\alpha_0v_1\alpha_1v_2\alpha_2v_3 \ldots v_i$ of M there exist an $\alpha_i \in \Sigma$ and a $v_{i+1} \in V$ such that $(v_i, \alpha_i, v_{i+1}) \in E$ holds.

Intuitively, this means that every finite trace can be extended to an infinite one. The stipulation that M has at least one trace rules out the pathological case that M has no trace at all. There are other possibilities to capture the idea of liveness (e.g., one could stipulate that every node reachable from v_0 has an outgoing edge) but the above definition describes exactly the desired behavior. Now we say that a plant automaton *satisfies* an LTL formula φ if it is live and every infinite trace fulfills φ. A plant automaton M is said to be *refineable with respect to* φ if there is a refinement of M satisfying φ. It is said to be *controllable with respect to* φ if there is a controller c such that $M|c$ satisfies φ. The fact whether refineability or controllability of a plant automaton can be decided by looking at bisimulation quotients is captured in the following definition:

Definition 2.9. Let φ be an LTL formula. We say that φ is *quotient compatible with respect to refinement (control)* if for all plant automata M and all bisimulation quotients M/B of M the equivalence

M is refineable (controllable) wrt. $\varphi \Leftrightarrow M/B$ is refineable (controllable) wrt. φ

holds.

There are indeed LTL formulae which are not quotient compatible with respect to refinement or control (cf. Lemmata 4.4 and 4.16). We call a plant automaton *irredundant* if every node has an outgoing edge iff it is reachable from v_0 (note that this implies that v_0 has an outgoing edge). A controller c is called *canonical* if $M|c$ is irredundant and $|c(v)| \leq 1$ holds for all nodes v. Reasoning about canonical controllers eases some proofs in the sequel of this work.

3   Basic Results

It is easy to see that a canonical controller induces a live plant automaton. Moreover, in the case of controllability, there always exists even a canonical controller:

Lemma 3.1. Let c be a controller such that $M|c$ satisfies some LTL formula φ. Then there is a canonical controller c' such that $M|c'$ satisfies φ.

Proof: We convert c into a canonical controller as follows: first, we restrict c in an arbitrary way to a controller $\hat{c}$ such that $|\hat{c}(v)| = 1$ holds for all nodes v with $|c(v)| \neq 0$ and $|\hat{c}(v)| = 0$ holds for all v with $|c(v)| = 0$. Note that this does not affect liveness: a node may lose outgoing edges if an edge is labelled by the empty set; however, it can still be left over the remaining edges (recall that for every $\alpha \in c(v)$ there is at least one w such that $(v, \alpha, w) \in E$ holds). Subsequently, we set $c(v) = \emptyset$ for all nodes v which are not reachable from v_0. It is straightforward to see that the resulting plant automaton corresponds to a canonical controller c' such that $M|c'$ satisfies φ ($M|c'$ may have less traces than $M|c$ but every trace in $M|c'$ is also a trace in $M|c$; hence every trace in $M|c'$ fulfills φ). ∎

Clearly, the existence of a canonical controller implies the existence of a controller so a plant automaton M is controllable with respect to φ iff there is a canonical controller c such that $M|c$ satisfies φ.

It is known since [3] that controllability of Rabin automata is NP-hard. An analogous result holds in our setting (note that this theorem only characterizes the hardness of the problem but not its exact complexity which may be NP-completeness):

Theorem 3.2. *In general, it is NP-hard to decide whether a plant automaton is refineable with respect to an LTL formula.*

Proof: We reduce the directed Hamiltonian cycle problem (see e.g. [4] for this topic) to refinement of plant automata. So, given a finite directed graph $G = (V, E)$ we pick an arbitrary $v_0 \in V$ and construct the plant automaton $M = (V, E, g, a)$ by $g(e) = \alpha_e$ for all $e \in E$ (i.e., we introduce a unique letter α_e for every edge $e \in E$), $a(v_0) = \mathsf{I}$ and $a(v) = \mathsf{F}$ for all $v \neq v_0$ (note that the node sets bearing F and I are disjoint). Now we consider the formula $\varphi =_{def}$ $(\bigwedge\limits_{i=1}^{|V|-1} \bigcirc^i \mathsf{F}) \wedge \bigcirc^{|V|} \mathsf{I}$ and a refinement M' satisfying φ. First we observe that M' can not contain a cycle of length $l < |V|$ containing a reachable node: if such a cycle contains v_0 this would contradict the conjugand $\bigcirc^l \mathsf{F}$ of φ (walking along this cycle starting in v_0 would give a trace fulfilling $\bigcirc^l \mathsf{I}$ contradicting the conjugand $\bigcirc^l \mathsf{F}$ due to the above mentioned disjointness). Otherwise, if such a cycle contains only nodes from $V \setminus \{v_0\}$ this would imply the existence of a trace $v_0 \alpha v_1 \alpha v_2 \ldots (v_i \alpha v_{i+1} \alpha v_j)^\omega$, contradicting the part $\bigcirc^{|V|} \mathsf{I}$ of φ. Hence, E' has to consist of one simple cycle (E' has to contain a cycle for both being live and satisfying $\bigcirc^{|V|} \mathsf{I}$) comprehending all nodes of V which is possible iff G admits a Hamiltonian cycle. ∎

In the plant automaton constructed in the above proof refineability and controllability are equivalent since every edge bears a unique label. Hence the claim holds also for controllability:

Corollary 3.3. *It is NP-hard to decide whether a plant automaton is controllable with respect to an LTL formula.*

4 Refineability and Controllability

In this section we will investigate the quotient compatibility of several LTL formulae. For each formula under consideration, we give an algorithm deciding refineability or controllability, resp., in order to evaluate usefulness of the quotient construction. The running times of the algorithms depend on the representation of a plant automaton, for an example data structure see [5,9]. In particular, $|E|$ may refer to the cardinality of the edge set (especially in algorithms not taking edge labels into account) or the number of labelled edges in a manner analogously to the conventions introduced immediately after Definition 2.1.

We investigate only a small amount of simple LTL formulae for which can both prove that they are quotient compatible and we can give algorithms for computing a refinement or a controller, resp. A more general investigation is the topic of future work.

4.1 Quotient Compatibility with Respect to Refinement

In this subsection we investigate quotient compatibility with respect to refinement (cf. Definition 2.9). Since refineability of the quotient implies refineability of the original plant automaton due to bisimilarity of $(M/B)'$ and $(M/B)'\backslash B$ (see the remark after Definition 2.7) it suffices to show that refineability of a plant automaton implies refineability of the quotient. The proof strategy will always be the same: we pick an arbitrary infinite trace (which by definition exists and fulfills the LTL formula under consideration) and use it to construct a refinement of the quotient.

Lemma 4.1. F is quotient compatible with respect to refinement.

Proof: Let M' be a refinement of M satisfying F. Then there is an infinite trace in M' of the form $p = v_0\alpha_0 v_1\alpha_1 \ldots$ with $F \in a(v_0)$. We define the plant automaton $(M/B)'$ as follows:

- $(V/B)' =_{def} V/B$
- $(a/B)' =_{def} a/B$
- $(v/B, \alpha, w/B) \in (E/B)' \Leftrightarrow_{def} \exists i : v \in v_i/B \wedge w \in v_{i+1}/B \wedge \alpha = \alpha_i$

Clearly, $(M/B)'$ is a live submodel of (M/B). Moreover, every trace fulfills F which shows the claim. ∎

Lemma 4.2. $\bigcirc$F is quotient compatible with respect to refinement.

Proof: Let M' be a refinement of M satisfying $\bigcirc$F. We define p and $(M/B)'$ analogously to above (here with $F \in a(v_1)$) and claim analogously that every trace of $(M/B)'$ fulfills $\bigcirc$F. To this end, it suffices to consider an arbitrary edge $(v_0/B, \alpha, w_0/B) \in E/B$ and to show that $F \in (a/B)(w_0/B)$ holds. However, if there is an edge $(v_0/B, \alpha, w_0/B) \in E/B$ then there is an edge $(v_0, \alpha, w_1) \in E'$ with $(w_0, w_1) \in B$. Because M' satisfies $\bigcirc$F this implies $F \in a(w_1)$ and hence by bisimulation properties also $F \in a/B(w_0/B)$. ∎

Lemma 4.3. $\bigcirc\bigcirc$ F is quotient compatible with respect to refinement.

Proof: Let M' be a refinement of M satisfying $\bigcirc\bigcirc$ F and define p analogously to the proof of Lemma 4.1 (but with F $\in a(v_2)$). We distinguish now several cases:

1. $v_0 = v_1$: Here we define $(M/B)'$ by $(V/B)' = V/B$, $(a/B)' = (a/B)$, $(E/B)' = \{(v_0/B, v_0/B)\}$ and $(g/B)(v_0/B, v_0/B) = \{\alpha_0\}$. Because $(v_0\alpha_0)^\omega$ is a trace in M' we have F $\in a(v_0)$ and hence F $\in (a/B)(v_0/B)$. Now it easy to see that $(M/B)'$ satisfies $\bigcirc\bigcirc$ F since it admits only the trace $(v_0\alpha_0)^\omega$.

2. $v_0 = v_2$: Here we define $(M/B)'$ by $(V/B)' = V/B$, $(a/B)' = (a/B)$, $(E/B)' = \{(v_0/B, v_1/B), (v_1/B, v_0/B)\}$, $(g/B)(v_0/B, v_1/B) = \{\alpha_0\}$ and $(g/B)(v_1/B, v_0/B) = \{\alpha_1\}$. Because $(v_0\alpha_0 v_1\alpha_1)^\omega$ is a trace in M' (which satisfies $\bigcirc\bigcirc$ F) we have F $\in a(v_0)$ and hence F $\in (a/B)(v_0/B)$. Since the only trace in $(M/B)'$ has the form $((v_0/B)\alpha_0(v_1/B)\alpha_1)^\omega$ we conclude that $(M/B)'$ satisfies $\bigcirc\bigcirc$ F.

3. $v_0 \neq v_1 \land v_0 \neq v_2 \land v_1 \in v_2/B$: Here we define $(M/B)'$ by $(V/B)' = V/B$, $(a/B)' = a/B$, $(E/B)' = \{(v_0/B, v_1/B), (v_1/B, v_1/B)\}$, $(g/B)(v_0/B, v_1/B) = \{\alpha_0\}$ and $(g/B)(v_1/B, v_1/B) = \{\alpha_1\}$ (note that this is well defined due to bisimulation properties; also, $v_1 \in v_2/B$ and $(v_1, \alpha_1, v_2) \in$ imply $(v_1/B, \alpha_1, v_1/B) \in E/B$). Because M' satisfies $\bigcirc\bigcirc$ F we have F $\in a(v_2)$ and hence F $\in a(v_1)$ as well as F $\in (a/B)(v_1/B)$ by bisimulation properties. Now the only trace in $(M/B)'$ is $(v_0/B)\alpha_0((v_1/B)\alpha_1)^\omega$ so $(M/B)'$ satisfies $\bigcirc\bigcirc$ F.

4. Otherwise: Now we have the situation that v_0, v_1 and v_2 as well as their equivalence classes under B are pairwise disjoint. If p visits v_0 only once at the beginning and does not visit a node in the equivalence class of v_1 except the first occurrence of v_1 we can apply the construction from the proof of Lemma 4.1. Otherwise, the immediate application of this construction could lead to uncontrollable side effects in the form of edges emanating from v_0/B or v_1/B not leading into v_1/B or v_2/B if p returns to v_0 or some $v_k \in v_1/B$. So let $k \geq 3$ be minimal such that $v_k = v_0$ or $v_k \in v_1/B$ holds, and define $(M/B)'$ as follows:

- $(V/B)' =_{def} V/B$
- $(a/B)' =_{def} a/B$
- $(v/B, \alpha, w/B) \in (E/B)' \Leftrightarrow_{def} \exists i : i < k \land v \in v_i/B \land w \in v_{i+1}/B \land \alpha = \alpha_i$

By choice of k, v_k/B equals either v_0/B or v_1/B and has an outgoing edge to v_1/B or v_2/B, resp. So $(M/B)'$ is live and all its traces start with $(v_0/B)\alpha_0(v_1/B)\alpha_1(v_2/B)$. Similar considerations as above show also F $\in (a/B)'(v_2/B)$ so $(M/B)'$ satisfies $\bigcirc\bigcirc$ F. ∎

The next lemma may come a little bit as a surprise:

Lemma 4.4. $\bigcirc\bigcirc\bigcirc$ F is not quotient compatible with respect to refinement.

Proof: To this end let us take a look at Fig. 1. We use the convention to omit set braces, to write edge labels next to an edge, node names in the interior of a node and node labels next to a node. The plant automaton at the left is refineable with respect to $\bigcirc\bigcirc\bigcirc F$ (if we keep the edges (v_0, v_1), (v_1, v_2), (v_2, w_1), (w_1, w_2) and (w_2, w_2)). In contrast, in its coarsest quotient on the right (note that v_1 and v_2 are indeed bisimilar by Definition 2.3; for simplicity we omitted set braces in the node labels and abbreviated $\{v_1, v_2\}$ by v_{12}) we have the choice whether we keep or remove the loop at v_{12} while keeping the resulting plant automaton live. After removing the loop at v_{12} there is no trace fulfilling $\bigcirc\bigcirc\bigcirc F$. Otherwise, if we keep the loop at v_{12} the resulting plant automaton has the trace $v_0\alpha(v_{12}\beta)^\omega$ which does not fulfill $\bigcirc\bigcirc\bigcirc F$. ∎

Remark: Lemma 4.4 can be generalized to $\bigcirc^i F$ for $i \geq 3$. For the proof one has simply to replace the edge (v_0, α, v_1) by a path $v_0 w_0 w_1 \ldots v_1$ (all edges labelled by α and all nodes labelled with the empty set) of length $4 - i$. □

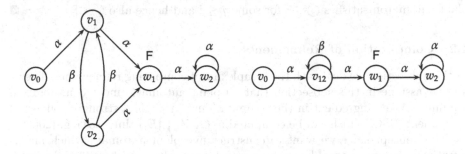

Fig. 1. An illustration of Lemmata 4.4 and 4.16

Lemma 4.5. $F \mathcal{U} G$ is quotient compatible with respect to refinement.

Proof: Let M' be a refinement of M satisfying $F \mathcal{U} G$ and define p analogously to the previous proofs (such that p fulfills $F \mathcal{U} G$). By definition, there is an index i such that $F \in a(v_j)$ for all $j < i$ and $G \in a(v_i)$ hold. The case $i = 0$ can be handled analogously to Lemma 4.1 so let us assume that $i > 0$ holds. First, we consider the trace $(v_0/B)\alpha_0(v_1/B)\alpha_1(v_2/B)\alpha_2 \ldots (v_i/B)$ in M/B. From this trace, we remove all cycles by standard construction (if there are indices $j < k$ with $(v_j/B) = (v_k/B)$ we simply remove all v_i and α_i with $j \leq i < k$) and obtain a trace $(w_0/B)\beta_0(w_1/B)\beta_1 \ldots (w_j/B)$ in (M/B) with $w_0/B = v_0/B$ and $w_j/B = v_i/B$. The trace $p' =_{def} w_0\beta_0 w_1\beta_1 \ldots w_j\alpha_i v_{i+1}\alpha_{i+1}v_{i+2}$ fulfills $F \mathcal{U} G$, however, it may be the case that there are indices k_1 and k_2 such that $w_{k_1} = v_{k_2}$ holds so putting all the edges from p' into $(E/B)'$ could lead to "outbreaks" in the prefix from w_0 till w_{i-1} leading to a node v with $F \notin a/B(v)$ and $G \notin a/B(v)$. To overcome this deficiency, we apply the following steps repeatedly:

1. If there is some v_k such that $v_k = w_l$ for some $l \leq i - 1$ we choose a minimal k with this property, otherwise we are done and terminate.
2. We construct the new trace $w_0\beta_0 w_1\beta_1 \ldots w_l\alpha_k v_{k+1}\alpha_{k+1}v_{k+2} \ldots$ and set $i =_{def} l$.

This procedure terminates since the length of the prefix containing the w's is strictly decreasing. Also, every such trace fulfills $\mathsf{F}\mathcal{U}\mathsf{G}$ (where $\mathsf{F} \in a/B(w_j)$ for all $j < i$ and $\mathsf{G} \in a/B(w_i)$ holds), and the trace obtained after termination has exactly one outgoing edge for each w_j with $j < i$. So we can use the set of edges from this trace as the edge set $(E/B)'$ of $(M/B)'$.   ∎

We saw a dichotomy in the formulae of the form $\bigcirc^i\mathsf{F}$ between $i = 2$ and $i = 3$; however, formulae of the form $\bigcirc^i\Box\mathsf{F}$ show a more uniform behavior:

Lemma 4.6. $\bigcirc^i\Box\mathsf{F}$ is quotient compatible with respect to refinement.

Proof: Let M' be a refinement of M satisfying $\bigcirc^i\Box\mathsf{F}$ and define p as in the proof of Lemma 4.1. By definition, we have $\mathsf{F} \in a(v_j)$ for all $j \geq i$ and hence also $\mathsf{F} \in a/B(v_j/B)$ for all $j \geq i$. We consider now the trace $(v_0/B)\alpha_0(v_1/B)\alpha_1(v_2/B)\ldots$ and apply the same construction as in the proof of Lemma 4.5. The obtained plant automaton satisfies $\bigcirc^j\Box\mathsf{F}$ for some $j \leq i$ and hence also $\bigcirc^i\Box\mathsf{F}$.   ∎

4.2 Computation of Refinements

In this subsection we deal with the complexity of computing refinements. Therefore we assume in this subsection that all plant automata under consideration are finite. A key ingredient in the computations is that of a strongly connected component (SSC) which can be computed in $\mathcal{O}(|V|+|E|)$ time; see e.g. [20]. We use this concept since we want to construct live plant automata which means intuitively that every trace has to be trapped eventually in such an SSC. Depending on the LTL formula we are interested in, every path leading into such an SSC or every reachable SSC has to fulfill certain properties.

Lemma 4.7. It can be decided in $\mathcal{O}(|V|+|E|)$ time whether M can be refined with respect to F. A corresponding refinement can also be computed in $\mathcal{O}(|V|+|E|)$ time.

Proof: It suffices to test whether $\mathsf{F} \in a(v_0)$ holds and there is an SSC of $G = (V, E)$ with at least one edge and that is reachable from v_0 (in order to construct a live refinement). The test $\mathsf{F} \in a(v_0)$ can be done in constant time. For the rest, we determine all SSCs of G. After this, we contract all SSCs (i.e., we define an equivalence relation S on $V \times V$ by $(v_1, v_2) \in S$ iff v_1 and v_2 lie in the same SSC and consider the quotient graph G/S defined analogously as in Definition 2.4) and test whether we can reach an SSC with at least on edge from v_0. A refinement (if one exists) can be obtained from a cycle-free path from v_0 so such an SSC together with all edges from this SSC.   ∎

Lemma 4.8. It can be decided in $\mathcal{O}(|V|+|E|)$ time whether M can be refined with respect to $\bigcirc\mathsf{F}$. A corresponding refinement can also be computed in $\mathcal{O}(|V|+|E|)$ time.

Proof: If $\mathsf{F} \in a(v_0)$ holds and there is a loop on v_0 we simply keep this loop as the only edge of the refinement and we are done. Next, we remove all edges (v_0, v_i)

where v_i is not labelled with F (obviously, such an edge can never be present in a refinement ensuring $\bigcirc$F). In the emerging graph (V, E') we compute again the SSCs and test whether an SSC with at least one edge is reachable from a node v_i with $(v_0, v_i) \in E'$ (note that such a node v_i is labelled with F). A refinement can be determined analogously to above; similarly, all described computations can be carried out in $\mathcal{O}(|V| + |E|)$ total time.   ∎

Lemma 4.9. It can be decided in $\mathcal{O}(|V| + |E|)$ time whether M can be refined with respect to $\bigcirc \bigcirc$ F. A corresponding refinement can also be computed in $\mathcal{O}(|V| + |E|)$ time.

Proof: If $F \in a(v_0)$ and $(v_0, v_0) \in E$ hold we proceed as in the previous lemma. Similarly, if there is a node v_1 bearing label F and both $(v_0, v_1) \in E$ and $(v_1, v_1) \in E$ hold we simply keep the edges (v_0, v_1) and (v_1, v_1). Otherwise, we compute all nodes reachable from v_0 in exactly two steps on paths without loops (i.e., immediately consecutive nodes are disjoint) bearing label F. If one such node is contained in an SSC we know that M is refineable with respect to $\bigcirc \bigcirc$ F, and a refinement can be obtained by the (two stepped) path and a suitable cycle of the SSC.   ∎

Lemma 4.10. For every i it can be decided in $\mathcal{O}(|V| + |E|)$ time whether M can be refined with respect to $\bigcirc^i \Box$F. A corresponding refinement can also be computed in $\mathcal{O}(|V| + |E|)$ time.

Proof: In contrast to the previous cases we have to construct a refinement with an SSC fulfilling a certain property (namely that all its nodes bear the label F) instead of a path fulfilling a certain property leading into a random SSC. So we compute first the SSCs of the graph $(V, E \cap \{(v_1, v_2) \mid F \in a(v_1) \land F \in a(v_2)\})$. Subsequently, we contract these SSCs in the original graph and test whether an SSC with at most one edge can be reached in i or less than i steps (if $i > |V|$ holds we simply set $i = |V|$). A refinement can be obtained from the edges of a shortest path in (V, E) to a computed SSC plus the edges of a cycle in the respective SSC. As above, all operations can be carried out in $\mathcal{O}(|V|+|E|)$ overall time.   ∎

Lemma 4.11. It can be decided in $\mathcal{O}(|V|+|E|)$ time whether M can be refined with respect to $\Diamond$F, $\Box$F, $\Diamond\Box$F and $\Box\Diamond$F. A corresponding refinement can also be computed in $\mathcal{O}(|V| + |E|)$ time.

Proof: Given the previous proofs we will only sketch the ideas. In the case of $\Diamond$F, we test whether a SSC can be reached from a reachable node labelled with F. To deal with $\Box$F we test whether v_0 is contained in a SSC of $(V, E \cap \{(v_1, v_2) \mid F \in a(v_1) \land F \in a(v_2)\})$. The case $\Diamond\Box$F can be handled analogously to Lemma 4.10 (here we have to test only reachability of an SSC instead of reachability in a certain number of steps). Finally, confronted with $\Box\Diamond$F we test whether we can reach from v_0 an SSC containing a node labelled with F.   ∎

Lemma 4.12. It can be decided in $\mathcal{O}(|V|+|E|)$ time whether M can be refined with respect to $F\mathcal{U}G$. A corresponding refinement can also be computed in $\mathcal{O}(|V|+|E|)$ time.

Proof: We compute the set of all nodes V_F reachable from v_0 by paths whose nodes bear all label F. Then we test whether node with label G in an SSC can be reached from a node in V_F. The actual refinement can be done analogously to the previous cases. ∎

The results of this Subsect. 4.2 show that one can not expect a speed-up for computing refinements using bisimulation quotients in the case of the formulae considered there. However, the results from Subsect. 4.1 apply also to infinite plant automata and may be used to derive refinements for this kind of plant automata.

4.3   Quotient Compatibility with Respect to Control

In this section we investigate analogously to Subsect. 4.1 selected LTL formulae and their compatibility with respect to control. Note that we have here a harder task since we can - intuitively spoken - enable or disable only certain groups of transitions but not arbitrary edges and hence we consider less formulae.

Lemma 4.13. F is quotient compatible with respect to control.

Proof: Assume that M is controllable with respect to F, and let c be a controller such that $M|c$ satisfies F. To construct a controller c/B on M/B we set $(c/B)(v/B) =_{def} \cup_{v_i \in v/B} c(v_i)$. By definition we have $F \in (a/B)(v_0/B)$ so it remains to show that $(M/B)|(c/B)$ is live. To this end, we observe that v_0/B has an outgoing edge in $(M/B)|(c/B)$ since v_0 has an outgoing edge in $M|c$ (note that $M|c$ has to be live). Consider now a node v_i/B which has an incoming edge $(v_j/B, \alpha, v_i/B)$ in $(M/B)|(c/B)$. Then there are $w_j \in v_j/B$, $w_i \in v_i/B$ such that (w_j, α, w_i) is an edge in $M|c$; hence there is an edge (w_i, β, w_k) in $M|c$ which implies the existence of the edge $(w_i/B\beta, w_k/B)$ in $(M/B)|(c/B)$. From this it follows easily that $(M/B)|(c/B)$ is live. ∎

Lemma 4.14. $\bigcirc$F is quotient compatible with respect to control.

Proof: Assume that M is controllable with respect to $\bigcirc$F, and let c be a controller such that $M|c$ satisfies $\bigcirc$F. Analogously to the proof of the previous Lemma 4.13 we define $(c/B)(v/B) =_{def} \bigcup_{v_i \in v/B} c(v_i)$ and argue as there that $(M/B)|(c/B)$ is live so we pick an arbitrary trace $(v_0/B)\alpha_0(v_1/B)\alpha_1(v_2/B)\dots$ in $(M/B)|(c/B)$. Because $M|c$ satisfies $\bigcirc$F every v_1' with $(v_0, \alpha, v_1') \in E$ bears the label F, and so does v_1/B by definition of bisimulation. Hence, $(M/B)|(c/B)$ satisfies $\bigcirc$F. ∎

Lemma 4.15. $\bigcirc\bigcirc$F is quotient compatible with respect to control.

Proof: Assume that M is controllable with respect to $\bigcirc\bigcirc$ F, and let c be a canonical controller such that $M|c$ satisfies $\bigcirc\bigcirc$ F. We define a controller c/B on M/B as follows:

- We choose $(c/B)(v_0/B) =_{def} c(v_0)$.
- For all $v/B \in V/B$ with $v \neq v_0$ such that there is a $w \in v/B$ which is reachable in one step from v_0 in $M|c$ we pick an arbitrary $w \in v/B$ and set $(c/B)(v/B) =_{def} c(w)$.
- For all other $v/B \in V/B$ we set $(c/B)(v/B) =_{def} \bigcup_{w \in v/B} c(w)$.

First we observe that in $(M/B)|(c/B)$ the node v_0/B has an outgoing edge because v_0 has an outgoing edge in $M|c$. An argument similar to the proof of Lemma 4.13 shows now that $(M/B)|(c/B)$ is live (note that all nodes considered in the second case of the definition of c/B keep at least one outgoing edge) so let us consider an arbitrary infinite trace $t = (v_0/B)\alpha_0(v_1/B)\alpha_1(v_2/B)\ldots$ in $(M/B)|(c/B)$. We distinguish the following cases:

1. $v_0/B = v_1/B = v_2/B$: Here $(v_0\alpha_0)^\omega$ is a trace in $M|c$, hence we have $\mathsf{F} \in a(v_0)$ which implies $\mathsf{F} \in (a/B)(v_0/B)$, and due to the assumption $v_0/B = v_2/B$ we conclude that t fulfills $\bigcirc\bigcirc$ F.
2. $v_0/B = v_1/B \wedge v_0/B \neq v_2/B$: In this case there exists a $w_2 \in v_2/B$ such that $v_0\alpha_0v_0\alpha_1w_2\ldots$ is a trace in $M|c$. Arguments analogous to the above case show that t fulfills $\bigcirc\bigcirc$ F.
3. $v_0/B \neq v_1/B$: Under these circumstances, there is a $w_1 \in v_1/B$ such that (v_0, α_0, w_1) is an edge in $M|c$, i.e., w_1 is reachable in $M|c$ in one step. By construction of c/B there is a $w_2 \in v_2/B$ such that (w_1, α_1, w_2) is an edge in $M|c$ which means that there is a trace $v_0\alpha_0w_1\alpha_1w_2\ldots$ in $M|c$. Similarly to above we conclude now that t fulfills $\bigcirc\bigcirc$ F.

In all cases, t fulfills $\bigcirc\bigcirc$ F which shows the claim together with the liveness of $(M/B)|(c/B)$. ∎

Lemma 4.16. $\bigcirc\bigcirc\bigcirc$F is not quotient compatible with respect to control.

Proof: This can be shown also by the plant automaton from Fig. 1: the plant automaton at the left can be controlled to satisfy $\bigcirc\bigcirc\bigcirc$F by the controller c, defined by $c(v_0) = c(v_2) = c(w_1) = c(w_2) = \{\alpha\}$ and $c(v_1) = \{\beta\}$. However, it is easy to see that the coarsest quotient at the right of Fig. 1 is not controllable with respect to $\bigcirc\bigcirc\bigcirc$F (cf. also the proof of Lemma 4.4). ∎

Remark: Analogously to the remark after Lemma 4.4, Lemma 4.16 can be generalized to $\bigcirc^i$F for $i \geq 3$. □

Lemma 4.17. $\Diamond$F, $\Box$F, $\Diamond\Box$F and $\Box\Diamond$F are quotient compatible with respect to control.

Proof: See 42ff. in [5,9]. Note that the proofs given there use different techniques than the ones presented here by employing sets of controllable predecessors arising during the construction of a controller. ∎

4.4   Computation of Controllers

In this subsection we will take a short look at the computation of controllers. Therefore we will again consider only finite plant automata. As an abbreviation we use $|E|$ instead of the correct $|\{(v, \alpha, w) \in E\}|$.

Lemma 4.18. It can be decided in $\mathcal{O}(|V| + |E|)$ time whether M can be controlled with respect to F. A corresponding controller can also be computed in $\mathcal{O}(|V| + |E|)$ time.

Proof: If v_0 does not bear the label F we know immediately that there is no controller with respect to F. If $\mathsf{F} \in a(v_0)$ holds we have to determine whether there is a controller ensuring liveness. Clearly, if M does not contain a node without outgoing edges we have a live plant automaton with an controller keeping all edges. Otherwise, we set $M_1 =_{def} M$ and construct a series $M_1, M_2, \ldots$ of plant automata as follows:

1. If M_i has no edges or no nodes without outgoing edges, we terminate the procedure.
2. Pick an arbitrary node v without outgoing but with incoming edges.
3. For every edge (w, α, v) in M_i, remove all edges (w, α, w').
4. Call the resulting plant automaton M_{i+1} and resume at step 1.

This procedure terminates with an overall number of $(|V| + |E|)$ steps (where removing an edge (w, α, w') is counted as one step). Moreover, M_{i+1} can be controlled to be live iff M_i can be controlled to be live. In the plant automaton obtained after the last iteration, every node with an incoming edge has also an outgoing edge so we only need to check whether v_0 has an outgoing edge. A controller can easily be derived from the edges of M'.   ∎

Lemma 4.19. It can be decided in $\mathcal{O}(|V| + |E|)$ time whether M can be controlled with respect to $\bigcirc$F. A corresponding controller can also be computed in $\mathcal{O}(|V| + |E|)$ time.

Proof: By an argument similar to the proof of Lemma 3.1, M is controllable with respect to $\bigcirc$F iff there is controller c that such $|c(v_0)| = 1$ holds and $M|c$ satisfies $\bigcirc$F. This is equivalent to the conditions that $c(v_0)$ is a singleton set $\{\alpha\}$, $\mathsf{F} \in a(v)$ holds for all $(v_0, \alpha, v) \in E$ and $M|c$ is live. A first approach to test the existence of such a controller is the following: first, we determine the set $A =_{def} \{\alpha \mid (v_0, \alpha, v) \in E \Rightarrow \mathsf{F} \in a(v)\}$. Next, for every $\alpha \in A$, we remove in M all edges emerging from v_0 which do not bear the label α and obtain the plant automaton M_α. To this plant automaton, we apply the construction from the proof of Lemma 4.18 to M_α and proceed accordingly. Since we do this for every $\alpha \in A$ this leads to a worst case running time of $\mathcal{O}(|E| \cdot (|V| + |E|))$ (if, for example, half of the edges emerge from v_0).

However, removing edges emerging from v_0 not bearing a label α and applying the construction from the proof of Lemma 4.18 commute. So we can first apply the construction from the proof of Lemma 4.18 and test subsequently whether there is an α such that all nodes reachable from v_0 via en edge labelled with α bear label F. This can be done in $\mathcal{O}(|V| + |E|)$, and a controller can be derived in an obvious way in the same time.   ∎

Lemma 4.20. It can be decided in $\mathcal{O}(|V| + |E|^2)$ time whether M can be controlled with respect to $\bigcirc\bigcirc$ F. A corresponding controller can also be computed in $\mathcal{O}(|V| + |E|^2)$ time.

Proof: By an argument similar to above, M is controllable with respect to $\bigcirc\bigcirc$F iff there is controller c that such $|c(v_0)| = 1$, $|c(v)| = 1$ holds for all v reachable from v_0 in exactly one step in $M|c$ (note that this set may contain v_0 without leading to an inconsistency), and $M|c$ satisfies $\bigcirc\,\bigcirc$ F. Before doing the actual computation, we execute first the construction from the proof of Lemma 4.18 to obtain a plant automaton M' and continue working on M'. On M' we test the existence of a controller with the properties from above by brute force:

```
for all α ∈ g(v₀) do
    controllable := FALSE
    if !((v₀, α, v₀) ∈ E ∧ F ∉ a(v₀)) then
        controllable := TRUE
        precont(v₀) := α
        for all v₁ with (v₀, α, v₁) ∈ E do
            if ∃β ∈ g(v₁)∀(v₁, β, v₂) ∈ E : F ∈ a(v₂) then
                precont(v₁) := β
            else
                controllable := FALSE
            end if
        end for
        if controllable = TRUE then return precont
    end if
end if
end for
```

If this algorithm returns a partial unction $precont : V \to \Sigma$ we modify M' in an obvious way by restrincting the edges emanating from nodes in the domain of $precont$. Otherwise, there is no controller ensuring $\bigcirc\,\bigcirc$ F.

Lemma 4.21. It can be decided in $\mathcal{O}(|V|^2 \cdot |\Sigma|)$ $(\mathcal{O}(|V|^3 \cdot |\Sigma|))$ time whether M can be controlled with respect to $\Diamond$F and $\Box$F ($\Diamond\Box$F and $\Box\Diamond$F). A corresponding controller can also be computed in $\mathcal{O}(|V|^2 \cdot |\Sigma|)$ $(\mathcal{O}(|V|^3 \cdot |\Sigma|))$ time.

Proof: A proof can be found in [5,9] pp. 52ff. Note that the algorithm given there (derived from [10]) uses other techniques than the previous constructions, compare also the remark made in the proof of Lemma 4.17.   ∎

The results of this Subsect. 4.4 show that one can expect a speed-up for computing controllers using bisimulation quotients in the case of the formulae from Lemmata 4.20 and 4.21 but not in the case of the formulae from the other lemmata of this subsection (recall again that the coarsest quotient can be computed in $\mathcal{O}(|E| \cdot log(|V|))$ time by an algorithm introduced in [13]). However, the results from Subsect. 4.1 apply also to infinite plant automata and may be used to derive controllers for this kind of plant automata.

5 Conclusion and Future Work

As one of the first results we showed NP-hardness of refinement and control. A sharper characterization of their complexity seems to be possible since model checking with respect to LTL formulae is PSPACE-complete as demonstrated in [2]. As main work, we investigated selected LTL formulae with respect to quotient compatibility. For every formula under consideration, we gave an algorithm computing a corresponding refinement or controller. In the case of refinement, the detour over the coarsest quotient did not lead to a speed-up, however, the results enable under certain circumstances the treatment of infinite systems. Dealing with the computation of controllers, a speed-up can be expected under favourable circumstances (note also that the algorithms given in Lemmata 4.20 and 4.21 may not be optimal; the hunt for faster algorithms could also be a possible topic of future research).

The formulae considered in Sect. 4 represent a rather erratic collection without any intrinsic structure. It would be interesting to find general criteria for quotient compatibility (which had to explain also the surprising dichotomy between $\bigcirc \bigcirc F$ and $\bigcirc \bigcirc \bigcirc F$ from Lemmata 4.3, 4.4, 4.15 and 4.16). This had to consider also formulae with more than one variable (the only one here was $F \mathcal{U} G$) and Boolean connectives (which were used only in the proof of Theorem 3.2).

The present work concentrated on LTL formulae only; a nearby idea would be to consider CTL or - even more general - CTL* formulae.

Acknowledgments. The author is grateful to Bernhard Möller and the anonymous reviewers for valuable hints and remarks which helped to improve quality and readability of the paper.

References

1. Bacci, G., Bouyer, P., Fahrenberg, U., Larsen, K.G., Markey, N., Reynier, P.-A.: Optimal and robust controller synthesis using energy timed automata with uncertainty. Form. Asp. Comput. **33**(1), 3–25 (2020). https://doi.org/10.1007/s00165-020-00521-4
2. Baier, C., Katoen, J.P.: Principles of Model Checking. MIT Press, Cambridge (2008)
3. Emerson, E.A., Jutla, C.S.: The complexity of tree automata and logics of programs. In: 29th Annual Symposium on Foundations of Computer Science, White Plains, New York, USA, 24–26 October 1988, pp. 328–337. IEEE Computer Society (1988)

4. Garey, M.R., Johnson, D.S.: Computers and Intractability: A Guide to the Theory of NP-Completeness. WH Freeman and Company, NY (1979)
5. Glück, R.: Bisimulations and model refinement (preprint). http://rolandglueck.de/html/PreprintDiss.pdf
6. Glück, R.: Using bisimulations for optimality problems in model refinement. In: de Swart, H. (ed.) RAMICS 2011. LNCS, vol. 6663, pp. 164–179. Springer, Heidelberg (2011). https://doi.org/10.1007/978-3-642-21070-9_14
7. Glück, R., Möller, B., Sintzoff, M.: A semiring approach to equivalences, bisimulations and control. In: Berghammer, R., Jaoua, A.M., Möller, B. (eds.) RelMiCS 2009. LNCS, vol. 5827, pp. 134–149. Springer, Heidelberg (2009). https://doi.org/10.1007/978-3-642-04639-1_10
8. Glück, R., Möller, B., Sintzoff, M.: Model refinement using bisimulation quotients. In: Johnson, M., Pavlovic, D. (eds.) AMAST 2010. LNCS, vol. 6486, pp. 76–91. Springer, Heidelberg (2011). https://doi.org/10.1007/978-3-642-17796-5_5
9. Glück, R.: Bisimulations and model refinement. PhD thesis, University of Augsburg. Pro Business (2015)
10. Maler, O., Pnueli, A., Sifakis, J.: On the synthesis of discrete controllers for timed systems. In: Mayr, E.W., Puech, C. (eds.) STACS 1995. LNCS, vol. 900, pp. 229–242. Springer, Heidelberg (1995). https://doi.org/10.1007/3-540-59042-0_76
11. Myhill, J.: Finite automata and the representation of events. WADD Tech. Rep. **57**, 112–137 (1957)
12. Nerode, A.: Linear automaton transformations. Proc. Am. Math. Soc. **9**(4), 541–544 (1958)
13. Paige, R., Tarjan, R.: Three partition refinement algorithms. SIAM J. Comput. **16**(6), 973–989 (1987)
14. Pous, D.: Complete lattices and up-to techniques. In: Shao, Z. (ed.) APLAS 2007. LNCS, vol. 4807, pp. 351–366. Springer, Heidelberg (2007). https://doi.org/10.1007/978-3-540-76637-7_24
15. Rabin, M.O.: Decidability of second-order theories and automata on infinite trees. Trans. Am. Math. Soc. **141**, 1–35 (1969)
16. Ramadge, P.J., Wonham, W.M.: The control of discrete event systems. Proc. IEEE **77**, 81–98 (1989)
17. Rey de Souza, F.G., Hirata, C.M., Nadjm-Tehrani, S.: Synthesis of a controller algorithm for safety-critical systems. IEEE Access **10**, 76351–76375 (2022)
18. Schmidt, G., Ströhlein, T.: Relations and Graphs: Discrete Mathematics for Computer Scientists. Springer, Heidelberg (1993). https://doi.org/10.1007/978-3-642-77968-8
19. Sintzoff, M.: Synthesis of optimal control policies for some infinite-state transition systems. In: Audebaud, P., Paulin-Mohring, C. (eds.) MPC 2008. LNCS, vol. 5133, pp. 336–359. Springer, Heidelberg (2008). https://doi.org/10.1007/978-3-540-70594-9_18
20. Tarjan, R.: Depth-first search and linear graph algorithms. In: 12th Annual Symposium on Switching and Automata Theory (SWAT 1971), pp. 114–121 (1971)
21. Thistle, J.G., Wonham, W.M.: Control of ω-automata, Church's problem, and the emptiness problem for tree ω-automata. In: Börger, E., Jäger, G., Kleine Büning, H., Richter, M.M. (eds.) CSL 1991. LNCS, vol. 626, pp. 367–381. Springer, Heidelberg (1992). https://doi.org/10.1007/BFb0023782

22. Thistle, J.G., Wonham, W.M.: Control of infinite behavior of finite automata. SIAM J. Control Optim. **32**(4), 1075–1097 (1994)
23. Bochmann, G.V., Hilscher, M., Linker, S., Olderog, E.-R.: Synthesizing and verifying controllers for multi-lane traffic maneuvers. Form. Asp. Comput. **29**(4), 583–600 (2017). https://doi.org/10.1007/s00165-017-0424-4
24. Winter, M.: A relation-algebraic theory of bisimulations. Fundam. Inform. **83**(4), 429–449 (2008)

Dependences Between Domain Constructions in Heterogeneous Relation Algebras

Walter Guttmann[✉]

Department of Computer Science and Software Engineering,
University of Canterbury, Christchurch, New Zealand
walter.guttmann@canterbury.ac.nz

Abstract. We show the following dependences between relational domain constructions in the framework of heterogeneous relation algebras. If all power sets and subsets exist and objects are comparable, then all sums exist. If all sums exist and atoms are rectangular, then all products exist. If all atoms are rectangular, then all subsets exist if and only if all quotients exist. We give models with rectangular atoms which rule out further dependences between these constructions.

1 Introduction

Applications of relations often need to work with structured data. To facilitate this, extensions of relation algebras by various domain constructions have been studied in the literature [2–6,11–14,20]. Examples include power sets, products, sums, quotients and subsets: well-known basic ingredients for the construction of more complex data types.

A typical way to extend relation algebras by a domain construction is to introduce operations axiomatically and to prove that the axioms characterise the intended domain uniquely up to isomorphism. Hence it is natural to ask about the independence of the axioms used for the various domain constructions. Studying their dependences is the topic of this paper.

We work in the framework of heterogeneous relation algebras and contribute the following main results:

- If all power sets and subsets exist and objects are comparable, then all sums exist.
- If all sums exist and atoms are rectangular, then all products exist.
- If all atoms are rectangular, then all subsets exist if and only if all quotients exist.
- There are models with rectangular atoms which rule out further dependences between these constructions.

We first recall basic definitions and properties of heterogeneous relation algebras, and the domain constructions of power set, product, sum, quotient and subset. Sections 3–6 provide the dependence results. This is followed by models for the independence results in Sect. 7.

© The Author(s), under exclusive license to Springer Nature Switzerland AG 2023
R. Glück et al. (Eds.): RAMiCS 2023, LNCS 13896, pp. 105–121, 2023.
https://doi.org/10.1007/978-3-031-28083-2_7

2   Heterogeneous Relation Algebras and Domain Constructions

In this section we define heterogeneous relation algebras and the domain constructions of power set, product, sum, quotient and subset. Heterogeneous relation algebras are a typed version of Tarski's relation algebras [15]. Related frameworks in which the dependence of domain constructions could be studied are allegories and Dedekind categories [4,6,10].

2.1   Heterogeneous Relation Algebras

The following definition is from [12] which provides a good overview of heterogeneous relation algebras and some domain constructions; also see [13].

Definition 1. A *heterogeneous relation algebra* is a locally small category with objects Obj, morphisms $\mathrm{Mor}(A, B)$ for $A, B \in$ Obj, composition ;, identities I_A, and the following additional structure.

- For each $A, B \in$ Obj there is a transposition $^\mathsf{T}_{A,B} : \mathrm{Mor}(A, B) \to \mathrm{Mor}(B, A)$.
- Each $\mathrm{Mor}(A, B)$ is a complete atomic Boolean algebra with join $\sqcup_{A,B}$, meet $\sqcap_{A,B}$, complement $^-_{A,B}$, order $\sqsubseteq_{A,B}$, least element $O_{A,B}$ and greatest element $L_{A,B}$, where $O_{A,B} \neq L_{A,B}$.
- Each $Q \in \mathrm{Mor}(A, B)$ and $R \in \mathrm{Mor}(B, C)$ and $S \in \mathrm{Mor}(A, C)$ satisfy the Schröder equivalences $Q ; R \sqsubseteq_{A,C} S \Leftrightarrow Q^\mathsf{T} ; \overline{S} \sqsubseteq_{B,C} \overline{R} \Leftrightarrow \overline{S} ; R^\mathsf{T} \sqsubseteq_{A,B} \overline{Q}$.
- The Tarski rule $R \neq O_{A,B} \Leftrightarrow L_{C,A} ; R ; L_{B,D} = L_{C,D}$ holds for each $R \in \mathrm{Mor}(A, B)$ and $C, D \in$ Obj.

We usually omit subscripts specifying type information and abbreviate composition $R;S$ as RS. Morphisms $R \in \mathrm{Mor}(A, B)$ are called *relations* and denoted $R : A \leftrightarrow B$.

An example of a heterogenous relation algebra is REL, which has all nonempty sets as objects, all (set-theoretic) binary relations $R \subseteq A \times B$ as morphisms $R : A \leftrightarrow B$, and the usual operations on binary relations. Further examples appear throughout this paper.

Relation R is *univalent* if $R^\mathsf{T} R \sqsubseteq I$, *total* if $I \sqsubseteq RR^\mathsf{T}$, a *mapping* if R is univalent and total, and *injective/surjective/bijective* if R^T is a univalent/total/a mapping. Relation R is *reflexive* if $I \sqsubseteq R$, *symmetric* if $R^\mathsf{T} = R$, *transitive* if $RR \sqsubseteq R$, a *partial equivalence* if R is symmetric and transitive, and an *equivalence* if R is reflexive and a partial equivalence. Relation R is a *partial identity* if $R \sqsubseteq I$. Relation R is a *vector* if $R = RL$ and *rectangular* if $RLR \sqsubseteq R$.

Partial identities are symmetric and they form a Boolean algebra in which composition coincides with meet and complement is given by $\neg R = \overline{RL} \sqcap I$.

A number of residual operations will be useful in particular for the construction of power sets. The *left residual* of relations $Q : B \leftrightarrow A$ and $R : C \leftrightarrow A$ is $Q/R = \overline{\overline{Q}R^\mathsf{T}}$. The *right residual* of relations $Q : A \leftrightarrow B$ and $R : A \leftrightarrow C$

is $Q\backslash R = \overline{Q^{\mathsf{T}}\overline{R}}$. Their *symmetric quotient* is $Q \div R = (Q\backslash R) \sqcap (Q^{\mathsf{T}}/R^{\mathsf{T}})$. The following table summarises the logical interpretation of these operations in REL:

$$(x,y) \in Q/R \Leftrightarrow (\forall z \in A : (y,z) \in R \Rightarrow (x,z) \in Q)$$
$$(x,y) \in Q\backslash R \Leftrightarrow (\forall z \in A : (z,x) \in Q \Rightarrow (z,y) \in R)$$
$$(x,y) \in Q \div R \Leftrightarrow (\forall z \in A : (z,x) \in Q \Leftrightarrow (z,y) \in R)$$

The following lemma collects properties of the above operations used in this paper. Here we only prove Lemma 2.7. The other properties are known from the literature or simple consequences; in particular, see [11–13].

Lemma 2

1. $QR \sqcap S \sqsubseteq (Q \sqcap SR^{\mathsf{T}})(R \sqcap Q^{\mathsf{T}}S)$.
2. $PQR \sqsubseteq S \Leftrightarrow P^{\mathsf{T}}\overline{S}R^{\mathsf{T}} \sqsubseteq \overline{Q}$.
3. $(QL \sqcap R)S = QL \sqcap RS$.
4. $(R \sqcap LQ)S = R(Q^{\mathsf{T}}L \sqcap S)$.
5. R is total if and only if $RL = L$.
6. R is rectangular if and only if $RLR = R$.
7. $R^{\mathsf{T}}R \neq O$ if R is total.
8. $Q(R \sqcap S) = QR \sqcap QS$ if Q is univalent.
9. $RL \sqcap I = R = LR \sqcap I$ if R is a partial identity.
10. $I\backslash R = R = R/I$.
11. $\backslash$ reverses $\sqsubseteq$ in its first argument and preserves $\sqsubseteq$ in its second argument.
12. $(Q \sqcup R)\backslash S = (Q\backslash S) \sqcap (R\backslash S)$.
13. $Q(R\backslash S) = RQ^{\mathsf{T}}\backslash S$ if Q is a mapping.
14. $R\backslash QS \sqsubseteq Q^{\mathsf{T}}R\backslash S$ if Q is univalent.
15. $QR \sqsubseteq S \Leftrightarrow Q \sqsubseteq S/R$.
16. $Q^{\mathsf{T}}/R^{\mathsf{T}} = (R\backslash Q)^{\mathsf{T}}$.
17. $I \div I = I$.
18. $I \div O = O$.
19. $I \sqsubseteq R \div R$.
20. $(Q \div R)^{\mathsf{T}} = R \div Q$.
21. $(Q \div R)(R \div S) = (Q \div S) \sqcap (Q \div R)L$.
22. $Q(R \div S) = RQ^{\mathsf{T}} \div S$ if Q is a mapping.
23. $(R \div S)Q = R \div SQ$ if Q is a bijective.
24. $Q^{\mathsf{T}}R \div S \sqsubseteq R \div QS$ if Q is injective and total.
25. $R \div Q$ is a vector if Q is a vector.

Proof (of Lemma 2.7). Using Lemma 2.5, we have $R^{\mathsf{T}}R \sqsubseteq O \Leftrightarrow RL \sqsubseteq \overline{R} \Leftrightarrow L \sqsubseteq \overline{R} \Leftrightarrow R \sqsubseteq O \Leftrightarrow RL \sqsubseteq O \Leftrightarrow L \sqsubseteq O$, which is false.   □

2.2   Power Sets

Power sets are introduced in heterogeneous relation algebras by axiomatising the membership relation based on symmetric quotients [2]. The following axioms characterise power sets uniquely up to isomorphism; a similar remark holds for products, sums, quotients and subsets below.

Definition 3. The *power* of object A is an object 2^A with a relation $\varepsilon : A \leftrightarrow 2^A$ satisfying

- $\varepsilon \div \varepsilon \sqsubseteq I$ and
- $R \div \varepsilon$ is total for each object B and relation $R : A \leftrightarrow B$.

It follows that 2^A is a categorical power object [4,6]. In REL, 2^A is the usual power set of A and $(x, Y) \in \varepsilon \Leftrightarrow x \in Y$.

The following lemma collects properties of ε used in this paper. Here we only prove Lemmas 4.6–4.8. The other properties are known from the literature; in particular, see [8,11,12].

Lemma 4

1. $\varepsilon \div \varepsilon = I$.
2. $R \div \varepsilon$ is a mapping.
3. $\varepsilon(\varepsilon \div R) = R$.
4. $\varepsilon(\varepsilon \backslash R) = R$.
5. $(Q \div \varepsilon)(\varepsilon \div R) = Q \div R$.
6. $(\varepsilon \backslash I) \div (\varepsilon \backslash I) = I$.
7. $\varepsilon \backslash I = (\varepsilon \div O) \sqcup (\varepsilon \div I)$.
8. $(\varepsilon \backslash I) \div I = O$.

Proof (of Lemmas 4.6–4.8)

6. Using Lemmas 2.10, 2.11, 2.13, 2.19, 2.20, 4.2 and 4.3,

$$(\varepsilon \backslash I) \div (\varepsilon \backslash I) \sqsubseteq (\varepsilon \backslash I) \backslash (\varepsilon \backslash I) \sqsubseteq (\varepsilon \div I) \backslash (\varepsilon \backslash I) = (I \div \varepsilon)(I \backslash (\varepsilon \backslash I)) = (I \div \varepsilon)(\varepsilon \backslash I)$$
$$= \varepsilon(\varepsilon \div I) \backslash I = I \backslash I = I \sqsubseteq (\varepsilon \backslash I) \div (\varepsilon \backslash I)$$

7. $\varepsilon \div O \sqsubseteq \varepsilon \backslash O \sqsubseteq \varepsilon \backslash I$ using Lemma 2.11 and $\varepsilon \div I \sqsubseteq \varepsilon \backslash I$. For the converse we have
$$(\varepsilon \backslash I) \sqcap \overline{\varepsilon \div O} = (\varepsilon \backslash I) \sqcap \varepsilon^T L \sqsubseteq (\varepsilon \backslash I) \sqcap \varepsilon^T \varepsilon(\varepsilon \backslash I) = (\varepsilon \backslash I) \sqcap \varepsilon^T = (\varepsilon \backslash I) \sqcap (\varepsilon^T / I) = \varepsilon \div I$$
using Lemmas 2.1, 2.10 and 4.7. The result follows by shunting.
8. Using Lemmas 2.1, 2.10, 2.12, 2.13, 2.18, 2.20, 4.2, 4.5 and 4.4,

$$(\varepsilon \backslash I) \div I \sqsubseteq (\varepsilon \backslash I) \backslash I = ((\varepsilon \div O) \sqcup (\varepsilon \div I)) \backslash I = ((\varepsilon \div O) \backslash I) \sqcap ((\varepsilon \div I) \backslash I)$$
$$= (O \div \varepsilon)(I \backslash I) \sqcap (I \div \varepsilon)(I \backslash I) = (O \div \varepsilon) \sqcap (I \div \varepsilon) \sqsubseteq (I \div \varepsilon)(\varepsilon \div O)(O \div \varepsilon)$$
$$= (I \div O)(O \div \varepsilon) = O$$

$\square$

2.3 Products

Products are introduced in heterogeneous relation algebras by axiomatising their projections [13].

Definition 5. The *product* of objects A, B is an object $A \times B$ with relations $p_A : A \times B \leftrightarrow A$ and $p_B : A \times B \leftrightarrow B$ satisfying

- p_A and p_B are mappings,
- $p_A^\mathsf{T} p_B = \mathsf{L}$ and
- $p_A p_A^\mathsf{T} \sqcap p_B p_B^\mathsf{T} \sqsubseteq \mathsf{I}$.

It follows that p_A and p_B are surjective, $p_A^\mathsf{T} p_A = \mathsf{I}_A$ and $p_B^\mathsf{T} p_B = \mathsf{I}_B$ and $p_A p_A^\mathsf{T} \sqcap p_B p_B^\mathsf{T} = \mathsf{I}_{A \times B}$. In general, $A \times B$ is not a categorical product, but it is one in the wide subcategory of mappings in REL [4].

2.4   Sums

Sums model disjoint unions and are introduced in heterogeneous relation algebras by axiomatising their injections [5, 20].

Definition 6. The *sum* of objects A, B is an object $A + B$ with relations $i_A : A \leftrightarrow A \times B$ and $i_B : B \leftrightarrow A \times B$ satisfying

- i_A and i_B are injective mappings,
- $i_A i_B^\mathsf{T} = \mathsf{O}$ and
- $\mathsf{I} \sqsubseteq i_A^\mathsf{T} i_A \sqcup i_B^\mathsf{T} i_B$.

It follows that $i_A i_A^\mathsf{T} = \mathsf{I}_A$ and $i_B i_B^\mathsf{T} = \mathsf{I}_B$ and $i_A^\mathsf{T} i_A \sqcup i_B^\mathsf{T} i_B = \mathsf{I}_{A+B}$. Moreover $A + B$ is a categorical coproduct; in REL, it is also a categorical product [4].

2.5   Quotients

Quotients are based on equivalence relations and introduced in heterogeneous relation algebras by axiomatising the projection to equivalence classes [11].

Definition 7. The *quotient* of object A by equivalence $E : A \leftrightarrow A$ is an object A/E with a relation $p : A \leftrightarrow A/E$ satisfying

- $p p^\mathsf{T} = E$ and
- $p^\mathsf{T} p = \mathsf{I}$.

It follows that p is a surjective mapping. Hence A/E with p is a categorical quotient object [1].

2.6   Subsets

Subsets are based on partial identities and introduced in heterogeneous relation algebras by axiomatising the injection into the base set. We specialise the axioms of 'subset extrusion' given in [11] to subsets specified by a partial identity.

Definition 8. The *subset* of object A corresponding to non-zero partial identity $S : A \leftrightarrow A$ is an object S with a relation $i : S \leftrightarrow A$ satisfying

- $i^\mathsf{T} i = S$ and
- $i i^\mathsf{T} = \mathsf{I}$.

It follows that i is an injective mapping. Hence S with i is a categorical subobject [1]. We overload the name of the object S with the name of the partial identity S on which it is based, because the two closely correspond to each other.

We remark that the quotient and subset constructions are special cases of the construction of splittings [3,6,18]. Splittings are based on partial equivalence relations; they combine taking a subset to the domain of the partial equivalence and a quotient to its classes. Every partial identity is a partial equivalence relation. Since projections go from A to the quotient whereas injections go from the subset to A, one of the two directions has to be reversed if they are unified as splittings. In this paper, we study quotients and subsets separately; see Sect. 6 for their dependence.

3   Sums from Power Sets and Subsets

In this section we show that all sums exist if all power sets and subsets exist and objects are comparable. Note that both the power set construction and the subset construction are based on a single object, whereas sums are based on two objects. To combine two different objects we need a way to relate them. This is provided by the following concept.

Definition 9. Object A is *contained* in object B if there is an injective mapping $i : A \leftrightarrow B$. Objects A, B are *comparable* if A is contained in B or B is contained in A.

In REL, A is contained in B if and only if $|A| \leq |B|$, where $|\cdot|$ is the cardinality of a set, and any two objects are comparable (this is equivalent to the Axiom of Choice). For a heterogeneous relation algebra in which not all objects are comparable consider $\mathrm{Obj} = \{A, B\}$ where $A = \{1, 2\}$ and $B = \{1, 2, 3\}$ and $\mathrm{Mor}(A, A) = \{O, I, \overline{I}, L\}$ and $\mathrm{Mor}(B, B) = \{O, I, \overline{I}, L\}$ and $\mathrm{Mor}(A, B) = \{O, L\}$ and $\mathrm{Mor}(B, A) = \{O, L\}$. It is a (heterogeneous) subalgebra of REL but none of the morphisms between A and B are injective mappings: $OO^T = O \neq I$ and $LL^T = L \neq I$ for all well-typed instances of these (in)equations.

Containment is connected to the subset domain construction as the following result shows.

Lemma 10. *A is contained in B if and only if A is a subset of B corresponding to some non-zero partial identity S.*

Proof. For the forward implication, let $i : A \leftrightarrow B$ be the injective mapping arising from the containment. Define $S : B \leftrightarrow B$ by $S = i^T i$. Then $S \sqsubseteq I$ since i is univalent. Moreover $S \neq O$ by Lemma 2.7 since i is total. Finally $ii^T = I$ is equivalent to i being injective and total. The backward implication follows immediately as the subset construction gives the desired injective mapping.   □

The existence of all subsets does not imply that objects are comparable. This is shown by the above example, which contains all subsets since I is the only non-zero partial identity on each of the two objects. The converse implication also does not hold. The single-object relation algebra of all binary relations on a two-element set does not contain subsets corresponding to the two partial identities between O and I, but its single object is comparable with itself since I is injective.

After these preliminaries we turn to the main goal of constructing sums from power sets and subsets. The general idea is to represent a sum as the power set of a power set. To illustrate this, consider the set-theoretic example of constructing the disjoint union of $A = \{1, 2, 3\}$ and $B = \{a, b\}$. Elements of A will be represented by singleton sets of singleton sets: 1 by $\{\{1\}\}$ and 2 by $\{\{2\}\}$ and 3 by $\{\{3\}\}$. Elements of B will be represented by including the empty set to distinguish the source: a by $\{\emptyset, \{a\}\}$ and b by $\{\emptyset, \{b\}\}$.

We implement this construction for general heterogeneous relation algebras in three parts. The first theorem will be used to discard sets of sets that are not used in the construction, but we formulate it more generally. This is where subsets come into play.

Theorem 11. *Assume all subsets exist. Assume $i_A : A \leftrightarrow C$ and $i_B : B \leftrightarrow C$ are injective mappings with $i_A i_B^\mathsf{T} = \mathsf{O}$. Then $A + B$ exists.*

Proof. Define $S : C \leftrightarrow C$ by $S = i_A^\mathsf{T} i_A \sqcup i_B^\mathsf{T} i_B$. Then $S \sqsubseteq \mathsf{I}$ since i_A and i_B are univalent. Moreover $S \neq \mathsf{O}$ by Lemma 2.7 since i_A is total. Hence the subset S exists with injection $i : S \leftrightarrow C$ satisfying $i^\mathsf{T} i = S$ and $i i^\mathsf{T} = \mathsf{I}$. We show $S = A + B$. To this end, define $j_A : A \leftrightarrow S$ by $j_A = i_A i^\mathsf{T}$ and $j_B : B \leftrightarrow S$ by $j_B = i_B i^\mathsf{T}$. Then

- $j_A j_A^\mathsf{T} = i_A i^\mathsf{T} i i_A^\mathsf{T} = i_A S i_A^\mathsf{T} = i_A i_A^\mathsf{T} i_A i_A^\mathsf{T} \sqcup i_A i_B^\mathsf{T} i_B i_A^\mathsf{T} = \mathsf{I} \sqcup \mathsf{O} = \mathsf{I}$ since i_A is injective and total.
- $j_B j_B^\mathsf{T} = i_B i^\mathsf{T} i i_B^\mathsf{T} = i_B S i_B^\mathsf{T} = i_B i_A^\mathsf{T} i_A i_B^\mathsf{T} \sqcup i_B i_B^\mathsf{T} i_B i_B^\mathsf{T} = \mathsf{O} \sqcup \mathsf{I} = \mathsf{I}$ since i_B is injective and total.
- $j_A j_B^\mathsf{T} = i_A i^\mathsf{T} i i_B^\mathsf{T} = i_A S i_B^\mathsf{T} \sqsubseteq i_A i_B^\mathsf{T} = \mathsf{O}$.
- $j_A^\mathsf{T} j_A \sqcup j_B^\mathsf{T} j_B = i i_A^\mathsf{T} i_A i^\mathsf{T} \sqcup i i_B^\mathsf{T} i_B i^\mathsf{T} = i S i^\mathsf{T} = i i^\mathsf{T} i i^\mathsf{T} = \mathsf{I}$. $\square$

The next corollary instantiates the previous theorem to the singleton-set construction outlined above. This is where power sets come into play. In REL, $\mathsf{I} \div \varepsilon$ relates an element with the singleton set containing it, that is, $(x, Y) \in \mathsf{I} \div \varepsilon \Leftrightarrow Y = \{x\}$. Hence $(\mathsf{I} \div \varepsilon)(\mathsf{I} \div \varepsilon)$ constructs the desired doubly singleton sets. Note that $(\mathsf{I} \div \varepsilon)(\mathsf{I} \div \varepsilon) = (\varepsilon \div \mathsf{I}) \div \varepsilon$ by Lemmas 2.20, 2.22 and 4.2. To include the empty set we allow subsets of the singleton set inside the outer set by replacing the inner symmetric quotient with a right residual as in $(\varepsilon \backslash \mathsf{I}) \div \varepsilon$.

Corollary 12. *Assume all subsets and power sets exist. Then $A + A$ exists for each object A.*

Proof. Define $i_A, i_B : A \leftrightarrow 2^{2^A}$ by $i_A = (\mathsf{I} \div \varepsilon)(\mathsf{I} \div \varepsilon)$ and $i_B = (\varepsilon \backslash \mathsf{I}) \div \varepsilon$. Then the assumptions of Theorem 11 are satisfied since

- $i_A i_A^\mathsf{T} = (\mathsf{I} \div \varepsilon)(\mathsf{I} \div \varepsilon)(\varepsilon \div \mathsf{I})(\varepsilon \div \mathsf{I}) = (\mathsf{I} \div \varepsilon)(\mathsf{I} \div \mathsf{I})(\varepsilon \div \mathsf{I}) = (\mathsf{I} \div \varepsilon)(\varepsilon \div \mathsf{I}) = \mathsf{I} \div \mathsf{I} = \mathsf{I}$ using Lemmas 2.17, 2.20 and 4.5.
- $i_B i_B^\mathsf{T} = ((\varepsilon \backslash \mathsf{I}) \div \varepsilon)(\varepsilon \div (\varepsilon \backslash \mathsf{I})) = (\varepsilon \backslash \mathsf{I}) \div (\varepsilon \backslash \mathsf{I}) = \mathsf{I}$ using Lemmas 2.20, 4.5 and 4.6.

- $i_A i_B^\mathsf{T} = (\mathsf{I} \div \varepsilon)(\mathsf{I} \div \varepsilon)(\varepsilon \div (\varepsilon \backslash \mathsf{I})) = (\mathsf{I} \div \varepsilon)(\mathsf{I} \div (\varepsilon \backslash \mathsf{I})) = \mathsf{O}$ using Lemmas 2.20, 4.5 and 4.8.
- $i_A^\mathsf{T} i_A = (\varepsilon \div \mathsf{I})(\varepsilon \div \mathsf{I})(\mathsf{I} \div \varepsilon)(\mathsf{I} \div \varepsilon) \sqsubseteq (\varepsilon \div \mathsf{I})(\varepsilon \div \varepsilon)(\mathsf{I} \div \varepsilon) = (\varepsilon \div \mathsf{I})(\mathsf{I} \div \varepsilon) \sqsubseteq \varepsilon \div \varepsilon = \mathsf{I}$ using Lemmas 2.20, 2.21 and 4.1.
- $i_B^\mathsf{T} i_B = (\varepsilon \div (\varepsilon \backslash \mathsf{I}))((\varepsilon \backslash \mathsf{I}) \div \varepsilon) \sqsubseteq \varepsilon \div \varepsilon = \mathsf{I}$ using Lemmas 2.20, 2.21 and 4.1. □

The following corollary generalises this to sums of different objects. This is where comparability comes into play. The proof reuses calculations from the proof of Corollary 12, keeping i_A and modifying i_B by composing the injection available through comparability.

Corollary 13. *Assume all subsets and power sets exist and objects are comparable. Then $A + B$ exists for each object A, B.*

Proof. Without loss of generality assume B is contained in A using injection $i : B \leftrightarrow A$ (a symmetric argument applies in the other case). Define $i_A : A \leftrightarrow 2^{2^A}$ by $i_A = (\mathsf{I} \div \varepsilon)(\mathsf{I} \div \varepsilon)$ and $i_B : B \leftrightarrow 2^{2^A}$ by $i_B = i((\varepsilon \backslash \mathsf{I}) \div \varepsilon)$. Then the assumptions of Theorem 11 (not already covered by the proof of Corollary 12) are satisfied since

- $i_B i_B^\mathsf{T} = ii^\mathsf{T} = \mathsf{I}$ since i is injective and total.
- $i_A i_B^\mathsf{T} = \mathsf{O}i^\mathsf{T} = \mathsf{O}$.
- $i_B^\mathsf{T} i_B = (\varepsilon \div (\varepsilon \backslash \mathsf{I}))i^\mathsf{T} i((\varepsilon \backslash \mathsf{I}) \div \varepsilon) \sqsubseteq (\varepsilon \div (\varepsilon \backslash \mathsf{I}))((\varepsilon \backslash \mathsf{I}) \div \varepsilon) \sqsubseteq \mathsf{I}$ as i is univalent. □

4   Products from Power Sets and Subsets

In this section we show that all products exist if all power sets and subsets exist and atoms are rectangular.

We recall concepts related to atoms; for example, see [7]. As usual, Q is an atom if $Q \neq \mathsf{O}$ and, for each $R \sqsubseteq Q$, either $R = Q$ or $R = \mathsf{O}$. Each $\mathrm{Mor}(A, B)$ is atomic, which means every $R \neq \mathsf{O}$ contains an atom $Q \sqsubseteq R$. Every atomic Boolean algebra is also atomistic, that is, every element is the supremum of the atoms below it. We denote the atomic partial identities of object A by $\mathrm{at}_1(A) = \{Q : A \leftrightarrow A \mid Q \text{ is an atom} \wedge Q \sqsubseteq \mathsf{I}\}$. Two atoms are either equal or their meet is O.

We remark about representability, since some of the following results assume that all atoms are rectangular. In a single-object relation algebra, this condition implies that the algebra is point-dense and therefore representable [9]. This consequence is not surprising as single-object relation algebras in which products exist are known to be representable by having conjugated quasi-projections [16].

The following result relates rectangular atoms to comparability.

Theorem 14. *Assume all atoms are rectangular. Then all objects are comparable.*

Proof. Let A, B be objects. Without loss of generality assume $|\mathrm{at}_1(A)| \leq |\mathrm{at}_1(B)|$ (otherwise swap A, B). Hence there is an injective function $g : \mathrm{at}_1(A) \to \mathrm{at}_1(B)$. Define $i : A \leftrightarrow B$ by $i = \bigsqcup_{a \in \mathrm{at}_1(A)} a\mathsf{L}g(a)$. Then

$$ii^{\mathsf{T}} = \bigsqcup_{a,b \in \mathrm{at}_1(A)} aLg(a)g(b)Lb \qquad \text{composition is completely distributive}$$

$$= \bigsqcup_{a \in \mathrm{at}_1(A)} aLg(a)La \qquad g(a)g(b) \neq \mathsf{O} \Leftrightarrow g(a) = g(b) \Leftrightarrow a = b$$

$$= \bigsqcup_{a \in \mathrm{at}_1(A)} aLa \qquad \text{Tarski rule}$$

$$= \bigsqcup_{a \in \mathrm{at}_1(A)} a \qquad \text{rectangular atoms, Lemma 2.6}$$

$$= \mathsf{I} \qquad \text{atomistic lattice}$$

$$i^{\mathsf{T}}i = \bigsqcup_{a,b \in \mathrm{at}_1(A)} g(a)LabLg(b) \qquad \text{composition is completely distributive}$$

$$= \bigsqcup_{a \in \mathrm{at}_1(A)} g(a)LaLg(a) \qquad ab \neq \mathsf{O} \Leftrightarrow a = b$$

$$= \bigsqcup_{a \in \mathrm{at}_1(A)} g(a)Lg(a) \qquad \text{Tarski rule}$$

$$= \bigsqcup_{a \in \mathrm{at}_1(A)} g(a) \qquad \text{rectangular atoms, Lemma 2.6}$$

$$\sqsubseteq \mathsf{I} \qquad g(a) \text{ partial identity} \qquad\qquad \square$$

The converse implication does not hold. Any non-representable single-object relation algebra is a counterexample.

After these preliminaries we turn to the main goal of constructing products from power sets, subsets and sums. The general idea is to represent a product as a power set of a sum. To illustrate this, consider the set-theoretic example of constructing the Cartesian product of $A = \{1, 2, 3\}$ and $B = \{a, b\}$. Every pair will be represented by a two-element set: for example, $(3, a)$ by $\{3, a\}$. Even if $A = B$ the sum construction will tag the components so that the set representing a pair does not collapse to a singleton set.

We implement this construction for general heterogeneous relation algebras in three parts. The first theorem will be used to discard sets that are not used in the construction, but we formulate it more generally. This is where subsets come into play.

Theorem 15. *Assume all subsets exist. Assume $p_A : C \leftrightarrow A$ and $p_B : C \leftrightarrow B$ are univalent with $p_A^{\mathsf{T}} p_B = \mathsf{L}$ and $p_A p_A^{\mathsf{T}} \sqcap p_B p_B^{\mathsf{T}} \sqsubseteq \mathsf{I}$. Then $A \times B$ exists.*

Proof. Define $S : C \leftrightarrow C$ by $S = p_A \mathsf{L} p_B^{\mathsf{T}} \sqcap \mathsf{I}$. Then $S \sqsubseteq \mathsf{I}$ and $S \neq \mathsf{O}$ since otherwise $p_A \mathsf{L} p_B^{\mathsf{T}} \sqsubseteq \overline{\mathsf{I}}$, which is equivalent to $p_A^{\mathsf{T}} \mathsf{I} p_B \sqsubseteq \mathsf{O}$ using Lemma 2.2, whence $\mathsf{L} \sqsubseteq \mathsf{O}$. Hence the subset S exists with injection $i : S \leftrightarrow C$ satisfying $i^{\mathsf{T}} i = S$ and $ii^{\mathsf{T}} = \mathsf{I}$. We show $S = A \times B$. To this end, define $q_A : S \leftrightarrow A$ by $q_A = i p_A$ and $q_B : S \leftrightarrow B$ by $q_B = i p_B$. Then

- $q_A^{\mathsf{T}} q_A = p_A^{\mathsf{T}} i^{\mathsf{T}} i p_A = p_A^{\mathsf{T}} S p_A \sqsubseteq p_A^{\mathsf{T}} p_A \sqsubseteq \mathsf{I}$ since p_A is univalent.
- $q_B^{\mathsf{T}} q_B = p_B^{\mathsf{T}} i^{\mathsf{T}} i p_B = p_B^{\mathsf{T}} S p_B \sqsubseteq p_B^{\mathsf{T}} p_B \sqsubseteq \mathsf{I}$ since p_B is univalent.

- $\mathsf{I} = ii^\mathsf{T} ii^\mathsf{T} = iSi^\mathsf{T} \sqsubseteq ip_A(\mathsf{L} p_B^\mathsf{T} \sqcap p_A^\mathsf{T})i^\mathsf{T} \sqsubseteq ip_A p_A^\mathsf{T} i^\mathsf{T} = q_A q_A^\mathsf{T}$ using Lemma 2.1.
- $\mathsf{I} = iSi^\mathsf{T} \sqsubseteq i(p_A \mathsf{L} \sqcap \mathsf{I} p_B)p_B^\mathsf{T} i^\mathsf{T} \sqsubseteq ip_B p_B^\mathsf{T} i^\mathsf{T} = q_B q_B^\mathsf{T}$ using Lemma 2.1.
- $q_A^\mathsf{T} q_B = p_A^\mathsf{T} i^\mathsf{T} ip_B = p_A^\mathsf{T} Sp_B = p_A^\mathsf{T}(p_A \mathsf{L} \sqcap \mathsf{L} p_B^\mathsf{T} \sqcap \mathsf{I})p_B = (p_A^\mathsf{T} \sqcap \mathsf{L} p_A^\mathsf{T})(p_B \sqcap p_B \mathsf{L}) = p_A^\mathsf{T} p_B = \mathsf{L}$ using Lemmas 2.3 and 2.4.
- $q_A q_A^\mathsf{T} \sqcap q_B q_B^\mathsf{T} = ip_A p_A^\mathsf{T} i^\mathsf{T} \sqcap ip_B p_B^\mathsf{T} i^\mathsf{T} = i(p_A p_A^\mathsf{T} \sqcap p_B p_B^\mathsf{T})i^\mathsf{T} \sqsubseteq ii^\mathsf{T} = \mathsf{I}$ using Lemma 2.8 since i is univalent. $\qquad\square$

The next lemma carries out the two-element-set construction outlined above and establishes some of the properties of products. This is where power sets and sums come into play. The sets used in the construction contain a single element from A and a single element from B. In REL, $\mathsf{I} \div i_A \varepsilon$ relates an element of A with the sets over $A + B$ containing that element and arbitrary elements of B. Similarly, $\mathsf{I} \div i_B \varepsilon$ relates an element of B with sets over $A + B$ containing it and arbitrary elements of A. Hence the intermediate sets in the composition $(\mathsf{I} \div i_A \varepsilon)(i_B \varepsilon \div \mathsf{I})$ contain exactly one element of A and exactly one element of B.

Lemma 16. *Assume all power sets and sums exist. Let A, B be objects. Then there are an object C and univalent surjective relations $p_A : C \leftrightarrow A$ and $p_B : C \leftrightarrow B$ with $p_A p_A^\mathsf{T} \sqcap p_B p_B^\mathsf{T} \sqsubseteq \mathsf{I}$.*

Proof. Define $p_A : 2^{A+B} \leftrightarrow A$ by $p_A = i_A \varepsilon \div \mathsf{I}$ and $p_B : 2^{A+B} \leftrightarrow B$ by $p_B = i_B \varepsilon \div \mathsf{I}$. Then

- $p_A^\mathsf{T} p_A = (\mathsf{I} \div i_A \varepsilon)(i_A \varepsilon \div \mathsf{I}) = (\mathsf{I} \div \mathsf{I}) \sqcap (\mathsf{I} \div i_A \varepsilon)\mathsf{L} = \mathsf{I} \sqcap \mathsf{L} = \mathsf{I}$ using Lemmas 2.17, 2.20 and 2.21 and that $\mathsf{L} = (i_A^\mathsf{T} \div \varepsilon)\mathsf{L} \sqsubseteq (\mathsf{I} \div i_A \varepsilon)\mathsf{L}$ using Lemma 2.24.
- Similarly, $p_B^\mathsf{T} p_B = \mathsf{I}$.
- Finally,

$$
\begin{aligned}
p_A p_A^\mathsf{T} \sqcap p_B p_B^\mathsf{T} &= (i_A \varepsilon \div \mathsf{I})(\mathsf{I} \div i_A \varepsilon) \sqcap (i_B \varepsilon \div \mathsf{I})(\mathsf{I} \div i_B \varepsilon) \\
&\sqsubseteq (i_A \varepsilon \div i_A \varepsilon) \sqcap (i_B \varepsilon \div i_B \varepsilon) \\
&= (i_A \varepsilon \backslash i_A \varepsilon) \sqcap (i_A \varepsilon \backslash i_A \varepsilon)^\mathsf{T} \sqcap (i_B \varepsilon \backslash i_B \varepsilon) \sqcap (i_B \varepsilon \backslash i_B \varepsilon)^\mathsf{T} \\
&\sqsubseteq (i_A^\mathsf{T} i_A \varepsilon \backslash \varepsilon) \sqcap (i_A^\mathsf{T} i_A \varepsilon \backslash \varepsilon)^\mathsf{T} \sqcap (i_B^\mathsf{T} i_B \varepsilon \backslash \varepsilon) \sqcap (i_B^\mathsf{T} i_B \varepsilon \backslash \varepsilon)^\mathsf{T} \\
&= ((i_A^\mathsf{T} i_A \varepsilon \sqcup i_B^\mathsf{T} i_B \varepsilon) \backslash \varepsilon) \sqcap ((i_A^\mathsf{T} i_A \varepsilon \sqcup i_B^\mathsf{T} i_B \varepsilon) \backslash \varepsilon)^\mathsf{T} \\
&= (\varepsilon \backslash \varepsilon) \sqcap (\varepsilon \backslash \varepsilon)^\mathsf{T} = \varepsilon \div \varepsilon = \mathsf{I}
\end{aligned}
$$

using Lemmas 2.12, 2.14, 2.16, 2.20, 2.21 and 4.1. $\qquad\square$

The following corollary completes the assumptions of Theorem 15 by establishing $p_A^\mathsf{T} p_B = \mathsf{L}$. This is where rectangular atoms come into play.

Corollary 17. *Assume all subsets and power sets exist and atoms are rectangular. Then $A \times B$ exists for each object A, B.*

Proof. By Theorem 14 all objects are comparable. Hence by Corollary 13 all sums exist. Hence by Theorem 15 and the proof of Lemma 16, it remains to show $p_A^\mathsf{T} p_B = \mathsf{L}$ reusing $p_A = i_A \varepsilon \div \mathsf{I}$ and $p_B = i_B \varepsilon \div \mathsf{I}$. Since

$$
\mathsf{L} = \mathsf{ILI} = \Big(\bigsqcup_{a \in \mathrm{at}_1(A)} a \Big) \mathsf{L} \Big(\bigsqcup_{b \in \mathrm{at}_1(B)} b \Big) = \bigsqcup_{\substack{a \in \mathrm{at}_1(A) \\ b \in \mathrm{at}_1(B)}} a \mathsf{L} b
$$

it suffices to show $a \mathsf{L} b \sqsubseteq p_A^\mathsf{T} p_B$ for each $a \in \mathrm{at}_1(A)$ and $b \in \mathrm{at}_1(B)$. Consider such a and b, and let $v = i_A^\mathsf{T} a \mathsf{L} \sqcup i_B^\mathsf{T} b \mathsf{L}$. Since a is rectangular, $a \mathsf{L} a \sqsubseteq a \sqsubseteq \mathsf{I}$, whence $a \sqsubseteq \mathsf{I}/\mathsf{L} a$ using Lemma 2.15. Therefore

$$a \sqsubseteq a \mathsf{L} \sqcap (\mathsf{I}/\mathsf{L} a) = (\mathsf{I} \backslash a \mathsf{L}) \sqcap (\mathsf{I}/\mathsf{L} a) = \mathsf{I} \div a \mathsf{L} = \mathsf{I} \div i_A v$$
$$= \mathsf{I} \div i_A \varepsilon (\varepsilon \div v) = (\mathsf{I} \div i_A \varepsilon)(\varepsilon \div v) = p_A^\mathsf{T}(\varepsilon \div v)$$

using Lemmas 2.10, 2.20, 2.23, 4.2 and 4.3. Similarly, $b \sqsubseteq p_B^\mathsf{T}(\varepsilon \div v)$. Hence

$$a \mathsf{L} b = a \mathsf{L} b^\mathsf{T} \sqsubseteq p_A^\mathsf{T}(\varepsilon \div v) \mathsf{L}(v \div \varepsilon) p_B = p_A^\mathsf{T}(\varepsilon \div v)(v \div \varepsilon) p_B \sqsubseteq p_A^\mathsf{T}(\varepsilon \div \varepsilon) p_B = p_A^\mathsf{T} p_B$$

using Lemmas 2.20, 2.21, 2.25 and 4.1 since v is a vector.   □

A referee noted that the construction of Lemma 16 was done in [17] and mentioned the following alternative to Corollary 17. If a heterogeneous relation algebra has powers, sums and products, then the product of A and B is isomorphic to the subset of 2^{A+B} in the above construction. Hence, if a heterogeneous relation algebra has powers and sums and is representable, then it can be embedded into REL, which has products, and the subset of 2^{A+B} is a product of A and B.

5   Products from Sums

In this section we show that all products exist if all sums exist and atoms are rectangular. This provides an alternative way to establish Corollary 17, but the proof in this section uses different ideas. We consider two cases, depending on whether $\mathrm{at}_1(B)$ is finite or infinite. In the finite case we represent $A \times B$ by an iterated sum $A + A + \cdots + A$ with as many summands as there are elements in $\mathrm{at}_1(B)$. In the infinite case we use that the cardinality of $\mathrm{at}_1(A) \times \mathrm{at}_1(B)$ is the same as the cardinality of $\mathrm{at}_1(A)$ or $\mathrm{at}_1(B)$ to obtain a bijection, so A or B will be the product.

Theorem 18. *Assume all sums exist and atoms are rectangular. Then $A \times B$ exists for each object A, B.*

Proof (if $\mathrm{at}_1(B)$ is finite). Let $b_1, \ldots, b_n$ be the atomic partial identities of B. Define objects $A_1, \ldots, A_n$ by $A_1 = A$ and $A_k = A_{k-1} + A$ with injections $i_k : A_{k-1} \leftrightarrow A_k$ and $j_k : A \leftrightarrow A_k$, for $2 \leq k \leq n$. Moreover, let $j_1 = \mathsf{I}$. We show $A_n = A \times B$.

Below, $i_{x..y}$ denotes the composition $i_x i_{x+1} \cdots i_{y-1} i_y$ for indices $x \leq y$; we also admit $i_{y+1..y} = \mathsf{I}$. The transposition is denoted $i_{y..x}^\mathsf{T} = i_y^\mathsf{T} i_{y-1}^\mathsf{T} \cdots i_{x+1}^\mathsf{T} i_x^\mathsf{T}$.

Let $p_k : A_n \leftrightarrow A$ with $p_k = (j_k i_{k+1..n})^\mathsf{T}$ for $1 \leq k \leq n$. Define $p_A : A_n \leftrightarrow A$ by $p_A = \bigsqcup_{1 \leq k \leq n} p_k$ and $p_B : A_n \leftrightarrow B$ by $p_B = \bigsqcup_{1 \leq k \leq n} p_k \mathsf{L} b_k$.

We first show that $p_k^\mathsf{T} p_l$ is I if $k = l$ and O otherwise.

- If $k < l$, then $p_k^\mathsf{T} p_l = j_k i_{k+1..n} i_n^\mathsf{T} ._{l+1} j_l^\mathsf{T} = j_k i_{k+1..l} j_l^\mathsf{T} = \mathsf{O}$.
- If $k > l$, then $p_k^\mathsf{T} p_l = (p_l^\mathsf{T} p_k)^\mathsf{T} = \mathsf{O}^\mathsf{T} = \mathsf{O}$.
- If $k = l$, then $p_k^\mathsf{T} p_l = j_k i_{k+1..n} i_{n..k+1}^\mathsf{T} j_k^\mathsf{T} = j_k j_k^\mathsf{T} = \mathsf{I}$.

The product axioms follow by

$$- \ p_A^\mathsf{T} p_A = \bigsqcup_{1\le k,l \le n} p_k^\mathsf{T} p_l = \bigsqcup_{1\le k \le n} p_k^\mathsf{T} p_k = \mathsf{I}.$$

$$- \ p_B^\mathsf{T} p_B = \bigsqcup_{1\le k,l \le n} b_k^\mathsf{T} \mathsf{L} p_k^\mathsf{T} p_l \mathsf{L} b_l = \bigsqcup_{1\le k \le n} b_k \mathsf{L} p_k^\mathsf{T} p_k \mathsf{L} b_k = \bigsqcup_{1\le k \le n} b_k \mathsf{L} b_k = \bigsqcup_{1\le k \le n} b_k = \mathsf{I} \text{ since } b_k \text{ is}$$

rectangular.

$$- \ p_A^\mathsf{T} p_B = \bigsqcup_{1\le k,l \le n} p_k^\mathsf{T} p_l \mathsf{L} b_l = \bigsqcup_{1\le k \le n} p_k^\mathsf{T} p_k \mathsf{L} b_k = \bigsqcup_{1\le k \le n} \mathsf{L} b_k = \mathsf{L} \bigsqcup_{1\le k \le n} b_k = \mathsf{L}.$$

- Finally,

$$p_A p_A^\mathsf{T} \sqcap p_B p_B^\mathsf{T} = (\bigsqcup_{1\le k,l \le n} p_k p_l^\mathsf{T}) \sqcap (\bigsqcup_{1\le k,l \le n} p_k \mathsf{L} b_k b_l \mathsf{L} p_l^\mathsf{T})$$

$$= (\bigsqcup_{1\le k,l \le n} p_k p_l^\mathsf{T}) \sqcap (\bigsqcup_{1\le k \le n} p_k \mathsf{L} b_k \mathsf{L} p_k^\mathsf{T}) \qquad b_k b_l \ne \mathsf{O} \Leftrightarrow b_k = b_l$$

$$= (\bigsqcup_{1\le k,l \le n} p_k p_l^\mathsf{T}) \sqcap (\bigsqcup_{1\le k \le n} p_k \mathsf{L} p_k^\mathsf{T}) \qquad \text{Tarski rule}$$

$$= \bigsqcup_{1\le k,l,m \le n} p_k p_l^\mathsf{T} \sqcap p_m \mathsf{L} p_m^\mathsf{T} = \bigsqcup_{1\le k \le n} p_k p_k^\mathsf{T} = \mathsf{I} \qquad \text{see below}$$

For the second last equality we have $p_k p_l^\mathsf{T} \sqcap p_m \mathsf{L} p_m^\mathsf{T} \sqsubseteq p_k \mathsf{L} \sqcap p_m \mathsf{L} \sqsubseteq p_k p_k^\mathsf{T} p_m \mathsf{L} = \mathsf{O}$ using Lemma 2.1 if $k \ne m$. Similarly, $p_k p_l^\mathsf{T} \sqcap p_m \mathsf{L} p_m^\mathsf{T} \sqsubseteq \mathsf{L} p_l^\mathsf{T} \sqcap \mathsf{L} p_m^\mathsf{T} \sqsubseteq \mathsf{L} p_l^\mathsf{T} p_m p_m^\mathsf{T} = \mathsf{O}$ using Lemma 2.1 if $l \ne m$. Hence it suffices to take the join over indices $k = l = m$. The last equality is a consequence of $\bigsqcup_{1\le k \le l} p_k p_k^\mathsf{T} = i_{n..l+1}^\mathsf{T} i_{l+1..n}$, which we show by induction over l. The base case holds since $p_1 p_1^\mathsf{T} = i_{n..2}^\mathsf{T} j_1^\mathsf{T} j_1 i_{2..n} = i_{n..2}^\mathsf{T} i_{2..n}$ since $j_1 = \mathsf{I}$. The inductive case holds since

$$\bigsqcup_{1\le k \le l+1} p_k p_k^\mathsf{T} = \bigsqcup_{1\le k \le l} p_k p_k^\mathsf{T} \sqcup p_{l+1} p_{l+1}^\mathsf{T} = i_{n..l+1}^\mathsf{T} i_{l+1..n} \sqcup i_{n..l+2}^\mathsf{T} j_{l+1}^\mathsf{T} j_{l+1} i_{l+2..n}$$

$$= i_{n..l+2}^\mathsf{T} (i_{l+1}^\mathsf{T} i_{l+1} \sqcup j_{l+1}^\mathsf{T} j_{l+1}) i_{l+2..n} = i_{n..l+2}^\mathsf{T} i_{l+2..n}$$

$\square$

Proof (of Theorem 18 if $\mathrm{at}_1(B)$ *is infinite).* Let C be the 'bigger' of A and B; formally, let $C = A$ if $|\mathrm{at}_1(A)| \ge |\mathrm{at}_1(B)|$ and $C = B$ otherwise. We show $C = A \times B$.

We have $|\mathrm{at}_1(C)| \le |\mathrm{at}_1(A)| \cdot |\mathrm{at}_1(B)| = |\mathrm{at}_1(A) \times \mathrm{at}_1(B)|$. Conversely, $|\mathrm{at}_1(A)| \cdot |\mathrm{at}_1(B)| \le |\mathrm{at}_1(C)|^2 = |\mathrm{at}_1(C)|$ since $\mathrm{at}_1(C)$ is infinite. By the Cantor-Schröder-Bernstein theorem, there is a bijective function $g : \mathrm{at}_1(C) \to \mathrm{at}_1(A) \times \mathrm{at}_1(B)$. Define $p_A : C \leftrightarrow A$ and $p_B : C \leftrightarrow B$ by

$$p_A = \bigsqcup_{\substack{a \in \mathrm{at}_1(C) \\ g(a)=(b,c)}} a \mathsf{L} b \qquad\qquad p_B = \bigsqcup_{\substack{a \in \mathrm{at}_1(C) \\ g(a)=(b,c)}} a \mathsf{L} c$$

Then

$$- \ p_A^\mathsf{T} p_A = \bigsqcup_{\substack{a,d \in \mathrm{at}_1(C) \\ g(a)=(b,c) \\ g(d)=(e,f)}} b \mathsf{L} a d \mathsf{L} e = \bigsqcup_{\substack{a \in \mathrm{at}_1(C) \\ g(a)=(b,c)}} b \mathsf{L} a \mathsf{L} b \sqsubseteq \bigsqcup_{\substack{a \in \mathrm{at}_1(C) \\ g(a)=(b,c)}} b \mathsf{L} b = \bigsqcup_{\substack{a \in \mathrm{at}_1(C) \\ g(a)=(b,c)}} b \sqsubseteq \mathsf{I} \text{ since } b \text{ is rectangular.}$$

$-\; p_B^\mathsf{T} p_B = \bigsqcup_{\substack{a,d\in\mathrm{at}_1(C)\\ g(a)=(b,c)\\ g(d)=(e,f)}} c\mathsf{L}ad\mathsf{L}f = \bigsqcup_{\substack{a\in\mathrm{at}_1(C)\\ g(a)=(b,c)}} c\mathsf{L}a\mathsf{L}c \sqsubseteq \bigsqcup_{\substack{a\in\mathrm{at}_1(C)\\ g(a)=(b,c)}} c\mathsf{L}c = \bigsqcup_{\substack{a\in\mathrm{at}_1(C)\\ g(a)=(b,c)}} c \sqsubseteq \mathsf{I}$ since c is rectangular.

$-\; p_A^\mathsf{T} p_B = \bigsqcup_{\substack{a,d\in\mathrm{at}_1(C)\\ g(a)=(b,c)\\ g(d)=(e,f)}} b\mathsf{L}ad\mathsf{L}f = \bigsqcup_{\substack{a\in\mathrm{at}_1(C)\\ g(a)=(b,c)}} b\mathsf{L}a\mathsf{L}c = \bigsqcup_{\substack{a\in\mathrm{at}_1(C)\\ g(a)=(b,c)}} b\mathsf{L}c = \bigsqcup_{\substack{b\in\mathrm{at}_1(A)\\ c\in\mathrm{at}_1(B)}} b\mathsf{L}c = \Big(\bigsqcup_{b\in\mathrm{at}_1(A)} b\Big)\mathsf{L}\Big(\bigsqcup_{c\in\mathrm{at}_1(B)} c\Big) = \mathsf{I}\mathsf{L}\mathsf{I} = \mathsf{L}$

using the Tarski rule and that g is bijective.

$-$ Finally,

$$p_A p_A^\mathsf{T} \sqcap p_B p_B^\mathsf{T} = \Big(\bigsqcup_{\substack{a,d\in\mathrm{at}_1(C)\\ g(a)=(b,c)\\ g(d)=(e,f)}} a\mathsf{L}be\mathsf{L}d\Big) \sqcap \Big(\bigsqcup_{\substack{a,d\in\mathrm{at}_1(C)\\ g(a)=(b,c)\\ g(d)=(e,f)}} a\mathsf{L}cf\mathsf{L}d\Big) = \Big(\bigsqcup_{\substack{a,d\in\mathrm{at}_1(C)\\ g(a)=(b,c)\\ g(d)=(b,f)}} a\mathsf{L}b\mathsf{L}d\Big) \sqcap \Big(\bigsqcup_{\substack{a,d\in\mathrm{at}_1(C)\\ g(a)=(b,c)\\ g(d)=(e,c)}} a\mathsf{L}c\mathsf{L}d\Big)$$

$$= \Big(\bigsqcup_{\substack{a,d\in\mathrm{at}_1(C)\\ g(a)=(b,c)\\ g(d)=(b,f)}} a\mathsf{L}d\Big) \sqcap \Big(\bigsqcup_{\substack{a,d\in\mathrm{at}_1(C)\\ g(a)=(b,c)\\ g(d)=(e,c)}} a\mathsf{L}d\Big)$$

$$= \bigsqcup_{\substack{a,d,a',d'\in\mathrm{at}_1(C)\\ g(a)=(b,c),g(a')=(b',c')\\ g(d)=(b,f),g(d')=(e',c')}} a\mathsf{L}d \sqcap a'\mathsf{L}d' = \bigsqcup_{\substack{a,d\in\mathrm{at}_1(C)\\ g(a)=g(d)}} a\mathsf{L}d = \bigsqcup_{a\in\mathrm{at}_1(C)} a\mathsf{L}a = \bigsqcup_{a\in\mathrm{at}_1(C)} a = \mathsf{I}$$

using the Tarski rule, that g is bijective, that a is rectangular and Lemma 2.6. For the fourth last equality note that $a\mathsf{L}d \sqcap a'\mathsf{L}d' \sqsubseteq aa^\mathsf{T} a'\mathsf{L}d' = aa'\mathsf{L}d' = \mathsf{O}$ using Lemma 2.1 if $a \neq a'$. Similarly $a\mathsf{L}d \sqcap a'\mathsf{L}d' \sqsubseteq a\mathsf{L}dd'^\mathsf{T} d' = a\mathsf{L}dd' = \mathsf{O}$ using Lemma 2.1 if $d \neq d'$. Hence it suffices to take the join over indices $a = a'$ and $d = d'$. Moreover $g(a) = g(d)$ since $g(a)$ and $g(d)$ agree in their first components and $g(a')$ and $g(d')$ agree in their second components. □

6 Subsets from Quotients and Vice Versa

In this section we show that all subsets exist if and only if all quotients exist, if all atoms are rectangular. We start with a lemma about atoms.

Lemma 19. *Assume all atoms are rectangular. Let A, B be objects and let $a \in \mathrm{at}_1(A)$ and $b \in \mathrm{at}_1(B)$. Then $a\mathsf{L}b$ is an atom.*

Proof. First, $a\mathsf{L}b \neq \mathsf{O}$. Otherwise, $a \sqsubseteq a\mathsf{L} = a\mathsf{L}b\mathsf{L} = \mathsf{O}\mathsf{L} = \mathsf{O}$ using the Tarski rule, which would contradict that a is an atom. Hence there is an atom $c \sqsubseteq a\mathsf{L}b$. Then $c\mathsf{L} \sqcap \mathsf{I} \neq \mathsf{O}$ since otherwise $c\mathsf{L} \sqsubseteq \overline{\mathsf{I}}$, which is equivalent to $c^\mathsf{T} \sqsubseteq \mathsf{O}$, whence $c = \mathsf{O}$. Similarly $\mathsf{L}c \sqcap \mathsf{I} \neq \mathsf{O}$. Moreover $c\mathsf{L} \sqcap \mathsf{I} \sqsubseteq a\mathsf{L}b\mathsf{L} \sqcap \mathsf{I} \sqsubseteq a\mathsf{L} \sqcap \mathsf{I} = a$ and $\mathsf{L}c \sqcap \mathsf{I} \sqsubseteq \mathsf{L}a\mathsf{L}b \sqcap \mathsf{I} \sqsubseteq \mathsf{L}b \sqcap \mathsf{I} = b$ using Lemma 2.9. Hence $c\mathsf{L} \sqcap \mathsf{I} = a$ and $\mathsf{L}c \sqcap \mathsf{I} = b$ since a, b are atoms. Thus $a\mathsf{L}b = (c\mathsf{L} \sqcap \mathsf{I})\mathsf{L}(\mathsf{L}c \sqcap \mathsf{I}) \sqsubseteq c\mathsf{L}\mathsf{L}\mathsf{L}c \sqsubseteq c\mathsf{L}c \sqsubseteq c$ since c is rectangular. It follows that $a\mathsf{L}b = c$ is an atom. □

Theorem 20. *Assume all atoms are rectangular. Then all quotients exist if and only if all subsets exist.*

Proof (of forward implication). Let $S : A \leftrightarrow A$ be a non-zero partial identity. Hence there is an atom $a \sqsubseteq S$. Define $E : A \leftrightarrow A$ by $E = S \sqcup aL\neg S \sqcup \neg SLa \sqcup \neg SL\neg S$. Then E is an equivalence:

- $I = S \sqcup \neg S \sqsubseteq S \sqcup \neg SL\neg S \sqsubseteq E$.
- $E^\mathsf{T} = E$ since S, $\neg S$ and a are partial identities and hence symmetric.
- Since $aS = Sa = a$ and $a\neg S = \neg Sa \sqsubseteq \neg SS = O$ and a is rectangular, we obtain

$$EE = S \sqcup aL\neg S \sqcup aL\neg SLa \sqcup aL\neg SL\neg S \sqcup \neg SLa \sqcup \neg SLaL\neg S \sqcup$$
$$\neg SL\neg SLa \sqcup \neg SL\neg SL\neg S$$
$$\sqsubseteq S \sqcup aL\neg S \sqcup aLa \sqcup \neg SLa \sqcup \neg SL\neg S = E \sqcup aLa = E \sqcup a = E$$

Hence $p : A \leftrightarrow A/E$ exists with $pp^\mathsf{T} = E$ and $p^\mathsf{T}p = I$. We show that A/E is a subset of A corresponding to S. To this end, define $i : A/E \leftrightarrow A$ by $i = p^\mathsf{T}S$. Then

- $ii^\mathsf{T} = p^\mathsf{T}SS^\mathsf{T}p = p^\mathsf{T}Sp \sqsubseteq p^\mathsf{T}p = I$. Conversely, $p^\mathsf{T}p \sqsubseteq p^\mathsf{T}pp^\mathsf{T}Spp^\mathsf{T}p = p^\mathsf{T}Sp$ since $I \sqsubseteq E = pp^\mathsf{T}Spp^\mathsf{T}$ using the Tarski rule:

$$pp^\mathsf{T}Spp^\mathsf{T} = ESE = (S \sqcup \neg SLa)E = S \sqcup aL\neg S \sqcup \neg SLa \sqcup \neg SLaL\neg S = E$$

- $i^\mathsf{T}i = S^\mathsf{T}pp^\mathsf{T}S = SES = S(S \sqcup \neg SLa) = S.$ □

Proof (of backward implication of Theorem 20). Let $E : A \leftrightarrow A$ be an equivalence. Consider the relation $\sim$ on $\mathrm{at}_1(A)$ defined by $a \sim b \Leftrightarrow aLb \sqsubseteq E$. It is an equivalence relation:

- $a \sim a$ since $aLa = a \sqsubseteq I$ using that a is rectangular.
- $a \sim b$ implies $b \sim a$ since $aLb \sqsubseteq E$ implies $bLa = (aLb)^\mathsf{T} \sqsubseteq E^\mathsf{T} = E$.
- $a \sim b$ and $b \sim c$ imply $a \sim c$ since $aLc = aLbLc = aLbbLc \sqsubseteq EE = E$ using the Tarski rule.

Let I be the equivalence classes of $\mathrm{at}_1(A)/\sim$ and let a_i be a representative of class $i \in I$. Define $S : A \leftrightarrow A$ by $S = \bigsqcup_{i \in I} a_i$. Then S is a non-zero partial identity since $\sim$ has at least one class. Hence $i : S \leftrightarrow A$ exists with $i^\mathsf{T}i = S$ and $ii^\mathsf{T} = I$. We show that S is A/E. To this end, define $p : A \leftrightarrow S$ by $p = Ei^\mathsf{T}$. Then

- $pp^\mathsf{T} = Ei^\mathsf{T}iE^\mathsf{T} = ESE \sqsubseteq EE = E$. Conversely, $E = EE \sqsubseteq EESEE = ESE$ since $I \sqsubseteq ESE$. To obtain the latter we show $a \sqsubseteq ESE$ for each $a \in \mathrm{at}_1(A)$. Since $ESE = E(\bigsqcup_{i \in I} a_i)E = \bigsqcup_{i \in I} Ea_iE$ it suffices to show $a \sqsubseteq Ea_iE$ using the representative a_i with $a \sim a_i$. This holds since $a \sqsubseteq aLa = aLa_iLa = aLa_ia_ia_iLa \sqsubseteq Ea_iE$ using the Tarski rule.
- $I = ii^\mathsf{T} \sqsubseteq iEi^\mathsf{T} = iE^\mathsf{T}Ei^\mathsf{T} = p^\mathsf{T}p$. Conversely, we have $iEi^\mathsf{T} = ii^\mathsf{T}iEi^\mathsf{T}ii^\mathsf{T} = iSESi^\mathsf{T} \sqsubseteq ii^\mathsf{T}$ since

$$SES = (\bigsqcup_{i \in I} a_i)E(\bigsqcup_{j \in I} a_j) = \bigsqcup_{i,j \in I} a_iEa_j \sqsubseteq I$$

for which it remains to show $a_i E a_j \sqsubseteq \mathsf{I}$. If $i = j$, then $a_i E a_i \sqsubseteq a_i \mathsf{L} a_i = a_i \sqsubseteq \mathsf{I}$ since a_i is rectangular. If $i \neq j$ we show $a_i E a_j = \mathsf{O}$. This is equivalent to $a_i \mathsf{L} a_j \sqsubseteq \overline{E}$ using Lemma 2.2. Since $a_i \mathsf{L} a_j$ is an atom by Lemma 19, the latter is equivalent to $a_i \mathsf{L} a_j \not\sqsubseteq E$, that is, to $a_i \not\sim a_j$, which holds since a_i and a_j represent different equivalence classes in this case. $\qquad\square$

7 Independence of Domain Constructions

In this section we give models which show that there are no dependences between the studied domain constructions apart from those proved in the previous sections, under the assumption that all atoms are rectangular. All of the following models are (heterogeneous) subalgebras of REL, where the objects are down-closed subsets of the natural numbers $\mathbb{N}$, and the morphisms are all relations between these sets. For $n \in \mathbb{N}$ let $\mathbf{n}$ denote the set $\{0, 1, 2, \ldots, n-1\}$ of numbers smaller than n; for example $\mathbf{1} = \{0\}$, $\mathbf{2} = \{0,1\}$ and $\mathbf{5} = \{0,1,2,3,4\}$.

The following table gives the models or why there are none, for each combination of having all power sets, products, sums and subsets. By Theorem 20 we do not distinguish between subsets and quotients. Due to lack of space we omit proofs that the models have or do not have the indicated domain constructions.

Power	Product	Sum	Subset	Objects
No	No	No	No	$\mathbf{2}$
No	No	No	Yes	$\mathbf{1}, \mathbf{2}$
No	No	Yes	No	No model by Theorem 18
No	No	Yes	Yes	No model by Theorem 18
No	Yes	No	No	$\mathbf{1}, \mathbb{N}$
No	Yes	No	Yes	$\mathbf{1}$
No	Yes	Yes	No	$\mathbb{N}$
No	Yes	Yes	Yes	$\mathbf{1}, \mathbf{2}, \mathbf{3}, \ldots, \mathbb{N}$
Yes	No	No	No	$\mathbf{2^i}, \mathbf{3^i}$ for $i \in \mathbb{N}$
Yes	No	No	Yes	No model by Corollary 13 or Corollary 17
Yes	No	Yes	No	No model by Theorem 18
Yes	No	Yes	Yes	No model by Corollary 17 or Theorem 18
Yes	Yes	No	No	$\mathbf{2^i}$ for $i \in \mathbb{N}$
Yes	Yes	No	Yes	No model by Corollary 13
Yes	Yes	Yes	No	$\mathbf{2}, \mathbf{3}, \mathbf{4}, \ldots$
Yes	Yes	Yes	Yes	$\mathbf{1}, \mathbf{2}, \mathbf{3}, \ldots$

8   Conclusion

We have shown a number of dependences between the domain constructions of power sets, products, sums, quotients and subsets in heterogeneous relation algebras. Some results assumed that objects are comparable or that atoms are rectangular. This raises questions for further study:

- Can sums be constructed without assuming comparability?
- Can products be constructed without assuming rectangular atoms?
- How are subsets and quotients related without assuming rectangular atoms?

The second question refers to products as axiomatised in this paper, which implies representability. A weaker version of relational products that does not imply representability was investigated in [18,19]. These papers also relate the relational product to the categorical product in the subcategory of mappings.

A referee noted that [17] shows 'every (small) heterogeneous relation algebra can be faithfully embedded into a relation algebra that has relational sums and powers so that these constructions can always be generated'. The present paper does not embed into another algebra but shows the existence of domain constructions within a heterogeneous relation algebra under certain assumptions.

Acknowledgement. I thank the anonymous referees for pointing out related works and other helpful comments.

References

1. Adámek, J., Herrlich, H., Strecker, G.E.: Abstract and Concrete Categories: The Joy of Cats, 2nd edn. Dover Publications (2009)
2. Berghammer, R., Schmidt, G., Zierer, H.: Symmetric quotients and domain constructions. Inf. Process. Lett. **33**(3), 163–168 (1989)
3. Berghammer, R., Winter, M.: Embedding mappings and splittings with applications. Acta Inf. **47**(2), 77–110 (2010)
4. Bird, R., de Moor, O.: Algebra of Programming. Prentice Hall, Hoboken (1997)
5. Desharnais, J.: Abstract relational semantics. Ph.D. thesis, McGill University (1989)
6. Freyd, P.J., Sčedrov, A.: Categories, Allegories, North-Holland Mathematical Library, vol. 39. Elsevier Science Publishers (1990)
7. Givant, S., Halmos, P.: Introduction to Boolean Algebras. Springer, Heidelberg (2009)
8. Guttmann, W.: Multirelations with infinite computations. J. Logical Algebraic Methods Program. **83**(2), 194–211 (2014)
9. Maddux, R.D.: Pair-dense relation algebras. Trans. Am. Math. Soc. **328**(1), 83–131 (1991)
10. Olivier, J.P., Serrato, D.: Catégories de Dedekind: Morphismes transitifs dans les catégories de Schröder. Comptes rendus hebdomadaires des séances de l'Académie des Sciences, Série A **290**, 939–941 (1980)
11. Schmidt, G.: Relational Mathematics. Cambridge University Press, Cambridge (2011)

12. Schmidt, G., Hattensperger, C., Winter, M.: Heterogeneous relation algebra. In: Brink, C., Kahl, W., Schmidt, G. (eds.) Relational Methods in Computer Science, pp. 39–53. Springer, Wien (1997). https://doi.org/10.1007/978-3-7091-6510-2_3

13. Schmidt, G., Ströhlein, T.: Relationen und Graphen. Springer, Heidelberg (1989)

14. Schmidt, G., Winter, M.: Relational Topology. LNM, vol. 2208. Springer, Cham (2018). https://doi.org/10.1007/978-3-319-74451-3

15. Tarski, A.: On the calculus of relations. J. Symb. Logic **6**(3), 73–89 (1941)

16. Tarski, A., Givant, S.: A Formalization of Set Theory Without Variables, Colloquium Publications, vol. 41. American Mathematical Society (1987)

17. Winter, M.: Strukturtheorie heterogener Relationenalgebren mit Anwendung auf Nichtdeterminismus in Programmiersprachen. Ph.D. thesis, Universität der Bundeswehr München (1998)

18. Winter, M.: Weak relational products. In: Schmidt, R.A. (ed.) RelMiCS/AKA 2006. LNCS, vol. 4136, pp. 417–431. Springer, Heidelberg (2006). https://doi.org/10.1007/11828563_28

19. Winter, M.: Products in categories of relations. J. Logic Algebraic Program. **76**(1), 145–159 (2008)

20. Zierer, H.: Relation algebraic domain constructions. Theor. Comput. Sci. **87**, 163–188 (1991)

Normal Forms for Elements of the *-continuous Kleene Algebras $K \otimes_{\mathcal{R}} C_2'$

Mark Hopkins[1] and Hans Leiß[2]($\boxtimes$)

[1] The Federation Archive, Munich, Germany
federation2005@netzero.net
[2] Centrum für Informations- und Sprachverarbeitung,
Ludwig-Maximilians-Universität München, Munich, Germany
leiss@cis.uni-muenchen.de, h.leiss@gmx.de
https://github.com/FederationArchive

Abstract. The tensor product $K \otimes_{\mathcal{R}} C_2'$ of the *-continuous Kleene algebra K with the polycyclic *-continuous Kleene algebra C_2' over two bracket pairs contains a copy of the fixed-point closure of K: the centralizer of C_2' in $K \otimes_{\mathcal{R}} C_2'$. As a next step, establishing a calculus for context-free expressions, we prove a representation of elements of $K \otimes_{\mathcal{R}} C_2'$ by automata à la Kleene and refine it by normal form theorems that restrict the occurrences of brackets on paths through the automata.

Keywords: *-continuous Kleene algebra · Polycyclic Kleene algebra · Centralizer · Normal form of automaton · Context-free expression

1 Introduction

A Kleene algebra $K = (K, +, \cdot, {}^*, 0, 1)$ is *-continuous, if

$$a \cdot c^* \cdot b = \sum \{ a \cdot c^n \cdot b \mid n \in \mathbb{N} \}$$

for all $a, b, c \in K$, where $\sum$ is the least upper bound with respect to the natural partial order $\leq$ on K given by $a \leq b$ iff $a + b = b$. Well-known examples of *-continuous Kleene algebras are the algebras $\mathcal{R}M = (\mathcal{R}M, +, \cdot, {}^*, 0, 1)$ of regular subsets of a monoid $M = (M, \cdot^M, 1^M)$, where $0 := \emptyset$, $1 := \{1^M\}$ and $+$ is union, $\cdot$ elementwise product, and * is iteration or "monoid closure", i.e. for $A \in \mathcal{R}M$, A^* is the least $B \supseteq A$ that contains 1^M and is closed under $\cdot^M$.

We will make use of two other kinds of *-continuous Kleene algebras: quotients K/ρ of a *-continuous Kleene algebras K under $\mathcal{R}$-congruences ρ on K, i.e. semiring congruences that behave well with respect to suprema of regular subsets, and tensor products $K \otimes_{\mathcal{R}} K'$ of *-continuous Kleene algebras K, K'.

Let Δ_m be a set of m pairs of "brackets", p_i, q_i, $i < m$, and $\mathcal{R}\Delta_m^*$ the *-continuous Kleene algebra of regular subsets of Δ_m^*. In [3] we considered the $\mathcal{R}$-congruence ρ_m on $\mathcal{R}\Delta_m^*$ generated by the equation set

© The Author(s), under exclusive license to Springer Nature Switzerland AG 2023
R. Glück et al. (Eds.): RAMiCS 2023, LNCS 13896, pp. 122–139, 2023.
https://doi.org/10.1007/978-3-031-28083-2_8

$$\{ p_i q_j = \delta_{i,j} \mid i,j < m \} \cup \{ q_0 p_0 + \ldots + q_{m-1} p_{m-1} = 1 \} \qquad (1)$$

and the finer $\mathcal{R}$-congruence ρ_m' generated by the equations

$$\{ p_i q_j = \delta_{i,j} \mid i,j < m \}, \qquad (2)$$

where $\delta_{i,j}$ is the Kronecker δ. These $\mathcal{R}$-congruences give rise to the *bra-ket* and the *polycyclic* *-continuous Kleene algebra $C_m = \mathcal{R}\Delta_m^* / \rho_m$ and $C_m' = \mathcal{R}\Delta_m^* / \rho_m'$, respectively. Equations (2) allow us to algebraically distinguish matching brackets, where $p_i q_j = 1$, from non-matching ones, where $p_i q_j = 0$.[1] For $m > 2$, C_m can be coded in C_2 and C_m' in C_2', so we focus on the case $m = 2$.

Two *-continuous Kleene algebras K and C can be combined to a *tensor product* $K \otimes_{\mathcal{R}} C$ which, intuitively, is the smallest common *-continuous Kleene algebra extension of K and C in which elements of K commute with those of C.

In an earlier incarnation of this result, the first author showed that for any *-continuous Kleene algebra K, the tensor product $K \otimes_{\mathcal{R}} C_2$ contains an isomorphic copy of the fixed-point closure of K. In particular, for finite alphabets X, each context-free set $L \subseteq X^*$ is represented in $\mathcal{R}X^* \otimes_{\mathcal{R}} C_2$ as the value of a regular expression over $X \cup \Delta_2$. In fact, the *centralizer of C_2* in $K \otimes_{\mathcal{R}} C_2$, i.e. the set of those elements of $K \otimes_{\mathcal{R}} C_2$ that commute with every element of C_2, consists of exactly the representations of context-free subsets of the multiplicative monoid of K. These results constitute a generalization of the well-known theorem by Chomsky and Schützenberger [1] in formal language theory; they were later shown to hold with the simpler algebra $K \otimes_{\mathcal{R}} C_2'$ instead of $K \otimes_{\mathcal{R}} C_2$ by the second author [6].

It is therefore of some interest to understand the structure of $K \otimes_{\mathcal{R}} C_2$ and $K \otimes_{\mathcal{R}} C_2'$. In this article, we show that some unpublished results and conjectures on $K \otimes_{\mathcal{R}} C_2$ by the first author, in particular the existence of normal forms, hold for the simpler algebra $K \otimes_{\mathcal{R}} C_2'$.

Section 2 recalls the definitions of *-continuous Kleene algebras and $\mathcal{R}$-dioids, bra-ket and polycyclic *-continuous Kleene algebras, and the tensor product of two $\mathcal{R}$-dioids. We then show a Kleene representation theorem, i.e. that each element φ of $K \otimes_{\mathcal{R}} C_2'$ is the value $L(\mathcal{A}) = SA^*F$ of an automaton $\mathcal{A} = \langle S, A, F \rangle$, where $S \in \{0,1\}^{1 \times n}$ resp. $F \in \{0,1\}^{n \times 1}$ code the set of initial resp. accepting states of the n states of $\mathcal{A}$ and $A \in Mat_{n,n}(K \otimes_{\mathcal{R}} C_2')$ is a transition matrix.

Section 3 refines the representation $\varphi = L(\mathcal{A})$ to a "normal form" where brackets on paths through the automaton $\mathcal{A}$ occur "mostly" in a balanced way. Section 3.1 identifies, in any *-continuous Kleene algebra with elements u, x, v, the value $(u + x + v)^*$ with the value $(Nv)^*N(uN)^*$, provided the algebra has a least solution N of the inequation $y \geq (x + uyv)^*$ defining Dyck's language $D(x) \subseteq \{u,x,v\}^*$ with "bracket" pair u, v. We then show that for any *-continuous Kleene algebra K and $n \geq 1$, $Mat_{n,n}(K \otimes_{\mathcal{R}} C_2')$ has such a solution N, for matrices U of 0's and opening brackets from C_2', X of elements of K and V of 0's and closing brackets from C_2', and that entries of N belong to the centralizer of C_2' in $K \otimes_{\mathcal{R}} C_2'$.

---

[1] In $\mathcal{R}\Delta_m^*$, elements of Δ_m^* are interpreted by their singleton sets, 0 by the empty set.

Section 3.2 refines the representation $\varphi = L(\mathcal{A})$ to the sketched "normal form": the transition matrix A can be split as $A = (U + X + V)$ into transitions $X \in K^{n \times n}$ by elements of K, transitions $U \in \{0, b, p\}^{n \times n}$ by opening brackets b, p, and transitions $V \in \{0, d, q\}^{n \times n}$ by closing brackets d, q of C'_2. Then A^* can be normalized to $(NV)^* N(UN)^*$, where N is balanced in U and V and all other occurrences of closing brackets V are in front of all other occurrences of opening brackets U. The Conclusion draws relations to previous work and mentions an open problem. An Appendix sketches how to combine normal forms by regular operations, which forms the core of a calculus of context-free expressions.

2   *-continuous Kleene Algebras and $\mathcal{R}$-diods

A *Kleene algebra* is an idempotent semiring or dioid $(K, +, \cdot, 0, 1)$ with a unary operation $* : K \to K$ such that for all $a, b \in K$

$$a \cdot a^* + 1 \leq a^* \wedge \forall x(a \cdot x + b \leq x \to a^* \cdot b \leq x),$$
$$a^* \cdot a + 1 \leq a^* \wedge \forall x(x \cdot a + b \leq x \to b \cdot a^* \leq x).$$

The boolean algebra $\mathbb{B} = (\{0, 1\}, +, \cdot, *, 0, 1)$ with boolean addition and multiplication and $*$ given by $0^* = 1^* = 1$ is a Kleene subalgebra of K.

A Kleene algebra $K = (K, +, \cdot, *, 0, 1)$ is *-*continuous*, if

$$a \cdot c^* \cdot b = \sum \{a \cdot c^n \cdot b \mid n \in \mathbb{N}\}$$

for all $a, b, c \in K$, where $\sum$ is the least upper bound with respect to the natural partial order $\leq$ on K given by $a \leq b$ iff $a + b = b$. Well-known *-continuous Kleene algebras are the algebras $\mathcal{R}M = (\mathcal{R}M, +, \cdot, *, 0, 1)$ of regular subsets of monoids $M = (M, \cdot^M, 1^M)$, where $0 := \emptyset$, $1 := \{1^M\}$ and for $A, B \in \mathcal{R}M$,

$$A + B = A \cup B, \qquad A \cdot B = \{a \cdot^M b \mid a \in A, b \in B\},$$
$$A^* = \bigcup \{A^n \mid n \in \mathbb{N}\} \qquad \text{with } A^0 = 1, A^{n+1} = A \cdot A^n.$$

If K is an idempotent semiring $(K, +^K, \cdot^K, 0^K, 1^K)$ or a Kleene algebra, by $\mathcal{R}K$ we mean the Kleene algebra $\mathcal{R}M$ of its multiplicative monoid $M = (K, \cdot^K, 1^K)$.

An $\mathcal{R}$-*dioid* is a dioid $K = (K, +^K, \cdot^K, 0^K, 1^K)$ where each $A \in \mathcal{R}K$ has a least upper bound $\sum A \in K$, and where $\sum(AB) = (\sum A)(\sum B)$ for all $A, B \in \mathcal{R}K$. It can be expanded to a *-continuous Kleene algebra by putting $c^* := \sum \{c\}^*$ for $c \in K$. It is *-continuous by

$$a \cdot c^* \cdot b = \left(\sum \{a\}\right) \cdot \left(\sum \{c\}^*\right) \cdot \left(\sum \{b\}\right) = \sum (\{a\}\{c\}^*\{b\}).$$

Conversely, the dioid reduct of a *-continuous Kleene algebra is an $\mathcal{R}$-dioid, since, by induction, every regular set C has a least upper bound $\sum C \in K$ satisfying $a \cdot (\sum C) \cdot b = \sum(aCb)$, which implies the $\mathcal{R}$-distributivity property $\sum(AB) = (\sum A)(\sum B)$ for $A, B, \in \mathcal{R}K$ (see [2]).

The *-continuous Kleene algebras, with Kleene algebra homomorphisms (semiring homomorphisms that preserve *), form a category. In the following, we often use the shorter term $\mathcal{R}$-dioids, cf. [2,3], and write $\mathbb{D}\mathcal{R}$ for this category; $\mathbb{D}$ stands for the category of dioids or idempotent semirings. The $\mathcal{R}$-dioids of the form $\mathcal{R}M$ with monoid M form the Kleisli subcategory of $\mathbb{D}\mathcal{R}$.

2.1   The Bra-ket and Polycyclic $\mathcal{R}$-diods

We will make use of two kinds of $\mathcal{R}$-dioids which do not belong to the Kleisli subcategory, but are quotients of the regular sets $\mathcal{R}\Delta^*$ by suitable $\mathcal{R}$-congruence relations ρ on $\mathcal{R}\Delta^*$, where Δ is an alphabet of "bracket" pairs.

Let ρ be a dioid-congruence on an $\mathcal{R}$-dioid D. The set D/ρ of congruence classes is a dioid under the operations defined by $(d/\rho)(d'/\rho) := (dd')/\rho$, $1 := 1/\rho$, $d/\rho + d'/\rho := (d+d')/\rho$, $0 := 0/\rho$. Let $\leq$ be the partial order on D/ρ derived from $+$. For $U \subseteq D$, put $U/\rho := \{ d/\rho \mid d \in U \}$ and

$$(U/\rho)^{\downarrow} = \{ e/\rho \mid e/\rho \leq d/\rho \text{ for some } d \in U, e \in D \}.$$

An $\mathcal{R}$-*congruence* on D is a dioid-congruence ρ on D such that for all $U, U' \in \mathcal{R}D$, if $(U/\rho)^{\downarrow} = (U'/\rho)^{\downarrow}$, then $(\sum U)/\rho = (\sum U')/\rho$.

Proposition 1. *If D is an $\mathcal{R}$-dioid and ρ an $\mathcal{R}$-congruence on D, then D/ρ is an $\mathcal{R}$-dioid. For every $R \subseteq D \times D$ there is a least $\mathcal{R}$-congruence $\rho \supseteq R$ on D.*

Proof. See [3].

Let $\Delta_m = P_m \,\dot\cup\, Q_m$ be a set of m "opening brackets" $P_m = \{ p_i \mid i < m \}$ and m "closing brackets" $Q_m = \{ q_i \mid i < m \}$, with $P_m \cap Q_m = \emptyset$. The *bra-ket* $\mathcal{R}$-*dioid* C_m is the quotient $\mathcal{R}\Delta_m^*/\rho_m$ of $\mathcal{R}\Delta_m^*$ by the $\mathcal{R}$-congruence ρ_m generated by the relations

$$\{ p_i q_j = \delta_{i,j} \mid i, j < m \} \cup \{ q_0 p_0 + \ldots + q_{m-1} p_{m-1} = 1 \}.$$

Remarkably, C_m is isomorphic to its own matrix semiring $Mat_{m,m}(C_m)$, with $a \mapsto (p_i a q_j)$ in one and $A \mapsto \sum_{i,j<m} q_i A_{i,j} p_j$ in the other direction.

The *polycyclic* $\mathcal{R}$-*dioid* C_m' is the quotient $C_m' = \mathcal{R}\Delta_m^*/\rho_m'$ of $\mathcal{R}\Delta_m^*$ by the $\mathcal{R}$-congruence ρ_m' generated by the relations

$$\{ p_i q_j = \delta_{i,j} \mid i, j < m \}, \tag{3}$$

where $\delta_{i,j}$ is the Kronecker δ. The latter equations allow us to algebraically distinguish matching brackets, where $p_i q_j = 1$, from non-matching ones, where $p_i q_j = 0$. While the "completeness condition"[2] $1 = \sum_{i<m} q_i p_i$ is a semiring

---

[2] The name is explained in terms of "complete" prefix/suffix codings in [6], Remark 7. It can also be understood in matrix terms: if p_i is the i-th unit (column) vector of size m and q_i its transpose, then $\sum_{i<m} q_i p_i$ is the $m \times m$ unit matrix 1. C_m arises by way of analogy with the bra-ket algebra in physics. We will not use the equation.

equation, the match relations $p_i q_j = \delta_{i,j}$ can be interpreted in monoids with an annihilating element 0. The *polycyclic monoid* P'_m of m generators is the quotient of $(\Delta_m \cup \{0\})^*$ by the monoid congruence σ_m generated by

$$\{\, p_i q_j = \delta_{i,j} \mid i, j < m \,\} \cup \{\, x0 = 0 \mid x \in \Delta_m \cup \{0\} \,\} \cup \{\, 0x = 0 \mid x \in \Delta_m \,\}.$$

Each element $w \in (\Delta_m \cup \{0\})^*$ has a *normal form* $nf(w) \in Q_m^* P_m^* \cup \{0\}$, obtained by using the equations to shorten w, that represents $w/\sigma_m \in P'_m$. Hence,

$$P'_m \simeq (Q_m^* P_m^* \cup \{0\}, \cdot, 1) \quad \text{with } v \cdot w = nf(vw).$$

This can be extended from P'_m to monoid extensions $P'_m[X]$ of P'_m in which elements of X are supposed to commute with elements of P'_m. Formally, let $Y = \Delta_m \cup \{0\} \cup X$ and $P'_m[X]$ the quotient Y^*/τ_m under the congruence τ_m generated by (i) the matching rules $\{\, p_i q_j = \delta_{i,j} \mid i, j < m \,\}$, (ii) the annihilation rules $y0 = 0$ and $0y = 0$ for $y \in Y$, and (iii) the commutation rules $\{\, xd = dx \mid x \in X, d \in \Delta_m \,\}$. Since Y^* can be disjointly decomposed into

$$Y^* = Y^*(P_m X \cup P_m Q_m \cup X Q_m) Y^* \cup Q_m^* X^* P_m^* \cup \{0\},$$

a normal form $nf(w) \in Q_m^* X^* P_m^* \cup \{0\}$ for strings $w \in Y^*$ can be obtained: use the commutation rules to move opening brackets $p_i \in P_m$ to the right and closing brackets $q_i \in Q_m$ to the left of elements of X^*, then use the matching rules to shorten $u p_i q_j v$ to uv or $u0v$ and the annihilation rules to replace $u0v$ by 0, and repeat this process. I.e. for $i, j < m, i \neq j$ and $x \in X, u, v \in Y^*$ we put

$$nf(u p_i x v) := nf(u x p_i v), \quad nf(u0v) := 0, \quad nf(u p_i q_i v) := nf(uv),$$
$$nf(u x q_i v) := nf(u q_i x v), \quad nf(1) := 1, \quad nf(u p_i q_j v) := 0.$$

We leave it to the readers to convince themselves that this amounts to a confluent rewriting system, so that $nf : Y^* \to Q_m^* X^* P_m^* \cup \{0\}$ is well-defined, and that

$$P'_m[X] \simeq (Q_m^* X^* P_m^* \cup \{0\}, \cdot, 1), \quad \text{where } u \cdot v := nf(uv).$$

The polycyclic $\mathcal{R}$-dioid C'_m can be seen as the quotient of $\mathcal{R} P'_m$ modulo the $\mathcal{R}$-congruence $\langle\langle \{0\} = 0 \rangle\rangle$ generated by identifying $\{0\}$ with the empty set. Hence, the elements of C'_m are represented by sets $A \setminus \{0\} \subseteq Q_m^* P_m^*$ with $A \in \mathcal{R} P'_m$.

   The normal form nf on $P'_m[X]$ is the motivating idea behind the normal form theorem (Theorem 5) for elements of the tensor product $\mathcal{R} X^* \otimes_{\mathcal{R}} C'_m$ to be introduced in the next section. On the tensor product, regular sets $A \in \mathcal{R} X^*$ and (congruence classes of) regular sets $B \in \mathcal{R} \Delta_m$ commute with each other, and the tensor product is an $\mathcal{R}$-dioid structure, not just a monoid structure.

   Notice that $m \geq 2$ bracket pairs can be coded by two, say b, d and p, q, by putting $p_i := b p^i$ and $q_i := q^i d$ for $i < m$, which extends to an embedding of C'_m in C'_2. We therefore formulate some results only for $m = 2$, using $\Delta_2 = P_2 \cup Q_2$ with $P_2 = \{b, p\}$ and $Q_2 = \{d, q\}$, unless stated otherwise.

2.2   The Tensor Product $K \otimes_{\mathcal{R}} C$ of $\mathcal{R}$-dioids K and C

In a category whose objects have a monoid structure, a *tensor product* of two objects M_1 and M_2 is an object $M_1 \otimes M_2$ with two relatively commuting morphisms $\top_1 : M_1 \rightarrow M_1 \otimes M_2 \leftarrow M_2 : \top_2$ such that for any pair $f : M_1 \rightarrow M \leftarrow M_2 : g$ of relatively commuting morphisms there is a unique morphism $h_{f,g} : M_1 \otimes M_2 \rightarrow M$ with $f = h_{f,g} \circ \top_1$ and $g = h_{f,g} \circ \top_2$. Intuitively, the tensor product $M_1 \otimes M_2$ is the free extension of M_1 and M_2 in which elements of M_1 commute with those of M_2.

In the category of monoids, $M_1 \otimes M_2$ is the cartesian product $M_1 \times M_2$ with componentwise unit and product, and $h_{f,g}(m_1, m_2) = f(m_1) \cdot g(m_2)$. The category $\mathbb{DR}$ of *-continuous Kleene algebras also has tensor products:

Theorem 1. *Let K_1, K_2 be $\mathcal{R}$-dioids and M_1, M_2 their monoid reducts. The tensor product of K_1, K_2 is*

$$K_1 \otimes_{\mathcal{R}} K_2 := \mathcal{R}(M_1 \times M_2)/_{\equiv},$$

the quotient of the regular sets $\mathcal{R}(M_1 \times M_2)$ of the monoid product $M_1 \times M_2$ by the $\mathcal{R}$-congruence $\equiv$ generated by the "tensor product" equations

$$\{\{(\textstyle\sum A, \sum B)\} = A \times B \mid A \in \mathcal{R}M_1, B \in \mathcal{R}M_2\}.$$

The embeddings $\top_1 : K_1 \rightarrow K_1 \otimes_{\mathcal{R}} K_2 \leftarrow K_2 : \top_2$ are $\top_1(a) := \{(a,1)\}/_{\equiv}$ for $a \in K_1$ and $\top_2(b) = \{(1,b)\}/_{\equiv}$ for $b \in K_2$. For a pair of commuting $\mathcal{R}$-morphisms $f : K_1 \rightarrow K \leftarrow K_2 : g$ to an $\mathcal{R}$-dioid K, the induced map is

$$h_{f,g}(R/_{\equiv}) := \textstyle\sum\{f(a)g(b) \mid (a,b) \in R\}, \quad R \in \mathcal{R}(M_1 \times M_2).$$

Proof. See [3], Theorem 4.

Concerning the definition of $\equiv$, we remark that with $A \in \mathcal{R}M_1$ and $B \in \mathcal{R}M_2$,

$$A \times B = \bigcup\{\{(a,b)\} \mid a \in A, b \in B\} \in \mathcal{R}(M_1 \times M_2).$$

In the following, for $\mathcal{R}$-dioids K_1, K_2, we also write $K_1 \times K_2$ for the product of their underlying monoids, and for $R \in \mathcal{R}(K_1 \times K_2)$, write $[R]$ instead of $R/_{\equiv}$. For $R, S \in \mathcal{R}(K_1 \times K_2)$, one has $[R] + [S] = [R \cup S]$, $[R][S] = [RS]$, and

$$[R]^* = \textstyle\sum\{[R]^n \mid n \in \mathbb{N}\} = \sum\{[R^n] \mid n \in \mathbb{N}\} = [\bigcup\{R^n \mid n \in \mathbb{N}\}] = [R^*].$$

We will mainly consider tensor products $K \otimes_{\mathcal{R}} C$ where $K = \mathcal{R}X^*$ and C is a polycyclic $\mathcal{R}$-dioid C_m' (or bra-ket $\mathcal{R}$-dioid C_m). In a monoid M, the *centralizer* $Z_C(M)$ of a set $C \subseteq M$ in M consists of those elements that commute with every element of C, i.e. the submonoid

$$Z_C(M) := \{m \in M \mid mc = cm \text{ for all } c \in C\}.$$

For example, the centralizer of Δ_m in $P_m'[X]$ is $X^* \cup \{0\}$.

The interest in Kleene algebras $\mathcal{R}X^* \otimes_\mathcal{R} C_2'$ comes from the fact that $\mathcal{C}X^*$, the algebra of context-free languages over X, is isomorphic to $Z_{C_2'}(\mathcal{R}X^* \otimes_\mathcal{R} C_2')$ cf. [6]. Since all elements of $\mathcal{R}X^* \otimes_\mathcal{R} C_2'$ can be denoted by regular expressions over $X \mathbin{\dot\cup} \Delta_2$, every *context-free* set $L \subseteq X^*$ is the value of a *regular* expression.

Example 1. Suppose $a, b \in X$ and write $\langle 0|, |0\rangle$ and $\langle 1|, |1\rangle$ for the two bracket pairs of Δ_2. Then $L = \{\, a^n b^n \mid n \in \mathbb{N} \,\} \in \mathcal{C}X^*$ is represented in $\mathcal{R}X^* \otimes_\mathcal{R} C_2'$ by the value of the regular expression $r_L := \langle 0|(a\langle 1|)^*(|1\rangle b)^*|0\rangle$ over $X \mathbin{\dot\cup} \Delta_2$:

$$
\begin{aligned}
r_L &= \textstyle\sum\{\, \langle 0|(a\langle 1|)^n(|1\rangle b)^m|0\rangle \mid n, m \in \mathbb{N} \,\} && \text{(*-continuity)} \\
&= \textstyle\sum\{\, a^n \langle 0|\langle 1|^n|1\rangle^m|0\rangle b^m \mid n, m \in \mathbb{N} \,\} && \text{(relative commutativity)} \\
&= \textstyle\sum\{\, a^n b^n \mid n \in \mathbb{N} \,\} && \text{(bracket match } \langle i||j\rangle = \delta_{i,j}).
\end{aligned}
$$

By Theorem 17 of [6], $L \mapsto \sum L$ embeds $\mathcal{C}X^*$ in $\mathcal{R}X^* \otimes_\mathcal{R} C_2'$.   ◁

2.3 Automata over a Kleene Algebra

A *finite automaton* $\mathcal{A} = \langle S, A, F\rangle$ *with n states* over a Kleene algebra K consists of a transition matrix $A \in Mat_{n,n}(K)$ and two vectors $S \in Mat_{1,n}(\mathbb{B})$ and $F \in Mat_{n,1}(\mathbb{B})$, coding the initial and final states. The 1-step transitions from state $i < n$ to state $j < n$ are represented by $A_{i,j}$, and paths from i to j of finite length by $A_{i,j}^*$, where A^* is the iteration of A. The sum of paths leading from initial to final states defines an element of K,

$$ L(\mathcal{A}) = S \cdot A^* \cdot F \ \in K. $$

The iteration M^* of $M \in Mat_{n,n}(K)$ is defined by induction on n: for $n = 1$ and $M = (k)$, $M^* = (k^*)$, and for $n > 1$,

$$ M^* = \begin{pmatrix} A & B \\ C & D \end{pmatrix}^* = \begin{pmatrix} F^* & F^*BD^* \\ D^*CF^* & D^*CF^*BD^* + D^* \end{pmatrix}, \tag{4} $$

where $F = A + BD^*C$ and $M = \begin{pmatrix} A & B \\ C & D \end{pmatrix}$ is any splitting of M in which A and D are square matrices of dimensions $n_1, n_2 < n$ with $n = n_1 + n_2$.

Theorem 2. *If K is a *-continuous Kleene algebra, so is $Mat_{n,n}(K)$, for $n \geq 1$.*

Proof. See [4], Chap. 7.1.   □

By Kleene's representation theorem, the set $\mathcal{R}X^*$ of regular subsets of X^* consists of the languages

$$ L(\mathcal{A}) = S \cdot A^* \cdot F \subseteq X^* $$

of finite automata $\mathcal{A} = \langle S, A, F\rangle$, where for some $n \in \mathbb{N}$, $A \in Mat_{n,n}(\mathcal{F}X^*)$, $S \in \mathbb{B}^{1\times n}$, $F \in \mathbb{B}^{n\times 1}$ and $\mathcal{F}X^*$ is the set of finite subsets of X^*.

For $\mathcal{R}$-dioid K, we next prove Kleene's representation theorem for $K \otimes_\mathcal{R} C_2'$: an element of $K \otimes_\mathcal{R} C_2'$ is the "language" $L(\mathcal{A}) = SA^*F$ of a finite automaton $\mathcal{A} = \langle S, A, F\rangle$ over $K \otimes_\mathcal{R} C_2'$. For $a \in K$ and $c \in C_2'$, we write a and c also for their images in $K \otimes_\mathcal{R} C_2'$, likewise ac for $[\{(a,c)\}]$, the product of their images.

Theorem 3. *Let K be an $\mathcal{R}$-dioid, i.e. a *-continuous Kleene-algebra, and C_2' the polycyclic Kleene algebra over $\Delta_2 = P_2 \cup Q_2$ with $P_2 = \{b, p\}$, $Q_2 = \{d, q\}$. For each $\varphi \in K \otimes_{\mathcal{R}} C_2'$ there are $n \in \mathbb{N}$, $S \in \mathbb{B}^{1 \times n}$, $F \in \mathbb{B}^{n \times 1}$, $U \in \{0, b, p\}^{n \times n}$, $V \in \{0, d, q\}^{n \times n}$ and $X \in K^{n \times n}$ such that*

$$\varphi = S(U + X + V)^* F.$$

Proof. Since $\varphi = [R]$ for some $R \in \mathcal{R}(K \times C_2')$, by induction on the construction of R we build an automaton $\mathcal{A}_R = \langle S, A, F \rangle$ over $K \otimes_{\mathcal{R}} C_2'$ such that $L(\mathcal{A}_R) = [R]$ and A splits as $U + X + V$ as in the claim.

- We leave the cases $R = \emptyset$, $R = R_1 \cup R_2$ and $R = R_1 \cdot R_2$ to the reader.
- $R = \{(k, c)\}$ with $k \in K, c \in C_2'$: Let $\mathcal{A}_R = \langle S, A, F \rangle$ consist of

$$S = (1 \; 0), \quad A = \begin{pmatrix} 1 & kc \\ 0 & 1 \end{pmatrix}, \quad F = \begin{pmatrix} 0 \\ 1 \end{pmatrix}.$$

Then $A^* = A$, since $A^0 \leq A = A^2$, hence $L(\mathcal{A}_R) = A_{1,2} = kc = [\{(k, c)\}]$. For the splitting of A into $U + X + V$, $\{(k, c)\} = \{(k, 1)\} \cdot \{(1, c)\}$ is a product of singleton factors, so we need only consider the cases $k = 1$ and $c = 1$. For $R = \{(1, c)\}$, we can further reduce the representing set $A \in \mathcal{R}\Delta_2^*$ of c to a singleton containing an element of Δ_2. If $c \in Q_2 = \{d, q\}$, split

$$A = \begin{pmatrix} 1 & c \\ 0 & 1 \end{pmatrix} = \begin{pmatrix} 0 & 0 \\ 0 & 0 \end{pmatrix} + \begin{pmatrix} 1 & 0 \\ 0 & 1 \end{pmatrix} + \begin{pmatrix} 0 & c \\ 0 & 0 \end{pmatrix} = U + X + V.$$

If $c \in P_2 = \{b, p\}$, switch the roles of U and V. For $R = \{(k, 1)\}$, keep $k1$ in the matrix X.

- $R = R_1^*$: Suppose $\mathcal{A}_{R_1} = (S_1, A_1, F_1)$, is an automaton such that

$$L(\mathcal{A}_{R_1}) = S_1 A_1^* F_1 = [R_1].$$

Let $\mathcal{A}_{R^+} = \langle S, A, F \rangle$ be $\langle S_1, A_1 + F_1 S_1, F_1 \rangle$. By equalities in Kleene algebras,

$$\begin{aligned} L(\mathcal{A}_{R^+}) &= S_1 (A_1 + F_1 S_1)^* F_1 \\ &= S_1 A_1^* (F_1 S_1 A_1^*)^* F_1 \\ &= S_1 A_1^* F_1 (S_1 A_1^* F_1)^* \\ &= [R_1][R_1]^* \\ &= [R_1][R_1^*] = [R_1^+], \end{aligned}$$

and then put $\mathcal{A}_{R^*} = \mathcal{A}_{1+R^+}$. The splitting $A = U + X + V$ is obtained from the splitting of A_1 by adding entries of $F_1 S_1$ to X.   □

3 Normal Form Theorems for $K \otimes_{\mathcal{R}} C_2'$ with $\mathcal{R}$-dioid K

In the representation of elements φ of $K \otimes_{\mathcal{R}} C_2'$ by $\varphi = L(\mathcal{A}) = SA^*F$ for automata $\mathcal{A} = \langle S, A, F \rangle$ with $A = U + X + V$ in Theorem 3, $A^* = (U + X + V)^*$ admits arbitrary sequences of opening brackets U with closing brackets V. We aim at a normal form for $(U + X + V)^*$ where brackets are mainly occurring in a balanced way. To this end, we now look at ways to express a Dyck-language with a single bracket pair u, v in a *-continuous Kleene algebra.

3.1 Least Solutions in *-continuous Kleene Algebras

We first show that in any *-continuous Kleene algebra K, least solutions of two fixed-point inequations that might be used to define the Dyck-language $D_1(X)$ with $X = \{x_1, \ldots, x_n\} \subseteq K$, namely

$$y \geq (x_1 + \ldots + x_n + uyv)^* \quad \text{and} \quad y \geq 1 + x_1 + \ldots + x_n + uyv + yy,$$

are related, where $u, v \in K \setminus X$ represent a pair of brackets. It is then shown that $(u + X + v)^* = (Nv)^*N(uN)^*$, where $N \in K$ is the least solution of $y \geq (X + uyv)^*$ corresponding to $D(X)$. Except for the balanced bracket occurrences in N, all occurrences of the closing bracket v are to the left of all occurrences of the opening bracket u. This is similar to the normal form $nf(w) \in Q_1^* P_1^* \cup \{0\}$ in the polycyclic monoid P_1' (of Sect. 2.1) with $P_1 = \{u\}$ and $Q_1 = \{v\}$, i.e. the normal forms on $\{u, v\}^*$ modulo the congruence generated by $uv = 1$.

Proposition 2. *Let K be a Kleene algebra, and $u, x, v \in K$. If $z \geq x + uz^*v$ has a least solution Z, then $y \geq (x + uyv)^*$ has a least solution, namely Z^*. If $y \geq (x + uyv)^*$ has a least solution N, then $z \geq x + uz^*v$ has a least solution, namely $x + uNv$. If they exist, then $N \geq Z$.*

Proof. By monotonicity of $+, \cdot, ^*$, one has:

- (i) If $z \geq x + uz^*v$ for $z \in K$, then z^* solves $y \geq (x + uyv)^*$,
- (ii) If $y \geq (x + uyv)^*$ for $y \in K$, then $(x + uyv)$ solves $z \geq x + uz^*v$.

Therefore, if Z is the least solution of $z \geq x + uz^*v$, and y is any solution of $y \geq (x + uyv)^*$, then by (ii), $y \geq (x + uyv)^* \geq (x + uyv) \geq Z^*$, and by (i), $Z^* \geq (x + uZ^*v)^*$, so Z^* is the least solution of $y \geq (x + uyv)^*$. On the other hand, if N is the least solution of $y \geq (x + uyv)^*$ and z is any solution of $z \geq x + uz^*v$, then by (i), $z^* \geq N$, so $z \geq x + uNv$, and since by (ii), $(x + uNy)$ is a solution of $z \geq x + uz^*v$, it is the least solution. □

Proposition 3. *Let K be a Kleene algebra and $u, x, v \in K$. If $y \geq (x + uyv)^*$ has a least solution N, then $y \geq 1 + x + uyv + yy$ has a least solution, namely $(x + uNv)^*$, and $N = (x + uNv)^*$. If $y \geq 1 + x + uyv + yy$ has a least solution D, then D is the least solution of $y \geq (x + uyv)^*$.*

Proof. We show: (i) if $z \geq 1 + x + uzv + zz$ for $z \in K$, then z is a solution of $y \geq (x + uyv)^*$ and (ii) if $y \geq (x + uyv)^*$ for $y \in K$, then $(x + uyv)^*$ is a solution of $z \geq 1 + x + uzv + zz$. For (i), assume $z \geq 1 + x + uzv + zz$. From $1 + zz \leq z$, we have $z^* \leq z$. Using $(x + uzv) \leq z$ and monotonicity of *,

$$(x + uzv)^* \leq z^* \leq z,$$

so z is a solution of $y \geq (x + uyv)^*$. For (ii), assume $y \geq (x + uyv)^*$. By monotonicity,

$$u(x + uyv)^*v \leq uyv \leq x + uyv \leq (x + uyv)^*,$$

and obviously, $1 + x \leq (x + uyv)^*$ and $(x + uyv)^*(x + uyv)^* \leq (x + uyv)^*$. So $(x + uyv)^*$ is a solution of $z \leq 1 + x + uzv + zz$.

It follows that if N is the least solution of $y \geq (x + uyv)^*$, then by (ii), $(x + uNv)^*$ is a solution of $z \geq 1 + x + uzv + zz$, and by (i), any $z \in K$ with $z \geq 1 + x + uzv + zz$ satisfies $z \geq N \geq (x + uNv)^*$, so $(x + uNv)^*$ is the least solution of $z \geq 1 + x + uzv + zz$. In particular, $N = (x + uNv)^*$. If D is the least solution of $y \geq 1 + x + uyv + yy$, then by (i), $D \geq (x + uDv)^*$. If $y \geq (x + uyv)^*$, then by (ii), $(x + uyv)^*$ is a solution of $z \geq 1 + x + uzv + zz$, so $y \geq (x + uyv)^* \geq D$.   □

Theorem 4. Let K be a Kleene algebra and $x, u, v \in K$. If $y \geq (x + uyv)^*$ has a least solution N in K, then $(u + x + v)^* = (Nv)^*N(uN)^*$.

Proof. Let $N = \mu y.(x + uyv)^*$ and $n = (u + x + v)^*$. We first show $N \leq n$, by showing that n solves $(x + uyv)^* \leq y$. By monotonicity of $+, \cdot,$ and *,

$$x + un^*v \leq n + nn^*n = (1 + nn^*)n = n^*n \leq n^*,$$

hence $(x + un^*v)^* \leq n^{**} = n^*$. So $N \leq n$, from which

$$(Nv)^*N(uN)^* \leq (u + x + v)^*$$

follows using $u, v, N \leq n = nn = n^*$.

Now consider the reverse inequality, $(u + x + v)^* \leq (Nv)^*N(uN)^*$: As $(x + uNv)^* = N$, we have $(x + uNv)N + 1 \leq N$. With this we show that $(Nv)^*N(uN)^*$ solves $(u + x + v)z + 1 \leq z$ in z:

$$
\begin{aligned}
&(u + x + v)(Nv)^*N(uN)^* + 1 \\
&= (u + x + v)N(vN)^*(uN)^* + 1 \\
&= uN(vN)^*(uN)^* + xN(vN)^*(uN)^* + vN(vN)^*(uN)^* + 1 \\
&= uN(1 + vN(vN)^*)(uN)^* + xN(vN)^*(uN)^* + vN(vN)^*(uN)^* + 1 \\
&= uN(uN)^* + uNvN(vN)^*(uN)^* + xN(vN)^*(uN)^* + vN(vN)^*(uN)^* + 1 \\
&= (x + uNv)N(vN)^*(uN)^* + uN(uN)^* + vN(vN)^*(uN)^* + 1 \\
&= (x + uNv)N(vN)^*(uN)^* + (1 + vN(vN)^*)(uN)^* \\
&= (x + uNv)N(vN)^*(uN)^* + (vN)^*(uN)^* \\
&= ((x + uNv)N + 1)(vN)^*(uN)^* \\
&\leq N(vN)^*(uN)^* \\
&= (Nv)^*N(uN)^*.
\end{aligned}
$$

Since $(u + x + v)^*$ is the least solution of $(u + x + v)z + 1 \leq z$, the claim $(u + x + v)^* \leq (Nv)^*N(uN)^*$ is shown.   □

3.2 Normal Form Theorems

Let $\mathcal{A} = \langle S, A, F \rangle$ be an automaton with $A = U + X + V$ as in Theorem 3, representing an element $\varphi = L(\mathcal{A}) = SA^*F$ of $K \otimes_{\mathcal{R}} C_2'$. We first show that there is a least solution of $y \geq (UyV + X)^*$ in $Mat_{n,n}(K \otimes_{\mathcal{R}} C_2')$, which is related to Dyck's context-free language $D \subseteq \{U, X, V\}^*$ of balanced strings of matrices, with U as "opening bracket" and V as "closing bracket".

Lemma 1. *Let K be an $\mathcal{R}$-dioid, $n \in \mathbb{N}$ and $U \in \{0, b, p\}^{n \times n}$, $V \in \{0, d, q\}^{n \times n}$ and $X \in K^{n \times n}$. In $Mat_{n,n}(K \otimes_{\mathcal{R}} C_2')$,*

$$y \geq (UyV + X)^* \tag{5}$$

has a least solution, namely $N := b(Up + X + qV)^*d$, and $N \in (Z_{C_2'}(K \otimes_{\mathcal{R}} C_2'))^{n \times n}$.

When multiplying b, d, p, q with $n \times n$-matrices, we identify them with corresponding diagonal matrices.

Proof. Let D and D' be the Dyck languages over $\{U, X, V\}$ resp. $\{Up, X, qV\}$ with brackets U, V and Up, qV, respectively.

Claim 1. *For each $m \in \mathbb{N}$, $b(Up + X + qV)^m d = \sum(\{U, X, V\}^m \cap D) \in K^{n \times n}$.*

Proof. Notice that every $A \in D$ evaluates in $Mat_{n,n}(K \otimes_{\mathcal{R}} C_2')$ to an element of $K^{n \times n}$. This is clear for $A = 1_n$ and $A = X$, and if $A, B \in K^{n \times n}$, then $AB \in K^{n \times n}$. Finally, consider $A = UBV$ with $B \in K^{n \times n}$. Since elements of K and C_2' commute with each other in $K \otimes_{\mathcal{R}} C_2'$, we have

$$(UBV)_{ij} = \sum_{k,l=1}^{n} U_{ik}(B_{kl}V_{lj}) = \sum_{k,l=1}^{n} B_{kl}(U_{ik}V_{lj}),$$

and since $U_{ik} \in \{0, b, p\}$ and $V_{lj} \in \{0, d, q\}$, we obtain $U_{ik}V_{lj} \in \{0, 1\}$, hence $(UBV)_{ij} \in K$, and so $A \in K^{n \times n}$. It follows that $\sum(\{U, X, V\}^m \cap D) \in K^{n \times n}$.

Let $A' \in D'$ be the matrix obtained from $A \in D$ by replacing factors U by Up and factors V by qV. Then $A' = A$, because $pq = 1$: this is clear for $A = 1_n$ and $A = X$, and if it is true for A, B, then $(UAV)' = UpA'qV = UpAqV = UAV$, since here $A \in D$ belongs to $K^{n \times n}$, and $(AB)' = A'B' = AB$ by induction.

Moreover, if $A \in D \cap \{U, X, V\}^m$, then $b(Up + X + qV)^m d \geq bA'd = bAd = bdA = A$, hence

$$b(Up + X + qV)^m d \geq \sum(\{U, X, V\}^m \cap D).$$

Finally, let $A' \in \{Up, X, qV\}^m$ be a summand of $(Up + X + qV)^m$ that is not obtained from some $A \in \{U, X, V\}^m \cap D$ by this substitution. Then $bA'd = 0$, because $A' \in (D'qV)^*D'(UpD')^* \setminus D'$ and b, d commute with factors from D' (with values in $K^{n \times n}$), so in $bA'd$, b can be moved over factors to the right, until it meets q and gives $bq = 0$, or d can be moved over factors to the left until it meets p and gives $pd = 0$. It follows that $b(Up + X + qV)^m d = \sum(\{U, X, V\}^m \cap D)$. ◁

By *-continuity, Claim 1 implies

$$N = \sum\{D \cap \{U, X, V\}^m \mid m \in \mathbb{N}\} = \sum D.$$

Claim 2. N *is the least solution of* $y \geq (UyV + X)^*$ *in* $Mat_{n,n}(K \otimes_{\mathcal{R}} C_2')$.

Proof. We show that N is the least solution of $y \geq 1 + X + UyV + yy$ and apply Proposition 3. By the previous claim, we get $N \geq X$ and

$$Ub(Up + X + qV)^m dV \leq b(Up + X + qV)^{m+2}d,$$

and so by *-continuity, $N = b(Up + X + qV)^*d \geq UNV$. It remains to show $NN \leq N$. Let $T = (Up + X + qV)$. By *-continuity,

$$NN = \sum_{k \in \mathbb{N}} bT^k dN = \sum_{k,l \in \mathbb{N}} bT^k dbT^l d.$$

By Claim 1, $bT^k d \in K^{n \times n}$ commutes with b, so $bT^k dbT^l d = b(bT^k d)T^l d$, and $bT^k d$ is a sum of products $A \in \{U, X, V\}^k \cap D$ of length k. If $A' \in \{Up, X, qV\}^k \cap D'$ is obtained from A by substituting Up for U and qV for V, then $A = A' \leq T^k$. Hence, $bT^k d \leq T^k$ and $NN = \sum_{k,l \in \mathbb{N}} b(bT^k d)T^l d \leq \sum_{k,l \in \mathbb{N}} bT^k T^l d = N$.

Therefore, N is a solution of $y \geq 1 + X + UyV + yy$. To show that it is the least solution, suppose $y \in Mat_{n,n}(K \otimes_{\mathcal{R}} C_2')$ satisfies $y \geq 1 + X + UyV + yy$. As $N = \sum D$, it is sufficient to show $A \leq y$ for each $A \in D$. This is clear for 1_n and X, and if $A, B \in D$ satisfy $A, B \leq y$, then $UAV \leq UyV \leq y$ and $AB \leq yy \leq y$ by monotonicity. So y is an upper bound of D.   ◁

Claim 3. $N \in (Z_{C_2'}(K \otimes_{\mathcal{R}} C_2'))^{n \times n}$.

Proof. As N is the supremum of the $b(Up + X + qV)^m d$ and these are equal to $\sum(\{U, X, V\}^m \cap D) \in K^{n \times n}$, we have $c(b(Up + X + qV)^m d) = (b(Up + X + qV)^m d)c$ for $c \in C_2'$. The claim follows by an application of *-continuity.   ◁

By the three claims, the Lemma is proven. (For C_m' with $m \geq 2$ instead of C_2 and brackets of P_m in U and of Q_m in V, use any two different bracket pairs b, d and p, q of $\Delta_m = P_m \cup Q_m$ to define N.)   □

Example 2. In the most simple case $n = 1$, with $Mat_{n,n}(K \otimes_{\mathcal{R}} C_2') \simeq K \otimes_{\mathcal{R}} C_2'$, suppose $U = b, V = d$ and $X = x \in K$. Then $N = b(bp + x + qd)^*d = \sum D$ for Dyck's language $D \subseteq \{b, x, d\}^*$. The proof shows $N = \sum D \in Z_{C_2'}(K \otimes_{\mathcal{R}} C_2')$.   ◁

Theorem 5 (First Normal Form). *Let K be an $\mathcal{R}$-dioid. For each $\varphi \in K \otimes_{\mathcal{R}} C_2'$ there are $n \in \mathbb{N}$, $S \in \mathbb{B}^{1 \times n}$, $F \in \mathbb{B}^{n \times 1}$, $U \in \{0, b, p\}^{n \times n}$, $V \in \{0, d, q\}^{n \times n}$ and $X \in K^{n \times n}$ such that*

$$\varphi = S(NV)^* N(UN)^* F,$$

where $N \in (Z_{C_2'}(K \otimes_{\mathcal{R}} C_2'))^{n \times n}$ is the least solution of $y \geq (UyV + X)^*$ in $Mat_{n,n}(K \otimes_{\mathcal{R}} C_2')$.

For $n = 1$, N commutes with U and V, so $(NV)^k N(UN)^l = V^k NU^l$, and by *-continuity, $(NV)^* N(UN)^* = V^* NU^*$. This is related to the normal form for the extension $P_m'[X]$ of the polycyclic monoid P_m' in Sect. 2.1.

Proof. By definition of $K \otimes_{\mathcal{R}} C_2'$, there is $R \in \mathcal{R}(K \times C_2')$ such that $\varphi = [R]$. As in Theorem 3, by induction on R one constructs an automaton $\langle S, A, F \rangle$ with

$$\varphi = [R] = L(\langle S, A, F \rangle) = SA^*F$$

and a transition matrix $A \in (K \otimes_{\mathcal{R}} C_2')^{n \times n}$ of the form $A = U + X + V$ where $U \in \{0, b, d\}^{n \times n}$, $X \in K^{n \times n}$ and $V \in \{0, d, q\}^{n \times n}$, for some n. By Lemma 1, $y \geq (UyV + X)^*$ has a least solution N in $Mat_{n,n}(K \otimes_{\mathcal{R}} C_2')$, and

$$N \in (Z_{C_2'}(K \otimes_{\mathcal{R}} C_2'))^{n \times n}.$$

By Theorem 4, this N allows us to write A^* as

$$A^* = (U + X + V)^* = (NV)^*N(UN)^*$$

and obtain the normal form $\varphi = [R] = SA^*F = S(NV)^*N(UN)^*F.$ □

Example 3. Let $P_2 = \{\langle 0|, \langle 1| \}$ and $Q_2 = \{|0\rangle, |1\rangle\}$, $K = \mathcal{R}\{a, b\}^* \otimes_{\mathcal{R}} C_2'$. The element $\varphi = (a\langle 1|)^*(|1\rangle b)^* \in K$ is represented as $\varphi = L(\mathcal{M}) = SM^*F$ by the automaton $\mathcal{M} = \langle S, M, F \rangle$ of Fig. 1 with initial state 1 and accepting state 3 (Fig. 2).

$$\langle (1\,0\,0\,0), \begin{pmatrix} 0 & a & 1 & 0 \\ \langle 1| & 0 & 0 & 0 \\ 0 & 0 & 0 & |1\rangle \\ 0 & 0 & b & 0 \end{pmatrix}, \begin{pmatrix} 0 \\ 0 \\ 1 \\ 0 \end{pmatrix} \rangle$$

Fig. 1. $\mathcal{M} = \langle S, M, F \rangle$

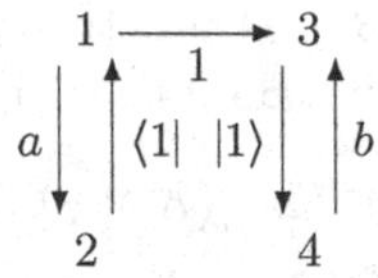

Fig. 2. Graph of M

The iteration M^* of M calculated using the formula (4) can be read off from the graph: the entry $(M^*)_{i,j}$ describes the labellings on paths from node i to node j. Hence, with $\bar{a} = a\langle 1|$ and $\bar{b} = |1\rangle b$, we have

$$M^* = \begin{pmatrix} \bar{a}^* & \bar{a}^*a & \bar{a}^*\bar{b}^* & \bar{a}^*\bar{b}^*|1\rangle \\ \langle 1|\bar{a}^* & 1 + \langle 1|\bar{a}^*a & \langle 1|\bar{a}^*\bar{b}^* & \langle 1|\bar{a}^*\bar{b}^*|1\rangle \\ 0 & 0 & \bar{b}^* & \bar{b}^*|1\rangle \\ 0 & 0 & b\bar{b}^* & 1 + b\bar{b}^*|1\rangle \end{pmatrix}.$$

To obtain the normal form $(NV)^*N(UN)^*$ of M^*, split M as $U + X + V$ with

$$U = \begin{pmatrix} 0 & 0 & 0 & 0 \\ \langle 1| & 0 & 0 & 0 \\ 0 & 0 & 0 & 0 \\ 0 & 0 & 0 & 0 \end{pmatrix}, \quad X = \begin{pmatrix} 0 & a & 1 & 0 \\ 0 & 0 & 0 & 0 \\ 0 & 0 & 0 & 0 \\ 0 & 0 & b & 0 \end{pmatrix}, \quad V = \begin{pmatrix} 0 & 0 & 0 & 0 \\ 0 & 0 & 0 & 0 \\ 0 & 0 & 0 & |1\rangle \\ 0 & 0 & 0 & 0 \end{pmatrix}.$$

To determine $N = \langle 0|(U\langle 1| + X + |1\rangle V)^*|0\rangle$, let $\widetilde{M} = (U\langle 1| + X + |1\rangle V)$ and read off $\widetilde{M}^*$ from the graph of $\widetilde{M}$, obtaining a copy of M^* with $\tilde{a} = a\langle 1|^2, \tilde{b} = |1\rangle^2 b, \langle 1|^2, |1\rangle^2$ instead of $\bar{a}, \bar{b}, \langle 1|, |1\rangle$, respectively. The entries of N are then

$$N_{i,j} = \langle 0|(\widetilde{M}^*)_{i,j}|0\rangle.$$

The resulting matrix is as follows, writing $\widehat{L}$ for $\sum L$ with $L = \{\, a^n b^n \mid n \in \mathbb{N} \,\}$,

$$N = \langle 0| \begin{pmatrix} \tilde{a}^* & \tilde{a}^* a & \tilde{a}^* \tilde{b}^* & \tilde{a}^* \tilde{b}^* |1\rangle^2 \\ \langle 1|^2 \tilde{a}^* & 1 + \langle 1|^2 \tilde{a}^* a & \langle 1|^2 \tilde{a}^* \tilde{b}^* & \langle 1|^2 \tilde{a}^* \tilde{b}^* |1\rangle^2 \\ 0 & 0 & \tilde{b}^* & \tilde{b}^* |1\rangle^2 \\ 0 & 0 & b\tilde{b}^* & 1 + b\tilde{b}^* |1\rangle^2 \end{pmatrix} |0\rangle = \begin{pmatrix} 1 & a & \widehat{L} & a\widehat{L} \\ 0 & 1 & \widehat{L}b & \widehat{L} \\ 0 & 0 & 1 & 0 \\ 0 & 0 & b & 1 \end{pmatrix}.$$

For example, $N_{1,3} = \langle 0|\tilde{a}^* \tilde{b}^*|0\rangle = \langle 0|(a\langle 1|^2)^*(|1\rangle^2 b)^*|0\rangle = \widehat{L}$ is calculated as in Example 1. It follows that

$$NV = \begin{pmatrix} 0\,0\,0\ \widehat{L}|1\rangle \\ 0\,0\,0\ \widehat{L}b|1\rangle \\ 0\,0\,0\ \ |1\rangle \\ 0\,0\,0\ b|1\rangle \end{pmatrix}, \qquad UN = \begin{pmatrix} 0 & 0 & 0 & 0 \\ \langle 1| & \langle 1|a & \langle 1|\widehat{L} & \langle 1|a\widehat{L} \\ 0 & 0 & 0 & 0 \\ 0 & 0 & 0 & 0 \end{pmatrix},$$

which imply $(NV)^* = 1 + NV(b|1\rangle)^*$ and $(UN)^* = 1 + (\langle 1|a)^* UN$. By matrix multiplication, one obtains the normal form $(NV)^* N (UN)^* = M^*$.

To determine N, one can also use that N is the least solution of $y \geq (UyV + X)^*$ in $Mat_{4,4}(K)$, hence $N = (UNV + X)^*$. Let e_i be the unit column vector with 1 in the i-th row, 0 else, e_i' its transpose row vector. Then $e_i e_j'$ is the 4×4-matrix with 1 at (i,j), 0 else, and $e_i' e_j$ the 1×1-matrix with entry $\delta_{i,j}$. Since

$$UNV = (e_2 \langle 1|e_1')(\sum_{1 \leq i,j \leq 4} e_i N_{i,j} e_j')(e_3|1\rangle e_4') = e_2 \langle 1|N_{1,3}|1\rangle e_4' = e_2 N_{1,3} e_4',$$

the graph of $X + UNV$ is that of X with additional edge $2 \xrightarrow{N_{1,3}} 4$, from which one can read off $(X + UNV)^*$ as

$$(X + UNV)^* = \begin{pmatrix} 1 & a & 1 + aN_{1,3}b & aN_{1,3} \\ 0 & 1 & N_{1,3}b & N_{1,3} \\ 0 & 0 & 1 & 0 \\ 0 & 0 & b & 1 \end{pmatrix} = N.$$

Since N is the least solution of $y \geq (UyV + X)^*$, $N_{1,3}$ is the least solution of $y_{1,3} \geq 1 + ay_{1,3}b$, i.e. $\mu x(1 + axb) = \sum L$ for $L = \{\, a^n b^n \mid n \in \mathbb{N} \,\} \in \mathcal{C}K$, leading to the matrix N shown above.   ◁

The normal forms of elements $\varphi_1, \varphi_2 \in K \otimes_{\mathcal{R}} C_2'$, determine the normal form of their combinations $\varphi_1 + \varphi_2$, $\varphi_1 \cdot \varphi_2$, and φ_1^*. A proof is sketched in the Appendix. For any $m \geq 2$, Theorem 5 holds as well with C_m' instead of C_2'.

Corollary 1. *Let Δ_m have the bracket pairs $\langle i|, |i\rangle$ for $i = 0, \ldots, m-1$. Suppose $\varphi = SA^*F \in K \otimes_{\mathcal{R}} C_m'$ is represented by an automaton $\langle S, A, F \rangle$ not using $\langle 0|, |0\rangle$, i.e. $U \in \{0, \langle 1|, \ldots \langle m-1|\}^{n \times n}$, $X \in K^{n \times n}$, $V \in \{0, |1\rangle, \ldots, |m-1\rangle\}^{n \times n}$ in $A = U + X + V$. If $S(NV)^* N (UN)^* F$ is the normal form of φ, then*

$$\langle 0|\varphi|0\rangle = SNF \in Z_{C_m'}(K \otimes_{\mathcal{R}} C_m').$$

If moreover $\varphi \in Z_{C_m'}(K \otimes_{\mathcal{R}} C_m')$, then $\varphi = SNF$.

Proof. By the assumption on U and V, we have $\langle 0|V = 0 = U|0\rangle$, and since N commutes with $\langle 0|$ and $|0\rangle$, we get $\langle 0|(NV)^* = \langle 0|$ and $(UN)^*|0\rangle = |0\rangle$. Hence

$$\langle 0|A^*|0\rangle = \langle 0|(NV)^*N(UN)^*|0\rangle = \langle 0|N|0\rangle = N,$$

and thus $\langle 0|\varphi|0\rangle = \langle 0|SA^*F|0\rangle = S\langle 0|A^*|0\rangle F = SNF \in Z_{C_2'}(K \otimes_{\mathcal{R}} C_2')$. If also $\varphi \in Z_{C_m'}(K \otimes_{\mathcal{R}} C_m')$, φ commutes with $\langle 0|$ and $|0\rangle$, so $\varphi = \langle 0|\varphi|0\rangle = SNF$.   □

Conjecture 1. Suppose we use the bra-ket $\mathcal{R}$-dioid C_2 instead of C_2'. Let $U \in \{0, b, p\}^{n \times n}$, $V \in \{0, d, q\}^{n \times n}$, $X \in K^{n \times n}$ and let N be the least solution of $y \geq (UyV + X)^*$ in $Mat_{n,n}(K \otimes_{\mathcal{R}} C_2)$. If $\varphi = S(NV)^*N(UN)^*F \in Z_{C_2}(K \otimes_{\mathcal{R}} C_2)$, then $\varphi = SNF$, i.e. the restriction on $\langle 0|, |0\rangle$ of Corollary 1 is unnecessary, leading to a simplified normal form for elements of $Z_{C_2}(K \otimes_{\mathcal{R}} C_2)$.

In some cases, we can characterize the elements of the centralizer of C_m':

Theorem 6. *For $m > 2$ and $\varphi \in \mathcal{R}X^* \otimes_{\mathcal{R}} C_m'$, we have $\varphi \in Z_{C_m'}(\mathcal{R}X^* \otimes_{\mathcal{R}} C_m')$ iff there is a regular expression r over $X \dot\cup (\Delta_m \setminus \{\langle 0|, |0\rangle\})$ such that $\varphi = \langle 0|r|0\rangle$.*

Proof. See Corollary 28 (and Lemma 31 for $Z_{C_2'}(K \otimes_{\mathcal{R}} C_2')$) of [6].   □

This generalizes the monoid case of $P_m'[X]$, where $Z_{\Delta_m}(P_m'[X]) = X^* \cup \{0\}$ and clearly $\varphi \in X^* \cup \{0\}$ iff $\varphi = \langle 0|w|0\rangle$ for some $w \in (X \cup \{0\} \cup (\Delta_m \setminus \{\langle 0|, |0\rangle\}))^*$.

Corollary 1 can be extended by admitting that $\varphi = SA^*F \in K \otimes_{\mathcal{R}} C_m'$ is given by an automaton $\langle S, A, F \rangle$ whose transition matrix A contains transitions by $|0\rangle\langle 0|$ in addition to those by elements of K and $\Delta_m \setminus \{\langle 0|, |0\rangle\}$. This is useful to combine representations $\langle 0|r_i|0\rangle = \sum L_i$ of $L_i \in \mathcal{C}X^*$, $i = 1, 2$, in $\mathcal{R}X^* \otimes_{\mathcal{R}} C_2'$ to a representation $\langle 0|r_1|0\rangle\langle 0|r_2|0\rangle = (\sum L_1)(\sum L_2) = \sum(L_1 L_2)$ of $L_1 L_2$.

Theorem 7 (Second Normal Form). *Let K be an $\mathcal{R}$-dioid, $m \geq 2$ and $\varphi \in K \otimes_{\mathcal{R}} C_m'$ be given in matrix form $\varphi = S(U + X + V + W\pi)^*F$, where $\pi = |0\rangle\langle 0|$ and for some $n \geq 0$,*

$$S \in \{0,1\}^{1 \times n}, \quad X \in K^{n \times n}, \quad U \in \{0, \langle 1|, \ldots, \langle m-1|\}^{n \times n},$$
$$F \in \{0,1\}^{n \times 1}, \quad W \in \{0,1\}^{n \times n}, \quad V \in \{0, |1\rangle, \ldots, |m-1\rangle\}^{n \times n}.$$

Then there is $N \in (Z_{C_m'}(K \otimes_{\mathcal{R}} C_m'))^{n \times n}$ such that $N = (UNV + X)^*$ and

$$\langle 0|\varphi|0\rangle = SN(WN)^*F \in Z_{C_m'}(K \otimes_{\mathcal{R}} C_m').$$

Proof. Let $A = U + X + V$. By Theorem 5, there is $N \in (Z_{C_m'}(K \otimes_{\mathcal{R}} C_m'))^{n \times n}$ such that

$$A^* = (U + X + V)^* = (NV)^*N(UN)^*.$$

As in the proof of Corollary 1, we obtain

$$\langle 0|A^*|0\rangle = \langle 0|(NV)^*N(UN)^*|0\rangle = \langle 0|N|0\rangle = N,$$

and therefore in the Kleene algebra $Mat_{n,n}(K \otimes_{\mathcal{R}} C_m')$,

$$\langle 0|(A + W\pi)^*|0\rangle = \langle 0|A^*(W\pi A^*)^*|0\rangle$$
$$= \langle 0|A^*(|0\rangle W \langle 0|A^*)^*|0\rangle$$
$$= \langle 0|A^*|0\rangle (W \langle 0|A^*|0\rangle)^*$$
$$= N(WN)^* \in (Z_{C_m'}(K \otimes_{\mathcal{R}} C_m'))^{n \times n}.$$

Because S, N, W, F commute with $\langle 0|$ and $|0\rangle$, it follows that

$$\langle 0|\varphi|0\rangle = S\langle 0|(A + W\pi)^*|0\rangle F = SN(WN)^*F \in Z_{C_m'}(K \otimes_{\mathcal{R}} C_m').$$

$\square$

4 Conclusion

The tensor product $\mathcal{R}X^* \otimes_{\mathcal{R}} C_m'$ of the algebra $\mathcal{R}X^*$ of regular sets of X^* with the polycyclic Kleene algebra C_m' based on $m \geq 2$ bracket pairs is a *-continuous Kleene algebra subsuming an isomorphic copy of the algebra $\mathcal{C}X^*$ of context-free sets of X^*, namely the centralizer $Z_{C_m'}(\mathcal{R}X^* \otimes_{\mathcal{R}} C_m')$ of C_m' (cf. [6]).

We have investigated $K \otimes_{\mathcal{R}} C_m'$ for arbitrary *-continuous Kleene algebras K. Every element $\varphi \in K \otimes_{\mathcal{R}} C_m'$ is the value SA^*F of an automaton $\langle S, A, F\rangle$ whose transition matrix $A = U + X + V$ splits into transitions by opening brackets (and 0's) in U, transitions by elements of K in X, and transitions by closing brackets (and 0's) in V. Our main result is a normal form theorem saying that $A^* = (NV)^*N(UN)^*$, where N is the least solution of $y \geq (UyV + X)^*$ in $Mat_{n,n}(K \otimes_{\mathcal{R}} C_m')$, corresponding to Dyck's language $D \subseteq \{U, X, V\}^*$ with brackets U and V, and N has entries in $Z_{C_m'}(K \otimes_{\mathcal{R}} C_m')$. The normal forms are the core of calculus of context-free expressions (without binders).

In an earlier incarnation, the first author considered establishing these results using the equation $1 = q_0 p_0 + \ldots + q_{m-1} p_{m-1}$ of C_m for the brackets $p_0, \ldots, q_{m-1}$. Here we showed that the match- and mismatch equations $p_i q_j = \delta_{i,j}$ of C_m' are sufficient. It remains open if the normal form specializes for elements of $Z_{C_m'}(K \otimes_{\mathcal{R}} C_m')$ similar to Conjecture 1; a description of these elements in different terms is given in [6], Lemma 31.

Applications of our results to parsing theory, where $\mathcal{R}X^* \otimes_{\mathcal{R}} C_2'$ provides an algebra for recognizers of context-free languages over the alphabet X and $\mathcal{R}X^* \otimes_{\mathcal{R}} \mathcal{R}Y^* \otimes_{\mathcal{R}} C_2'$ an algebra for translations between X^* and Y^*, will be defered to future publications.

Appendix: Combination of Normal Forms

The normal form for $\varphi \in K \otimes_{\mathcal{R}} C_2'$ can be obtained by induction on the construction of φ. We leave it to the reader to provide $\langle S, A, F\rangle$ and N for atomic elements $\varphi \in K$ or $\varphi \in C_2'$. Suppose for $i = 1, 2$, $\varphi_i = S_i A_i^* F_i$ and $A_i = U_i + X_i + V_i$ as in the normal form theorem, and $N_i = \mu y.(U_i y V_i + X_i)^*$, so that $A_i^* = (N_iV_i)^*N_i(U_iN_i)^*$. For a regular combination φ of φ_1 and φ_2, we define an automaton $\langle S, A, F \rangle$, a splitting $A = U + X + V$ and a matrix N such that $SA^*F = \varphi$ and N is the least solution of $y \geq (UyV + X)^*$, hence $A^* = (NV)^*N(UN)^*$. Space allows us only to prove the claims for φ_1^*.

$\varphi = (\varphi_1 + \varphi_2)$: Put $S = (S_1 \ S_2)$, $F = \begin{pmatrix} F_1 \\ F_2 \end{pmatrix}$, $A = U + X + V$ with

$$U = \begin{pmatrix} U_1 & 0 \\ 0 & U_2 \end{pmatrix}, X = \begin{pmatrix} X_1 & 0 \\ 0 & X_2 \end{pmatrix}, V = \begin{pmatrix} V_1 & 0 \\ 0 & V_2 \end{pmatrix}, \text{ and } N = \begin{pmatrix} N_1 & 0 \\ 0 & N_2 \end{pmatrix}.$$

$\varphi = (\varphi_1 \cdot \varphi_2)$: Put $S = (S_1 \ 0)$, $F = \begin{pmatrix} 0 \\ F_2 \end{pmatrix}$, $A = U + X + V$ with

$$U = \begin{pmatrix} U_1 & 0 \\ 0 & U_2 \end{pmatrix}, X = \begin{pmatrix} X_1 & F_1S_2 \\ 0 & X_2 \end{pmatrix}, V = \begin{pmatrix} V_1 & 0 \\ 0 & V_2 \end{pmatrix}, \text{ and } N = \begin{pmatrix} N_1 & \alpha \\ 0 & N_2 \end{pmatrix}$$

with $\alpha = \mu z.(N_1U_1zV_2N_2 + N_1F_1S_2N_2)$. The existence of α follows from the fact (shown in [6]) that $Z_{C_2'}(K \otimes_\mathcal{R} C_2')$ is a C-dioid, i.e. its context-free subsets A, B have sups $\sum A, \sum B$ and $\sum(AB) = (\sum A)(\sum B)$, in which systems of polynomial inequations $p(z) \leq z$ have least solutions (cf. [5]).

$\varphi = \varphi_1^*$: Since $\varphi_1^* = 1 + \varphi_1^+$, it is sufficient to treat φ_1^+. Put $S = S_1$, $F = F_1$, $A = U + X + V$ with $U = U_1$, $X = X_1 + FS$, $V = V_1$. Then

$$SA^*F = S(A_1 + FS)^*F = SA_1^*(FSA_1^*)^*F = SA_1^*F(SA_1^*F)^* = \varphi\varphi^* = \varphi^+.$$

Moreover, let $N = \mu z.(UzV + N_1 + FS)^*$. The existence of N follows from a generalization of Lemma 1 in which $X \in K^{n \times n}$ is replaced by $X \in (Z_{C_2'}(K \otimes_\mathcal{R} C_2'))^{n \times n}$. Since in Kleene algebra, $(a + b)^* = (a^* + b)^*$, from $N_1 \leq N$ we get

$$\begin{aligned}(UNV + X)^* &= (U(N_1 + N)V + X_1 + FS)^* \\ &= (UN_1V + X_1 + UNV + FS)^* \\ &= ((UN_1V + X_1)^* + UNV + FS)^* \\ &= (N_1 + UNV + FS)^* \leq N,\end{aligned}$$

hence N is a solution of $(UyV + X)^* \leq y$. To show that it is least, suppose $(UyV + X)^* \leq y$. Then $(UyV + X_1)^* \leq y$, hence $N_1 \leq y$, and therefore

$$\begin{aligned}(UyV + N_1 + FS)^* &= (UyV + (UN_1V + X_1)^* + FS)^* \\ &= (UyV + UN_1V + X_1 + FS)^* \\ &\leq (UyV + X)^* \leq y.\end{aligned}$$

Since N is the least solution of $(UzV + N_1 + FS)^* \leq z$, this shows $N \leq y$.

References

1. Chomsky, N., Schützenberger, M.: The algebraic theory of context free languages. In: Braffort, P., Hirschberg, D. (eds.) Computer Programming and Formal Systems, pp. 118–161 (1963)
2. Hopkins, M.: The algebraic approach I: the algebraization of the Chomsky hierarchy. In: Berghammer, R., Möller, B., Struth, G. (eds.) RelMiCS 2008. LNCS, vol. 4988, pp. 155–172. Springer, Heidelberg (2008). https://doi.org/10.1007/978-3-540-78913-0_13
3. Hopkins, M., Leiß, H.: Coequalizers and tensor products for continuous idempotent semirings. In: Desharnais, J., Guttmann, W., Joosten, S. (eds.) RAMiCS 2018. LNCS, vol. 11194, pp. 37–52. Springer, Cham (2018). https://doi.org/10.1007/978-3-030-02149-8_3
4. Kozen, D.: The Design and Analysis of Algorithms. Springer, New York (1991). https://doi.org/10.1007/978-1-4612-4400-4
5. Leiß, H.: The matrix ring of a μ-continuous Chomsky algebra is μ-continuous. In: Regnier, L., Talbot, J.M. (eds.) 25th EACSL Annual Conference on Computer Science Logic (CSL 2016), pp. 1–16. Leibniz International Proceedings in Informatics, Leibniz-Zentrum für Informatik, Dagstuhl Publishing (2016)
6. Leiß, H.: An algebraic representation of the fixed-point closure of *-continuous Kleene algebras – a categorical Chomsky-Schützenberger-theorem. Math. Struct. Comput. Sci. **32**(6), 686–725 (2022). https://doi.org/10.1017/S0960129522000329

Representable and Diagonally Representable Weakening Relation Algebras

Peter Jipsen[1] and Jaš Šemrl[2](✉)

[1] Chapman University, Orange, USA
jipsen@chapman.edu
[2] UCL (University College London), London, UK
j.semrl@cs.ucl.ac.uk
https://www1.chapman.edu/~jipsen/,
http://www0.cs.ucl.ac.uk/staff/jsemrl/

Abstract. A binary relation defined on a poset is a weakening relation if the partial order acts as a both-sided compositional identity. This is motivated by the weakening rule in sequent calculi and closely related to models of relevance logic. For a fixed poset the collection of weakening relations is a subreduct of the full relation algebra on the underlying set of the poset. We present a two-player game for the class of representable weakening relation algebras akin to that for the class of representable relation algebras. This enables us to define classes of abstract weakening relation algebras that approximate the quasivariety of representable weakening relation algebras. We give explicit finite axiomatisations for some of these classes. We define the class of diagonally representable weakening relation algebras and prove that it is a discriminator variety. We also provide explicit representations for several small weakening relation algebras.

Keywords: Weakening relation algebra · Relevance frames · Sugihara monoids · Representation games

1 Introduction

The *full algebra of binary relations on* X is

$$\mathbf{Rel}(X) = (\mathcal{P}(X^2), \cap, \cup, \emptyset, \top, ;, id_X, \neg, \smallsmile)$$

where $\top = X^2$, $R;S$ is the composition of R, S, $\neg R = X^2 \backslash R$, and $R^\smallsmile = \{(x,y) \mid (y,x) \in R\}$. The class RRA of *representable relation algebras* $= \mathbb{SP}\{\mathbf{Rel}(X) \mid X$ is a set$\}$. Tarski [22] proved that RRA is a variety and Monk [17] proved that RRA is not finitely axiomatisable. For more details see the books by Givant [5,6] and Maddux [11].

This work was supported by the Engineering and Physical Sciences Research Council EP/S021566/1.

R. Glück et al. (Eds.): RAMiCS 2023, LNCS 13896, pp. 140–157, 2023.
https://doi.org/10.1007/978-3-031-28083-2_9

The set of *weakening relations* on a poset $\mathbf{X} = (X, \leq)$ is $\mathcal{W}(\mathbf{X}) = \{R \subseteq X^2 \mid \leq;R;\leq = R\}$. The *full algebra of weakening relations on a poset* $\mathbf{X}$ is

$$\mathbf{wk}(\mathbf{X}) = (\mathcal{W}(X, \leq), \cap, \cup, \emptyset, \top, ;, 1, \sim)$$

where $1 = \leq$ and $\sim R = \neg R^{\smile}$ is the complement-converse operation. The class of **representable weakening relation algebras** is

$$\mathsf{RwkRA} = \mathbb{SP}\{\mathbf{wk}(X, \leq) \mid (X, \leq) \text{ is a poset}\}.$$

Weakening relations are the analogue of binary relations when the category **Set** of sets and functions is replaced by the category **Pos** of partially ordered sets and order-preserving functions. Since sets can be considered as discrete posets (i.e. antichains, ordered by the identity relation), **Pos** contains **Set** as a full subcategory, which implies that weakening relations are a substantial generalisation of binary relations. However, weakening relations do not allow $\neg$ or $\smile$ as operations.

They have applications in sequent calculi [2], quasi-proximity lattices/spaces [19], order-enriched categories [10], mathematical morphology [21], and program semantics, e.g. via separation logic [18].

The closely related algebras $\mathbf{Wk}(\mathbf{X})$ are defined as the expansions of $\mathbf{wk}(\mathbf{X})$ by the Heyting implication $R \to S = \{(x, y) \mid \forall u, v(u \leq x \ \& \ y \leq v \ \& \ uRv \Rightarrow uSv)\}$. The $\mathbb{SP}$-closure of these algebras is denoted by RWkRA and has been studied in [3,4,9,20,21]. It is a discriminator variety that has RRA as a proper subvariety. The algebras in RWkRA are generalised bunched implication algebras, and the algebras in RwkRA are all the subreducts of algebras in RWkRA, hence RwkRA is a quasivariety. We show that it is not a variety, but with respect to representability the two classes behave the same way.

In Sect. 2 we define a representation game for RwkRA (which can be extended to a game for RWkRA) and use it to give an explicit universal axiomatisation for the class. Section 3 defines (Kripke) frames for weakening relation algebras and adapts the game to this setting. From an n-pebble version of this frame game we define a sequence of classes wkRA_n that approximate RwkRA from above, similar to the sequence RA_n that converges the RRA. In the next section we find finite axiomatisation for wkRA_2 and wkRA_3. In Sect. 5 we define the class of representable diagonal weakening relation algebras and show that is a discriminator variety. Finally, in the last section we show that all associative algebras in wkRA_3 with 6 elements or fewer are representable.

2 Representation Game

In this section we present a representation game for weakening relation algebras similar to those defined for relation algebras, defined in [8]. We begin by defining some notation.

Definition 1. A *bounded cyclic involutive unital distributive lattice-ordered magma* $\mathcal{A} = (A, \cdot, +, \perp, \top, ;, 1, \sim)$ is an algebra such that

(1) $(A, \cdot, +, \perp, \top)$ is a bounded distributive lattice
(2) $(s + t);(u + v) = s; u + s; v + t; u + t; v$
(3) $s; \perp = \perp = \perp; s$
(4) $s; 1 = s = 1; s$
(5) $\sim(\sim s) = s$
(6) $\sim(s \cdot t) = \sim s + \sim t$

for all $s, t, u, v \in A$. A *representation* of $\mathcal{A}$ is an injective homomorphism $h : \mathcal{A} \to$ **wk(X)** for some poset $\mathbf{X} = (X, \leq)$ such that $h(\top)$ is an equivalence relation on X.

Note that $s \leq t$ if and only if $s + t = t$, or equivalently $\sim s \cdot \sim t = \sim t$ which can be rewritten as $\sim t \leq \sim s$, hence $\sim$ is order reversing. The adjective "cyclic" is included in the name to contrast it to the non-cyclic general case where there are two unary operations $\sim, -$ in the language that satisfy $\sim -s = s = -\sim s$. In the cyclic case $\sim, -$ have the same interpretation.

Distributive lattice-ordered magmas are abbreviated as $d\ell$-magmas. Let $\mathcal{A}$ be a bounded cyclic involutive unital $d\ell$-magma. Additionally we define $0 = \sim 1$.

Definition 2. A *network (for $\mathcal{A}$)* is a tuple $\mathcal{N} = (N, \lambda)$ where N is a set of *nodes* and $\lambda : N^2 \to \wp(A)$ is a *labelling function* such that for all $x, y \in N$, $1 \in \lambda(x, x)$ and $\top \in \lambda(x, y)$. Such a network is *consistent* if and only if for all $x, y \in N$ we have that

$$\lambda(x, y) \cap \{\sim a \mid a \in \lambda(y, x)\} = \emptyset.$$

A network $\mathcal{N} = (N, \lambda)$ is a *prenetwork* of $\mathcal{N}' = (N', \lambda')$ – denoted $\mathcal{N} \subseteq \mathcal{N}'$ – if and only if $N \subseteq N'$ and for all $x, y \in N$ we have $\lambda(x, y) \subseteq \lambda'(x, y)$.

Observe that the prenetwork predicate is a partial order and that inconsistency is inherited from prenetworks.

We now have the tools to define a two player game and prove that the existence of a winning strategy for one of the players coincides with $\mathcal{A}$'s membership in the class of RWkRA.

Definition 3. An *n-round representation game*, denoted $\Gamma_n(\mathcal{A})$, for some $n \leq \omega$ is a two player game played between the challenger $\forall$ (Abelard) and the responder $\exists$ (Héloïse) over $n + 1$ moves. After the ith move for $0 \leq i \leq n$, $\exists$ will return a network $\mathcal{N}_i$ such that $\mathcal{N}_0 \subseteq \mathcal{N}_1 \subseteq ... \subseteq \mathcal{N}_n$. The game is won by $\forall$ if $\exists$ returns an inconsistent network. Otherwise $\exists$ wins.

On the *initialisation move* $\forall$ picks a pair of elements $a \nleq b \in \mathcal{A}$ and $\exists$ must return a network $\mathcal{N}_0$ with some $(x, y) \in N_0^2$ such that $a \in \lambda(x, y)$ and $\sim b \in \lambda(y, x)$.

On the ith move for $0 < i \leq n$, $\forall$ may challenge $\exists$ with any of the following four moves.

join move: $\forall$ picks $x, y \in N_{i-1}$, some $a \in \lambda_{i-1}(x, y)$, and some $b, c \in \mathcal{A}$ such that $a \leq b + c$. $\exists$ must return a $\mathcal{N}_i$ with $b \in \lambda_i(x, y)$ or $c \in \lambda_i(x, y)$.

involution move: $\forall$ picks $x, y \in N_{i-1}$ and some $a, b \in \mathcal{A}$ such that $b = \sim a$. $\exists$ must return a $\mathcal{N}_i$ with $a \in \lambda_i(x, y)$ or $b \in \lambda_i(y, x)$.

composition move: $\forall$ picks $x, y, z \in N_{i-1}$ and $a \in \lambda_{i-1}(x, y), b \in \lambda_{i-1}(y, z)$. $\exists$ must return a $\mathcal{N}_i$ with $c \in \lambda_i(x, z)$ where $c = a; b$.

witness move: $\forall$ picks $x, y \in N_{i-1}$, $a \in \lambda_{i-1}(x, y)$, and $b, c \in \mathcal{A}$ such that $a = b; c$. $\exists$ must return a $\mathcal{N}_i$ with some $z \in N_i$ such that $b \in \lambda_i(x, z), c \in \lambda_i(z, y)$.

The proof for the following proposition is an outline. The argument is based on [8].

Proposition 1. $\mathcal{A}$ *is representable if and only if* $\exists$ *has a winning strategy for* $\Gamma_\omega(\mathcal{A})$.

Proof. If $\mathcal{A}$ is representable, then $\exists$ can take some representation h over X. Let $a \not\le b$ be the pair played on initialisation. There will exist some maximal $X' \subseteq X$ such that $\exists x, y \in X' : (x, y) \in h(a) \backslash h(b)$ and $\forall z, w \in X' : (z, w) \in h(\top)$. On initialisation, $\exists$ can return the network $\mathcal{N} = (X', \lambda)$ where $\lambda(x, y) = \{c \in \mathcal{A} \mid (x, y) \in h(c)\}$. Because h preserves all the operations in the language, all moves $\forall$ may call are trivially responded to by returning the same network after every move.

If $\mathcal{A}$ is countable then $\forall$ can schedule his moves in a way that every move will be called eventually. Let $\mathcal{N}_0^{a,b}, \mathcal{N}_1^{a,b}, \mathcal{N}_2^{a,b}, \ldots$ be the networks during an $\exists$-winning play of $\Gamma_\omega(\mathcal{A})$ where $\forall$ scheduled his moves in such a way and the initialisation move was called for the pair $a \not\le b$. Define $N_\omega^{a,b}$ as $\{x \mid \exists i < \omega : j \ge i \Rightarrow x \in N_j^{a,b}\}$, $\lambda_\omega^{a,b}(x, y)$ as $\{c \mid \exists i < \omega : j \ge i \Rightarrow (x, y \in N_j^{a,b} \wedge c \in \lambda_j^{a,b}(x, y))\}$, and a relation $\equiv$ as $\{(x, y) \in (N_i^{a,b})^2 \mid 1 \in \lambda_\omega^{a,b}(x, y), 1 \in \lambda_\omega^{a,b}(y, x)\}$. It is symmetric by definition, reflexive because networks are defined as having $1 \in \lambda_\omega^{a,b}(x, x)$ and transitive because all composition moves were called eventually and $1; 1 = 1$. Therefore, we can define $h^{a,b} : \mathcal{A} \to ((N_\omega / \equiv)^2)$ where for all $c \in \mathcal{A}$ we have $h^{a,b}(c) = \{([x]_\equiv, [y]_\equiv) \mid c \in \lambda_\omega^{a,b}(x, y)\}$.

The reader can check that $h^{a,b}$ is a homomorphism (because all moves were called eventually) for $\mathcal{A}$, discriminating $a \not\le b$ (because of initialisation). Thus let $h(c)$ for all $c \in \mathcal{A}$ be the disjoint union $\biguplus_{a \not\le b \in \mathcal{A}} h^{a,b}(c)$. Because h is a homomorphism that discriminates all $a \not\le b$ pairs, it is a representation.

This generalises to uncountable algebras by the downward Löwenheim Skolem Theorem since RWkRA is a pseudoelementary class. $\square$

Next we show that the existence of a winning strategy for $\exists$ can be expressed by a universal first-order sentence. For this result we define the following concepts.

Definition 4. A *term network* is a network $\mathcal{N} = (N, \lambda)$ where N is a finite set of nodes and λ is a labelling function that maps every pair of nodes to a finite set of terms. We also require that for all $x, y \in N$, $1 \in \lambda(x, x)$ and $\top \in \lambda(x, y)$.

For every term network $\mathcal{N} = (N, \lambda)$ we define a network $\mathcal{N}^{+, x, y, t} = (N \cup \{y\}, \lambda^\ell)$ where $x \in N, y \in N \uplus \{x^+\}$ (for some new node x^+), t is a term in the language of RWkRA and for all $z, w \in N \uplus \{x^+\}$.

$$
\lambda^\ell(z,w) = \begin{cases}
\{1, \top\} & \text{if } x^+ = z = w \\
\{\top\} & \text{if } x^+ = z \neq w \neq x \text{ or } x^+ = w \neq z \neq x, w \neq y \\
\{t, \top\} & \text{if } z = x, w = y = x^+ \\
\lambda(z,w) \cup \{t\} & \text{if } z = x, w = y \neq x^+ \\
\lambda(z,w) & \text{otherwise}
\end{cases}
$$

For variables a, b we define two initial term networks below.

$$
\mathcal{N}^{1,a,b} = (\{x\}, \{(x,x) \mapsto \{\top, 1, a, \sim b\}\})
$$
$$
\mathcal{N}^{2,a,b} = (\{x,y\}, \{(x,x) \mapsto \{\top, 1\}, (x,y) \mapsto \{\top, a\},
$$
$$
(y,x) \mapsto \{\top, \sim b\}, (y,y) \mapsto \{\top, 1\}\})
$$

Proposition 2. *For every $n < \omega$ there exists a first-order formula σ_n that corresponds to $\exists$ having a winning strategy for $\Gamma_n(\mathcal{A})$.*

Proof. We show by induction that there exists a formula $\phi_n(\mathcal{N})$ for every $0 \leq n < \omega$, defined for a finite term network $\mathcal{N}$, with all the variables universally quantified that signifies that the network can remain consistent for n more moves of the representation game where $\exists$ plays *conservatively*, i.e., only adds the requested labels. It is easy to see that she has a winning strategy for the game if and only if she also has one for the conservative play.

In the base case, $\phi_0(\mathcal{N})$ defined below signifies consistency (remaining consistent for zero moves)

$$
\phi_0(\mathcal{N}) = \bigwedge_{x,y \in N} \bigwedge_{t \in \lambda(x,y)} \bigwedge_{t' \in \lambda(y,x)} t \neq \sim t'.
$$

In the induction case, we assume that $\phi_n(\mathcal{N}')$, where $\mathcal{N}'$ is a term network with all variables universally quantified, is both necessary and sufficient for $\mathcal{N}'$ to be able to remain consistent for n moves. Then we show you can define $\phi_{n+1}(\mathcal{N})$ that extends the assumption to $n + 1$ moves. Although we use a, b here, the variable names should be unique when constructing these formulas.

$$
\phi_{n+1}(\mathcal{N}) = \bigwedge_{x,y \in N} \bigwedge_{t \in \lambda(x,y)} \forall a, b \big(t \leq a + b \implies (\phi_n(\mathcal{N}^{+,x,y,a}) \vee \phi_n(\mathcal{N}^{+,x,y,b})) \big)
$$
$$
\wedge \bigwedge_{x,y \in N} \bigwedge_{t \in \lambda(x,y)} \forall a \big(\phi_n(\mathcal{N}^{+,x,y,a}) \vee \phi_n(\mathcal{N}^{+,y,x,\sim a}) \big)
$$
$$
\wedge \bigwedge_{x,y,z \in N} \bigwedge_{t \in \lambda(x,y)} \bigwedge_{t' \in \lambda(y,z)} \phi_n(\mathcal{N}^{+,x,z,t;t'})
$$
$$
\wedge \bigwedge_{x,y \in N} \bigwedge_{t \in \lambda(x,y)} \forall a, b \big(t = a; b \implies \bigvee_{z \in N \uplus \{x^+\}} \phi_n((\mathcal{N}^{+,x,z,a})^{+,z,y,b}) \big)
$$

We now have a formula $\phi_n(\mathcal{N})$ for every $0 \leq n < \omega$ that ensures $\exists$ can keep a universally quantified term network $\mathcal{N}$ consistent. Hence the formula

$$\sigma_n = \forall a, b \ (a \not\leq b \implies (\phi_n(\mathcal{N}^{1,a,b}) \vee \phi_n(\mathcal{N}^{2,a,b})))$$

ensures that $\exists$ has a winning strategy for a conservative game of length n. $\square$

Corollary 3. $\Sigma = \{\sigma_1, \sigma_2, \ldots\}$ *together with the axioms for cyclic distributive involutive semirings is a recursively enumerable theory that axiomatises* RWkRA.

3   Frames, Frame Games, and Finite Pebble Games

In this section we present finite algebras as frames, similar to Routley-Meyer frames or relevance frames for relevance logic [1] and atom structures of atomic relation algebras [12]. We then define a modified version of the representation game that utilises frames.

Finally, we define an n-pebble versions of the frame game. Analogous to the abstract classes of relation algebras $RA_\omega \subseteq \ldots \subseteq RA_3 \subseteq RA_2$, this gives rise to classes of weakening relation algebras $wkRA_\omega \subseteq \ldots \subseteq wkRA_3 \subseteq wkRA_2$. Clearly $RA_\omega, wkRA_\omega$ are the classes of representable relation algebras and weakening algebras, respectively. Furthermore, similarly to RA_4, we say that $wkRA_4$ is the class of weakening relation algebras.

First, observe that the language of RwkRA does not include negation and hence the lattice need not be Boolean. As we will see in Sect. 6, the smallest representable non-Boolean algebra is a 4-element chain $\mathbf{S}_4$. Thus we cannot present finite weakening relation algebras using atoms. Instead, we make use of join-irreducibles.

Definition 5. A non-$\perp$ element a of a representable weakening relation algebra is *join-irreducible* if and only if for all b, c if $a = b + c$ then $a = b$ or $a = c$. It is *join-prime* if and only if for all b, c if $a \leq b + c$ then $a \leq b$ or $a \leq c$.

Because $\cdot$ distributes over $+$ we have that an element is join-irreducible if and only if it is join-prime. In the finite case every algebra will have join-irreducibles and every element is a join of join-irreducibles. (In general this is only true for *perfect* algebras. In fact, by definition, a distributive lattice is *join-perfect* if every element is a join of completely join-irreducible elements. This generalises the concept of *atomic* for Boolean algebras).

The element a in the result below is called the *join-irreducible label* of (x, y).

Proposition 4. *In a representation h of a finite representable weakening relation algebra $\mathcal{A}$, for any pair (x, y) there exists a join-irreducible $a \in \mathcal{A}$ such that*

$$\uparrow a = \{s \in \mathcal{A} \mid (x, y) \in h(s)\}.$$

Proof. A representation h maps joins to unions, hence the set $\{s \mid (x,y) \in h(s)\}$ is upward closed and if $(x,y) \in h(a+b)$ then it is also in $h(a)$ or $h(b)$. Hence the base set of the representation is itself a union of upward closures of join-primes. Now if it is above $\uparrow a$ and $\uparrow b$ then it must be the case that (y,x) is in neither $h(\sim a)$ nor $h(\sim b)$ and thus $(x,y) \in h(\sim(\sim a + \sim b)) = h(a \cdot b)$. Thus the meet of all such join-irreducibles must also be a non-$\bot$ element that is join-prime and below all elements in the set. □

Although the converse operation is not defined in our language, we can use the following trick to define a useful unary operation on the join-irreducibles.

Definition 6. For every join-irreducible a in a finite algebra, define $\hat{a} = \sim\sum_{a \not\leq s} s$ where $\sum$ is with respect to join ($+$).

The join $\sum_{s \not\leq t} t$, defined for all s in a finite algebra $\mathcal{A}$, is usually denoted $\kappa(s)$. If we take $s \leq s' \in \mathcal{A}$ we have $s \not\leq t \Rightarrow s' \not\leq t$ and thus $\kappa(s) \leq \kappa(s')$, hence κ is order preserving. Because $\sim$ is order reversing and κ is order preserving we have that $\hat{\ }$ is order reversing.

Proposition 5. *In any finite bounded distributive involutive additive algebra $\mathcal{A}$, if a is a join-irreducible, so is $\hat{a}$.*

Proof. It is well known that $\kappa(a)$ of a join-irreducible a in a lattice is meet irreducible and because $\sim$ is order reversing, that means that $\hat{a} = \sim\kappa(a)$ is a join-irreducible. □

Proposition 6. *If a pair (x,y) in a representation has the join-irreducible label a, then (y,x) has label $\hat{a}$. Moreover, $\hat{\hat{a}} = a$.*

Proof. $\sim s \in h(y,x)$ if and only if $s \notin h(x,y)$, i.e. $a \not\leq s$. Thus, by the argument from Proposition 4 the join-irreducible label of (y,x) can be written as $\prod_{a \not\leq s} \sim s$ where $\prod$ is with respect to meet ($\cdot$) and this is equivalent to $\sim\sum_{a \not\leq s} s$ by the De Morgan equivalence. □

Finally to characterise composition, we need to define a ternary predicate, similar to the set of allowed triangles in relation algebras.

Definition 7. Let $\mathcal{A}$ be a finite bounded cyclic involutive unital $d\ell$-magma and define a ternary relation R on the set of join-irreducibles of $\mathcal{A}$ by

$$R(a,b,c) \text{ if and only if } a \leq b; c.$$

For relation algebras with atoms a, b, c the Peircian triangle law says that

$$a \leq b; c \iff \hat{a} \leq \hat{c}; \hat{b} \iff b \leq a; \hat{c} \iff \hat{b} \leq c; \hat{a} \iff c \leq \hat{b}; a \iff \hat{c} \leq \hat{a}; b.$$

As we will see in the next section, this law does not hold for the class of representable weakening relation algebra frames. However, atom structures for relation algebras generalise to the weakening setting as follows.

Definition 8. A *relevance frame* $\mathcal{F} = (F, I, \leq, R, \hat{\ })$ is a structure with a carrier set F, a unary predicate I, a partial order predicate $\leq$, a ternary predicate R, and an order-reversing involution operation $\hat{\ }$ where for all a, b, c, d in F

(1) $a \leq b \Leftrightarrow \exists e : I(e) \wedge R(a, e, b)$
(2) $a \leq b \Leftrightarrow \exists e : I(e) \wedge R(a, b, e)$
(3) $a \leq b \wedge R(b, c, d) \Rightarrow R(a, c, d)$
(4) $b \leq c \wedge R(a, b, d) \Rightarrow R(a, c, d)$
(5) $c \leq d \wedge R(a, b, c) \Rightarrow R(a, b, d)$

Proposition 7. *A relevance frame $\mathcal{F} = (F, I, \leq, R, \hat{\ })$ defines a bounded involutive unital dℓ-magma $\mathcal{A} = (A, \cdot, +, \bot, \top, ;, 1, \sim)$ by taking $(F, \leq)$ as the join-irreducibles of the lattice with their partial order and for all $s, t \in A$*

$$1 = \sum_{I(a)} a, \qquad \sim s = \sum_{\hat{a} \nleq s} a, \qquad s; t = \sum_{b \leq s, c \leq t, R(a,b,c)} a$$

where $a, b, c \in F$.

Proof. A bounded distributive lattice can be defined by its join-irreducibles and their ordering. To show that the magma is unital, we can see that no term of the join defining the composition with the identity is above the identity by Definition 8(1)(2) and because $\leq$ is reflexive, there will exist, for every join-irreducible a term in the composition with the identity (on either side) equal to that join-irreducible. Thus 1 is precisely the identity. Composition is additive by definition. $\sim$ is an involution because a join-irreducible $a \leq \sim(\sim s)$ if and only if $\hat{a} \nleq \sim s$ which is true if and only if $a = \hat{\hat{a}} \leq s$. For the De Morgan equivalence, $a \leq \sim(\sim s + \sim t)$ if and only if $\hat{a} \nleq \sim s + \sim t$, or equivalently $\hat{a} \nleq \sim s \wedge \hat{a} \nleq \sim t$ which by definition is true if and only if $a = \hat{\hat{a}} \leq s$ and $a = \hat{\hat{a}} \leq t$, or simply $a \leq s \cdot t$. $\square$

Proposition 8. *Every finite bounded cyclic involutive unital dℓ-magma has a unique equivalent relevance frame.*

Proof. Finite distributive lattices are determined by their poset of join-irreducibles, and from Proposition 5 they have a unique $\hat{\ }$ defined on the join-irreducibles. The mapping to R is unique as Definition 8(3)(4)(5) ensure that R is downward closed in the first argument and upward closed in the other arguments. $\square$

Proposition 9. *For finite algebras and finite frames, the mappings described in the previous two lemmas are inverses of each other.*

Proof. Finite distributive lattices correspond uniquely to their posets of join-irreducibles. The preservation of identity and the composition follow trivially from the definition. For $\sim$, $\hat{\ }$ observe that $\sim \kappa(a) = \hat{a} \nleq s$ if and only if $\sim s \nleq \kappa(a) = \sum_{a \nleq t} t$, or equivalently $a \leq \sim s$. For the converse note that $\sim \kappa(a) = \sum_{\hat{b} \nleq \kappa(a)} b = \sum_{a \leq \hat{b}} b = \sum_{b \leq \hat{a}} b = \hat{a}$. $\square$

Although we have defined these frames for finite algebras, we can say that a possibly infinite algebra is *frame-definable* if it can be defined by a relevance frame. In the context of relation algebras, this corresponds to complete and atomic relation algebras. Similarly to that class, we will show that every non-frame definable algebra embeds into a frame definable algebra with equivalent representability.

Proposition 10. *Let $\mathcal{A}$ be a bounded cyclic involutive unital $d\ell$-magma. Then a frame $\mathcal{F}(\mathcal{A})$ can be defined by taking the carrier set of all prime filters $U \subseteq A$, with $U \leq V$ if and only if $V \subseteq U$, $\hat{U} = \{\sim s \mid s \in A\backslash U\}$, $I(U) \Leftrightarrow 1 \in U$ and $R(U,V,W)$ if and only if for all $v \in V, w \in W$ we have $v;w \in U$.*

Proof. $\leq$ is clearly a partial order, for closure of $\hat{\ }$ note that $A\backslash U$ is a prime ideal, so by the order reversing property of $\sim$, $\hat{U}$ is a prime filter. Furthermore, all the unitality conditions are trivially preserved and by $U' \leq U$ if and only if $U \subseteq U'$ we have downward closure of R in the first argument and upward closure in the other two. □

Proposition 11. *$\mathcal{A}$ is representable if and only if the algebra defined by $\mathcal{F}(\mathcal{A})$ is a representable weakening relation algebra. This algebra is called the* canonical extension *of $\mathcal{A}$.*

Proof. Because $\mathcal{A}$ is a subalgebra of the algebra defined by $\mathcal{F}(\mathcal{A})$, we know that the right to left implication is true. For the other direction, if $\mathcal{A}$ is representable, every (x,y) will have a prime filter U such that $(x,y) \in h(a)$ if and only if $a \in U$ to represent the lattice correctly. The prime filter defining (y,x) will be exactly $\hat{U}$. The identity is also correctly represented as it is only above those prime filters that include it. Finally for ; we have shown that it suffices for $\exists$ to have a winning strategy for a game of any finite length. Thus at any point we need to show that the compositions are correctly represented if and only if all compositions finite meets are properly included in the relevant prime filter. □

Definition 9. A *frame network* $\mathcal{N} = (N,\lambda)$ is defined for a frame $\mathcal{F} = (F,I,\leq,R,\hat{\ })$ with N being the set of nodes and $\lambda : N^2 \to F$ is the labelling function. The network is said to be *consistent* if and only if for all $x,y \in N$ we have $\lambda(x,y) = \widehat{\lambda(y,x)}$ and for all $x,y,z \in N$ we have $R(\lambda(x,y),\lambda(x,z),\lambda(z,y))$.

We say for two frame networks $\mathcal{N} = (N,\lambda)$, $\mathcal{N}' = (N',\lambda')$ that $\mathcal{N} \subseteq \mathcal{N}'$ if and only if $N \subseteq N'$ and $\lambda = \lambda' \upharpoonright_{N^2}$ where $\upharpoonright$ denotes the restriction of the function to the domain in the subscript.

Definition 10. An infinite length *frame game* $G(\mathcal{F})$ where $\mathcal{F} = (F,I,\leq,R,\hat{\ })$ is a relevance frame is defined for two players $\forall$ and $\exists$.

The game starts with $\forall$ picking a join-irreducible a and $\exists$ must return a frame network $\mathcal{N}_0 = (N_0,\lambda_0)$ such that there exists $x,y \in N_0$ such that $\lambda_0(x,y) = a$.

At the ith move for $0 < i < \omega$ $\forall$ picks a pair $x,y \in N_{i-1}$ and a pair of join-irreducibles a,b such that $R(\lambda(x,y),a,b)$ and for all $a' \leq a, b' \leq b \in N_{i-1}$ if

$R(\lambda(x,y), a', b')$ then $a = a'$ and $b = b'$. $\exists$ must return a network $\mathcal{N}_i = (N_i, \lambda_i)$ such that $\mathcal{N}_{i-1} \subseteq \mathcal{N}_i$ and $\exists z \in N_i$ such that $\lambda(x, z) = a, \lambda(z, y) = b$.

$\forall$ wins if and only if $\exists$ returns an inconsistent network at any point in the game.

Proposition 12. $\exists$ *has a winning strategy for* $G(\mathcal{F}(\mathcal{A}))$ *if and only if she has a winning strategy for* $\Gamma_\omega(\mathcal{A})$.

Proof. It suffices to prove that she has a winning strategy for the play where all moves are called eventually. Thus if she has a winning strategy for $\Gamma_\omega(\mathcal{A})$, we know that the limit network will have the relevant prime filters as labels. Thus if a is the initial join-irreducible $\exists$ can map all her moves from the limit network of the play where the initialisation pair was $a \not\leq \kappa(a)$.

For the converse, assume she has a winning strategy for $G(\mathcal{F}(\mathcal{A}))$. To respond to the initialisation move with $s \not\leq t$ there will exist a join-irreducible a such that $a \leq s$ but $a \not\leq t$ or rather $t \leq \kappa(a)$ so returning the initial network for a will ensure that $a \leq s$ and $\hat{a} \leq \tilde{t}$. Any witness move called can be responded to by minimal join-irreducible pairs, which makes any other witness moves called by $\forall$ redundant. $\qquad\square$

We now define for every $2 \leq n \leq \omega$ the n-pebble equivalent version of the frame game as follows.

Definition 11. The n-pebble infinite move game $G^n(\mathcal{F})$ for a frame $\mathcal{F}$ is defined exactly as $G(\mathcal{F})$, except before $\forall$ calls a witness move, he takes $N' \subseteq \mathcal{N}_{i-1}$ such that $|N'| \leq n$ and then proceeds to call the witness move.

In particular, the frame game G is equivalent to G^ω. Next we define wkRA$_n$ and wkRA analogous to RA$_n$, the variety of all n-dimensional relation algebras, and RA, the variety of all (4-dimensional) relation algebras [8,13].

Definition 12. The class wkRA$_n$ is the class of all bounded cyclic involutive unital $d\ell$-magmas $\mathcal{A}$ for which $\exists$ has a winning strategy for $G^n(\mathcal{F}(\mathcal{A}))$. The class of weakening relation algebras wkRA is defined as wkRA$_4$.

It follows that wkRA$_\omega$ is equivalent to RwkRA and wkRA$_\omega \subseteq \ldots \subseteq$ wkRA$_4 \subseteq$ wkRA$_3 \subseteq$ wkRA$_2$.

4   Axiomatisation of the Abstract Classes

In this section we provide finite axiomatisations for wkRA$_2$ and wkRA$_3$. We leave open the problem of whether, similarly to RA$_4$ the axiomatisation for wkRA$_4$ consists of axioms of wkRA$_3$ and associativity of ;.

We begin by axiomatising wkRA$_2$. This will be done using the axiomatisation of bounded cyclic involutive unital $d\ell$-magmas together with the theory Φ_2, defined below.

Definition 13. Let Φ_2 be the first order theory given by the following quasiequations:

(1) $s \cdot \sim s \leq 0$
(2) $s \leq t \Leftrightarrow s; \sim t \cdot 1 \leq 0$
(3) $s \leq t; u \wedge s; t \leq \sim u \Rightarrow s \cdot 1 \leq 0$
(4) $s \leq t; u \wedge u; s \leq \sim t \Rightarrow s \cdot 1 \leq 0$
(5) $s \leq t; u \wedge (s \cdot 1 \cdot t; v) + (1 \cdot s \cdot \sim v; u) \leq 0 \Rightarrow s \cdot 1 \leq 0$

Before we prove the soundness and completeness, we introduce a ternary predicate for the language of frames $R^{\min}$ from the equivalence below.

$$R^{\min}(a, b, c) \Leftrightarrow R(a, b, c) \wedge \forall b', c' : (R(a, b', c') \wedge b' \leq b \wedge c' \leq c \Rightarrow b' = b \wedge c' = c)$$

Note that since the union of a chain of prime filters is again a prime filter, frames of the form $\mathcal{F}(\mathcal{A})$ have the property that $R(a, b, c)$ can be refined to $R^{\min}(a, b', c')$ for some prime filters $b' \leq b$ and $c' \leq c$.

Lemma 13. *Let $\mathcal{A}$ be a bounded cyclic involutive unital dℓ-magma. $\mathcal{A} \models \Phi_2$ if and only if $\mathcal{F}(\mathcal{A})$ satisfies*

(1) $\forall a \exists b : I(b) \wedge \hat{b} = b \wedge R(b, a, \hat{a})$
(2) $\forall a, b : I(a) \wedge \hat{a} = a \wedge R(a, b, \hat{b}) \Rightarrow R(b, a, b)$
(3) $\forall a, b : I(a) \wedge \hat{a} = a \wedge R(a, b, \hat{b}) \Rightarrow R(\hat{b}, \hat{b}, a)$
(4) $\forall a, b, c : I(a) \wedge \hat{a} = a \wedge R^{\min}(a, b, c) \Rightarrow b = \hat{c}$

Proof. For the left to right implication, observe that for any join-irreducible a, we know that $a \not\leq \kappa(a)$ so $(a; \sim \kappa(a) \cdot 1) \not\leq 0$ by $\Phi_2(2)$. Thus there must exist a join-irreducible $b \leq 1, b \not\leq 0, b \leq a; \hat{a}$. Suppose $b \neq \hat{b}$. Then there would exist some $b \leq s, \hat{b} \not\leq s$. Because $\hat{\hat{b}} = b$ we know that $b \leq \sim s$ and thus $b \leq s \cdot \sim s \leq 0$, contradicting $\Phi_2(1)$ and we have proven (1) follows from Φ_2. For (2) assume we have $a \leq 1, \hat{a} = a$ then $a \not\leq \sim 1 = 0$. Thus $a = a \cdot 1 \not\leq 0$ and $a \leq b; \sim \kappa(b)$ implies $a; b \not\leq \sim \sim \kappa(b)$ or simply $a \leq a; b$. By a similar argument we get (3). Finally if $a \leq b; c$ and $a = a \cdot 1 \leq b; c \cdot 1 \not\leq 0$ we have $b \not\leq \sim c$. We also have that $a \cdot 1 = a \not\leq 0$ and $a \leq b; c$ so $a \cdot \kappa(b); c \not\leq 0$ or $a \cdot b; \hat{b} \not\leq 0$. In the former case that means that $\kappa(b); c \not\leq 0$ and thus $\kappa(b) \leq \sim c$ or $c \leq \hat{b}$ and we are done. In the latter case it means that there exists a join-irreducible $a' \leq a$ such that $a \not\leq 0$ and thus $a' = \hat{a}'$ as well as $a' \leq b; \hat{b} \leq b; c$ by monotinicity. Because $a' \leq a \leq \hat{a}' = a'$ we have $a = a'$ and by minimality $\hat{b} = c$.

For the right to left implication note that if $s \cdot \sim s \not\leq 0$ then there exists some $a \leq s \cdot \sim s$ not below 0. Thus $\hat{a} \leq 1$ and we know there exists a join-irreducible $b = \hat{b}, b \leq a; \hat{a} \leq (s \cdot \sim s); 1 = s \cdot \sim s$ and that contradicts $b = \hat{b}$. Assume $s \not\leq t$. That is true if there exists a join-irreducible a such that $a \leq s, \hat{a} \leq \sim t$. Thus there exists a join-irreducible $b = b \cdot 1 \not\leq 0$ below $1 \cdot a; \hat{a} \leq 1 \cdot s; \sim t$ and we conclude $1 \cdot s; \sim t \not\leq 0$. If $1 \cdot s; \sim t \not\leq 0$ then there exist join-irreducibles a, b, c such that $I(a), a = \hat{a}, b \leq s, c \leq \sim t$ and b, c also being minimal and hence $\hat{b} = c$. Therefore

$\hat{b} \leq \sim t$ or simply $s \not\leq t$. $s \cdot 1 \not\leq 0 \wedge s \leq t; u \Rightarrow s; t \not\leq \sim u$ follows directly from (2) and its dual directly from (3). Finally $s \cdot 1 \not\leq 0$ and $s \leq t$; u iff there exist some a, b, c in the corresponding frame such that $a \leq s \cdot 1, b \leq t, c \leq u, I(a), \hat{a} = a, R^{\min}(a, b, c)$ and thus $\hat{b} = c$ by (4). Observe that for every v either $b \leq v$ or $\hat{b} \leq \sim v$ and thus $a \leq t; v$ or $a \leq \sim v; u$ and the join of the two terms is not below 0.   □

Theorem 14. wkRA$_2$ *is axiomatised by the basic axioms for bounded cyclic involutive unital dℓ-magmas and Φ_2.*

Proof. By Lemma 13 this axiomatisation is equivalent to the frame conditions, enumerated (1)–(4). First we show these are sound for the two pebble game. If there existed a join-irreducible a with no $b, I(b), \hat{b} = b$ with $R(b, a, \hat{a})$, then $\forall$ would win on initialisation with a because if $\lambda(x, y) = a$, no consistent b would exist for $\lambda(x, x)$. We show (4) next. If this didn't hold for some a, b, c then $\forall$ could start by asking a on initial move. By order reversing of ˆ and the identity, a is the only join-irreducible to be set as $\lambda(x, x)$ where $\lambda(x, y) = a$. For the second move, $\forall$ calls the witness $b; c$ on (x, x) and we get an inconsistency because $b \neq \hat{c}$. For (2) and (3) see that if $R(a, b, \hat{b})$ we have $R^{\min}(a, b, \hat{b})$ by order reversing properties of ˆ and (4). Thus if $\forall$ again starts by forcing $\lambda(x, x) = a$ then calling the witness $b; \hat{b}$ then both $R(b, a, b), R(\hat{b}, \hat{b}, a)$ must hold to keep the network consistent.

To show completeness, it suffices to say that $\exists$ can respond to any initialisation with a by returning a network with two nodes x, y with $\lambda(x, y) = a, \lambda(y, x) = \hat{a}$ and by (1) there exists a b for a and b' for $\hat{a}$ to be set as $\lambda(x, x)$ and $\lambda(y, y)$ respectively and by (2)(3) all other triangles are also consistent. A witness move can only be called on a reflexive node (x, x) and that means that by (2)(3)(4) any witness will be consistent and by the same reasoning as with initialisation, $\exists$ can put a label on $\lambda(y, y)$ and keep the network consistent.   □

In order to axiomatise wkRA$_3$ we only need to add two well known axioms as well as a set of quasiequations. The first axiom is called *rotation* for involutive semirings and the second one was found by Maddux in [15] as an axiom that holds for binary relations, but not for relevance logic frames.

Definition 14. Let Φ_3 be the first order theory containing all the formulas in Φ_2 as well as

(1) $s; t \leq \sim u \Rightarrow t; u \leq \sim s$
(2) $s \cdot t; u \leq ((s; v) \cdot t); u + t; (u \cdot \sim v)$
(3) $1 \cdot \sim s'; s \cdot t; \sim t' \leq 0 \Rightarrow s; t \leq (s \cdot s'); t + s; (t \cdot t')$
(4) $1 \cdot s \cdot 0 = \bot \Rightarrow (s \cdot 1); (t; u) \leq ((s \cdot 1); t); u$
(5) $1 \cdot u \cdot 0 = \bot \Rightarrow (s; t); (u \cdot 1) \leq s; (t; (u \cdot 1))$

Lemma 15. *Let $\mathcal{A}$ be a bounded cyclic involutive unital dℓ-magma. $\mathcal{A} \models \Phi_3$ if and only if for $\mathcal{F}(\mathcal{A})$ all the formulas from Lemma 13 hold as well as*

(1) $\forall a, b, c : R^{\min}(a, b, c) \Rightarrow R(b, a, \hat{c})$
(2) $\forall a, b, c : R(a, \hat{b}, \hat{c}) \Rightarrow R(b, \hat{c}, \hat{a})$

(3) $\forall a,b,c : R^{\min}(a,b,c) \Rightarrow \exists d : d = \hat{d} \wedge I(d) \wedge R(d,\hat{b},b) \wedge R(d,c,\hat{c})$

(4) $\forall a,b,c,d : d = \hat{d} \wedge I(d) \wedge R(a,d,a) \wedge R^{\min}(a,b,c) \Rightarrow R(b,d,b)$

(5) $\forall a,b,c,d : d = \hat{d} \wedge I(d) \wedge R(a,a,d) \wedge R^{\min}(a,b,c) \Rightarrow R(c,c,d)$

Proof. For the left to right implication of (1) if a,b,c are join-irreducibles with $a \le b;c$ as well as the minimality condition for b,c then see that $a = a \cdot b;c \le (a;\hat{c} \cdot b);c + b;(c \cdot \kappa(c))$. $c \cdot \kappa(c)$ is strictly below c and due to minimality of b,c for this composition $a \not\le b;(c \cdot \kappa(c))$. Thus $a \le (a;\hat{c} \cdot b);c$ and again by minimality $a;\hat{c} \cdot b = b$ or simply $R(b,a,\hat{c})$. For (2) observe that $a \not\le \hat{b};\hat{c}$ is the same as $\sim\kappa(b);\sim\kappa(c) \le \kappa(a)$ and by rotate we get $\sim\kappa(c);\sim\kappa(a) \le \kappa(b)$ and $\sim\kappa(a);\sim\kappa(b) \le \kappa(c)$ so $R(a,\hat{b},\hat{c}), R(b,\hat{c},\hat{a}), R(c,\hat{a},\hat{b})$ are equivalent. For (3) if $R^{\min}(a,b,c)$ then we know $b;c \not\le (b\cdot\kappa(b));c + b;(c\cdot\kappa(b))$ and thus $1\cdot\sim s';s\cdot t;\sim t' \not\le 0$ and we can find a d satisfying $I(d), \hat{d} = d, R(d,\hat{b},b), R(d,c,\hat{c})$. For (4) see that $1 \cdot d \cdot 0 = \bot$ and thus $a \le d;a \le d;(b;c) \le (d;b);c$. By minimality $b = d;b$. By a similar argument we get (5).

For the right to left implication, if $s;t \le \sim u$ observe that for all join-irreducibles a,b,c such that $a \le s, b \le t, c \le u$ we have $a;b \le \kappa(\hat{c})$ and thus $\neg R(\hat{c},a,b)$ and by (2) we have $\neg R(\hat{a},b,c)$ and thus $b;c \le \kappa(\hat{a}) = \sim a$. If for all join-irreducibles a,b,c below s,t,u respectively that holds then $t;u \le \sim s$. To show $s \cdot t;u \le ((s;v) \cdot t);u + t;(u \cdot \sim v)$ take any $a \le s \cdot t;u$ and some minimal b,c witnessing the $t;u$ composition. Then all v will either have $c \le \sim v$ or $\hat{c} \le v$, in either case the term is above a by monotonicity. Finally if $s;t \not\le (s \cdot s');t \cdot s;(t \cdot t')$ it means that $s;t$ is non-empty and as such there exists some $a \le s;t$ and some $R^{\min}(a,b,c)$ and as such $b \not\le s'$ and $c \not\le t'$ and thus $\hat{b};b \le \sim s';s$ and $c;\hat{c} \le t;\sim t'$ and there exists a $d \not\le 0$ such that $d \le 1 \cdot \sim s';s \cdot t;\sim t'$ and therefore the term cannot be below 0. Take any join-irreducible $a \le (s \cdot 1);t;u$. There will exist a self-^ join-irreducible $d \le s \cdot 1$ such that $d \le d;a$ and a minimal b,c below t,u such that $a \le b;c$ and so we have by (4) $b \le d;b$ and thus $a \le b;c \le (d;b);c \le ((s \cdot 1);t);u$. The dual is shown similarly from (5). $\qquad\square$

Theorem 16. wkRA$_3$ *is axiomatised by the basic axioms for bounded cyclic involutive unital dℓ-magmas and Φ_3.*

Proof. First we show that all the formulas from Lemma 15 are sound. If we have a,b,c such that $R^{\min}(a,b,c)$ then $\forall$ calls a on initialisation and calls the witness $R^{\min}(a,b,c)$ on the $\lambda(x,y) = a$ and $\exists$ must return such a network where $\lambda(x,z) = a, \lambda(y,z) = \hat{c}$ so $R(b,a,\hat{c})$ must hold for consistency and we have (1). For (2) assume without loss that we have $R(a,\hat{b},\hat{c})$ so there must be some minimal $\hat{b}' \le \hat{b}, \hat{c}' \le \hat{c}$ to call the witness on the initial pair a. Observe that for consistency $b \le b' \le \hat{c}'; \hat{a} \le \hat{c}; \hat{a}$ by monotonicity. For (3) if $\forall$ initialises with a and calls the b,c witness, $\exists$ needs a join-irreducible d to put on the reflexive edge of the added node.

From Lemma 13, Theorem 16 we have that $\exists$ can survive the initial move and we only need to examine the two possible witness moves, that on a non-reflexive edge in a two-node network and that on a reflexive edge. If a witness move $R^{min}(a,b,c)$ is called on a non-reflexive edge (x,y), check that all Peircian transformations of this triangle hold. By (1) we have $R(b, a, \hat{c})$ and through (2) we get $R(\hat{b}, c, \hat{a}), R(\hat{c}, \hat{a}, b)$ from $R(a, b, c)$ and $R(\hat{a}, \hat{c}, \hat{b}), R(c, \hat{b}, a)$ from $R(b, a, \hat{c})$. For the reflexive edge on (z, z) you can see that $\exists$ can add $\lambda(z, z) = d$ from (3) and by similar reasoning to Theorem 16 all triangles including (z, z) are consistent. Finally let $\lambda(x, x) = d$. By (4) $R(b, d, b)$ and by (2) $R(\hat{d} = d, b, \hat{b})$. The consistency of other triangles follows from formulas in Lemma 13. Similarly we get consistency for $\lambda(y, y)$. For the reflexive witness $R^{min}(d, a, \hat{a})$ on (x, x) observe due to order reversing of $\hat{}$, $\exists$ can either find a join-irreducible c such that $R^{min}(\lambda(x, y), a, c)$ or $R^{min}(\lambda(y, x), c, \hat{a})$ and $\exists$ can use the same strategy as for the non-reflexive witness move. $\square$

To axiomatise the class wkRA = wkRA$_4$ we would at least need to add associativity for composition. For RA, it is precisely the axioms for RA$_3$ and composition that axiomatise RA$_4$, however, whether this also holds for wkRA remains open.

Problem 17. What axioms are necessary to axiomatise wkRA? Is it finitely axiomatisable?

Problem 18. Let $n > 4$. RA$_n$ is not finitely axiomatisable [8]. Is the same true for wkRA$_n$?

5   Representable Diagonal Weakening Relation Algebras Form a Discriminator Variety

In this section we define *representable diagonal weakening relation algebras* as those relation algebras where 1 can be represented as an antichain. Thus in this section when we talk about the concrete binary relation 1, we mean the diagonal on X. The algebras with this property are the members of RwkRA that satisfy the identity $1 \cdot 0 = \perp$.

We show that the simple representable diagonal relation algebras have a discriminator term. A neat consequence is that, unlike representable weakening relation algebras, representable diagonal weakening relation algebras can be defined by an equational theory.

Lemma 19. *For all $R \subseteq X^2$ we have $1 \cdot (R;(R \cdot {\sim}R)) = \perp = 1 \cdot (\sim R;(R \cdot {\sim}R))$.*

Proof. Suppose there exists $(x, x') \in 1 \cdot (R;(R \cdot {\sim}R))$. Because $(x, x') \in 1$ we have $x = x'$. Thus there must exist a y to witness the composition by having $(x, y) \in R, (y, x) \in R \cdot {\sim}R$. This means that $(x, y) \in R$ and $(y, x) \in {\sim}R$ and we have reached a contradiction.

The second equation can be proven by a similar argument or by substitution of R with $\sim R$, the involution law, and the commutativity of meet. $\square$

Let $d_1(R, S) = 1 \cdot (R;(S \cdot {\sim}S))$ and $d_2(R, S) = 1 \cdot (\sim S;(R \cdot {\sim}S))$.

Lemma 20. *If $R \backslash S \neq \emptyset$ for $R, S \subseteq X^2$ then $d_1(R, S) \mid d_2(R, S) \neq \perp$.*

Proof. Assume $(x, y) \in R \backslash S$ and consider the two cases, $(y, x) \in S$ and $(y, x) \notin S$. In the first case, because $(x, y) \notin S$ we also have $(y, x) \in \sim S$ and consequently $(y, x) \in S \cdot \sim S$. Hence $(x, x) \in R; (S \cdot \sim S)$ and also by definition in 1 and thus $(x, x) \in d_1(R, S)$.

In the second case $(y, x) \notin S$ and therefore $(x, y) \in \sim S$. Because $(x, y) \notin S$, $(y, x) \notin \sim S$. By composition $(y, y) \in \sim S; (R \cdot \sim S)$ and by reflexivity of 1 we also have $(y, y) \in 1 \cdot (\sim S; (R \cdot \sim S))$.

In either case we have that at least one of $d_1(R, S), d_2(R, S)$ is nonempty and thus their join is always nonempty given $R \backslash S \neq \emptyset$.   □

Theorem 21. *Simple diagonal weakening relation algebras have a term $d(a, b, c)$ such that $d(a, b, c) = c$ if $a = b$ and $d(a, b, c) = a$ otherwise.*

Proof. It is easy to see that in simple weakening relation algebras $\top; s; \top = \top$ if $s \neq \bot$ and $\top; s; \top = \bot$ otherwise. By the lemmas above, we have for representable simple algebras that $a = b$ if and only $d_1 + d_2 = \bot$, where $d_i = d_i(a, b) + d_i(b, a)$ for $i = 1, 2$. Thus $d(a, b, c) = \top; (d_1 + d_2); \top \cdot a + \sim(\top; (d_1 + d_2); \top) \cdot c$ will equal to c if $a = b$ and a otherwise.   □

Corollary 22. *Representable diagonal weakening relation algebras form a discriminator variety.*

Proof. The representation game defined for weakening relation algebras only needs an additional move where $\exists$ is requested add 1 to $\lambda(y, x)$ if $1 \in \lambda(x, y)$ and this game gives rise to a similar style of a recursive axiomatisation as presented in Proposition 2. If all variables are given unique names, the universal quantifiers can also be moved to the begining of all these formulas. Observe that although these formulas apply to all algebras, the game is played on the homomorphic image of the algebra where $\top$ maps to $\top; a; \top$ where $\sigma_n = s \not\leq t \Rightarrow (\phi_n(\mathcal{N}^{1,s,t}) \vee \phi_n(\mathcal{N}^{2,s,t}))$. Thus we can construct a term from any universally quantified first order formula that is equal to $\top; a; \top$ if and only if the formula is true and $\bot$ otherwise. For equations $t = t'$ we take $\top; a; \top \cdot \sim d(t, t', \top; a; \top)$. If a term t corresponds to a formula, then $\sim t \cdot \top; a; \top$ corresponds to its negation and for disjunctions we can take the join of the corresponding terms. Thus every formula σ_n has an equivalent equation.   □

6   Representing Associative Members of **wkRA$_3$** with Weakening Relations

Sugihara monoids are commutative distributive idempotent involutive residuated lattices. This variety is semilinear, i.e., generated by linearly ordered algebras, and the structure of these algebras is well known. In particular, the Sugihara monoid S_n is a chain with n elements $\{a_{-k}, a_{-k+1}, \ldots, a_{-1}, a_0, a_1, \ldots, a_{k-1}, a_k\}$ if $n = 2k + 1$ is odd, and otherwise for even n, $S_n = S_{n+1} \backslash \{a_0\}$. The involution operation is given by $\sim a_i = a_{-i}$ and the multiplication is $a_i; a_j = a_{-\max|i|, |j|}$. It follows that in the odd case the identity element is $1 = a_0$ and in the even case it is $1 = a_1$.

Note that S_2 is the 2-element Boolean algebra and that for even n, there is a surjective homomorphism from S_n to S_{n-1} that identifies a_1 and a_{-1}.

It is proved in [14] that the even Sugihara chains can be represented by algebras of weakening relations. For S_2 this is clear since $S_2 \cong \text{Rel}(1)$. For S_4 an infinite base set is needed with a dense order. E.g., we can take $(\mathbb{Q}, \leq)$ be the poset of rational numbers with the standard order and check that $S_4 \cong \{\emptyset, <, \leq , \mathbb{Q}^2\}$ is a representation in $\text{Wk}(\mathbb{Q}, \leq)$.

It follows from the consistency of networks that no nontrivial member of wkRA_2 has an element that satisfies $a = \sim a$. Hence any finite member of wkRA has an even number of elements. In particular, the odd Sugihara chains do not have a representation by weakening relations. However they are in the variety generated by all algebras of weakening relations since they are homomorphic images of even Sugihara chains. This shows that RwkRA is not closed under homomorphic images, so it is a proper quasivariety.

Let $\mathbf{2} = \{0, 1\}$ be the two element chain with $0 < 1$. The algebra $\textbf{wk(2)}$ is shown in Fig. 2, and it has the following six elements: $\emptyset$, $\{(0,1)\}$, $\{(0,0),(0,1)\}$, $\{(0,1),(1,1)\}$, $\leq$, $\mathbf{2} \times \mathbf{2}$.

The *point algebra* $\mathbf{P}$ shown in Fig. 1 (see also [7]) is a representable relation algebra with 3 atoms $id_{\mathbb{Q}}, <, >$ where $<$ is the strict order on the rational numbers $\mathbb{Q}$. It has two weakening subalgebras: $\mathbf{S_4} = \{\emptyset, <, \leq, \top\}$ and $\mathbf{W_{6,1}} = \{\emptyset, id_{\mathbb{Q}}, <, \leq, <\text{U}>, \top\}$. Like the point algebra, both of these algebras can only be represented on an infinite set. Note that $\mathbf{W_{6,1}}$ is diagonally representable, while $\mathbf{S_4}$ is not.

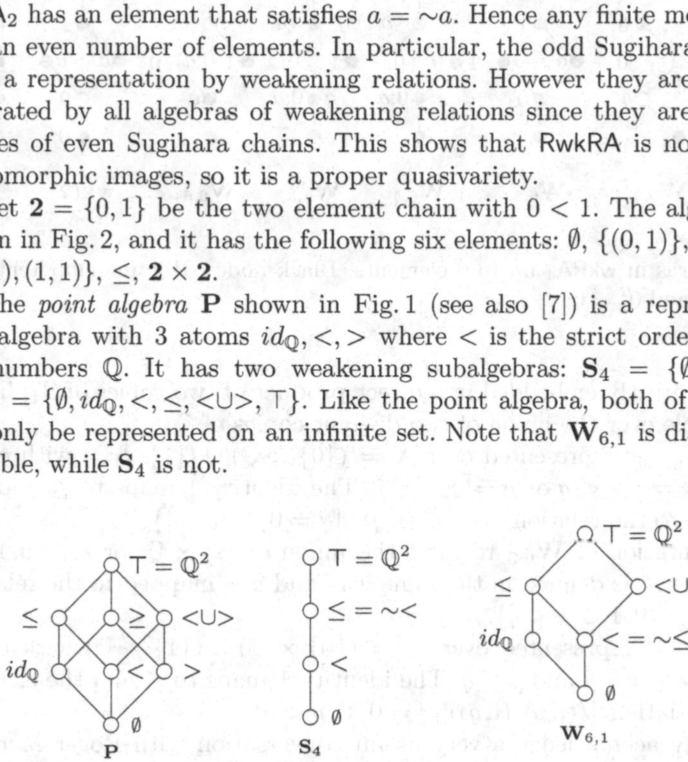

Fig. 1. The point algebra $\mathbf{P}$, the weakening subalgebra $\mathbf{S_4}$ and the diagonally representable weakening subalgebra $\mathbf{W_{6,1}}$.

Since wkRA_3 is finitely axiomatised, one can use a model finder such as Mace4 [16] to compute all members of cardinality n for small values of n. Up to isomorphism there are 14 algebras with 6 elements or fewer in wkRA_3 such that ; is associative, shown in Fig. 2. We now briefly describe their representations by weakening relations.

The first 5 are symmetric representable relation algebras, hence they are diagonally representable weakening relation algebras.

As mentioned above, the Sugihara algebra $\mathbf{S_4}$ and the algebra $\mathbf{W_{6,1}}$ are representable as subalgebras of the $\sim$-reduct of the point algebra (Fig. 1). The algebra $\mathbf{W_{6,2}}$ is representable as $\sim$-subreduct of the complex algebra of $\mathbb{Z}_7$, where the element $a = \{1, 2, 4\}$ and $1 = \{0\}$.

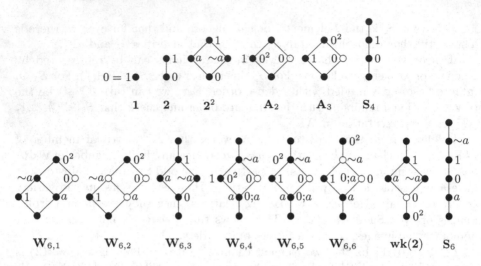

Fig. 2. All algebras in wkRA$_4$ up to 6 elements. Black nodes denote idempotent elements ($x;x = x$) and $0^2 = 0;0$.

$\mathbf{W}_{6,3}$ is subdirectly embedded in a direct product of two copies of $\mathbf{S}_4$, hence it is representable over the union of two disjoint copies of $\mathbb{Q}$.

Similarly $\mathbf{W}_{6,4}$ is represented over $X = (\{0\} \times \mathbb{Q}) \cup (\{1\} \times \mathbb{Q})$ with order $(i,p) \leq (j,q) \iff p < q$ or $p = q, i = j$. The identity 1 maps to $\leq$ and the element a maps to the relation $\{((i,p),(i,q)) \mid i = 0,1, p < q\}$.

The representation of $\mathbf{W}_{6,5}$ requires the union of $\{i\} \times \mathbb{Q}$ for $i \in \{0,1,2\}$. The partial order $\leq$ is defined in the same way and a is mapped to the relation $\{((i,p),(i,q)) \mid i = 0,1,2, p < q\}$.

Finally $\mathbf{W}_{6,6}$ is represented over $X = (\{0\} \times \mathbb{Q}) \cup (\{1\} \times \mathbb{Q})$ with order $(i,p) \leq (j,q) \iff i = j$ and $p \leq q$. The identity 1 maps to $\leq$ and the element a maps to the relation $\{((i,p),(i,q)) \mid i = 0,1, p < q\}$.

We gratefully acknowledge a very useful conversation with Roger Maddux regarding relevance frames, relevance logic and its connections with relation algebras. In particular, formulas (3.101), (3.102) in [15] provided key insights into the axiomatisation of wkRA$_3$.

References

1. Bimbó, K., Dunn, J.M., Maddux, R.D.: Relevance logics and relation algebras. Rev. Symb. Log. **2**(1), 102–131 (2009). https://doi.org/10.1017/S17550203090090145
2. Galatos, N., Jipsen, P.: Distributive residuated frames and generalized bunched implication algebras. Algebra Univers. **78**(3), 303–336 (2017). https://doi.org/10.1007/s00012-017-0456-x
3. Galatos, N., Jipsen, P.: The structure of generalized BI-algebras and weakening relation algebras. Algebra Univers. **81**(3), 1–35 (2020). https://doi.org/10.1007/s00012-020-00663-9

4. Galatos, N., Jipsen, P.: Weakening relation algebras and FL^2-algebras. In: Fahrenberg, U., Jipsen, P., Winter, M. (eds.) RAMiCS 2020. LNCS, vol. 12062, pp. 117–133. Springer, Cham (2020). https://doi.org/10.1007/978-3-030-43520-2_8
5. Givant, S.: Advanced Topics in Relation Algebras, vol. 2. Springer, Heidelberg (2017). https://doi.org/10.1007/978-3-319-65945-9
6. Givant, S.: Introduction to Relation Algebras, vol. 1. Springer, Heidelberg (2017). https://doi.org/10.1007/978-3-319-65235-1
7. Hirsch, R.: Relation algebras of intervals. Artif. Intell. **83**(2), 267–295 (1996). https://doi.org/10.1016/0004-3702(95)00042-9
8. Hirsch, R., Hodkinson, I.: Relation Algebras by Games. Elsevier, Amsterdam (2002)
9. Jipsen, P.: Relation algebras, idempotent semirings and generalized bunched implication algebras. In: Höfner, P., Pous, D., Struth, G. (eds.) RAMICS 2017. LNCS, vol. 10226, pp. 144–158. Springer, Cham (2017). https://doi.org/10.1007/978-3-319-57418-9_9
10. Kurz, A., Velebil, J.: Relation lifting, a survey. J. Log. Algebraic Methods Program. **85**(4), 475–499 (2016)
11. Maddux, R.: Relation Algebras, vol. 13. Elsevier, Amsterdam (2006)
12. Maddux, R.: Some varieties containing relation algebras. Trans. Amer. Math. Soc. **272**(2), 501–526 (1982). https://doi.org/10.2307/1998710
13. Maddux, R.: A sequent calculus for relation algebras. Ann. Pure Appl. Log. **25**(1), 73–101 (1983). https://doi.org/10.1016/0168-0072(83)90055-6
14. Maddux, R.D.: Relevance logic and the calculus of relations. Rev. Symb. Log. **3**(1), 41–70 (2010). https://doi.org/10.1017/S1755020309990293
15. Maddux, R.D.: Tarskian classical relevant logic. In: Düntsch, I., Mares, E. (eds.) Alasdair Urquhart on Nonclassical and Algebraic Logic and Complexity of Proofs. OCL, vol. 22, pp. 67–161. Springer, Cham (2022). https://doi.org/10.1007/978-3-030-71430-7_3
16. McCune, W.: Prover9 and Mace4 (2005–2010). www.cs.unm.edu/ mccune/prover9/
17. Monk, D.: On representable relation algebras. Michigan Math. J. **11**, 207–210 (1964). projecteuclid.org/euclid.mmj/1028999131
18. Reynolds, J.C.: Separation logic: a logic for shared mutable data structures. In: Proceedings 17th Annual IEEE Symposium on Logic in Computer Science, pp. 55–74. IEEE (2002)
19. Smyth, M.: Stable compactification i. J. Lond. Math. Soc. **2**(2), 321–340 (1992)
20. Stell, J.G.: Relations on hypergraphs. In: Kahl, W., Griffin, T.G. (eds.) RAMICS 2012. LNCS, vol. 7560, pp. 326–341. Springer, Heidelberg (2012). https://doi.org/10.1007/978-3-642-33314-9_22
21. Stell, J.G.: Symmetric Heyting relation algebras with applications to hypergraphs. J. Log. Algebr. Methods Program. **84**(3), 440–455 (2015). https://doi.org/10.1016/j.jlamp.2014.12.001
22. Tarski, A.: Contributions to the theory of models. J. Symb. Log. **21**(4) (1956)

Completeness and the Finite Model Property for Kleene Algebra, Reconsidered

Tobias Kappé[1,2]([✉]) [iD]

[1] Open University of the Netherlands, Heerlen, The Netherlands
tobias.kappe@ou.nl
[2] ILLC, University of Amsterdam, Amsterdam, The Netherlands

Abstract. Kleene Algebra (KA) is the algebra of regular expressions. Central to the study of KA is Kozen's (1994) *completeness* result, which says that any equivalence valid in the language model of KA follows from the axioms of KA. Also of interest is the *finite model property* (FMP), which says that false equivalences always have a finite counterexample. Palka (2005) showed that, for KA, the FMP is equivalent to completeness.

We provide a unified and elementary proof of both properties. In contrast with earlier completeness proofs, this proof does not rely on minimality or bisimilarity techniques for deterministic automata. Instead, our approach avoids deterministic automata altogether, and uses Antimirov's derivatives and the well-known transition monoid construction.

Our results are fully verified in the Coq proof assistant.

1 Introduction

Kleene Algebra (KA) [10,18] provides an algebraic perspective on the equivalence of regular expressions. It is the foundation for Kleene Algebra with Tests (KAT) [9,19,20,24], which has been applied to reason about equivalence of programs in general [22,25], and programming languages such as NetKAT [1,34].

Central to Kleene Algebra and its extensions is the *completeness* property, which says that every equivalence valid in the language model can be proved using the laws of KA. Salomaa showed an important precursor to this result [32], and other authors [6,10,13,26] have studied alternative axiomatizations.

The axiomatization most commonly used today is due to Kozen [18], and has the advantage of being *algebraic*, i.e., it allows one to define a "Kleene algebra" as a model that may verify or falsify equations. A number of alternative proofs of the same result have been proposed [12,14,21,23]; notably, it was shown that one of the quasi-equations can be dropped from Kozen's axioms [12,23].

Another phenomenon of interest is the *finite model property* (FMP) [5]. For KA, the FMP states that any *invalid* equivalence is witnessed by some finite Kleene algebra where it does not hold—contrapositively, equivalences valid in any finite Kleene algebra are also valid in any (possibly infinite) Kleene algebra.

The author was supported by the EU's Horizon 2020 research and innovation programme under Marie Skłodowska-Curie grant agreement No. 101027412 (VERLAN).

R. Glück et al. (Eds.): RAMiCS 2023, LNCS 13896, pp. 158–175, 2023.
https://doi.org/10.1007/978-3-031-28083-2_10

Palka [29] showed that the FMP is a consequence of completeness for KA, and moreover that completeness can be recovered if one assumes the FMP. This equivalence raises a question: can we provide an elementary proof of the FMP for KA, i.e., one that does not rely on completeness? Indeed, Palka writes that "an independent proof of [the FMP] would provide a quite different proof of the Kozen completeness theorem, based on purely logical tools" [29].

Our main contribution is a positive answer to this question, providing a proof of the FMP for KA. More specifically, our argument weaves together considerations from Palka's proof as well as classical facts from automata theory, in such a way that both the FMP and completeness can be concluded.

In contrast with earlier completeness proofs, our method does not center on minimality [18], bisimilarity [14,23] or the construction of a cyclic proof system [12]. Instead, we rely purely on the fact that KA allows one to find least solutions to linear systems [3,10,19], or in our case, to automata. The arguments towards our main result exploit this property in concert with various ideas around automata, such as the transition monoid [27], and Antimirov's construction [2], eventually building a particular finite Kleene algebra with sufficient structure to conclude both completeness and the finite model property.

The remainder of this paper is organized as follows. Section 2 provides an overview of the context, and defines fundamental notions. Section 3 recalls the notion of *solutions to an automaton*, a technique that will be leveraged repeatedly. Sections 4 and 5 provide an algebraic perspective on transformation automata [27], and Antimirov's construction [2] respectively. Section 6 shows how to construct a particularly useful Kleene algebra, and Sect. 7 shows how to conclude completeness and the FMP using the notions discussed up to that point. Section 8 concludes with some discussion and suggestions for future work.

To save space, proofs of auxiliary facts appear in the full version [15]. Our formalization of the proofs in Coq is available online [16].

2   Overview

Our primary objects of study are *Kleene algebras*. The equations that hold in a Kleene algebra correspond well to properties expected of program composition, which makes them a suitable semantic domain for programs.

Definition 2.1. *A* (weak[1]) *Kleene algebra* (KA) *is a tuple* $(K, +, \cdot, {}^*, 0, 1)$, *where K is a set (the* carrier*), $+$ and $\cdot$ are binary operators on K, * is a unary operator on K, and $0, 1 \in K$ are constants, satisfying the following for all $x, y, z \in K$:*

$$x + 0 = x \qquad x + x = x \qquad x + y = y + x \qquad x + (y + z) = (x + y) + z$$

$$x \cdot (y \cdot z) = (x \cdot y) \cdot z \qquad x \cdot (y + z) = x \cdot y + x \cdot z \qquad (x + y) \cdot z = x \cdot z + y \cdot z$$

$$x \cdot 1 = x = 1 \cdot x \qquad x \cdot 0 = 0 = 0 \cdot x \qquad 1 + x \cdot x^* = x^* \qquad x + y \cdot z \leq z \implies y^* \cdot x \leq z$$

---

[1] This is a "weak" KA in the sense that it does not require the right unrolling and right-fixpoint laws from [19]. As it turns out, this does not change the equational theory [12,23]. For the sake of brevity, we omit "weak" in the remainder of this paper.

Here, we use ≤ to denote the natural order *induced by +, that is, $x \leq y$ if and only if $x + y = y$; it is straightforward to verify that this makes ≤ a partial order on K, and that all operators are monotone w.r.t. this order.*

We often denote a generic KA $(K, +, \cdot, ^*, 0, 1)$ by its carrier K, and simply write $+$, $\cdot$, etc. for the operators and constants when there is no risk of ambiguity.

Typically, the additive operator $+$ is used to implement nondeterministic composition, the multiplicative operator $\cdot$ corresponds to sequential composition, the *Kleene star* operator * implements iteration, 0 represents a program that fails immediately, and 1 is the program that does nothing and terminates successfully. The equations of KA correspond well to what might be expected of such operators on programs—for instance, and iteration is characterized as a least fixpoint.

One very natural instance of Kleene algebras, which we will connect to the interpretation of programs shortly, is given by the *relational model*.

Example 2.2 (KA of relations). Let X be a set. The set of relations on X, i.e., $\mathcal{P}(X \times X)$, can be equipped with a KA $\mathcal{R}_X = (\mathcal{P}(X \times X), \cup, \circ, ^*, \emptyset, \mathrm{id}_X)$, in which $\circ$ is relational composition; * is the reflexive-transitive closure operator on relations; and id_X is the diagonal or identity relation on X given by $\{(x, x) : x \in X\}$.

When interpreting programs in $\mathcal{R}_X$, we think of the relations on X as a way of representing how a program may transform the machine states represented by X. To make this more precise, we need a syntax and semantics for programs.

Definition 2.3 (Expressions). *We fix a set of* actions $\Sigma = \{\mathsf{a}, \mathsf{b}, \mathsf{c}, \dots\}$ *called the* alphabet. *The set of* regular expressions $\mathbb{E}$ *is given by*

$$e, f ::= 0 \mid 1 \mid \mathsf{a} \in \Sigma \mid e + f \mid e \cdot f \mid e^*$$

Given a KA K and a function $h : \Sigma \to K$, we can define $\widehat{h} : \mathbb{E} \to K$ inductively:

$$\widehat{h}(0) = 0 \qquad \widehat{h}(\mathsf{a}) = h(\mathsf{a}) \qquad \widehat{h}(e \cdot f) = \widehat{h}(e) \cdot \widehat{h}(f)$$

$$\widehat{h}(1) = 1 \qquad \widehat{h}(e + f) = \widehat{h}(e) + \widehat{h}(f) \qquad \widehat{h}(e^*) = \widehat{h}(e)^*$$

Example 2.4. Consider a programming language with integer variables $\mathsf{Var} = \{\mathsf{x}, \mathsf{y}, \dots\}$, and statements Σ comprised of (for all $\mathsf{x}, \mathsf{y} \in \mathsf{Var}$, $n \in \mathbb{N}$ and $v \in \mathsf{Var} \cup \mathbb{N}$) *assignments* $\mathsf{x} \leftarrow n$, *increments* $\mathsf{x} \leftarrow \mathsf{x} + v$, and *comparisons* $\mathsf{x} < \mathsf{y}, \mathsf{x} \geq \mathsf{y}$.

The state of the machine is defined by the value of each variable, and so we choose $S = \{\sigma : \mathsf{Var} \to \mathbb{N}\}$ as the state space. The semantics of the actions are relations that represent their effect, i.e., we define $h : \Sigma \to \mathcal{P}(S \times S)$ by

$$h(\mathsf{x} \leftarrow n) = \{(\sigma, \sigma[n/\mathsf{x}]) : \sigma \in S\}$$
$$h(\mathsf{x} \leftarrow \mathsf{x} + n) = \{(\sigma, \sigma[\sigma(\mathsf{x}) + n/\mathsf{x}]) : \sigma \in S\}$$
$$h(\mathsf{x} \leftarrow \mathsf{x} + \mathsf{y}) = \{(\sigma, \sigma[\sigma(\mathsf{x}) + \sigma(\mathsf{y})/\mathsf{x}]) : \sigma \in S\}$$
$$h(\mathsf{x} < \mathsf{y}) = \{(\sigma, \sigma) : \sigma \in S, \sigma(\mathsf{x}) < \sigma(\mathsf{y})\}$$
$$h(\mathsf{x} \geq \mathsf{y}) = \{(\sigma, \sigma) : \sigma \in S, \sigma(\mathsf{x}) \geq \sigma(\mathsf{y})\}$$

Here, $\sigma[n/\mathsf{x}]$ is denotes the function that assigns n to x, and $\sigma(\mathsf{y})$ to all $\mathsf{y} \neq \mathsf{x}$.

This gives us a semantics $\hat{h} : \mathbb{E} \to \mathcal{P}(S \times S)$ for regular expressions over Σ, and allows us to express and interpret programs like the following:

$$\mathsf{x} \leftarrow 1 \cdot \mathsf{y} \leftarrow 0 \cdot \mathsf{i} \leftarrow 0 \cdot (\mathsf{i} < \mathsf{n} \cdot \mathsf{y} \leftarrow \mathsf{y} + \mathsf{x} \cdot \mathsf{x} \leftarrow \mathsf{x} + 2 \cdot \mathsf{i} \leftarrow \mathsf{i} + 1)^* \cdot (\mathsf{i} \geq \mathsf{n})$$

which will compute the square of n and store it in y.

Of course, one can build more involved programming languages based on KA; one elaborate and well-studied instance is NetKAT [1], a programming language for specifying and reasoning about software-defined networks.

Let $e, f \in \mathbb{E}$. When $\hat{h}(e) = \hat{h}(f)$ for all $h : \Sigma \to K$, we write $K \models e = f$. If $\mathfrak{C}$ is a class of KAs and $K \models e = f$ for each KA K in $\mathfrak{C}$, then we write $\mathfrak{C} \models e = f$. We use $\equiv$ for the smallest congruence on $\mathbb{E}$ that satisfies the axioms of KA, and use $e \leq f$ as shorthand for $e + f \equiv f$. One can show that all operators are monotone w.r.t. the preorder $\leq$, and that $e \leq f$ and $f \leq e$ together imply $e \equiv f$.

We use $[\phi]$ for some logical condition ϕ to denote $1 \in \mathbb{E}$ when ϕ holds, and $0 \in \mathbb{E}$ otherwise. We also use the familiar $\sum$ notation to generalize $+$. The empty sum is defined to be 0, the unit of $+$. Note that the sum notation is well-defined up to $\equiv$, because $+$ is associative, commutative and idempotent in any KA.

The following is a standard fact of universal algebra—see, e.g., [8].

Lemma 2.5. *Let $e, f \in \mathbb{E}$. We have $e \equiv f$ iff $K \models e = f$ for all KAs K.*

Given that KA provides such a suitable semantic domain, can we characterize the *equational theory* of KA, i.e., the equations valid in all models (programming languages) captured by $\equiv$, as the equations that hold for a particular model or class of models? Conversely, can we guarantee any properties of countermodels (pathological programming languages) that witness invalid equations?

Kozen [19] answered these questions by showing that the valid equations of KA are characterized by the *language model*. Intuitively, this model assigns to each expression the set of possible sequences of actions that may be executed by the program it represents. We will now make this more precise.

A *word* is a finite sequence of actions $\mathsf{a}_1 \cdots \mathsf{a}_n$; the *empty word* (with no letters) is denoted ϵ. We write Σ^* for the set of words, and denote its elements by $w, x, y, \ldots$. We can *concatenate* words by juxtaposing them, i.e., if $w = \mathsf{a}_1 \cdots \mathsf{a}_n$ and $x = \mathsf{b}_1 \cdots \mathsf{b}_m$, then wx is the word given by $\mathsf{a}_1 \cdots \mathsf{a}_n \mathsf{b}_1 \cdots \mathsf{b}_m$.

A set of words $L, K, M, \ldots$ is called a *language*. We can combine languages as one would combine sets (e.g., by taking their union). The concatenation of words can also be lifted to languages in a pointwise manner, writing $L \cdot K$ for the set $\{wx : w \in L, x \in K\}$. Finally, the *Kleene star* of a language L, denoted L^*, is the set $\{w_1 \cdots w_n : w_1, \ldots, w_n \in L\}$. Note that L^* includes the empty word.

Definition 2.6. *The* KA *of languages* $\mathcal{L}$ *is given by* $(\mathcal{P}(\Sigma^*), \cup, \cdot, ^*, \emptyset, \{\epsilon\})$, *where* $\cdot$ *is language concatenation, and* * *is the Kleene star of a language as above. We furthermore define the function* $\ell : \Sigma \to \mathcal{P}(\Sigma^*)$ *by* $\ell(\mathsf{a}) = \{\mathsf{a}\}$.

Remark 2.7. Readers familiar with regular languages will recognize that $\hat{\ell} : \mathbb{E} \to \mathcal{P}(\Sigma^*)$ is the standard language interpretation of regular expressions.

Algebraically, Kozen's theorem can now be stated as follows.

Theorem 2.8 (Kozen [19]). *For all $e, f \in \mathbb{E}$, we have that $e \equiv f$ iff $\mathcal{L} \models e = f$.*

One of the payoffs of this result is that we can decide $e \equiv f$ by checking whether $\mathcal{L} \models e = f$, which turns out to be a PSPACE-complete problem [36].

Another property, known as the *finite model property* and proved by Palka [29], states that the equational theory of KA can also be characterized by the class of *finite* KAs, denoted $\mathfrak{F}$. Her result can be stated as follows:

Theorem 2.9 (Palka [29]). *For all $e, f \in \mathbb{E}$, we have that $e \equiv f$ iff $\mathfrak{F} \models e = f$.*

In her proof of the above, Palka applied Kozen's theorem. The central contribution of this paper is that *both theorems can be proved independently of each other*, by a generic construction that allows one to conclude either result.

3 Solutions to Automata

In this section, we recall *automata* as a way of defining a language, as well as the notion of the *least solution* to an automaton. Both of these are well-known, but since they play such a central role for our results we discuss them in detail.

Definition 3.1 (Automaton). *A (non-deterministic finite) automaton A is a tuple (Q, δ, I, F) where Q is a finite set of* states, *$\delta : Q \times \Sigma \to \mathcal{P}(Q)$ is the* transition function *and $I \subseteq Q$ (resp. $F \subseteq Q$) holds the* initial *(resp.* final*) states.*

For $\mathsf{a} \in \Sigma$, we write δ_{a} for the relation given by $\{(q, q') : q' \in \delta(q, \mathsf{a})\}$. This family of relations can be extended to words, as follows:

$$\delta_\epsilon = \mathsf{id}_Q \qquad\qquad \delta_{w\mathsf{a}} = \delta_w \circ \delta_{\mathsf{a}}$$

The language *of a state $q \in Q$, denoted $L(A, q)$, is the set of words w such that q can reach a final state through δ_w, given by $L(A, q) = \{w \in \Sigma^* : q \, \delta_w \, q_f \in F\}$. The* language *of A is defined by its initial states, i.e., $L(A) = \bigcup_{q_i \in I} L(A, q_i)$.*

It is well known that the set of languages defined by regular expressions is the same as the set of languages described by (finite) automata [17]. In fact, the translations that demonstrate this equivalence will play an important role in the remainder of this paper, and we will outline one of these now.

Definition 3.2 (Solutions). *Let $A = (Q, \delta, I, F)$ be an automaton, and $e \in \mathbb{E}$. An e-solution to A is a function $s : Q \to \mathbb{E}$ s.t. the following hold for all $q, q' \in Q$:*

$$q \in F \implies e \leq s(q) \qquad\qquad q' \in \delta(q, \mathsf{a}) \implies \mathsf{a} \cdot s(q') \leq s(q)$$

A 1-solution to A is simply called a solution *to A. We say that s is the* least *e-solution to A if for all e-solutions s' it holds for all $q \in Q$ that $s(q) \leq s'(q)$.*

Least e-solutions are unique up to the laws of KA; this explains why we can speak of *the* least e-solution to an automaton.

3.1 Computing Solutions

It is well-known that least e-solutions always exist for (finite) automata; the process to compute these [10,18] closely resembles the *state elimination* technique from [17,28], which computes a regular expression representing the language accepted by an automaton.

Theorem 3.3 (Computing solutions). *Let $A = (Q, \delta, I, F)$ be an automaton, and let $e \in \mathbb{E}$. We can compute the least e-solution to A, denoted $\overline{A}^e$.*

In fact, the above statement can be strengthened: as it turns out, the least solution to A gives rise to *all* of the least e-solutions [18], in the following sense.

Theorem 3.4 (Relating solutions). *Let $A = (Q, \delta, I, F)$ be an automaton, and let $e \in \mathbb{E}$. For all $q \in Q$, it holds that $\overline{A}^1(q) \cdot e \equiv \overline{A}^e(q)$.*

The two results above form the technical nexus of this paper, and will be applied repeatedly throughout the coming three sections. The second result in particular, which connects solutions to e-solutions, will prove to be rather useful.

To lighten notation, we will simply write $\overline{A}$ for $\overline{A}^1$, which we call the *least solution to A*. We also write $\lfloor A \rfloor$ for the expression $\sum_{q \in I} \overline{A}(q)$.

3.2 Properties of Solutions

We conclude this section by recording three more properties of solutions. For the remainder of this section, we fix two automata $A_i = (Q_i, \delta_i, I_i, F_i)$ for $i \in \{1, 2\}$.

For the first property, we need to define *morphisms* of automata.

Definition 3.5. *A morphism from A_1 to A_2 is a function $h : Q_1 \to Q_2$ where (1) if $q \in F_1$ then $h(q) \in F_2$, and (2) if $q' \in \delta_1(q, \mathsf{a})$, then $h(q') \in \delta_2(h(q), \mathsf{a})$. Furthermore, h is strong when for all $q \in I_1$ we have that $h(q) \in I_2$.*

Morphisms between automata relate their least solutions, as follows.

Lemma 3.6. *Let $h : Q_1 \to Q_2$ be a morphism from A_1 to A_2. For all $q \in Q$, it holds that $\overline{A_1}(q) \leq \overline{A_2}(h(q))$. Furthermore, if h is strong, then $\lfloor A_1 \rfloor \leq \lfloor A_2 \rfloor$.*

For the second property, we need the notion of a subautomaton.

Definition 3.7. *We say A_1 is a subautomaton of A_2 when $Q_1 \subseteq Q_2$, and furthermore for all $\mathsf{a} \in \Sigma$ we have that $\delta_1(q, \mathsf{a}) = \delta_2(q, \mathsf{a})$.*

Unsurprisingly, the least solution to a subautomaton coincides with the least solution of the automaton that contains it, on the states where they overlap.

Lemma 3.8. *If A_1 is a subautomaton of A_2 and $q \in Q_1$, then $\overline{A_1}(q) \equiv \overline{A_2}(q)$.*

The third and last property that we will use connects the least solution of an automaton to the languages of that automaton.

Lemma 3.9. *Both of the following hold for all* $q \in Q_1$:

$$\overline{A_1}(q) \equiv [q \in F] + \sum_{q' \in \delta(q,\mathsf{a})} \mathsf{a} \cdot \overline{A_1}(q') \qquad \widehat{\ell}(\overline{A_1}(q)) = L(A_1, q)$$

Here, $[q \in F]$ *is shorthand for* 1 *if* $q \in F$, *and* 0 *otherwise.*

4   Transformation Automata

Throughout this section, we fix an automaton $A = (Q, \delta, I, F)$.

We now turn our attention to *transformation automata* [27]. Intuitively, the states of a transformation automaton A' obtained from A are relations on Q, with the intention that reading $w \in \Sigma^*$ starting from a state R in A' leads (uniquely) to $R \circ \delta_w$. In particular, reading w in A' from id_Q takes us to δ_w, which is why we will pay special attention to the solutions to id_Q in transformation automata.

Definition 4.1. *We define* $\delta^\tau : \mathcal{P}(Q \times Q) \times \Sigma \to \mathcal{P}(\mathcal{P}(Q \times Q))$ *by setting*

$$\delta^\tau(R, \mathsf{a}) = \{R \circ \delta_\mathsf{a}\}$$

For each $R \subseteq Q \times Q$, *write* $A[R]$ *for the* R-*transformation automaton*

$$(\mathcal{P}(Q \times Q), \delta^\tau, \{\mathrm{id}_Q\}, \{R\})$$

Note that the above still fits our definition of an automaton, since if Q is finite then so is the set of relations on Q, i.e., $\mathcal{P}(Q \times Q)$. It is also useful to point out that transformation automata are *deterministic*, in that each state leads to one (and only one) next state for a given letter.

Remark 4.2. Readers familiar with formal language theory may recognize transformation automata as the construction used to show that each language accepted by an automaton can also be recognized by a (finite) monoid [27].

In the remainder of this section, we characterize the solution to A in terms of solutions to its transformation automata. To this end, we first analyze the solutions to transformation automata in general. A useful first observation is that, for each $\mathsf{a} \in \Sigma$, words read from id_Q to δ_a in the transformation automaton include a. This gives rise to the following property on the level of solutions.

Lemma 4.3. *For all* $\mathsf{a} \in \Sigma$, *it holds that* $\mathsf{a} \leq \lfloor A[\delta_\mathsf{a}] \rfloor$.

Furthermore, if R_1, R_2 and R_3 are relations, and if we can read w by moving from R_1 to R_2, then we can also read w by moving from $R_3 \circ R_1$ to $R_3 \circ R_2$. This can be expressed in terms of solutions to transformation automata, as follows.

Lemma 4.4. *For all $R_1, R_2, R_3 \subseteq Q \times Q$, it holds that*

$$\overline{A[R_2]}(R_1) \leqq \overline{A[R_3 \circ R_2]}(R_3 \circ R_1)$$

Proof Sketch. Let's fix R_2 and R_3. We choose $s : \mathcal{P}(Q \times Q) \to \mathbb{E}$ by setting $s(R) = \overline{A[R_3 \circ R_2]}(R_3 \circ R)$. Now show that s is a solution to $A[R_2]$. $\qquad\square$

We can think of the least solution to id_Q in the R-transformation automaton of A as an expression representing all words w such that $\delta_w = R$. This explains the next property, which is an algebraic encoding of the fact that if $w_1, w_2 \in \Sigma^*$ are such that $\delta_{w_1} = R_1$ and $\delta_{w_2} = R_2$, then $\delta_{w_1 \cdot w_2} = \delta_{w_1} \circ \delta_{w_2} = R_1 \circ R_2$.

Lemma 4.5. *For all $R_1, R_2 \subseteq Q$ it holds that $\lfloor A[R_1] \rfloor \cdot \lfloor A[R_2] \rfloor \leqq \lfloor A[R_1 \circ R_2] \rfloor$*

Proof Sketch. Using Lemmas 4.3 and 4.4, one can show that $\overline{A[R_1 \circ R_2]}$ is an $\lfloor A[R_2] \rfloor$-solution to $A[R_1]$, which implies the claim. $\qquad\square$

With this property in hand, we can now express the least solution to A in terms of the least solutions to its transformation automata, as follows.

Lemma 4.6. *For all $q \in Q$ it holds that $\overline{A}(q) \equiv \sum_{qRq_f \in F} \lfloor A[R] \rfloor$*

Proof (Proof sketch). To show that the left-hand side is contained in the right-hand side, we use the preceding lemmas to show that it constitutes a solution to A. For the converse containment, we argue that the solution of A gives rise to a solution to each of the automata $A[R]$ that appear on the right-hand side. $\qquad\square$

5   Antimirov's Construction

We now discuss the least solution to an automaton A_e that accepts $\widehat{\ell}(e)$, for each $e \in \mathbb{E}$. Many methods to obtain such an automaton exist (for instance [7,38]; see [39] for a good overview). We focus on Antimirov's construction [2], and show that an expression e can be recovered from its Antimirov automaton.

Remark 5.1. In a sense, the property we prove is analogous to the one shown by Kozen [21] (see also [14]), who proved that e can recovered from the solution to its Brzozowski automaton [7]. We diverge from this for two reasons.

1. Antimirov's construction produces non-deterministic automata, which makes it a bit easier to express than Brzozowski's construction, which uses *deterministic* automata. In particular, this saves us from having to consider the theory necessary to make Brzozowski's construction produce a *finite* automaton.

2. Kozen's result about Brzozowski's construction leverages the fact that *bisimilar automata have equivalent solutions*. This is a very powerful (and somewhat tricky to prove) observation, which also underlies the completeness proof in [14,21]. For Antimirov automata, however, it turns out that we can rely on *morphisms* of automata instead, which are fairly easy to establish.[2]

---

[2] Kozen's bisimilarity property would also not help us obtain a completeness proof using Antimirov automata instead of Brzozowski automata, because some non-deterministic automata are not bisimilar despite having the same language; see also Remark 5.13.

Having said that, the structure of the proof that follows is very much inspired by the strategy employed in [14, 21], especially when it comes to Lemma 5.15.

5.1   Recalling Antimirov's Automata

The main idea behind Antimirov's construction is that expressions are endowed with the structure of an automaton. The language of a state in this automaton is meant to be $\widehat{\ell}(e)$, the language denoted by its expression. From this perspective, the accepting states should be those representing expressions whose language contains the empty word. This set of expressions is fairly easy to describe.

Definition 5.2. *The set* $\mathbb{F}$ *is defined as the smallest subset of* $\mathbb{E}$ *satisfying:*

$$\frac{}{1 \in \mathbb{F}} \qquad \frac{e_1 \in \mathbb{F} \quad e_2 \in \mathbb{E}}{e_1 + e_2, e_2 + e_1 \in \mathbb{F}} \qquad \frac{e_1, e_2 \in \mathbb{F}}{e_1 \cdot e_2 \in \mathbb{F}} \qquad \frac{e_1 \in \mathbb{E}}{e_1^* \in \mathbb{F}}$$

Next, we recall Antimirov's transition function. The intuition is that an expression e has an a-transition to an expression e' when e' denotes remainders of words in $\widehat{\ell}(e)$ that start with a—i.e., if $w \in \widehat{\ell}(e')$, then $aw \in \widehat{\ell}(e)$. Together, the expressions reachable by a-transitions from e should describe *all* such words.

In the following, when $S \subseteq \mathbb{E}$ and $e \in \mathbb{E}$, we write $S \mathbin{\mathring{,}} e$ for $\{e' \cdot e : e' \in S\}$. We are now ready to define Antimirov's transition function, as follows.

Definition 5.3. *We define* $\partial : \mathbb{E} \times \Sigma \to \mathcal{P}(\mathbb{E})$ *recursively, as follows*

$$\begin{aligned} \partial(0, \mathsf{a}) &= \emptyset & \partial(e_1 + e_2, \mathsf{a}) &= \partial(e_1, \mathsf{a}) \cup \partial(e_2, \mathsf{b}) \\ \partial(1, \mathsf{a}) &= \emptyset & \partial(e_1 \cdot e_2, \mathsf{a}) &= \partial(e_1, \mathsf{a}) \mathbin{\mathring{,}} e_2 \cup e_1 \star \partial(e_2, \mathsf{a}) \\ \partial(\mathsf{b}, \mathsf{a}) &= \{1 : \mathsf{a} = \mathsf{b}\} & \partial(e_1^*, \mathsf{a}) &= \partial(e_1, \mathsf{a}) \mathbin{\mathring{,}} e_1^* \end{aligned}$$

Here, we use $e \star S$ *as a shorthand for* S *when* $e \in \mathbb{F}$, *and* $\emptyset$ *otherwise.*

Of course, the expression e could serve as the sole initial state in the automaton for e. However, our automata allow multiple initial states, and distributing this task among them will simplify some of the arguments that follow.

Definition 5.4. *We define* $\iota : \mathbb{E} \to \mathcal{P}(\mathbb{E})$ *recursively, as follows:*

$$\begin{aligned} \iota(0) &= \emptyset & \iota(e_1 + e_2) &= \iota(e_1) \cup \iota(e_2) \\ \iota(1) &= \{1\} & \iota(e_1 \cdot e_2) &= \iota(e_1) \mathbin{\mathring{,}} e_2 \\ \iota(\mathsf{a}) &= \{\mathsf{a}\} & \iota(e_1^*) &= \iota(e_1) \mathbin{\mathring{,}} e_1^* \cup \{1\} \end{aligned}$$

We could now try to package these parts into an automaton $(\mathbb{E}, \partial, \iota(e), \mathbb{F})$ for each expression e. Unfortunately, we have defined our automata to be finite, so a little more work is necessary to identify the expressions that are relevant (i.e., represented by reachable states) for a starting expression e.

Definition 5.5. *We define* $\rho : \mathbb{E} \to \mathcal{P}(\mathbb{E})$ *recursively, as follows:*

$$\rho(0) = \emptyset \qquad\qquad \rho(e_1 + e_2) = \rho(e_1) \cup \rho(e_2)$$
$$\rho(1) = \{1\} \qquad\qquad \rho(e_1 \cdot e_2) = \rho(e_1) \,\fatsemi\, e_2 \cup \rho(e_2)$$
$$\rho(\mathsf{a}) = \{\mathsf{a}, 1\} \qquad\qquad \rho(e_1^*) = \rho(e_1) \,\fatsemi\, e_1^* \cup \{1\}$$

With this function in hand, we can verify that it fits all of the requirements of the state space of an automaton with respect to the other parts identified above.

Lemma 5.6. *For all* $e \in \mathbb{E}$, *the set* $\rho(e)$ *is finite and closed under* ∂, *i.e., if* $e' \in \rho(e)$ *and* $e'' \in \partial(e', \mathsf{a})$, *then* $e'' \in \rho(e)$ *as well. Furthermore,* $\iota(e) \subseteq \rho(e)$.

In light of this, we write ∂_e for the function $\partial_e : \rho(e) \times \Sigma \to \mathcal{P}(\rho(e))$ obtained by restricting ∂. We can now define Antimirov automata, as follows.

Definition 5.7. *Let* $e \in \mathbb{E}$. *We write* A_e *for the* Antimirov *automaton*

$$(\rho(e), \partial_e, \iota(e), \mathbb{F} \cap \rho(e))$$

Antimirov's transition function can be used to decompose an expression e into several "derivatives" e', which can then reconstitute e. This validates the intuition that the derivatives collectively contain (only) the "tails" of words denoted by e. Similarly, the initial expressions $\iota(e)$ can also be used to reconstitute e.

Theorem 5.8. *Let* $e \in \mathbb{E}$. *The following two equivalences hold:*

$$e \equiv [e \in \mathbb{F}] + \sum_{e' \in \partial(e, \mathsf{a})} \mathsf{a} \cdot e' \qquad\qquad e \equiv \sum_{e' \in \iota(e)} e'$$

The first property above is usually referred to as the *fundamental theorem* of Kleene algebra [31], because of its close resemblance to the fundamental theorem of calculus. One caveat is that one needs to prove that the sums on the right-hand sides are in fact finite, but this turns out to be the case.

We end this subsection by recording two more useful properties of ι.

Lemma 5.9. *Let* $e \in \mathbb{E}$. *We have* $e \in \mathbb{F}$ *if and only if there exists an* $e' \in \iota(e)$ *such that* $e' \in \mathbb{F}$. *Also,* $e'' \in \partial(e, \mathsf{a})$ *if and only if* $e'' \in \partial(e', \mathsf{a})$ *for some* $e' \in \iota(e)$.

5.2 Solving Antimirov's Automata

Having fully described Antimirov's construction, we resume with the proof of the main technical point of this section, which is that the solution to A_e can be used to construct an expression equivalent to e.

More precisely, we will prove that e is equivalent to the sum of the solutions to its initial states, $\lfloor A_e \rfloor$, by showing that $\lfloor A_e \rfloor \leq e$ and $e \leq \lfloor A_e \rfloor$. The former property is easy to prove using the theory established up to this point.

Lemma 5.10. *For all $e \in \mathbb{E}$, it holds that $\lfloor A_e \rfloor \leq e$.*

Proof Sketch. By Theorem 5.8, the injection of $\rho(e)$ into $\mathbb{E}$ is a solution to A_e. $\qquad \square$

To show that $e \leq \lfloor A_e \rfloor$, we cannot exploit the fact that $\overline{A_e}$ is the least solution to A_e, as above. Our proof will instead operate by induction on e. First, we will need to develop some theory; the following abstraction is useful.

Definition 5.11. *Let $e, f \in \mathbb{E}$. We write $e \lesssim f$ when there exists a strong automaton morphism $h : \rho(e) \to \rho(f)$ from A_e to A_f.*

By Lemma 3.6, if we want to show that $\lfloor A_e \rfloor \leq \lfloor A_f \rfloor$, it is sufficient to prove $e \lesssim f$. We record the following instances of expressions being related by $\lesssim$.

Lemma 5.12. *The following hold for all $e_0, e_1, e_2 \in \mathbb{E}$:*

$$e_0 \lesssim e_0 \cdot 1 \qquad\qquad e_0 \lesssim e_0 + e_1 \qquad\qquad e_0 \lesssim e_1 \implies e_0 \cdot e_2 \lesssim e_1 \cdot e_2$$

$$e_0 \cdot e_0^* \lesssim e_0^* \qquad\qquad 1 \lesssim e_0^* \qquad\qquad e_0 \cdot (e_1 \cdot e_2) \lesssim (e_0 \cdot e_1) \cdot e_2$$

Proof Sketch. In all cases, a map can be gleaned from the structure of the relevant state spaces; checking that it is a term morphism is routine. $\qquad \square$

Remark 5.13. Kozen [21] and Jacobs [14] show that if $e, f \in \mathbb{E}$ are such that $e \leq f$, then the Brzozowski automaton of e is simulated by that of f, and hence these automata yield solutions e' and f' such that $e' \leq f'$.

It is tempting to try and prove a similar property for Antimirov automata, along the lines of "if $e \leq f$, then $e \lesssim f$". Unfortunately, this is not true. For instance, if $e = \mathsf{a} \cdot (\mathsf{b} + \mathsf{c})$ and $f = \mathsf{a} \cdot \mathsf{b} + \mathsf{a} \cdot \mathsf{c}$, then $e \leq f$, but there is no strong morphism from A_e to A_f. Fortunately, Lemma 5.12 is sufficient for our purposes.

The solutions to the automata for e, $e \cdot 1$ and $1 \cdot e$ are also related.

Lemma 5.14. *Let $e \in \mathbb{E}$. It holds that $\lfloor A_{e \cdot 1} \rfloor \leq \lfloor A_e \rfloor$ and $\lfloor A_e \rfloor \leq \lfloor A_{1 \cdot e} \rfloor$.*

Proof Sketch. We show that the solution to one automaton gives rise to a solution to the other automaton, using Lemmas 5.9 and 3.9 for the latter claim. $\qquad \square$

The next lemma is the main workhorse that we need to show that $e \leq \lfloor A_e \rfloor$. The proof is very similar to that of [21, Lemma 3].

Lemma 5.15. *Let $e, f \in \mathbb{E}$. It holds that $e \cdot \lfloor A_f \rfloor \leq \lfloor A_{e \cdot f} \rfloor$*

Proof Sketch. As in [21, Lemma 3], we proceed by induction on e; we use Lemmas 3.6 and 5.14 in the base, and Lemma 5.12 in the inductive cases. $\qquad \square$

With this in hand, we now have everything required to conclude the desired property of solutions to Antimirov automata, which we record below.

Lemma 5.16. *For all $e \in \mathbb{E}$, we have that $e \equiv \lfloor A_e \rfloor$.*

Proof. We already knew that $\lfloor A_e \rfloor \leq e$ by Lemma 5.10. To show that $e \leq \lfloor A_e \rfloor$, we derive using Lemmas 5.14 and 5.15, as follows:

$$e \equiv e \cdot 1 \leqq e \cdot \lfloor A_1 \rfloor \leq \lfloor A_{e \cdot 1} \rfloor \leq \lfloor A_e \rfloor$$

The second step is valid because $1 \in \iota(1)$ and $1 \in \mathbb{F}$, so $1 \leqq \overline{A_1}(1) \leq \lfloor A_1 \rfloor$. $\square$

6   From Monoids to Kleene Algebras

Recall that our objective was to derive a finite KA for two expressions, whose properties can then be used to conclude completeness and the FMP. We already saw how an expression gives rise to an automaton, which can then be turned into a transformation automaton. As it happens, the states of this transformation automaton have the internal structure of a monoid—indeed, this was the original motivation for the construction [27]—but we still do not have a KA.

In this section, we recall a straightforward translation from monoids to KAs proposed by Palka [29], and prove a useful property that we will leverage in the proof later on. Let us start by recalling the definition of a monoid.

Definition 6.1. *A monoid is a tuple $(M, \cdot, 1)$ where M is a set, $\cdot$ is a binary operator and $1 \in M$ such that the following hold for all $m_0, m_1, m_2 \in M$:*

$$m_1 \cdot (m_2 \cdot m_3) = (m_1 \cdot m_2) \cdot m_3 \qquad m_1 \cdot 1 = m_1 \qquad 1 \cdot m_1 = m_1$$

A function $h : \Sigma \to M$ gives rise to the function $\widetilde{h} : \Sigma^* \to M$, defined by

$$\widetilde{h}(\mathsf{a}_1 \cdots \mathsf{a}_n) = h(\mathsf{a}_1) \cdot \cdots \cdot h(\mathsf{a}_n)$$

As for KAs, we may identify a monoid $(M, \cdot, 1)$ with its carrier M, if the accompanying operator and unit are clear from context.

As stated above, if $A = (Q, \delta, I, F)$ is an automaton, then the state space of its transition automata is given by $\mathcal{P}(Q \times Q)$—i.e., the relations on Q—which has a monoidal structure: the operator is given by relational composition, and the unit is the identity relation on Q. In the sequel, we write M_A for this monoid.

The composition operator of a monoid can be lifted to sets of its elements, which can then be used to derive a fixed point operator, as follows.

Lemma 6.2 (Palka [29]). *If $(M, \cdot, 1)$ is a monoid, then $(\mathcal{P}(M), \cup, \otimes, ^{\circledast}, \emptyset, \{1\})$ is a KA, where $\otimes$ and $^{\circledast}$ are defined by choosing for $U, V \subseteq M$:*

$$U \otimes V = \{m \cdot n : m \in U, n \in V\} \qquad U^{\circledast} = \{u_1 \cdots u_n : u_1, \ldots, u_n \in U\}$$

As an example of this construction, note that applying this construction to the free monoid $(\Sigma^*, \cdot, \epsilon)$ precisely yields the free KA of languages.

Now, given an expression $e \in \mathbb{E}$, a monoid $(M, \cdot, 1)$, and a map $h : \Sigma \to M$, we have two ways of interpreting e inside of the KA that arises from this monoid:

1. We lift the map $\mathbf{a} \mapsto \{h(\mathbf{a})\}$ to obtain a map $\mathbb{E} \to \mathcal{P}(M)$.
2. We map each $w \in \widehat{\ell}(e)$ to an element of M via $\widetilde{h} : \Sigma \to M$.

The next lemma shows that these two interpretations of expressions inside the KA for $(M, \cdot, 1)$ are actually the same; it can be thought of as a generalization of [29, Lemma 3.1], which covers the special case for the syntactic monoid.

Lemma 6.3. *Let $(M, \cdot, 1)$ be a monoid and let $(\mathcal{P}(M), \cup, \otimes, ^{\circledast}, \emptyset, \{1\})$ be the KA obtained from it, per Lemma 6.2. Furthermore, let $h_1 : \Sigma \to M$ and $h_2 : \Sigma \to \mathcal{P}(M)$ be such that for all $\mathbf{a} \in \Sigma$, we have that $h_2(\mathbf{a}) = \{h_1(\mathbf{a})\}$. Then for $e \in \mathbb{E}$:*

$$\widehat{h_2}(e) = \{\widetilde{h_1}(w) : w \in \widehat{\ell}(e)\}$$

We conclude this section by leveraging the above to prove a pivotal lemma: the solution to a state q of an automaton A can be recovered by interpreting this expression inside of the KA obtained from the transformation automata of A, and looking at the solutions to the relations inside that interpretation.

Lemma 6.4. *Let $A = (Q, \delta, I, F)$ be an automaton and $q \in Q$. Furthermore, let $h : \Sigma \to \mathcal{P}(M_A)$ be given by $h(\mathbf{a}) = \{\delta_\mathbf{a}\}$. The following holds.*

$$\overline{A}(q) \equiv \sum_{R \in \widehat{h}(\overline{A}(q))} \lfloor A[R] \rfloor$$

Proof. We start by massaging the proof goal. Lemma 6.3 tells us that $R \in \widehat{h}(\overline{A}(q))$ if and only if there exists a $w \in \widehat{\ell}(\overline{A}(q))$ such that $R = \delta_w$. Using this observation and Lemmas 4.6 and 3.9, it suffices to show

$$\sum_{qRq' \in F} \lfloor A[R] \rfloor \equiv \sum_{w \in L(A,q)} \lfloor A[\delta_w] \rfloor$$

For the inclusion from left to right, let $R \subseteq Q \times Q$ and $q' \in F$ be such that $q \, R \, q'$. On the one hand, if $\widehat{\ell}(\lfloor A[R] \rfloor) = \emptyset$, then an easy inductive argument shows that $\lfloor A[R] \rfloor \equiv 0$, which means that the term $\lfloor A[R] \rfloor$ does not contribute to the sum. Otherwise, let $w \in \widehat{\ell}(\lfloor A[R] \rfloor)$. By Lemma 3.9, we know that $w \in L(A[R], \mathrm{id}_Q)$, and thus $\delta_w = R$. Since $q \, \delta_w \, q' \in F$, we also know that $w \in L(\overline{A}(q))$. Therefore $\lfloor A[R] \rfloor = \lfloor A[\delta_w] \rfloor$ appears in the sum on the right-hand side.

For the inclusion from right to left, let $w \in L(A, q)$. In that case, $q\delta_w q'$ for some $q' \in F$. Thus $\lfloor A[\delta_w] \rfloor$ appears in the sum on the left-hand side. $\square$

7   Completeness and the FMP

We are now ready to prove our main claims. In a nutshell, our proof will take two expressions e and f, apply the transformation automaton construction to A_{e+f}, and then use the resulting state space monoid to obtain a KA. The previously derived facts connect e and f to their interpretation inside this KA, which we will use in two different ways to conclude both completeness and the FMP.

Throughout this section, we fix $e, f \in \mathbb{E}$. For brevity, we also write ∂ for the transition function of A_{e+f}. We fix $h : \Sigma \to \mathcal{P}(M_{A_{e+f}})$ by $h(\mathbf{a}) = \{\partial_{\mathbf{a}}\}$. Note that $M_{A_{e+f}}$ is a finite monoid, and hence $\mathcal{P}(M_{A_{e+f}})$ is a finite KA by Lemma 6.2.

The next lemma puts the results of the previous sections together to connect e and f to their interpretations inside $\mathcal{P}(M_{A_{e+f}})$, in the following way.

Lemma 7.1. *The following two equivalences hold:*

$$ e \equiv \sum_{R \in \widehat{h}(e)} \lfloor A_{e+f}[R] \rfloor \qquad f \equiv \sum_{R \in \widehat{h}(f)} \lfloor A_{e+f}[R] \rfloor $$

Proof. Without loss of generality, we prove the first property by deriving:

$$ e \equiv \lfloor A_e \rfloor \qquad\qquad\qquad\qquad\qquad \text{(Lemma 5.16)} $$

$$ \equiv \sum_{e' \in \iota(e)} \overline{A_e}(e') \qquad\qquad\qquad\qquad \text{(def.}\lfloor A_e \rfloor) $$

$$ \equiv \sum_{e' \in \iota(e)} \overline{A_{e+f}}(e') \qquad\qquad\qquad\qquad \text{(Lemma 3.8)} $$

$$ \equiv \sum_{e' \in \iota(e)} \sum_{R \in \widehat{h}(\overline{A_{e+f}}(e'))} \lfloor A_{e+f}[R] \rfloor \qquad\quad \text{(Lemma 6.4)} $$

$$ \equiv \sum_{R \in \widehat{h}(e)} \lfloor A_{e+f}[R] \rfloor \qquad\qquad\qquad\quad \text{(see below)} $$

The last equivalence holds because by Lemmas 3.8 and 5.16, we have that:

$$ e \equiv \lfloor A_e \rfloor \equiv \sum_{e' \in \iota(e)} \overline{A_e}(e') \equiv \sum_{e' \in \iota(e)} \overline{A_{e+f}}(e') $$

and thus $\widehat{h}(e) = \bigcup_{e' \in \iota(e)} \widehat{h}(\overline{A_{e+f}}(e'))$ by Lemma 2.5. $\qquad\qquad\square$

We are now ready to conclude our first main claim: the finite model property holds for KA. Recall that $\mathfrak{F}$ denotes the class of all *finite* Kleene algebras, to which $\mathcal{P}(M_{A_{e+f}})$ belongs. This allows us to apply Lemma 7.1, as follows.

Theorem 7.2 (Finite model property). *If $\mathfrak{F} \models e = f$, then $e \equiv f$.*

Proof. By the premise, we have that $\widehat{h}(e) = \widehat{h}(f)$, and so by Lemmas 7.1 and 2.5:

$$ e \equiv \sum_{R \in \widehat{h}(e)} \lfloor A_{e+f}[R] \rfloor = \sum_{R \in \widehat{h}(f)} \lfloor A_{e+f}[R] \rfloor \equiv f $$

$$\square$$

Finally, we note that a very similar proof also allows us to conclude that the axioms of KA are complete w.r.t. its language model, thanks to the connection between interpretations in lifted monoids given by Lemma 6.3.

Theorem 7.3 (Completeness). *If $\mathcal{L} \models e = f$, then $e \equiv f$.*

Proof. Let $h' : \Sigma \to M_{A_{e+f}}$ be given by $h'(\mathbf{a}) = \partial_{\mathbf{a}}$. By the premise $\widehat{\ell}(e) = \widehat{\ell}(f)$, and thus $\widehat{h}(e) = \widehat{h}(f)$ because we can use Lemma 6.3 to derive

$$\widehat{h}(e) = \{\widehat{h'}(w) : w \in \widehat{\ell}(e)\} = \{\widehat{h'}(w) : w \in \widehat{\ell}(f)\} = \widehat{h}(f)$$

We can then conclude by leveraging Lemma 7.1, as for Lemma 7.2.   □

8   Discussion

We leave the reader with some final considerations regarding our formalization and directions for possible further work.

Coq Formalization. We have formalized all of our results in Coq [4,11]. The trusted base comes down to (1) the axioms of the Calculus of Inductive Constructions, (2) injectivity of dependent equality (equivalent to Streicher's axiom K [37]), and (3) dependent functional extensionality. The latter is a result of our encoding of subsets, and can most likely be factored out with better data structures.

All proofs as presented here are faithful to the insights underlying the claim, although some encodings differ slightly. For instance, the definition of $\rho(e)$ in the development is more accurately rendered using disjoint union.

Possible Extensions. Guarded Kleene Algebra with Tests (GKAT) [33,35] is a fragment of KAT with favorable decidability properties. GKAT in particular admits a set of axioms that are complete w.r.t. its language (resp. relational, probabilistic) model, but this set is infinite as a result of an axiom scheme. We wonder whether the techniques discussed here could be applied to arrive at a more satisfactory completeness result. To start answering this question, one would first have to devise an analogue transformation automata and monoids for GKAT.

Relational Models. Pratt [30] connected the language model of KA to the relational model—essentially saying that if $\mathfrak{R}$ is the class of relational KAs (as in Example 2.2), then $\mathfrak{R} \models e = f$ if and only if $\mathcal{L} \models e = f$ for all $e, f \in \mathbb{E}$. By Theorem 7.3, this means that relational models are also complete for KA.

In light of Theorem 7.2, we wonder: can this form of completeness be strengthened to *finite* relational models? A positive answer would mean that the finite countermodel accompanying an invalid equation would correspond to an interpretation of the primitive actions as state transformers on a finite state space.

There is a tantalizing candidate for a canonical model that might be able to fill the role of $\mathcal{P}(M_{A_{e+f}})$ in the previous section: simply use the relational KA with the carrier $M_{A_{e+f}}$. For this to work, we would have to connect e and f with their interpretations inside this KA, which will require further research.

Acknowledgements. The author wishes to thank Nick Bezhanishvili for his suggestion to investigate the FMP for KA, Alexandra Silva for general advice about this manuscript, and the anonymous reviewers for their careful comments.

References

1. Anderson, C.J., et al.: NetKAT: semantic foundations for networks. In: POPL, pp. 113–126 (2014). https://doi.org/10.1145/2535838.2535862
2. Antimirov, V.M.: Partial derivatives of regular expressions and finite automaton constructions. Theor. Comput. Sci. **155**(2), 291–319 (1996). https://doi.org/10.1016/0304-3975(95)00182-4
3. Backhouse, R.: Closure algorithms and the star-height problem of regular languages. Ph.D. thesis, University of London (1975). https://hdl.handle.net/10044/1/22243
4. Bertot, Y., Castéran, P.: Interactive Theorem Proving and Program Development - Coq'Art: The Calculus of Inductive Constructions. Texts in Theoretical Computer Science. An EATCS Series (2004). https://doi.org/10.1007/978-3-662-07964-5
5. Blackburn, P., de Rijke, M., Venema, Y.: Modal Logic, Cambridge Tracts in Theoretical Computer Science, vol. 53. Cambridge University Press, Cambridge (2001). https://doi.org/10.1017/CBO9781107050884
6. Boffa, M.: Une remarque sur les systèmes complets d'identités rationnelles. RAIRO Theor. Informatics Appl. **24**, 419–423 (1990). https://doi.org/10.1051/ita/1990240404191
7. Brzozowski, J.A.: Derivatives of regular expressions. J. ACM **11**(4), 481–494 (1964). https://doi.org/10.1145/321239.321249
8. Burris, S., Sankappanavar, H.P.: A Course in Universal Algebra. Graduate Texts in Mathematics, Springer, Heidelberg (1981)
9. Cohen, E., Kozen, D., Smith, F.: The complexity of Kleene algebra with tests. Technical report TR96-1598 (1996). https://hdl.handle.net/1813/7253
10. Conway, J.H.: Regular Algebra and Finite Machines. Chapman and Hall Ltd., London (1971)
11. Coq Development Team: The Coq Reference Manual, version 8.15 (2022). https://coq.inria.fr/doc
12. Das, A., Doumane, A., Pous, D.: Left-handed completeness for Kleene algebra, via cyclic proofs. In: LPAR, pp. 271–289 (2018). https://doi.org/10.29007/hzq3
13. Foster, S., Struth, G.: On the fine-structure of regular Algebra. J. Autom. Reason. **54**(2), 165–197 (2014). https://doi.org/10.1007/s10817-014-9318-9
14. Jacobs, B.: A bialgebraic review of deterministic automata, regular expressions and languages. In: Futatsugi, K., Jouannaud, J.-P., Meseguer, J. (eds.) Algebra, Meaning, and Computation. LNCS, vol. 4060, pp. 375–404. Springer, Heidelberg (2006). https://doi.org/10.1007/11780274_20
15. Kappé, T.: Completeness and the finite model property for Kleene algebra, reconsidered (2022). https://doi.org/10.48550/arXiv.2212.10931. (Full version)
16. Kappé, T.: Completeness and the finite model property for Kleene algebra, reconsidered - Coq formalization (2022). https://doi.org/10.5281/zenodo.7467245
17. Kleene, S.C.: Representation of events in nerve nets and finite automata. Automata Stud. **34**, 3–41 (1956)
18. Kozen, D.: A completeness theorem for Kleene algebras and the algebra of regular events. Inf. Comput. **110**(2), 366–390 (1994). https://doi.org/10.1006/inco.1994.1037

19. Kozen, D.: Kleene algebra with tests and commutativity conditions. In: Margaria, T., Steffen, B. (eds.) TACAS 1996. LNCS, vol. 1055, pp. 14–33. Springer, Heidelberg (1996). https://doi.org/10.1007/3-540-61042-1_35

20. Kozen, D.: Kleene algebra with tests. ACM Trans. Program. Lang. Syst. **19**(3), 427–443 (1997). https://doi.org/10.1145/256167.256195

21. Kozen, D.: Myhill-Nerode relations on automatic systems and the completeness of Kleene algebra. In: Ferreira, A., Reichel, H. (eds.) STACS 2001. LNCS, vol. 2010, pp. 27–38. Springer, Heidelberg (2001). https://doi.org/10.1007/3-540-44693-1_3

22. Kozen, D., Patron, M.-C.: Certification of compiler optimizations using Kleene algebra with tests. In: Lloyd, J., et al. (eds.) CL 2000. LNCS (LNAI), vol. 1861, pp. 568–582. Springer, Heidelberg (2000). https://doi.org/10.1007/3-540-44957-4_38

23. Kozen, D., Silva, A.: Left-handed completeness. Theor. Comput. Sci. **807**, 220–233 (2020). https://doi.org/10.1016/j.tcs.2019.10.040

24. Kozen, D., Smith, F.: Kleene algebra with tests: completeness and decidability. In: van Dalen, D., Bezem, M. (eds.) CSL 1996. LNCS, vol. 1258, pp. 244–259. Springer, Heidelberg (1997). https://doi.org/10.1007/3-540-63172-0_43

25. Kozen, D., Tseng, W.-L.D.: The Böhm–Jacopini theorem is false, propositionally. In: Audebaud, P., Paulin-Mohring, C. (eds.) MPC 2008. LNCS, vol. 5133, pp. 177–192. Springer, Heidelberg (2008). https://doi.org/10.1007/978-3-540-70594-9_11

26. Krob, D.: A complete system of B-rational identities. In: Paterson, M.S. (ed.) ICALP 1990. LNCS, vol. 443, pp. 60–73. Springer, Heidelberg (1990). https://doi.org/10.1007/BFb0032022

27. McNaughton, R., Papert, S.: The syntactic monoid of a regular event. Algebraic Theory of Machines, Languages, and Semigroups, pp. 297–312 (1968)

28. McNaughton, R., Yamada, H.: Regular expressions and state graphs for automata. IRE Trans. Electron. Comput. **9**(1), 39–47 (1960). https://doi.org/10.1109/TEC.1960.5221603

29. Palka, E.: On finite model property of the equational theory of Kleene algebras. Fundam. Inform. **68**(3), 221–230 (2005). https://content.iospress.com/articles/fundamenta-informaticae/fi68-3-02

30. Pratt, V.R.: Dynamic algebras and the nature of induction. In: STOC, pp. 22–28 (1980). https://doi.org/10.1145/800141.804649

31. Rutten, J.J.M.M.: Universal coalgebra: a theory of systems. Theor. Comput. Sci. **249**(1), 3–80 (2000). https://doi.org/10.1016/S0304-3975(00)00056-6

32. Salomaa, A.: Two complete axiom systems for the algebra of regular events. J. ACM **13**(1), 158–169 (1966). https://doi.org/10.1145/321312.321326

33. Schmid, T., Kappé, T., Kozen, D., Silva, A.: Guarded Kleene algebra with tests: coequations, coinduction, and completeness. In: ICALP, pp. 142:1–142:14 (2021). https://doi.org/10.4230/LIPIcs.ICALP.2021.142

34. Smolka, S., Eliopoulos, S.A., Foster, N., Guha, A.: A fast compiler for NetKAT. In: ICFP, pp. 328–341 (2015). https://doi.org/10.1145/2784731.2784761

35. Smolka, S., Foster, N., Hsu, J., Kappé, T., Kozen, D., Silva, A.: Guarded Kleene algebra with tests: verification of uninterpreted programs in nearly linear time. In: POPL (2020). https://doi.org/10.1145/3371129

36. Stockmeyer, L.J., Meyer, A.R.: Word problems requiring exponential time: Preliminary report. In: STOC, pp. 1–9 (1973). https://doi.org/10.1145/800125.804029

37. Streicher, T.: Investigations into intensional type theory. Habilitiation thesis, Ludwig Maximilian Universität (1993). https://www2.mathematik.tu-darmstadt.de/streicher/HabilStreicher.pdf

38. Thompson, K.: Regular expression search algorithm. Commun. ACM **11**(6), 419–422 (1968). https://doi.org/10.1145/363347.363387
39. Watson, B.W.: A taxonomy of finite automata construction algorithms. Technical report, Technische Universiteit Eindhoven (1993). https://research.tue.nl/files/2482472/9313452

What Else is Undecidable About Loops?

Laura Kovács[iD] and Anton Varonka[(✉)][iD]

TU Wien, Vienna, Austria
anton.varonka@tuwien.ac.at

Abstract. We address algebraic aspects of invariant generation and proving termination, two central questions of program analysis, for non-deterministic loops with polynomial updates.

Our focus is sketching the boundary of (un-)decidability for both problems between different classes of programs. The first main contribution of this work is related to the question raised by Braverman in 2006: "How much non-determinism can be introduced in a linear loop before termination becomes undecidable?" We show that termination of loops with a purely non-deterministic choice between linear updates is undecidable in the presence of linear inequality loop conditions.

In the context of invariants, an algorithm is known that computes all polynomial relations among variables of loops with multiple linear updates. At the same time, allowing polynomial assignments of higher degrees was shown to result in algorithmic unsolvability. We highlight that negative results in fact do not exploit general polynomial updates. We show that no algorithm can find all polynomial relations for programs as soon as quadratic updates or updates guarded by affine equalities are involved.

1 Introduction

One prominent approach towards proving program correctness comes with showing partial program correctness and providing termination arguments [4]. Automating both tasks is hence a key challenge in verification, especially in the context of program loops.

While the undecidability of termination for general programs cannot be ignored, drawing inflexible pessimistic conclusions about proving termination in the special cases is inaccurate [8]. It becomes of importance to detect certain "core" restricted classes of programs, where the termination problem becomes decidable, or vice versa [2]. Most of the research in this direction focused on dealing with one simple loop at a time, establishing positive decidability results for some classes of loops [6,14,22,29] over the last years. Identifying limits of (un-)decidability by exposing classes of general programs for which termination remains undecidable is nevertheless an equally challenging task, see, e.g., [2,5,29].

Reasoning about partial correctness of programs and, in particular, loops is, in turn, inseparable from constructing (inductive) loop invariants at each program point, that is, assertions that evaluate to true on every program execution reaching that point [3]. When thinking of inductive invariants as relations among

© The Author(s), under exclusive license to Springer Nature Switzerland AG 2023
R. Glück et al. (Eds.): RAMiCS 2023, LNCS 13896, pp. 176–193, 2023.
https://doi.org/10.1007/978-3-031-28083-2_11

Table 1. Decidability (solvability) results in terms of multi-path loops

Problems / Loops	Invariants strongest algebraic	Termination inequality condition	Termination equality condition
multi-path affine	solv. [15]	**undec.**	dec. [22]
——″—— + equality guards	unsolv. [24]	**undec.**	dec. [22]
——″—— + inequality guards	unsolv. [24]	undec. [29]	undec. [29]
q-admissible	*unsolv.*	*undec.*	dec. [22]
polynomial	unsolv. [15]	undec. [5]	dec. [22]

New results are highlighted
[a] in bold: follows from our result given in Theorem 1;
[b] in italics: follows from Theorem 4, or due to techniques that simulate polynomial updates of arbitrary degree with those of degree at most 2.
Loop guards (incl. loop conditions) are assumed to be Boolean combinations of *linear* (in-)equalities in program variables. (The decidability result of [22], however, holds for polynomial equalities of arbitrary degree.)

variables that are preserved under all behaviours of the program, it is natural to ask about the quality of those relations. For example, questions may touch upon restricting to invariants defined using equalities vs. inequalities, or exploiting affine vs. polynomial equations of arbitrary degree. For some of such restricted classes of invariants, automated approaches computing all corresponding invariants of program loops exist, see, e.g., [15,16,24].

Given the intrinsic hardness of proving termination and generating invariants suitable for enforcing partial correctness, in this paper *we focus on imperative program loops and establish new limits of undecidability towards loop verification.*

Key to our approach is algebraic modelling of programs as abstract transition systems with associated variables that are subject to polynomial updates. To this end, we analyse actions of finitely generated semigroups of polynomial updates on the set $\mathbb{Q}^n$ of potential variables valuations. In the case of non-deterministic affine updates, this semigroup is the rational matrix semigroup generated by finitely many matrices originating from the updates of the loop. The termination question studied in Sect. 4 is inherently akin to the decision problems for matrix semigroups, such as vector reachability or matrix membership problems [1]. When addressing invariants in Sect. 5, we turn to considering the Zariski closures of those subsets of $\mathbb{Q}^n$ that, in our model, constitute the collecting semantics of program locations, much in the spirit of [15].

Related Work. Table 1 puts our results around program termination and invariant synthesis into the context of the related works, which we discuss next. *Program termination.* Total correctness properties assert that certain states are eventually reached during program execution [4]. Consider a simple single-path linear loop of the form

$$x := c; \ while \ Bx \geq b \ do \ x := Mx,$$

where for a fixed initial vector c there exists a unique (finite or not) program execution. Termination question is whether, for every initial c, such execution reaches the complement of the loop condition set. This problem, with loop conditions given by conjunctions of linear inequalities, has been shown decidable over reals [29], rationals [6] and integers [14]. We shift the discussion to the *multi-path loops*, where a non-deterministic choice between different linear updates is possible. We highlight that a loop with a fixed initial vector has multiple potential executions. Therefore, termination now asks about finiteness of them all, and that for all initial vectors from a set. In [29], the undecidability of termination for multi-path affine loops, each update *guarded by a linear inequality*, has been established. An undecidability result is also provided in [5], where the same problem is shown undecidable for non-deterministic loops with inequalities as *loop conditions*, but *polynomial* updates of arbitrary degree. Another work that considers multi-path polynomial loops is [22]. The crucial aspect of the model considered there is that the loop condition is only allowed to be a polynomial *equality*. Remarkably, this grants decidability, even with non-deterministic polynomial updates. Notice also that allowing non-linear loop conditions enriches the model in such way that the undecidable Hilbert's 10th Problem can be encoded in its termination [30] while basically ignoring the loop body. Our work goes beyond these results and *proves undecidability of loop termination for multi-path loops with affine updates and linear inequality conditions*, see Table 1.

Termination of single-path *non-linear loops* renders undecidable as well. In a natural deterministic generalisation of a simple single-update model, namely one of piecewise affine updates, there is again no algorithm to decide termination [2], see also the undecidability of the generalised Collatz Problem [21]. Furthermore, the *Halting Problem*, i.e., termination on a single input, poses additional challenges. Decidability is known for polynomial loops with equality conditions [22], or loops with general conditions but restricted spectral structure [13].

Invariant Synthesis. The work of [18] computes all affine equality relations among variables of a program with affine assignments, the so-called *strongest affine invariant*. Another seminal paper [9] generalises the domain to polyhedra and thus allows to determine valid affine inequalities in program variables. We refer to references in [4, Chapter 12] for discussions of other invariant domains.

Polynomial invariants provide greater expressiveness than just the affine and hence have been actively studied recently (see, e.g., [7] for the state-of-the-art and discussions). Formally, an algebraic inductive invariant assigns to each program location an algebraic set such that the resulting family is preserved under the transition relation of the program. This is equivalent to associating with every location an ideal of polynomial relations among program variables that hold every time the location is reached. The approach of [25] infers all polynomial relations up to a given degree d for programs with polynomial assignments.

An important result that motivated our investigations in Sect. 5 was attained in [15]. In that work, an algorithm is established that computes the *strongest algebraic invariant*, i.e., the ideal of *all* polynomial relations (regardless of the degree) among variables of an *affine* program. The same paper highlights the unsolvability of computing the strongest algebraic invariant for *polynomial* programs, see Sect. 7 therein. At the same time, the work of [16] presents a method to compute strongest algebraic invariants of polynomial programs which allow non-linear updates of certain kind. In connection to this line of research, *we show that synthesising the strongest algebraic invariant is unsolvable already for loops with quadratic updates* (which we call q-admissible), as stated in Table 1.

Our Contributions. We investigate imperative programs with polynomial updates both in terms of termination analysis and invariant generation. In our work, we consider refinements of polynomial programming models, such as guards and loop conditions of different types, and observe how these shape the decidability landscape of problems in question.

- We prove that termination of a multi-path loop with a linear inequality condition and a non-deterministic choice between several affine updates is undecidable (Theorem 1). There is no conditional branching in our programming model, so undecidability results for, say, piecewise affine updates [2,21], cannot be interpreted as a special case of our setting.

- We show the algorithmic unsolvability of computing the strongest algebraic invariant for programs with updates of degree at most 2 (Theorem 4).

2 Preliminaries

2.1 Computational Models

Polynomial Programs and Loops. We introduce the programming model and related notions in continuity with [15,25].

Definition 1 (Polynomial Program). *A polynomial program with n variables is a tuple $\mathcal{P} = (Q, E, q_0)$, where Q is a finite set consisting of program locations, $E \subseteq Q \times \mathbb{Q}[x_1, \ldots, x_n]^n \times Q$ is a set of edges (transitions) and $q_0 \in Q$ is the initial location.*

We think of $x = (x_1, \ldots, x_n)$ as a variables vector in $\mathbb{Q}^n$ and of an edge (q, f, q') as performing a simultaneous update $x := f(x)$ described by a vector f of n polynomials. We often use a, b, c, ... for variables valuations and so distinguish them from the variables vector, denoted by x throughout this work.

Definition 2 (Configuration). *A configuration (q, a) of a polynomial program $\mathcal{P}$ is a pair consisting of a location $q \in Q$ and a variables valuation $a \in \mathbb{Q}^n$. An edge (q, f, q') induces a transition from a configuration (q, a) to $(q', f(a))$.*

The *executions* of $\mathcal{P}$ are defined intuitively starting from the initial configuration $(q_0, 0)$.

Definition 3 (Collecting Semantics). *The collecting semantics of $\mathcal{P}$ associates with each $q \in Q$ a set $S_q^{\mathcal{P}}$ of all vectors $a \in \mathbb{Q}^n$ reachable in location q, i.e., such that an execution of $\mathcal{P}$ reaching configuration (q, a) exists.*

Definition 4 (Polynomial Loop). *A polynomial loop $\mathcal{L} = (Q, E, q_0)$ is a polynomial program with a location set $Q = \{q_0, q_1\}$ and a set of transitions $E = \{(q_0, f^{(0)}, q_1), (q_1, f^{(1)}, q_1), \ldots, (q_1, f^{(k)}, q_1)\}$. All components of $f^{(0)}$ are required to be constant functions.*

A polynomial loop is thus a special case of a polynomial program that only has two locations $\{q_0, q_1\}$ and a special transition from q_0 to q_1 with a *constant* update $f^{(0)}$. This *initialising transition* is interpreted as the initialisation of loop variables and is fixed for a particular loop. The location q_1 is the "looping" location; transitions from q_1 to itself correspond to k non-deterministic polynomial updates of variables in the loop body.

Polynomial programs are referred to as *affine* if all their update functions are vectors of polynomials of degree at most 1. Besides that, we consider a subclass of polynomial programs where the updates are *at most quadratic*. That is, for each edge $(q, (f_1, \ldots, f_n), q') \in E$, for all $i \in \{1, \ldots, n\}$, the polynomial f_i is of degree at most 2. We refer to such polynomial programs as *q-admissible programs*.

Guarded Updates. The definition of a polynomial program can be extended by incorporating *guarded updates*, or *guarded commands*. A transition $e \in Q \times \mathbb{Q}[x]^n \times Q$ of a *guarded polynomial program (GPP)* is modified by adding a guard predicate G which applies to the variables vector x.

For a transition that applies update f to x, we use Dijkstra's notation [11]: $G(x) \to x := f(x)$. The semantics is as follows. If (q, a) is the current configuration of a GPP execution, only transitions from q whose guards are satisfied, that is, $G(a)$ evaluates to *true*, can be taken. Notice that the choice between several transitions with satisfied guards is non-deterministic. We emphasise that a GPP where every guard is trivial, i.e., *true*, is exactly an unguarded polynomial program as in Definition 1.

A generic template of a *guarded polynomial loop* is shown in Program 1.

Program 1 A polynomial loop with guards

$x := f^{(0)}(x)$ $\hfill \{x := c \in \mathbb{Q}^n\}$
while *true* **do**
$\quad \sigma_1 : \qquad G_1(x) \to x := f^{(1)}(x)$
$\quad$ **or**
$\qquad \vdots$
$\quad$ **or**
$\quad \sigma_k : \qquad G_k(x) \to x := f^{(k)}(x)$

Guarded loops of the form

$$x := c; \text{ while } G(x) = true \text{ do } x := f^{(1)}(x) \text{ or } \ldots \text{ or } x := f^{(k)}(x)$$

are a special case of guarded polynomial loops, where essentially every update is guarded by the same predicate G. We refer to G as *loop condition* and use the notation $\mathcal{L}_G$ for the loop. In particular, we dedicate Sect. 4 to the studies of termination of loops with a loop condition. We proceed directly to defining the necessary preliminaries for the termination discussions.

Loop Termination. We call the variables valuation assigned by the (constant) initialising transition of a loop an *initial vector*. When a loop $\mathcal{L}$ has the initial vector c, we may emphasise this by writing $\mathcal{L}[c]$.

Definition 5 (Loop with Inputs). *Let $S \subseteq \mathbb{Q}^n$ be a set of vectors viewed as initial vectors. A family $\mathcal{L}_G[S] = \{\mathcal{L}_G[c] : c \in S\}$ of guarded polynomial loops is refered to as a* loop with inputs *from S, if all loops share the same loop condition G and the same body, while only differing in their initialising transitions.*

Termination is formally defined using the concept of execution tree, see also [23]. Crucially, an execution tree is defined for a loop-input pair.

Definition 6 (Loop Execution Tree). *The* execution tree *of a guarded polynomial loop $\mathcal{L}[c]$ is defined inductively as*

- *the root is $c = f^{(0)}(0)$;*
- *any node $a \in \mathbb{Q}^n$ has a child for every transition $(q_1, f^{(i)}, q_1)$ whose guard evaluates to true on a. There is a directed edge from a to the corresponding node $f^{(i)}(a)$, labeled by i.*

Definition 7 (Termination of Loop with Inputs). *A loop with inputs $\mathcal{L}_G[S]$ terminates (on S) if for every input $c \in S$ an execution tree of $\mathcal{L}_G[c]$ is finite.*

A combination of a loop condition, loop body and an input (initial vector) yields a single execution tree. A tree is finite if and only if there is no infinite execution starting with a given initial vector. When considering multiple potential inputs from a set S, the termination of the loop family $\mathcal{L}_G[S]$ means that every execution with an initial vector from S is finite.

Inductive Invariants. We use the notation $\overline{\mathbb{Q}}$ for the field of *algebraic* numbers. In order to work over an algebraically closed field, we consider the sets $S_{q_i}^{\mathcal{P}}$ as subsets of $\overline{\mathbb{Q}}^n$, when talking about invariants and Zariski closures.

Definition 8 (Inductive Invariant). *A family of sets $\mathcal{X} = \{X_q : q \in Q\}$, where $X_q \subseteq \overline{\mathbb{Q}}^n$, is referred to as an* inductive invariant *of the polynomial program $\mathcal{P}$ if it is a solution of the following system of inclusions:*

$$X_{q_0} \supseteq \{0\}$$
$$X_{q'} \supseteq f(X_q) \text{ for all } (q, f, q') \in E. \tag{1}$$

The inductive invariant is closed under applying the transitions of $\mathcal{P}$. This way, it is an overapproximation of the program's collecting semantics.

Definition 8 can be phrased for GPPs by only adding $f(a)$ to the invariant set $X_{q'}$ if the guard $G(a)$ of the edge (q, f, q') in question evaluates to *true*.

An invariant family as in Definition 8 is said to be an *algebraic* inductive invariant if each X_q is an algebraic subset of $\overline{\mathbb{Q}}^n$. For a given program location, there exists a minimal algebraic solution of Eq. (1), hence we can talk about the minimal algebraic inductive invariant. Following [15], we point out that the minimal algebraic inductive invariant can be characterised as the family

$\mathcal{X} = \{X_q : q \in Q\}$, where $X_q := \overline{S_q^{\mathcal{P}}}$ for all $q \in Q$, i.e., X_q is the *Zariski closure* of $S_q^{\mathcal{P}}$ in $\overline{\mathbb{Q}}^n$. We will refer to $\mathcal{X}$ as the *strongest algebraic invariant* of $\mathcal{P}$.

Given a polynomial $P \in \overline{\mathbb{Q}}[x_1, \ldots, x_n]$, we say that the equation $P = 0$ holds at $q \in Q$ if P vanishes on $S_q^{\mathcal{P}}$. The set of all polynomials that hold at $q \in Q$ constitutes an ideal I_q. The algebraic set that corresponds to this ideal is $V(I_q) = \overline{S_q^{\mathcal{P}}}$. Essentially, computing the strongest algebraic invariant is equivalent to computing a basis of ideal I_q for each $q \in Q$.

The strongest algebraic invariant of a polynomial loop $\mathcal{L}$ consists of two sets. The reachable set of the initial location q_0 is a single vector and thus a closed set of the Zariski topology. When discussing the strongest algebraic invariant of a polynomial loop, we are going to focus on the reachable set $S_{q_1}^{\mathcal{L}}$ (and its Zariski closure), for simplicity denoting it by $S^{\mathcal{L}}$ (and $\overline{S^{\mathcal{L}}}$, respectively).

From Programs to Loops. We observe that considering loops instead of programs with multiple locations does not limit the computational power. This is also inherited by the corresponding subclasses of both. In particular, there exists a technique (see also [15, Sect. 4]) that allows to omit additional locations at a cost of blowing up the number of variables in the program. Let $\mathcal{P}$ be an unguarded polynomial program with non-initial locations $q_1, \ldots, q_\ell$ and n variables. We sketch how a polynomial loop $\mathcal{L}$ simulates its behaviour by adding ℓ copies of each variable. That is, we associate a block of n variables in $\mathcal{L}$ with each program location. Only one block of copies is "active" in each configuration of the resulting loop $\mathcal{L}$, while the other variables are kept to 0. This indicates the active location in an execution of $\mathcal{P}$. In a self-loop of $\mathcal{L}$ that corresponds to a transition from q_i to q_j, the variables of j-th block are updated from the values of variables of i-th block. Let $\Pi_i : \overline{\mathbb{Q}}^{\ell \cdot n} \to \overline{\mathbb{Q}}^n$ be the projection to the i-th block of coordinates for some $i \in \{1, \ldots, \ell\}$.

Lemma 1. *Let $\mathcal{P}$ be a (q-admissible) polynomial program and $\mathcal{L}$ its (q-admissible) transformation as above. Then, for any $i \in \{1, \ldots, \ell\}$, the projection of $\mathcal{L}$'s collecting semantics relates to the collecting semantics of $\mathcal{P}$ as*

$$\Pi_i(S^{\mathcal{L}} \setminus \{\mathbf{0}\}) = S_{q_i}^{\mathcal{P}}.$$

2.2   Undecidable Problems

Post's Correspondence Problem. The *Post's Correspondence Problem* (PCP) is an undecidable decision problem [28]. Let Σ be a finite alphabet. As an input of a PCP instance, a finite set of pairwise distinct pairs $\{(x_i, y_i) : i = 1, \ldots, k\}$ is given, where each x_i and y_i is a finite word over Σ. A solution to the problem is a sequence of indices (i_ℓ), $\ell \in \{1, \ldots, L\}$, with each $i_\ell \in \{1, \ldots, k\}$, such that concatenating elements of the corresponding pairs results in the same word:

$$x_{i_1} \ldots x_{i_L} = y_{i_1} \ldots y_{i_L}.$$

The decision problem is to decide whether such a solution exists or not. It is easy to see by encoding symbols of the alphabet (one by one) in binary that PCP restricted to the alphabet of size 2 is undecidable as well.

Boundedness of Reset VASS. *Reset vector addition systems with states (reset VASS)* are finite state automata extended with non-negative counters. They can be viewed as unguarded polynomial programs with affine updates, however with a different semantics that only allows non-negative variable values. Alternatively, reset VASS are polynomial programs guarded by inequality guards–then no change of previously introduced semantics is needed. In a reset VASS $\mathcal{V} = (Q, E, q_1)$ with n counters (variables), each update f_i of an edge $(q, (f_1, \ldots, f_n), q')$ comes from the set $\{x_i - 1, x_i, x_i + 1, 0\}$. That is, the value of each counter can be decremented, preserved, incremented, or reset to 0. Crucially, each transition that tries to decrement a zero-valued counter **is blocked**.

A *configuration* $(q, \boldsymbol{a})$ of a reset VASS $\mathcal{V}$ is a pair consisting of a location $q \in Q$ and a *non-negative* valuation $\boldsymbol{a} \in \mathbb{N}^n$ of counters. The initial configuration is $(q_1, \boldsymbol{0})$. The *collecting semantics* of $\mathcal{V}$ associates with each $q \in Q$ a set $S_q^{\mathcal{V}}$ of all $\boldsymbol{a} \in \mathbb{N}^n$ reachable in location q, i.e., such that there exists an execution of $\mathcal{V}$ reaching configuration $(q, \boldsymbol{a})$ while avoiding blocked transitions.

In the *Boundedness Problem* for reset VASS an input consists of a reset VASS $\mathcal{V}$ and its location $\hat{q} \in Q$. The question is whether the set $S_{\hat{q}}^{\mathcal{V}}$ of reachable counter valuations in location $\hat{q}$ is *finite*. The Boundedness Problem is undecidable [12].

3 Polynomial Program Challenges

In this section, we fix the precise termination and invariant synthesis problems to be addressed in our work. In extension of Definition 7, we formulate the *Termination Problem* for loops with inputs and loop conditions, as follows.

Termination Problem for $\mathcal{L}_G$ – TERM
Given: a loop with inputs $\mathcal{L}_G[S]$ with condition G, the set of inputs S
Question: Does $\mathcal{L}_G[S]$ terminate?

Note that answering TERM requires reasoning about all execution trees of $\mathcal{L}_G[S]$. A positive answer means that **all** execution trees with roots from S and the shared updates of $\mathcal{L}_G[S]$ are finite. An instance of the problem is negative if there is **at least one** infinite execution tree or, equivalently, if an infinite execution with some initial vector $\boldsymbol{c} \in S$ exists.

Note that the question whether a loop $\mathcal{L}_G[\boldsymbol{c}]$ with a fixed initial vector terminates is a special case of TERM where S is guaranteed to be a singleton. This is *the Halting Problem*, denoted HALT, see Sect. 6 for a discussion.

The problem we address in Sect. 5 is about finding the minimal collection of algebraic sets that contains the collecting semantics of a polynomial program $\mathcal{P}$.

Unlike TERM and HALT, the problem of computing strongest (algebraic) invariants is not a decision problem.

Strongest Algebraic Invariant for $\mathcal{P}$ – INV
Given: a polynomial program $\mathcal{P}$ with location set Q
Compute: the strongest algebraic invariant $\mathcal{X} = \{\overline{S_q^{\mathcal{P}}} : q \in Q\}$

---

4   Termination with Inequality Conditions

In this section, we consider the Termination Problem TERM for polynomial loops with loop conditions defined by Boolean combinations of linear inequalities. Formally, such loop condition G can be expressed as $\bigvee_i \bigwedge_j L_{ij} \{>, \geq\} l_{ij}$, where each L_{ij} is a linear form of program variables $x_1, \ldots, x_n$ with integer coefficients and each $l_{ij} \in \mathbb{Z}$ is a constant. Equivalently, this can be written as $\bigvee_i \bigwedge_j A_{ij} \{>, \geq\} 0$, where each A_{ij} is an affine function of program variables. We refer to the loop conditions of this type as *linear inequality conditions*.

Program 2 A polynomial loop $\mathcal{L}_G$ with a linear inequality condition

$x := f^{(0)}(x)$ $\hfill \{x := c \in \mathbb{Q}^n\}$
while $\bigvee_i \bigwedge_j A_{ij}(x) \{>, \geq\} 0$ **do**
$\quad \sigma_1 : \qquad x := f^{(1)}(x)$
$\quad$ **or**
$\quad \vdots$
$\quad$ **or**
$\quad \sigma_k : \qquad x := f^{(k)}(x)$

As shown in [5], TERM is undecidable for *general* polynomial loops with linear inequality condition. The proof provided in [5] includes a reduction from the undecidable Hilbert's 10th Problem. The degrees of polynomial updates $f^{(i)}$ in the loops from that reduction are arbitrary, just as in the original problem of whether a polynomial has integer solutions.

In what follows, we provide a stronger result which shows *undecidability already for a substantially smaller subclass of polynomial loops* than in [5].

Multi-path Affine Loops. We proceed by considering a subclass of polynomial loops with **linear inequality conditions and affine updates** in the loop body.

Example 1 (multi-path affineloop).
$\quad (x, y, z) := (1, 1, 1)$
$\quad$ **while** $x + y + z > 0 \wedge x \leq 10$ **do**

$$\sigma_1: \quad (x, y, z) := (x + y, \tfrac{1}{2}y, -3z + \tfrac{11}{7})$$
or
$$\sigma_2: \quad (x, y, z) := (-2y + z + \tfrac{3}{2}, x - y, z)$$
or
$$\sigma_3: \quad (x, y, z) := (y, z, -x + 1)$$

Such loops are defined as in Program 2 with each update function $f^{(i)}$ being a vector of polynomials of degree at most 1. An example of such loop with $n = 3$ variables and $k = 3$ affine update functions is given in Example 1.

We refer to such loops as *multi-path affine loops* emphasising the non-determinism in their definition in contrast to the well-studied (single-path) linear/affine loops, cf. [14,17].

Observe that affine updates of a loop can be simulated by linear ones by virtue of introducing an additional variable which is constantly set to 1. Therefore, we can view k updates of a multi-path affine loop as k linear transformations in $\mathbb{Q}^{n+1}$ which, in turn, can be expressed by k rational square matrices of size $n + 1$.

For a single-path counterpart, i.e., $k = 1$, TERM with $S = \mathbb{Q}$ is known to be decidable [6] (see [14] for a similar result regarding termination on $\mathbb{Z}$ of loops with integer updates). In the multi-path case, mostly reachability problems have been studied, establishing undecidability in multiple situations [1,19]. We emphasise that in the non-deterministic setting termination on a given input is not equivalent to the reachability questions previously considered in literature. A loop is terminating iff all of its executions are finite, so we need to argue about reaching the negation of the loop condition on *all* branches of the execution tree.

We notice that the undecidability result of [29] also concerns termination of multi-path affine loops with linear inequality guards. Observe that the updates of that construction simulate finite control. It is thus crucial that different update functions are applied depending on different conditions. The result we prove next considers a less general class of loops: it can be thought of as a guarded polynomial loop where every guard is in fact the same inequality.

Theorem 1. TERM *is undecidable for multi-path affine loops with inputs.*

Proof. We reduce the undecidable Post's Correspondence Problem over the binary alphabet $\{0, 1\}$ to TERM for multi-path affine loops. More specifically, we describe an algorithm that for any PCP instance $\mathcal{I} = \{(x_i, y_i) : i = 1, \ldots, k\}$ yields a multi-path affine loop with inputs $\mathcal{L}_G[S]$ which terminates on S iff $\mathcal{I}$ has *no* solutions. Notice that the negation of PCP is not semi-decidable, so our reduction shows that TERM for multi-path affine loops is not semi-decidable as well, much in the spirit of the undecidability result in [5]. The multi-path affine loop with inputs $\mathcal{L}_G[S]$ operates over the variable set $\{c, x, y, z\}$:

$(x, y, z) := (1, 1, 1)$
while $c \geq 0 \land z > 0 \land z < 1$ **do**
  σ_1 or $\ldots$ or σ_k or σ_{k+1}

Note that we set $S := \{(c, 1, 1, 1) : c \in \mathbb{Q}\}$. That is, the reduction produces a family of loops with all possible initial values of the variable c. The loop condition is indeed a conjunction of linear inequalities.

Now for every pair (x_i, y_i) of words in $\mathcal{I}$, we introduce a *simultaneous update* σ_i of loop variables. First, define a function num : $\{0, 1\}^* \to \mathbb{N}$ that maps a binary word to its numerical value. Observe, in particular, that this function is not injective, as $\mathrm{num}(\varepsilon) = \mathrm{num}(0) = 0$ or $\mathrm{num}(11) = \mathrm{num}(011) = 3$. For each pair of words $(x_i, y_i) \in I$ with $x_i \in \{0, 1\}^{t_i}$ and $y_i \in \{0, 1\}^{b_i}$, we define an update σ_i (on the left); one more update is defined (on the right):

$$\sigma_i : x := 2^{t_i} \cdot x + \mathrm{num}(x_i); \qquad\qquad \sigma_{k+1} : x := x;$$
$$y := 2^{b_i} \cdot y + \mathrm{num}(y_i); \qquad\qquad\qquad\quad y := y;$$
$$z := 0; \qquad\qquad\qquad\qquad\qquad\quad z := -2z + x - y;$$
$$c := c - 1; \qquad\qquad\qquad\qquad\qquad\; c := 0;$$

The effect of an update σ_i with $i \neq k + 1$ is simply appending both words $x_i, y_i \in \{0, 1\}^*$ to the binary representation of the values of variables x and y accumulated so far. We furthermore notice that applying two updates σ_i and σ_j (with $i \neq k + 1$ and $j \neq k + 1$) to some valuation (x, y) results in the same valuation (x', y') only if $i = j$.

Let $L \geq 1$ and $(i_1, \ldots, i_L)$ be a sequence of I's elements, i.e., $i_j \in \{1, \ldots, k\}$ for any $j \in \{1, \ldots, L\}$. The structure of updates σ_i, $i = 1, \ldots, k$, guarantees that $x = y$ holds after applying the sequence of updates $\sigma_{i_1}, \ldots, \sigma_{i_L}$ starting from $x = y = 1$ iff $x_{i_1} \ldots x_{i_L} = y_{i_1} \ldots y_{i_L}$. We classify all possible executions of a loop from $\mathcal{L}_G[S]$. The following case distinction is complete:

1. Executions that start with σ_{k+1}. Termination after the first iteration due to $z = -2 < 0$.
2. Executions that after applying σ_{k+1} at iteration N, proceed with applying some σ_i, $i = 1, \ldots, k$, at iteration $N+1$. These terminate after iteration $N+1$ due to $c = -1 < 0$.
3. Executions without σ_{k+1}. Termination after $\max\{0, \lfloor c \rfloor + 1\}$ iterations at latest due to $c < 0$.
4. Executions where $N \geq 1$ (but finitely many) updates with indices from the set $\{1, \ldots, k\}$ are applied first, followed by finitely or infinitely many applications of σ_{k+1}.

In the rest of the proof we show that a non-terminating infinite execution (necessarily of type 4) exists if and only if a PCP instance $\mathcal{I}$ has a solution. Indeed, if $(i_1, \ldots, i_L)$ is a solution to $\mathcal{I}$, then consider the loop $\mathcal{L}_G[L, 1, 1, 1]$ with the initial value of c set to L. This program has a non-terminating execution $\sigma_{i_1} \ldots \sigma_{i_L} \sigma_{k+1}^\omega$. After applying the first L updates of it, a fixpoint of σ_{k+1} with $x = y \wedge c = z = 0$ is reached, hence the loop does not terminate. If the PCP instance $\mathcal{I}$, on the other hand, has no solutions, then after applying $N \geq 1$ updates with indices from the set $\{1, \ldots, k\}$, either $c \geq 0$ is violated (and the loop terminates) or $x \neq y$. Let $(\hat{x}, \hat{y})$ be the values of variables x and y after applying the last update different from σ_{k+1}. Then, in the next two iterations (if they exist), the value of z becomes $\hat{x} - \hat{y}$ and $-(\hat{x} - \hat{y})$ respectively. Since $\hat{x} \neq \hat{y}$, at least one of those values is not from the set $\{0, 1\}$–and the loop terminates due to the violation of $z \geq 0 \wedge z \leq 1$. Hence, all executions of type 4 terminate. The executions of other types are all terminating. $\square$

Remark 1. The update coefficients in the proof above are all integers. We observe that the input set S can be viewed as a subset of $\mathbb{Z}^n$, $\mathbb{Q}^n$ or $\mathbb{R}^n$ without affecting the proof. Therefore, TERM is in fact proven undecidable *not only over the domain*[1] $\mathbb{Q}$, *but also over integers or reals*. We emphasise that the three termination problems are not trivially equivalent, see also [6].

Remark 2. If PCP is undecidable when restricted to n_{PCP} word pairs, then TERM is undecidable for multi-path affine loops with $n_{PCP} + 1$ update functions. Since PCP has been proven undecidable with 5 pairs (see [26]), we argue that TERM is undecidable already for multi-path affine loops with 4 variables[2] and 6 updates.

Corollary 1. *The Termination Problem* TERM *is undecidable for multi-path linear loops with 2 update matrices.*

5 Strongest Algebraic Invariants

In this section we investigate the gap between two results on versions of INV for different classes of polynomial programs, with both results shown in [15]. The discussions of this section refer to unguarded programs, unless stated differently.

Theorem 2 ([15]). *There is an algorithm that given an affine program $\mathcal{P}$ with locations set Q computes its strongest algebraic invariant $\mathcal{X} = \{X_q : q \in Q\}$.*

Theorem 3 ([15]). *There is no algorithm that computes the strongest algebraic invariant of an arbitrary polynomial program.*

The unsolvability result of Theorem 3 holds when the degrees of updates are not bounded by an a priori constant. It is, however, of interest how far Theorem 2 can be pushed. That is, whether the strongest algebraic invariant can be computed algorithmically for programs with update degrees up to some $d > 1$.

It turns out that the solvability boundary in fact lies between linearity and non-linearity. That is, the problem cannot be solved algorithmically already for $d = 2$. We prove this using a modification of the reduction given in [15, Sect. 7]. Other than in the reference, we keep the degrees of update polynomials at most 2. Our reduction avoids a blow-up in the number of program variables while at the same time flattening the degrees of updates.

---

[1] TERM is defined over D when $S \subseteq D^n$ and the updates coefficients are from D.
[2] Note that TERM remains undecidable even for a fixed number of variables.

Theorem 4. *There is no algorithm that computes the strongest algebraic invariant of a q-admissible program.*

Proof. We reduce the Boundedness Problem for reset VASS to the problem of computing strongest algebraic invariants of q-admissible programs.

First, we can assume without loss of generality that every transition of a reset VASS $\mathcal{V}$ involves at most one decrement. Indeed, if $m > 1$ counters are decremented at once, such an edge can be substituted with a path of length m that visits new (finitely many) "dummy" locations and where the same counters are decremented one by one. The collecting semantics associated with the "old" locations of a VASS is preserved. Therefore, the Boundedness Problem for reset VASS without simultaneous decrements is still undecidable.

Reduction. In the next step, we give a reduction from the latter problem by exhibiting a q-admissible program $\mathcal{P}$ capable of simulating executions of $\mathcal{V}$. The assignments of $\mathcal{P}$ are manufactured so that they simulate the computation of $\mathcal{V}$, among other things handling the non-negativity requirement.

Let $\mathcal{V} = (Q, E, q_1)$ have n counters and let $E = E_0 \cup E_1$, where $E_0 \cap E_1 = \varnothing$ and E_i stands for the set of transitions which have exactly i decrements. In turn, $\mathcal{P} := (Q \cup \{q_0\}, E', q_0)$ operates over $n + 1$ variables $x_0, x_1, \ldots, x_n$. The updates $f^{(0)}$ of the edge $e_0 := (q_0, f^{(0)}, q_1) \in E'$ are assigning $x_0 := 1$ and $x_i := 0$ for all $i \neq 0$. Further, given an update polynomial $f_i(x_1, \ldots, x_n) = cx_i + d$, where $(c, d) \in \{(0, 0), (1, 1), (1, 0), (1, -1)\}$, we consider a homogeneous map

$$f_i^*(x_0, x_1, \ldots, x_n) := cx_i + dx_0.$$

We also refer to id as the identity map. The edges of $E' \setminus \{e_0\}$ are in one-to-one correspondence with the edges of the set E. We proceed by describing their polynomial assignments. For each non-decrementing transition $(q_i, (f_1, \ldots, f_n), q_j) \in E_0$ of VASS there is an edge $(q_i, (2 \cdot \text{id}, 2 \cdot f_1^*, \ldots, 2 \cdot f_n^*), q_j) \in E'$, where each f_i^* has the new variables $x_0, \ldots, x_n$ as arguments. We emphasise that the update polynomials of this first transition type are *at most linear*. For the edges of VASS where decrements take place (set E_1), the updates are slightly different. For each such edge $(q_i, (f_1, \ldots, f_n), q_j) \in E_1$, where x_d is the decremented variable, there is an edge $(q_i, \langle \ldots \rangle, q_j) \in E'$ with the assignments

$$\begin{cases} x_0 := 2x_d \cdot x_0, \\ x_i := 2x_d \cdot f_i^*(x_0, \ldots, x_n) \text{ for each } i \in \{1, \ldots, n\}. \end{cases}$$

The update polynomials in this case have *degree at most 2*. The result of the reduction is thus a q-admissible program $\mathcal{P}$.

The key idea to prove **correctness** of this reduction is to observe that the ratios $x_1/x_0, \ldots, x_n/x_0$ mimic the values of the n counters in $\mathcal{V}$. Furthermore, any decrement of $\mathcal{V}$'s zero-valued counter is caught in the corresponding transition of $\mathcal{P}$. If such an update happens in an execution of $\mathcal{V}$, all variables of $\mathcal{P}$, including x_0, are set to 0. This can be outlined as follows:

For any location q of $\mathcal{V}$, it holds that for any $\boldsymbol{a} \in S_q^{\mathcal{V}}$ there exists $\boldsymbol{b} \in S_q^{\mathcal{P}} \setminus \{\boldsymbol{0}\}$ with $a_i = b_i/b_0$ for all $i \in \{1, \ldots, n\}$, and vice versa.

The rest of the proof relates the finiteness of $S_{q_i}^{\mathcal{V}}$ to the dimension of the Zariski closure $\overline{S_{q_i}^{\mathcal{P}}}$. The argument is close to that of [15, Proposition 17]. First, if $S_{q_i}^{\mathcal{V}}$ is a finite set, then a finite collection of lines through the origin in $\mathbb{Z}^{n+1}$ contains $S_{q_i}^{\mathcal{P}}$, hence the dimension of the closure $\overline{S_{q_i}^{\mathcal{P}}}$ is at most one. Now assume $S_{q_i}^{\mathcal{V}}$ is infinite. Then, without loss of generality there exist infinitely many values b_1/b_0 for $\boldsymbol{b} \in S_{q_i}^{\mathcal{P}}$. Each transition of $\mathcal{P}$ increases the value of x_1/x_0 by at most one. At the same time, every transition at least doubles the value of x_0. An asymptotic argument factors out the existence of a non-zero polynomial that relates infinitely many values b_1/b_0 to the values b_0. We finally observe that a non-zero polynomial relating b_0 and b_1 for all reachable vectors exists if and only if some non-zero polynomial relates b_0 and b_1/b_0. Therefore, no polynomial that only binds variables x_1 and x_0 vanishes on $S_{q_i}^{\mathcal{P}}$ and hence, on its Zariski closure $\overline{S_{q_i}^{\mathcal{P}}}$. However, for a set of dimension at most 1, it is known (see [10, Chapter 9, Sect. 5, Corollary 4]) that for any choice of two variables, there exists a non-zero polynomial in them that vanishes on this set. We thus argue that the dimension of $\overline{S_{q_i}^{\mathcal{P}}}$ is at least 2.

Suppose we can compute the strongest algebraic invariant for $\mathcal{P}$ that simulates $\mathcal{V}$. Then, we can find the dimension of the set $\overline{S_q^{\mathcal{P}}}$ for any location q, since an algorithm exists that given a polynomial ideal, finds the dimension of its zero set (see, e.g., [10, Chapter 9, Sect. 3]). However, we can then deduce whether the dimension is at most one–and thus decide whether the set $S_q^{\mathcal{V}}$ is finite. But this is an undecidable problem, hence, computing the strongest algebraic invariant for a q-admissible program cannot be done algorithmically.   □

While Theorem 4 holds for programs with multiple locations, we want to show the unsolvability of a version of the problem restricted to loops, too. Let $\mathcal{L}$ be a q-admissible loop that simulates a q-admissible program $\mathcal{P}$, as in Lemma 1. Since projection Π_i is a Zariski-continuous map, it is easy to see the following from Lemma 1: $X := \Pi_i(S^{\mathcal{L}} \setminus \{\boldsymbol{0}\})$ not only contains $S_{q_i}^{\mathcal{P}}$ but also has the same Zariski closure $\overline{S_{q_i}^{\mathcal{P}}}$. Now, if we could compute $\overline{S^{\mathcal{L}}}$, then the set X would be constructible (see [15, Sect. 3.2] for definitions). In that case, an approach of e.g. [20, Theorem 1], could be used to find its Zariski closure $\overline{S_{q_i}^{\mathcal{P}}}$. This is, however, an algorithmically unsolvable problem. Therefore, $\overline{S^{\mathcal{L}}}$ cannot be computed.

Corollary 2. *There is no algorithm that computes the strongest algebraic invariant of a q-admissible loop.*

Equality Guards. While the Termination Problem TERM for multi-path affine programs renders undecidable, the result of Theorem 2 exhibits an algorithm to solve its safety counterpart INV for the same class of updates. Here, we briefly discuss how introducing guards changes the solvability landscape. The following result holds already in the presence of a single equality test

Theorem 5 ([24]). *The Constant Propagation problem, i.e., establishing whether a variable maintains a unique constant value at a given program location, is undecidable for guarded affine programs.*

This immediately implies that generating the strongest algebraic invariant of a GPP with affine updates cannot be done algorithmically. Indeed, whether an expression of the type $x_j = c$ for some $c \in \mathbb{Q}$ holds at location q can be easily checked, as long as the Zariski closure $\overline{S_q^{\mathcal{P}}}$ is computed. As in the q-admissible case, the uncomputability transfers to loops. It suffices to see that the executions of $\mathcal{P}$ can be mimicked by those of a guarded polynomial loop $\mathcal{L}$ with additional new variables $y_1, \ldots, y_\ell$, each of which stands for a location of $\mathcal{P}$. The idea is to force the variable y_i to have value 0 every time the location q_i is active in the execution of $\mathcal{P}$, while $y_j = 1, j \neq i$. The transitions from q_i are respectively guarded by "$y_i = 0$?" This way, only those executions of $\mathcal{L}$ are singled out that correspond to the proper executions of $\mathcal{P}$. Meanwhile, the loop variables $x_1, \ldots, x_n$ go through the same sequence of valuations as in $\mathcal{P}$. Finally, $x_j = c$ is valid at location q_i of $\mathcal{P}$ iff $\mathcal{L}$ has $(y_i - 1) \cdot (x_j - c) = 0$ as an invariant.

Corollary 3. *There is no algorithm that computes the strongest algebraic invariant of a guarded affine loop.*

We conclude that both TERM and INV become unsolvable for affine loops with guarded updates, just as they do for q-admissible loops due to Theorems 1 and 4.

6   Discussion and Conclusions

Termination. We put the undecidability result of Theorem 1 into the context of existing works in this direction and discuss some follow-up questions.

Both the abstraction of loops and the formulation of the termination problem considered in this work are similar to those in [5]. Taking into account a set of possible inputs as a parameter of the decision problem is natural for a verification framework: thus finiteness of executions is considered for all inputs that satisfy a precondition. An improvement with respect to the result of [5] is that we prove TERM to be undecidable already with updates of degree at most 1.

We emphasise once more that the model we consider in Theorem 1 is a restriction of Tiwari's model [29] (see also the discussion in Sect. 4). In particular, Theorem 1 implies that TERM is undecidable for the programs with several updates guarded, in general, by different predicates, as in [29, Sect. 6]. Moreover, results presented here witness that TERM becomes undecidable already with two linear updates guarded by the same inequality. In turn, restricting loop conditions in Theorem 1 to equalities makes TERM decidable since the set of non-terminating inputs is algorithmically computable for that model [22]. This remains true for multiple equality guards.

Further challenges arise in the study of the Halting Problem HALT:

Halting Problem for $\mathcal{L}_G$ – HALT
Given: a polynomial loop $\mathcal{L}_G$ with condition G and an initial vector $c \in \mathbb{Q}^n$
Question: Does $\mathcal{L}_G$ terminate (halt) on c?

Recall that termination on all rational (resp. integer) inputs is decidable for single-path affine loops with linear inequality conditions [6,14]. At the same time, HALT for these loops is equivalent to the *Positivity Problem*, a long-standing open problem in number theory [27]. Consider the problem HALT for the class of multi-path loops, where a loop $\mathcal{L}_G$ has several affine updates chosen non-deterministically. On the one hand, decidability of HALT would not contradict undecidability of TERM proved here and on the other hand, proving undecidability of HALT for multi-path loops would not show its special case, HALT for single-path affine loops (or, equivalently, the Positivity Problem) undecidable. This question remains open.

Invariants. With Theorem 4, we have shown that generalising the class of updates in a polynomial program from affine to q-admissible changes the solvability characterisation of the strongest algebraic invariant problem INV. We point out that the unsolvability result of Theorem 4 exploits non-determinism of updates, following [15]. The problem of computing the strongest algebraic invariant of an arbitrary *single-path* polynomial loop, for its part, remains open (see also [22]).

Acknowledgements. We thank the anonymous reviewers for their valuable feedback on our work. The work presented in this paper was partially supported by the ERC consolidator grant ARTIST 101002685, the ProbInG project of the Vienna Science and Technology Fund (WWTF [10.47379/ICT19018]), and the Marie Sklodowska-Curie Doctoral Network LogiCS@TU Wien.

References

1. Bell, P., Potapov, I.: On undecidability bounds for matrix decision problems. Theor. Comput. Sci. **391**(1), 3–13 (2008). https://doi.org/10.1016/j.tcs.2007.10.025

2. Ben-Amram, A.M., Genaim, S., Masud, A.N.: On the termination of integer loops. ACM Trans. Program. Lang. Syst. **34**(4), 1–24 (2012). https://doi.org/10.1145/2400676.2400679

3. Beyer, D., Henzinger, T.A., Majumdar, R., Rybalchenko, A.: Invariant synthesis for combined theories. In: Cook, B., Podelski, A. (eds.) VMCAI 2007. LNCS, vol. 4349, pp. 378–394. Springer, Heidelberg (2007). https://doi.org/10.1007/978-3-540-69738-1_27

4. Bradley, A.R., Manna, Z.: The Calculus of Computation - Decision Procedures with Applications to Verification. Springer, Heidelberg (2007). https://doi.org/10.1007/978-3-540-74113-8

5. Bradley, A.R., Manna, Z., Sipma, H.B.: Termination of polynomial programs. In: Cousot, R. (ed.) VMCAI 2005. LNCS, vol. 3385, pp. 113–129. Springer, Heidelberg (2005). https://doi.org/10.1007/978-3-540-30579-8_8
6. Braverman, M.: Termination of integer linear programs. In: Ball, T., Jones, R.B. (eds.) CAV 2006. LNCS, vol. 4144, pp. 372–385. Springer, Heidelberg (2006). https://doi.org/10.1007/11817963_34
7. Chatterjee, K., Fu, H., Goharshady, A.K., Goharshady, E.K.: Polynomial invariant generation for non-deterministic recursive programs. In: PLDI 2020, pp. 672–687 (2020). https://doi.org/10.1145/3385412.3385969
8. Cook, B., Podelski, A., Rybalchenko, A.: Proving program termination. Commun. ACM **54**(5), 88–98 (2011). https://doi.org/10.1145/1941487.1941509
9. Cousot, P., Halbwachs, N.: Automatic discovery of linear restraints among variables of a program. In: POPL 1978, pp. 84–96 (1978). https://doi.org/10.1145/512760.512770
10. Cox, D.A., Little, J., O'Shea, D.: Ideals, Varieties, and Algorithms. UTM, Springer, Cham (2015). https://doi.org/10.1007/978-3-319-16721-3
11. Dijkstra, E.W.: Guarded commands, nondeterminacy and formal derivation of programs. Commun. ACM **18**(8), 453–457 (1975). https://doi.org/10.1145/360933.360975
12. Dufourd, C., Finkel, A., Schnoebelen, P.: Reset nets between decidability and undecidability. In: Larsen, K.G., Skyum, S., Winskel, G. (eds.) ICALP 1998. LNCS, vol. 1443, pp. 103–115. Springer, Heidelberg (1998). https://doi.org/10.1007/BFb0055044
13. Hark, M., Frohn, F., Giesl, J.: Polynomial loops: beyond termination. In: LPAR-23, vol. 73, pp. 279–297 (2020). https://doi.org/10.29007/nxv1
14. Hosseini, M., Ouaknine, J., Worrell, J.: Termination of linear loops over the integers. In: ICALP 2019, pp. 118:1–118:13 (2019). http://drops.dagstuhl.de/opus/volltexte/2019/10694
15. Hrushovski, E., Ouaknine, J., Pouly, A., Worrell, J.: On strongest algebraic program invariants. J. ACM. (2019). https://people.mpi-sws.org/~joel/publications/strongest_algebraic_invariants19abs.html
16. Humenberger, A., Jaroschek, M., Kovács, L.: Invariant generation for multi-path loops with polynomial assignments. In: VMCAI 2018. LNCS, vol. 10747, pp. 226–246. Springer, Cham (2018). https://doi.org/10.1007/978-3-319-73721-8_11
17. Karimov, T., et al.: What's decidable about linear loops? In: POPL 2022, pp. 1–25 (2022). https://doi.org/10.1145/3498727
18. Karr, M.: Affine relationships among variables of a program. Acta Inf. **6**(2), 133–151 (1976). https://doi.org/10.1007/BF00268497
19. Ko, S., Niskanen, R., Potapov, I.: Reachability problems in low-dimensional non-deterministic polynomial maps over integers. Inf. Comput. **281**, 104785 (2021). https://doi.org/10.1016/j.ic.2021.104785
20. Koiran, P.: The complexity of local dimensions for constructible sets. J. Complex. **16**, 311–323 (2000). https://doi.org/10.1006/jcom.1999.0536
21. Kurtz, S.A., Simon, J.: The undecidability of the generalized Collatz problem. In: Cai, J.-Y., Cooper, S.B., Zhu, H. (eds.) TAMC 2007. LNCS, vol. 4484, pp. 542–553. Springer, Heidelberg (2007). https://doi.org/10.1007/978-3-540-72504-6_49
22. Li, Y., Zhan, N., Chen, M., Lu, H., Wu, G., Katoen, J.P.: On termination of polynomial programs with equality conditions (2015). https://doi.org/10.48550/ARXIV.1510.05201

23. Liu, J., Xu, M., Zhan, N., Zhao, H.: Discovering non-terminating inputs for multi-path polynomial programs. J. Syst. Sci. Complex. **27**, 1286–1304 (2014). https://doi.org/10.1007/s11424-014-2145-6
24. Müller-Olm, M., Seidl, H.: A note on Karr's algorithm. In: Díaz, J., Karhumäki, J., Lepistö, A., Sannella, D. (eds.) ICALP 2004. LNCS, vol. 3142, pp. 1016–1028. Springer, Heidelberg (2004). https://doi.org/10.1007/978-3-540-27836-8_85
25. Müller-Olm, M., Seidl, H.: Computing polynomial program invariants. Inf. Process. Lett. **91**(5), 233–244 (2004). https://doi.org/10.1016/j.ipl.2004.05.004
26. Neary, T.: Undecidability in binary tag systems and the post correspondence problem for five pairs of words. In: STACS 2015, pp. 649–661 (2015). https://doi.org/10.4230/LIPIcs.STACS.2015.649
27. Ouaknine, J., Worrell, J.: On linear recurrence sequences and loop termination. ACM SIGLOG News **2**(2), 4–13 (2015). https://doi.org/10.1145/2766189.2766191
28. Post, E.L.: A variant of a recursively unsolvable problem. Bull. Am. Math. Soc. **52**, 264–268 (1946). https://doi.org/10.2307/2267252
29. Tiwari, A.: Termination of linear programs. In: Alur, R., Peled, D.A. (eds.) CAV 2004. LNCS, vol. 3114, pp. 70–82. Springer, Heidelberg (2004). https://doi.org/10.1007/978-3-540-27813-9_6
30. Xia, B., Zhang, Z.: Termination of linear programs with nonlinear constraints. J. Symb. Comput. **45**(11), 1234–1249 (2010). https://doi.org/10.1016/j.jsc.2010.06.006

Implication Algebras and Implication Semigroups of Binary Relations

Andrew Lewis-Smith[1] and Jaš Šemrl[2(✉)]

[1] King's College London, London, UK
andrew.lewis-smith@kcl.ac.uk
[2] UCL (University College London), London, UK
j.semrl@cs.ucl.ac.uk
http://eecs.qmul.ac.uk/profiles/lewis-smithandrewstephen.html,
http://www0.cs.ucl.ac.uk/staff/jsemrl/

Abstract. Representable implication algebras are known to be axiomatised by a finite number of equations (making the representation and finite representation problems decidable here). We show that this also holds in the context of unary (and binary) relations and present a Stone-style representation theorem. We then show that the (finite) representation decision problem is undecidable for implication semigroups, in stark contrast with implication algebras.

Keywords: Implication algebras · Implication semigroups · Representability as binary relations

1 Introduction

The variety of *implication algebras*, so-named by Abbott [1] and studied by Rasiowa [12], Diego [2], and their students, forms the algebraic semantics of the implicational fragment of classical propositional logic. These are Boolean algebras restricted to one operation ($\rightarrow$) and a constant ($\top$ or 1). The variety of *Relation algebras*, alias residuated Boolean algebras with an additional involution operator ($x^\smile$ or 'converse of x', with x understood as a relation), forms the algebraic semantics of the calculus of relations. By a classical result of Korselt attributed in [10], the variety of relation algebras corresponds to the three-variable fragment of classical first-order logic, permitting a study of mathematical logic, particularly set theory [14], via a quantifier-free equational theory. Proofs in this theory consist of simple manipulations of identities, similar to proofs in abstract algebra. This situation contrasts with proofs of standard mathematical logic (or set theory) which can involve complex alternations of quantifiers. Meanwhile, implication algebras (having one operation, classical implication) yield an algebraic analysis of the entailment relation between propositions in classical logic. Although informed by different motivations, a certain elegance recommends the study of relation and implication algebras.

This work was supported by the Engineering and Physical Sciences Research Council EP/S021566/1.

The present paper considers a fragment of the signature of relation algebras we call *implication semigroups* based on adjoining a semigroup operation (;), i.e. relational composition, to the implication algebras of Abbott. When the carrier set of this structure is a set of binary relations, we obtain the fragment of relation algebras consisting of $(S, \rightarrow, ;)$, i.e. relation algebras with signature restricted to implication and composition. There are good reasons to examine this signature. For one, it has not been well-explored: practically speaking, most algebraic structures considered in algebraic logic are residuated lattices, groups, or at least monoids – this can be noticed already in a standard definition of relation algebras, as residuated lattices [9] – where the implication and monoidal operations interact via residuation. Algebras featuring implication and semigroup operations fall out of the mainstream substructural logic literature as the algebra at hand lacks the interdefinability present even in the case of a residuated monoid.

For algebraic logic (particularly relation algebras) the question of whether an algebra has a finite representation looms large. One typically asks whether a given logical system of interest is not just consistent but has finite models, i.e. models we can inspect within finite time or employing finitely many resources. The present paper demonstrates the (finite) representation problem for implication semigroups is undecidable. Our results are curious for two reasons. First, implication semigroups represent, in a sense, a limiting case of substructural logics of implication for which the question of decidability of finite representations, to our knowledge, has not been raised, and certainly not approached from the angle considered here. This suggests a track of further research in what one might call *substructural relation algebras*, exploring the effects of weakening the Boolean base in relation algebra into other algebras of residuation. This is already a current area of research by Peter Jipsen and Nikolaos Galatos [4,5], and has been broached from another angle in [13], where the signature considered there bears two residuals and a semigroup operation and is in fact a model of the famed Lambek Calculus (thus connecting that algebra to the base system for infinitely many substructural logics). Second, our results contribute to a research programme seeking a better grasp of the consequences for relation algebras when operating in a restricted signature. We are particularly motivated to understand the effect on representability when moving to subsignatures of the standard presentation of a relation algebra [6].

2 Preliminaries

In this section we present the definitions of the algebraic structures and operators for binary relations. We begin by defining Abbott's implication algebras.

Definition 1. *An* implication algebra[1] $\mathfrak{A}$ *is a pair* $(A, \rightarrow)$*, with A a set and $\rightarrow$ a binary operation on A satisfying the following properties:*

  (i) $(a \rightarrow b) \rightarrow a = a$ (Contraction)
  (ii) $(a \rightarrow b) \rightarrow b = (b \rightarrow a) \rightarrow a$ (Quasi-commutativity)
  (iii) $a \rightarrow (b \rightarrow c) = b \rightarrow (a \rightarrow c)$ (Exchange)

---
[1] Also known as Tarski algebras.

Trivially, because the class of implication algebras is equationally definable, it forms a *variety*. We shall refer to this class as **IA**. Abbott shows a neat property about these in [1].

Proposition 2 (Abbott). *Let* $\mathfrak{A} = (A, \rightarrow)$ *be an implication algebra. We can implicitly define a constant* 1 *as* $a \rightarrow a$ *such that* $b \rightarrow 1 = 1$ *and* $1 \rightarrow b = b$, *for all* $b \in A$.

This also gives us

Proposition 3. *For an implication algebra* $(A, \rightarrow)$, *we can define a partial order as*

$$a \leq b \Leftrightarrow (a \rightarrow b) = 1$$

Proof. Let $a + b = (a \rightarrow b) \rightarrow b$. It is commutative by quasi-commutativity, idempotent by contraction, has 1 as the top by Proposition 2, and can be shown associative (see [1][Theorem 12]). So $a + b = b$ forms a partial order. Observe that if $a + b = b$ then $a \rightarrow b = a \rightarrow ((a \rightarrow b) \rightarrow b) = (a \rightarrow b) \rightarrow (a \rightarrow b) = 1$. If $a \rightarrow b = 1$ then $(a \rightarrow b) \rightarrow b = 1 \rightarrow b = b$. □

Definition 4. *Let* $\top \subseteq X \times X$ *be a binary relation. Define* $\mathfrak{A}(\top) = (\wp(\top), \rightarrow)$ *where* $\rightarrow$ *is interpreted as proper Boolean implication defined below*

$$a \rightarrow b = (\top \setminus a) \cup b$$

One can check that $\mathfrak{A}(\top) \in$ **IA**. Although $\top$ is conventionally an arbitrary maximal relation, this is not the only possible interpretation of the $\rightarrow$ operation for binary relations. We say that the implication operator is *absolute* if we require $\top = X \times X$, else we say that it is *relative*.

We say that $\mathfrak{A} \in$ **IA** is *representable* if and only if it embeds into $\mathfrak{A}(\top)$ for some $\top \subseteq X \times X$. The embedding (usually denoted h) is called a *representation*. If $\mathfrak{A}$ embeds into $\mathfrak{A}(\top)$ and $\top$ is over a finite base X, then we say $\mathfrak{A}$ is also *finitely representable*.

Another standard presentation of implication algebras is $\mathfrak{A} = (A, \rightarrow, 1)$. However, the constant 1 can be defined as $a \rightarrow a$, for any a. Furthermore, the quasi-commutativity axiom is a consequence of the fact that $(a \rightarrow b) \rightarrow b$ is equivalent to the Boolean join of $a + b$.

Proposition 5. *Let* $\mathfrak{A} = (A, \rightarrow) \in$ **IA** *be representable via* h. *Then* $h((a \rightarrow b) \rightarrow b) = h(a) \cup h(b)$ *and* $h(1) = h(a \rightarrow a) = \top$, *for any* $a, b \in A$.

Proof. Since $1 = a \rightarrow a$ and h is a representation we get $h(1) = h(a \rightarrow a) = h(a) \rightarrow h(a) = (\top \setminus h(a)) \cup h(a) = \top$.

By h being a representation, DeMorgan's law, and $a \cap \top = a$ we also have $h((a \rightarrow b) \rightarrow b) = (h(a) \rightarrow h(b)) \rightarrow h(b) = \top \setminus ((\top \setminus h(a)) \cup b) \cup h(b) = (h(a) \cap (\top \setminus h(b))) \cup h(b) = (h(a) \cup h(b)) \cap ((\top \setminus h(b)) \cup h(b)) = (h(a) \cup h(b)) \cap \top = h(a) \cup h(b)$. □

We now direct our attention to what happens when we add a semigroup operation (;) to the signature.

Definition 6. *An* implication semigroup $\mathfrak{S}$ *is a tuple* $(S, \to, ;)$*, with a carrier set* S *and* $\to, ;$ *binary operations on* S *where*

(i) $(S, \to)$ *is an implication algebra*
(ii) $(S, ;)$ *is a semigroup*
(iii) $((a \to b) \to b); c = (a; c \to b; c) \to b; c$ *(Left quasi-additivity)*
(iv) $c; ((a \to b) \to b) = (c; a \to c; b) \to c; b$ *(Right quasi-additivity)*

The class of implication semigroups will be called **ISG**. Similarly to **IA** we also examine structures where the carrier set is a set of binary relations.

Definition 7. *Let* $\top \subseteq X \times X$ *be a transitive binary relation. Define* $\mathfrak{S}(\top) = (\wp(\top), \to, ;)$ *where* $\to$ *is interpreted as proper Boolean implication and* $;$ *as proper relational composition defined as*

$$a; b = \{(x, z) \mid \exists y \in X : (x, y) \in a, (y, z) \in b\}$$

Again checking $\mathfrak{S}(\top) \in$ **ISG** is relatively straightforward, note that they are closed under composition due to the transitivity of $\top$. Similarly to **IA**, $\mathfrak{S} \in$ **ISG** is (finitely) representable if it embeds into $\mathfrak{S}(\top)$ for some transitive $\top$ (over a finite base).

3   Basic Theory, Stone Representation, and Decidability for Implication Algebras

We now present the basic theory of implication algebras, the implicational fragment of the implication semigroups discussed in the previous section. We first consider the more general *positive* implication algebras, subsuming the implication algebras. This culminates in a representation theorem for implication algebras, informing our construction in Sect. 4.[2]

The axiomatics here are largely in [1] and [12] with some corrections and modifications. Their presentations of the implication algebras are quite different, Abbott preferring an equational presentation where Rasiowa utilises a quasiequational definition.

Definition 8 (Rasiowa 2). *A positive* implication algebra[3] *(Postive* **IA***) is a pair* $(A, \to, 1)$[4]*, a set* A *and* $\to$ *satisfying:*

(P1) $a \to (b \to a) = 1$
(P2) $(a \to (b \to c)) \to ((a \to b) \to (a \to c)) = 1$
(P3) if $a \to b = 1$ *and* $b \to a = 1$ *then* $a = b$
(P4) $a \to 1 = 1$

---

[2] The representation result for implication algebras appears to have been known to Diego [2], perhaps Abbott [1], but the proof is given in full by Rasiowa in [12]. It was probably known to several others throughout different traditions of algebraic logic.

[3] Also known as a Hilbert algebra.

[4] With this axiomatisation we cannot omit 1 from the signature. Alternatively, 1 could be replaced with $a \to a$ and an extra axiom added as $a \to a = b \to b$.

Without proof, we state the following lemmas. For proofs, refer to [12].

Proposition 9 (Rasiowa 2(1)) *In any positive implication algebra, the following condition is fulfilled: if $a \to b = 1$ and $a = 1$, then $b = 1$. Also, if $a = 1$, then $b \to a = 1$ for any $b \in A$.*

Proposition 10 (Rasiowa 2.2). *For any positive* **IA** $\mathfrak{A}$*, for all $a, b \in A$, we can define a partial order $\leq$ on A as*

$$a \leq b \Longleftrightarrow a \to b = 1$$

and $1 = c \to c$ for all maximal c in the poset $(A, \leq)$.

Proposition 11 (Rasiowa 2.3). *The following hold in any positive implication algebra:*

(1) If $a \leq b \to c$ then $b \leq a \to c$
(2) $a \leq (a \to b) \to b$
(3) $1 \to a = a$
(4) If $b \leq c$, then $a \to b \leq a \to c$
(5) If $a \leq b$ then $b \to c \leq a \to c$
(6) $a \to (b \to c) = b \to (a \to c)$

Proposition 12 (Distributivity). *In any (positive) implication algebra $\mathfrak{A} = (A, \to, 1)$, we have $a \to (b \to c) = (a \to b) \to (a \to c)$*

Proof We have $b \leq a \to b$ by (P1) and Proposition 10. Applying Proposition 11(5)(6), we get $(a \to b) \to (a \to c) \leq b \to (a \to c) = a \to (b \to c)$. So, $a \to (b \to c) = (a \to b) \to (a \to c)$ follows from (P2) and Proposition 10.

The proof that distributivity holds in implication algebras is found in [1, Theorem 5]. □

We now show that the class of implication algebras lies below the class of positive implication algebras. Although the following proposition is not in Abbott or Rasiowa, it is latent in the published results concerning implicative, positive implication, and implication algebras.

Proposition 13. *Any implication algebra $(A, \to)$ is a positive implication algebra.*

Proof. (P1) follows from the exchange axiom and Proposition 2, more specifically $a \to (b \to a) = b \to (a \to a) = b \to 1 = 1$. For (P2) follows from Proposition 12 and Proposition 2. For (P3) see that by Proposition 3 we have the anti-symmetry for the partial order in implication algebras. Finally (P4) follows directly from Proposition 2. □

Proposition 14. *Any positive implication algebra $(A, \to)$ satisfying*

$$(a \to b) \to a = a$$

for all $a, b \in A$ is an implication algebra.[5]

---

[5] Note that the contraction identity is not provable from the axioms (P1)–(P4), a counterexample can be found using Mace4.

Proof. To show the other direction, let $(A, \rightarrow, 1)$ be a positive implication algebra satisfying $(a \rightarrow b) \rightarrow a = a$. The first axiom of implication algebras $(a \rightarrow b) \rightarrow a = a$ we have already assumed adjoined to the algebra, and the third axiom, $a \rightarrow (b \rightarrow c) = b \rightarrow (a \rightarrow c)$, is found in Proposition 11(6). To show the second axiom: $(a \rightarrow b) \rightarrow b = (b \rightarrow a) \rightarrow a$, we note $a \rightarrow b \leq 1 = (b \rightarrow b) = (b \rightarrow a) \rightarrow (b \rightarrow b) = b \rightarrow ((b \rightarrow a) \rightarrow b)$ by Proposition 10 and Proposition 11(6). By Proposition 11(1) we have $b \leq (a \rightarrow b) \rightarrow ((b \rightarrow a) \rightarrow b)$ and thus by Proposition 11(1) and (3) we get $(a \rightarrow b) \rightarrow b \leq ((a \rightarrow b) \rightarrow (a \rightarrow b)) \rightarrow ((b \rightarrow a) \rightarrow b) = 1 \rightarrow ((b \rightarrow a) \rightarrow b) = ((b \rightarrow a) \rightarrow b)$. By a completely analogous argument, $(a \rightarrow b) \rightarrow b \leq (b \rightarrow a) \rightarrow a$. Hence $(a \rightarrow b) \rightarrow b = (b \rightarrow a) \rightarrow a$ as desired.   □

In anticipation of the Stone-style representation theorem, we define some required notions like that of an implicative filter.

Definition 15 (Abbott). *An implicative filter of a (positive) implication algebra* $\mathfrak{A} = (A, \rightarrow)$ *is a subset* $F \subseteq A$ *such that:*

(i) $1 \in F$
(ii) if $a \in F$ *and* $a \rightarrow b \in F$ *then* $b \in F$

Definition 16. *We say that an implicative filter* F *is* proper *if* $F \neq A$. *We say that a proper implicative filter is* irreducible *if it is not the intersection of two proper implicative filters distinct from it, or formally:* F *is irreducible if for any two proper implicative filters* F_1, F_2 *such that* $F = F_1 \cap F_2$, *either* $F = F_1$ *or* $F = F_2$. *Finally, a proper implicative filter* F *is said to be* prime *if* $a + b \in F$ *(or equivalently* $(a \rightarrow b) \rightarrow b \in F$) *implies that either* $u \in F$ *or* $b \in F$, *for all* $a, b \in A$.

The proof of the Stone-like Representation theorem follows the following steps. For proofs, refer to [12].

Proposition 17 (Rasiowa 1.8).[6] *If in any (positive) implication algebra* $\mathfrak{A} = (A, \rightarrow)$ *one of the following conditions is satisfied for all* $a, b, c \in A$:

(F1) $(a \rightarrow (b \rightarrow c)) \rightarrow a \rightarrow b) \rightarrow (a \rightarrow c)) = 1$
(F2) $(a \rightarrow b) \rightarrow (a \rightarrow (b \rightarrow c)) \rightarrow (a \rightarrow c)) = 1$

then for every implicative filter F *in* $\mathfrak{A}$ *and for every* $a \in A$, *the set* $F_{a^*} = \{x \in A : a \rightarrow x \in F\}$ *is an implicative filter. If, moreover, for all* $a, b \in A : a \rightarrow (b \rightarrow a) = 1$, *then* F_{a^*} *is the* least *implicative filter containing* F *and* a.

Proposition 18 (Rasiowa 3.4). *If* $(A, \rightarrow)$ *is a (positive) implication algebra, then for every implicative filter* F *and for every element* $a \in A$ *the set* $F_{a^*} = \{x \in A : a \rightarrow x \in F\}$ *is the* least *implicative filter containing* F *and* a.

---

[6] Rasiowa states this result for implicative algebras, the weakest algebra she considers in her text. Since all implication algebras are positive implication algebras, and all positive implication algebras are implicative algebras, we can specialise her result for the present case.

Proposition 19 (Rasiowa 6.1). *An implicative filter in an implication algebra is prime if and only if it is irreducible.*

Lemma 20 (Rasiowa 1.4). *If F_0 is an implicative filter in an implicative algebra $\mathfrak{A}$ such that $a_0 \notin F_0$ for some $a_0 \in \mathfrak{A}$ then there exists an irreducible implicative filter G such that $F_0 \subset G$ and $a_0 \notin G$.*

Immediately, by Lemma 20 and Proposition 19 we have:

Corollary 21. *If F is an implicative filter in an implicative algebra $\mathfrak{A}$ such that $a \notin F$ for some $a \in \mathfrak{A}$ then there exists a prime implicative filter G such that $F \subset G$ and $a \notin G$.*

This next corollary we prove, as it is not found in any of the literature cited above and is required for the representation theorem.

Corollary 22. *Let F be an implicative filter of an implication algebra $\mathfrak{A} = (A, \rightarrow)$ such that $a \rightarrow b \notin F$ for some $a, b \in A$. Then there exists a prime implicative filter $G : F \subseteq G$ such that $a \in G$ and $b \notin G$.*

Proof. Let F_a* be the implicative filter generated by the filter F and a. Suppose that $a \rightarrow b \notin F$. If $b \in F_a*$, then we have $a \rightarrow b \in F$ by the definition of F_a*. This contradicts our assumption that $a \rightarrow b \notin F$; hence $b \notin F_a*$, and applying Corollary 21 for F_a* and b we have a prime filter G such that $F_a* \subseteq G$ and $b \notin G$. Clearly, $a \in G$ and $F \subseteq G$.   □

We have then, as an immediate corollary from Corollary 22 and Proposition 19, the following:

Corollary 23. *Let F be an implicative filter of an implication algebra $\mathfrak{A} = (A, \rightarrow)$ such that $a \rightarrow b \notin F$ for some $a, b \in A$. Then there exists an irreducible implicative filter $G : F \subseteq G$ such that $a \in G$ and $b \notin G$.*

Finally, the culminating representation theorem. Rasiowa presents this for irreducible implicative filters [12], which given her equivalence result, one can also state using prime implicative filters, or maximal implicative filters.

Theorem 24 (Rasiowa 7.1). *For any implication algebra $\mathfrak{A} = (A, \rightarrow)$, there is a monomorphism h from $\mathfrak{A}$ to $(\wp(X), \rightarrow)$ of an arbitrary space X with $|X| \geq A$.*

From this it follows that every implication algebra is isomorphic to an implication algebra of sets. Since the focus of the present paper is on representations, we note a corollary from this last result [1,2,12]:

Corollary 25. *For any implication algebra $\mathfrak{A}$, if A is finite, then $\mathfrak{A}$ has a finite representation.*

Proof. Let $\mathfrak{A}$ be a finite implication algebra. Then by Theorem 24 $\mathfrak{A}$ is monomorphic to the algebra $\mathfrak{A}\prime$ under h, where $\mathfrak{A}\prime = (\wp(X), \rightarrow)$, an implication algebra of sets. Now if $|X| = A$ then $|\wp(X)| = 2^{|A|}$, and thus finite. That means $h(\mathfrak{A})$, the subalgebra of $\mathfrak{A}\prime$ under h, is finite. So we have $h(\mathfrak{A})$ is a finite implication algebra (induced by h and $\mathfrak{A}$), and hence h is a finite representation of $\mathfrak{A}$.   □

Now, the focus of the rest of the paper revolves around the (finite) representation decision problem for implication semigroups. In the case of **IA**, this is defined as follows:

Definition 26. The (finite) representation decision problem for implication algebras *is a decision problem that takes an implication algebra with a (finite) carrier set as input. The algebra is a yes instance if and only if it is (finitely) representable.*

Closing this section, we note:

Corollary 27. IA *is finitely axiomatisable.*

Corollary 28. *The (finite) representation problem for* **IA** *is decidable.*

4 Undecidability Results for Implication Semigroups

In this section we build on results from [7,8,11] to show undecidability of some decision problems for. We begin by defining the representation and the finite representation decision problems.

Definition 29. *The* (finite) *representation decision problem for implication semigroups is a decision problem that takes an implication semigroup with a finite carrier set as input. The semigroup is a yes instance if and only if it is (finitely) representable.*

As we mention in Sect. 2, whether a structure is representable, also depends on our interpretation of the constant 1. Here we show that the (finite) decision problem is undecidable in both cases.

4.1 Representation Problem with Absolute Implication

We begin by examining the case with absolute implication, i.e. we require $\top = X \times X$ for some (finite) base X.

Definition 30. *An* implication monoid $\mathfrak{M} = (M, 1', \to, ;)$ *is an algebra where* $(M, \to, ;)$ *is an implication semigroup and* $1'$ *is the monoidal identity for* $;$. *For some transitive and reflexive* $\top \subseteq X \times X$, *we define* $\mathfrak{M}(\top) = (\wp(\top), 1', \to, ;)$ *where* $\to, ;$ *are proper relational implication and composition respectively and* $1'$ *is the proper relational identity for* X *defined as* $1' = \{(x, x) \mid x \in X\}$.

In [8, Section 4] a construction of a Boolean monoid from a square cancellative partial group $\mathfrak{G}$ is given. Its implication monoid reduct is denoted $\mathfrak{M}(\mathfrak{G}) = (M, 1', \to, ;)$. By [8, Proposition 5.1, Example 6.2] $\mathfrak{M}(\mathfrak{G})$ is representable (over a finite base) if and only if $\mathfrak{G}$ embeds into a (finite) group.

From the fact that both the group and the finite group embedding problems are undecidable [3] for finite structures it follows that the (finite) representation decision problem is undecidable. Thus if we prove that the **ISG** reduct of $\mathfrak{M}(\mathfrak{G})$

is (finitely) representable if and only if $\mathfrak{M}(\mathfrak{G})$ is representable, we have shown that the (finite) representability is undecidable. The right to left implication is trivial. But we must examine the case where we relax the requirement where we represent $1'$ as the true relational identity, and show that this is still sufficient for the structure to remain (finitely) representable with $1'$ taken as the true relational identity.

Suppose we have an embedding h from $\mathfrak{M}(\mathfrak{G})$ to $\mathfrak{S}(\mathsf{T})$, i.e. an injective mapping that preserves $\rightarrow, ;$, but not necessarily $1'$.

Lemma 31. *If $(x, y) \in h(1')$ then $(y, x) \in h(1')$.*

Proof. Suppose $(y, x) \notin h(1')$. That means that $(y, x) \in h(1' \rightarrow 0) = h(\overline{1'})$. By composition of $(x, y) \in h(1')$ and $(y, x) \in h(\overline{1'})$ we get that $(x, x) \in h(\overline{1'})$ and by composing that with $(x, y) \in h(1')$ we have that $(x, y) \in h(\overline{1'})$. As $(x, y) \in h(1')$ and $(x, y) \in h(\overline{1'}) = h(1' \rightarrow 0)$, we also have $(x, y) \in h(0)$. By a series of compositions we also get that $(y, x) \in h(0)$ and because $0 \leq 1'$ we also get $(y, x) \in h(1')$ and we've reached a contradiction. $\square$

Lemma 32. *$h(1')$ is an equivalence relation.*

Proof. By Lemma 31 we have that $h(1')$ is symmetric. Furthermore, since all $(x, x) \in h(\mathsf{T})$ there must exist a z witnessing $1'; \mathsf{T} = \mathsf{T}$. Thus $(x, z) \in h(1')$ and $(z, x) \in h(1')$ and we compose that to get $(x, x) \in h(1')$, so $h(1')$ is reflexive. Finally, as $1' = 1'; 1'$ we also have that $h(1')$ is transitive. $\square$

Lemma 33. *For all $x, x', y, y' \in X$ where $(x', x), (y, y') \in h(1')$ we have for all $a \in \mathfrak{M}(\mathfrak{G})$ that $(x, y) \in h(a) \Leftrightarrow (x', y') \in h(a)$.*

Proof. If $(x, y) \in h(a)$ we have $(x, y') \in h(a)$ by $(x', x), (y, y') \in h(1')$ and the composition of $1'; a; 1' = a$. By Lemma 31, we also have $(x, x'), (y', y) \in h(1')$ so similarly if $(x, y') \in h(a)$ then $(x, y) \in h(a)$. $\square$

Theorem 34. *The (finite) representation decision problem for* **ISG** *is undecidable when $\rightarrow$ is interpreted as absolute implication.*

Proof. As $h(1')$ is an equivalence relation by Lemma 32, so we can define $h' : \mathfrak{M}(\mathfrak{G}) \rightarrow X/h(1')$ where

$$h'(a) = \{([x]_{h(1')}, [y]_{h(1')}) \mid (x, y) \in h(a)\}$$

and show that h' is indeed an embedding of $\mathfrak{M}(\mathfrak{G})$ into $\mathfrak{M}(X \times X)$.

By Lemma 33, we know that if $(x, y) \in h(a)$ then for any $x' \in [x]_{h(1')}, y' \in [y]_{h(1')}$ we have $(x', y') \in h(a)$.

Take any $a \leq b$. Then there exists $(x, y) \in h(a) \setminus h(b)$. From this follows $([x]_{h(1')}, [y]_{h(1')}) \in h'(a)$ and if it were the case that $([x]_{h(1')}, [y]_{h(1')}) \in h'(b)$ that would mean that there exist some $(x', y') \in h(b)$ with $(x, x') \in h(1')$ and $(y', y) \in h(1')$ and that would also means that $(x, y) \in h(b)$. Thus h' is injective.

Every composition is witnessed by the equivalence class of the witness for the composition in h and if $(x, y) \in h(a)$ and $(y', z) \in h(b)$ with $y' \in [y]_{h(1')}$ we also have $(y, y') \in h(1')$ and thus we have the composition $(x, z) \in h(a; 1'; b) = h(a; b)$. Thus h' represents ; correctly. Finally $1'$ is represented correctly as a pair of equivalence classes is in $h'(1')$ if and only if they are the same equivalence class.

Thus we have shown that if we have an embedding of $\mathfrak{M}(\mathfrak{G})$ into $\mathfrak{G}(X \times X)$ then we also have an embedding of $\mathfrak{M}(G)$ into $\mathfrak{M}(X' \times X')$ where $X' = X/h(1')$. Furthermore if X is finite, so is X'. Trivially if $\mathfrak{M}(G)$ embeds into $\mathfrak{M}(X \times X)$ it also embeds into $\mathfrak{G}(X \times X)$ via the same embedding. This, together with the results presented in [8] shows that the (finite) representation decision problem for **ISG** is undecidable.   □

4.2 Representation Problem with Relative Implication

Now we show the same result for relative implication.

Definition 35. *A Boolean semigroup is a tuple* $\mathfrak{B} = (B, 0, 1, -, +, ;)$ *is an algebraic structure where S is a carrier set*

(i) $(B, 0, 1, -, +)$ *is a Boolean algebra*
(ii) $(B, ;)$ *is a semigroup*
(iii) ; *is additive over* +
(iv) $0; a = a; 0 = 0$

Similarly to **ISG**, we denote the class of Boolean semigroups **BSG** and we say that a Boolean semigroup is representable if and only for some transitive $\top \subseteq X \times X$ it embeds into $\mathfrak{B}(\top) = (\wp(X \times X), \emptyset, \top, -, +, ;)$ where $-a$ is interpreted as proper Boolean negation $\top \setminus a$, $+$ is interpreted as proper Boolean join $\cup$ and ; is interpreted as proper relational composition.

The (finite) representation problem for Boolean semigroups is defined analogous to that for implication semigroups. [7, Theorem 11.2] shows that the representation problem for Boolean semigroups is undecidable and [11, Theorem 2.5] shows that the finite representation problem for Boolean semigroups is undecidable. From this we show that the (finite) representation problem for implication semigroups is also undecidable.

Note that the above results require an operation · to be defined in the signature, but much like $\top$ in **IA**, · is term definable for **BSG** as $a \cdot b = -(-a + -b)$.

Lemma 36. *Let* $\mathfrak{G} = (S, \rightarrow, ;)$ *be an implication semigroup that contains some element 0 such that for all $a \in S$ we have $0 \leq a$ and $0; a = a; 0 = 0$. If $\mathfrak{G}$ is representable via some representation h then there exists a representation h' of $\mathfrak{G}$ where $h'(0) = \emptyset$. If h is defined over a finite base, so is h'.*

Proof. Let $h(\mathfrak{G})$ be the proper structure defined by h for some $\top \subseteq X \times X$. As h is a representation, there exists for every pair $a \not\leq b \in S$ a *discriminator pair* $(\iota, o) \in \top$ such that $(\iota, o) \in h(a) \setminus h(b)$.

Define $X^{\iota, o}$ as

$$X^{\iota, o} = \left\{ x \in X \mid \left(x = \iota \vee (\iota, x) \in \top \right) \wedge \left(y = o \vee (y, o) \in \top \right) \right\}$$

$\mathsf{T}^{\iota,o}$ as $\mathsf{T} \cap (X^{\iota,o} \times X^{\iota,o})$, and a mapping $h^{\iota,o} : \mathfrak{S} \to \mathfrak{S}(\mathsf{T}^{\iota,o})$ where $h^{\iota,o}(a) = h(a) \cap \mathsf{T}^{\iota,o}$.

First observe that $h^{\iota,o}(0) = \emptyset$. Suppose that there was a pair $(x,y) \in h^{\iota,o}(0)$. If $x = \iota$, we have $(\iota,y) \in h^{\iota,o}(0)$, else $(\iota,x) \in h(1) = \mathsf{T}$ and thus $(\iota,y) \in h(0)$ since $1;0 = 0$ and h preserves composition. Similarly if $y = o$ we get $(\iota,o) \in h(0)$, else by composing $(\iota,y) \in h(0)$ with $(y,z) \in h(1)$ we get $(\iota,o) \in h(0)$. Since $b \geq 0$ that would mean $(\iota,o) \in h(b)$ that contradicts the fact that (ι,o) is a discriminator pair for $a \not\leq b$.

Now let us check that $h^{\iota,o}$ preserves composition. Suppose $(x,y) \in h^{\iota,o}(a;b)$. This means that there exists $z \in X$ such that $(x,z) \in h(a)$ and $(y,z) \in h(b)$. If $x = \iota$, we trivially have $(x,\iota) \in h(1) = \mathsf{T}$. Else, by composing $(\iota,x) \in h(1)$ and $(x,y) \in h(a)$ we get $(\iota,y) \in h(1) = \mathsf{T}$ as $1;a \leq 1$. Similarly $(y,o) \in \mathsf{T}$ and thus $y \in X^{\iota,o}$. Thus we have $(x,z) \in h^{\iota,o}(a), (y,z) \in h^{\iota,o}(b)$ and we have shown $h^{\iota,o}(a;b) \subseteq h^{\iota,o}(a); h^{\iota,o}(b)$. The fact that $h^{\iota,o}(a;b) \supseteq h^{\iota,o}(a); h^{\iota,o}(b)$ follows from $(x,y),(y,z) \in \mathsf{T}^{\iota,o}$ then $x,z \in X^{\iota,o}$ and we have $(x,z) \in \mathsf{T}^{\iota,o}$. Thus $h^{\iota,o}$ preserves composition.

We have $h^{\iota,o}(a) \neq h^{\iota,o}(b)$ as $(\iota,o) \in \mathsf{T}^{\iota,o}$. The operation $\to$ is preserved by $h^{\iota,o}$ as for all $(x,y) \in \mathsf{T}^{\iota,o}$ it holds $(x,y) \in h(a) \iff (x,y) \in h^{\iota,o}(a)$. Finally, $|X^{\iota,o}| \leq |X|$. Thus we conclude that $h^{\{\iota,o\}}$ is a homomorphism for $\mathfrak{S}$ that discriminates the pair $a \not\leq b$.

Now let us pick for every $a \not\leq b$ a $\delta(a,b) = (\iota,o)$ such that $(\iota,o) \in \mathsf{T}$ is a discriminator pair for $a \not\leq b$ and let $\dot\cup$ denote a disjoint union. A mapping

$$ h' : S \to \wp \left(\dot\bigcup_{a \not\leq b \in S} \mathsf{T}^{\delta(a,b)} \right) $$

$$ h'(c) = \dot\bigcup_{a \not\leq b \in S} h^{\delta(a,b)}(c) $$

still represents $;, \to$ correctly, discriminates all pairs $a \not\leq b$ (i.e. is injective), which makes it a representation. Furthermore, $h'(0) = \emptyset$ and the size of its base is bounded by $|S|^2|X|$. $\qquad\square$

Theorem 37. *The (finite) representation decision problem for implication semigroups is undecidable.*

Proof. We show this by proving that $\mathfrak{B} \in \mathbf{BSG}$ is representable if and only if its $\langle \to, ; \rangle$-reduct $\mathfrak{S}$ is representable. The left to right implication is trivial as $a \to b \in S$ is term-definable as $(-a) + b \in B$. For the right to left implication, $+, 1$ are term-definable in $\langle \to, ; \rangle$ (see Proposition 5). By Lemma 36 the $\langle \to, ; \rangle$-reduct of $\mathfrak{B}$ is representable if and only if it has a representation where $h(0) = \emptyset$. See how in that representation $a \to 0$ (corresponding to $-a + 0 = -a$ in the Boolean semigroup) is represented as $\mathsf{T} \setminus a \cup \emptyset = \mathsf{T} \setminus a$.

As the (finite) representation decision problem is undecidable for Boolean semigroups, we conclude the same for implication semigroups. $\qquad\square$

5   Problems

In this section we outline some open problems. It follows from the results in Sect. 4 that the class of representable implication semigroups is not finitely axiomatisable, nor does it have the finite representation property, i.e. not every finite representable structure in the class is finitely representable. However, another decision problem of interest is posed below.

Problem 38. Is membership in the equational theory generated by the class of representable implication semigroups decidable?

The reader can see that if we add the bottom element 0 to the signature, the undecidability follows from the undecidability of the equational theory of the Boolean semigroups as described in [7]. This is because all negations of terms $-t$ can be rewritten as $t \to 0$ and all joins $t + t'$ as $(t \to t') \to t'$ where t, t' are terms.

The problem remains open for the class of representable implication semigroups without the bottom element. One of the possible ways to prove undecidability is by using discriminator terms, defined below.

Definition 39. *A discriminator term $d(a, b, c)$ is a term defined in terms of elements of algebra a, b, c such that for all representable algebras $d(a, b, c) = c$ if $a = b$ and a otherwise.*

Although the existence of discriminator terms is not a guarantee for the undecidability of the equational theory membership decision problem, it is an interesting open question in its own right.

Problem 40. Is it possible to define a discriminator term in the language of implication semigroups?

It is well known that subreducts of representable relation algebras form quasivarieties. As such, the class of implication semigroups can be characterised by quasiequations. However, some open questions about the equational theory generated by the class of representable implication semigroups are listed below.

Problem 41. Is the class of representable implication semigroups a variety?

Problem 42. Is the equational theory generated by the class of representable implication semigroups finitely axiomatisable?

We continue by looking at the alternative interpretations of $\to$ operation for binary relations. An interesting example, as mentioned in the introduction section is that of a weakening relation defined below.

Definition 43. *Let $\mathbf{P} = (X, \leq)$ be a poset. $R \subseteq X \times X$ is a weakening relation if and only if $\leq; R; \leq \subseteq R$.*

In the context of the weakening relation algebras as described in [4], the $\rightarrow$ operation can be given in first order terms as

$$R \rightarrow S = \{(x,y) \mid \forall x', y' : ((x' \leq x \land y \leq y' \land (x',y') \in R) \Rightarrow (x',y') \in S)\}$$

where R, S are weakening relations over a poset $\mathbf{P} = (X, \leq)$.

This interpretation of the $\rightarrow$ operation gives rise to the class of representable weakening implication semigroups, for which the following properties remain open.

Problem 44. Is the (finite) representation decision problem decidable for the class of representable weakening implication semigroups? Is the class finitely axiomatisable and does it have the finite representation property?

Problem 45. Is the class of representable weakening implication semigroups a (discriminator) variety? Is the equational theory generated by the class finitely axiomatisable/decidable?

Finally, we note that it can be checked that all results presented in this paper can be generalised to the dual operation $\leftarrow$ by presenting dual axioms for the class of implication algebras and defining negation as $0 \leftarrow a$.

References

1. Abott, J.C.: Implicational algebras. Bulletin mathématique de la Société des Sciences Mathématiques de la République Socialiste de Roumanie **11** (59)(1), 3–23 (1967). http://www.jstor.org/stable/43679502
2. Diego, A.: Sobre álgebras de Hilbert. Notas de lógica matemática, Instituto de Matemática, Universidad Nacional del Sur (1965). https://books.google.co.uk/books?id=AfcSAQAAMAAJ
3. Evans, T.: Embeddability and the word problem. J. Lond. Math. Soc. **1**(1), 76–80 (1953)
4. Galatos, N., Jipsen, P.: The structure of generalized bi-algebras and weakening relation algebras. Algebra Universalis **81**(3), 1–35 (2020)
5. Galatos, N., Jipsen, P.: Weakening relation algebras and FL^2-algebras. In: Fahrenberg, U., Jipsen, P., Winter, M. (eds.) RAMiCS 2020. LNCS, vol. 12062, pp. 117–133. Springer, Cham (2020). https://doi.org/10.1007/978-3-030-43520-2_8
6. Hirsch, R., Hodkinson, I.: Relation Algebras by Games. Elsevier, Amsterdam (2002)
7. Hirsch, R., Hodkinson, I., Jackson, M.: Undecidability of algebras of binary relations. In: Madarász, J., Székely, G. (eds.) Hajnal Andréka and István Németi on Unity of Science. OCL, vol. 19, pp. 267–287. Springer, Cham (2021). https://doi.org/10.1007/978-3-030-64187-0_11
8. Hirsch, R., Jackson, M.: Undecidability of representability as binary relations. J. Symb. Log. **77**(4), 1211–1244 (2012)
9. Jónsson, B., Tsinakis, C.: Relation algebras as residuated Boolean algebras. Algebra Universalis **30**, 469–478 (1993)
10. Löwenheim, L.: Über möglichkeiten im relativkalkül. Math. Ann. **76**, 447–470 (1915)

11. Neuzerling, M.: Undecidability of representability for lattice-ordered semigroups and ordered complemented semigroups. Algebra Universalis **76**(4), 431–443 (2016). https://doi.org/10.1007/s00012-016-0409-9
12. Rasiowa, H.: An Algebraic Approach to Non-Classical Logics. Warszawa, Pwn - Polish Scientific Publishers, Amsterdam, Netherlands (1974)
13. Rogozin, D.: The finite representation property for some reducts of relation algebras (2020). https://doi.org/10.48550/ARXIV.2007.13079, https://arxiv.org/abs/2007.13079
14. Tarski, A., Givant, S.R.: A Formalization of Set Theory Without Variables. American Mathematical Soc., Ann Arbor (1987)

On the Complexity of Kleene Algebra
with Domain

Igor Sedlár[✉][iD]

The Czech Academy of Sciences, Institute of Computer Science,
Prague, Czech Republic
sedlar@cs.cas.cz

Abstract. We prove that the equational theory of Kleene algebra with
domain is EXPTIME-complete. Our proof makes essential use of Hollenberg's equational axiomatization of program equations valid in relational
test algebra. We also show that the equational theory of Kleene algebra
with domain coincides with the equational theory of *-continuous Kleene
algebra with domain.

Keywords: Complexity · Kleene algebra · Kleene algebra with
domain · Propositional dynamic logic · Test algebra

1 Introduction

Kleene algebra with domain [5] is an expansion of Kleene algebra [15] with a
negation-like antidomain operator a. The image of a Kleene algebra with domain
under a is a Boolean algebra, and so every Kleene algebra with domain gives
rise to a Kleene algebra with tests [16,18]. In fact, it can be shown that the
equational theory of Kleene algebra with tests embeds into the equational theory
of Kleene algebra with domain. However, Kleene algebra with domain is more
expressive than Kleene algebra with tests [24]. In this paper we approach the
difference between Kleene algebra with tests and Kleene algebra with domain
from the viewpoint of computational complexity. In particular, we prove that
the (problem of deciding membership in the) equational theory of Kleene algebra
with domain is EXPTIME-complete. On the other hand, it is well-known that
the equational theory of Kleene algebra with tests is PSPACE-complete [2].

While this result is not quite unexpected, it fills a certain gap in the literature
on Kleene algebra with domain, where only limited attention has been paid to
complexity-theoretic questions. To the best of our knowledge, the only explicit
result is that the equational theory of left-inductive extensional Kleene algebra
with domain is in EXPTIME [21].[1]

We will use the well-known fact that the validity problem for Propositional
Dynamic Logic (PDL) is EXPTIME-complete [6,23]. In particular, our proof

---

[1] Extensionality is closely related to the property of *separability*, well known from the
literature on dynamic algebra [12,22].

© The Author(s), under exclusive license to Springer Nature Switzerland AG 2023
R. Glück et al. (Eds.): RAMiCS 2023, LNCS 13896, pp. 208–223, 2023.
https://doi.org/10.1007/978-3-031-28083-2_13

uses an algebraic formulation of PDL in terms of *test algebras* [22,25] and Hollenberg's equational axiomatization of program equations valid in relational test algebra (RTA) [10]. We establish our EXPTIME-completeness result by showing that the equational theory of test algebras embeds into the equational theory of Kleene algebra with domain (lower bound) and vice versa (upper bound). The embedding result entails that PDL, RTA and KAD have essentially the same expressive power. Our proof of the embedding result also shows that the equational theory of KAD coincides with the equational theory of *-continuous KAD. The main result is established in Sect. 5. The necessary prerequisites are given in Sects. 2 (Kleene algebra and Kleene algebra with tests), Sect. 3 (Kleene algebra with domain) and 4 (test algebra). The concluding section summarizes the paper and lists some interesting open problems.[2]

2 Kleene Algebra with Tests

This section gives some necessary background information on Kleene algebra and Kleene algebra with tests. For more detail we refer the reader to [15,16].

Definition 1. *A* Kleene algebra *is an algebra of the form*

$$\mathcal{K} = (K, \cdot, +, {}^*, 1, 0)$$

where $(K, \cdot, +, 1, 0)$ *is an idempotent semiring and* $^* : K \to K$ *satisfies the following, for all* $x, y, z \in K$ *(where* $x \le y$ *iff* $x + y = y$*):*

$$1 + xx^* \le x^* \tag{1}$$
$$1 + x^*x \le x^* \tag{2}$$
$$y + xz \le z \to x^*y \le z \tag{3}$$
$$y + zx \le z \to yx^* \le z \tag{4}$$

A Kleene algebra is *-continuous *iff*

$$xy^*z = \sup_{\le}\{xy^n z \mid n \in \omega\} \tag{5}$$

for all $x, y, z \in K$*, where* $y^0 = 1$ *and* $y^{n+1} = y^n y$*.*

---

[2] We note that the the main result of our original submission was an EXPTIME-completeness result on the equational theory of *-continuous KAD. The proof was essentially the same as the one given here, but we relied on the assumption of *-continuity in the proof of Theorem 4 (in particular, the argument showing that the translation of T8 is valid). An anonymous reviewer showed that the proof of Theorem 4 can be carried out without assuming *-continuity and that the resulting argument establishes in conjunction with Lemma 4 that the equational theory of KAD coincides with the equational theory of *-continuous KAD. We gratefully acknowledge the input of the reviewer and we suggest that the results of this paper be considered joint work of the official author and the anonymous reviewer.

It follows from the definition that $\leq$ is a partial order on K and x^* is the smallest reflexive transitive element above x with respect to $\leq$; that is, $1 \leq x^*$, $x^*x^* \leq x^*$ and $x \leq x^*$, and if $1 \leq y$, $yy \leq y$ and $x \leq y$, then $x^* \leq y$. In (5), it is assumed that each set of the form $\{xy^n z \mid n \in \omega\}$ has a supremum with respect to $\leq$ and that the supremum is identical to xy^*z.

The standard examples of Kleene algebras are the algebra of regular languages over a finite alphabet [11], the Kleene algebra expanding the (min,+)-semiring where * maps each element to the multiplicative identity [14], and the *relational Kleene algebra* over a set S, consisting of a collection of binary relations on S such that 1 is the identity relation on S (1_S), 0 is the empty set, $\cdot$ is relational composition, + is set union, and * is reflexive transitive closure. The class (quasivariety) of Kleene algebras will be denoted as KA and the subclass of *-continuous Kleene algebras as KA*. The class of relational Kleene algebras is denoted as RKA.

Let $\mathsf{P} = \{\mathsf{p}_n \mid n \in \omega\}$ be a set of variables ("program variables"). The set of *Kleene algebra terms*, Tm_{KA}, is generated by the following grammar:

$$p, q := \mathsf{p} \mid 1 \mid 0 \mid p \cdot q \mid p + q \mid p^*.$$

(We follow the standard approach and we do not distinguish between the syntactic operators corresponding to – or interpreted by – Kleene algebra operators and the operators themselves.) The set of Kleene algebra terms can be seen as an algebra of the Kleene algebra type (the absolutely free algebra of the Kleene algebra type). A *Kleene algebra equation* is an ordered pair of Kleene algebra terms, written as $p \approx q$. A *Kleene algebra model* is a pair $(\mathcal{K}, v)$ where $\mathcal{K} \in$ KA and v is a homomorphism from Tm_{KA} to $\mathcal{K}$. An equation $p \approx q$ is *valid* in $(\mathcal{K}, v)$ iff $v(p) = v(q)$; it is valid in $\mathcal{K}$ iff it is valid in all $(\mathcal{K}, v)$; and it is valid in a class of Kleene algebras K iff it is valid in all members of the class. We usually write "$\ldots \models p \approx q$" instead of "$p \approx q$ is valid in $\ldots$". The *equational theory* of a class of Kleene algebras K, denoted as $Eq(\mathsf{K})$, is the set of all Kleene algebra equations valid in K. Later on we will use similar definitions and notation for other classes of algebras and other languages.

Definition 2. *A* Kleene algebra with tests *is an algebra of the form*

$$\mathcal{K} = (K, B, \cdot, +, {}^*, {}^-, 1, 0)$$

where

- $(K, \cdot, +, {}^*, 1, 0)$ *is a Kleene algebra;*
- $B \subseteq K$;
- $^-$ *is a unary operation on B;*
- $(B, \cdot, +, {}^-, 1, 0)$ *is a Boolean algebra.*

*A Kleene algebra with tests is *-continuous iff its underlying Kleene algebra is *-continuous.*

Every Kleene algebra gives rise to a Kleene algebra with tests where $B = \{1, 0\}$ and $^-$ is defined as the complement in the two-element Boolean algebra. A standard example of a (*-continuous) Kleene algebra with tests is the full relational Kleene algebra with tests over a set S where K is the set of all binary relations on S, B is the power set of 1_S (the identity relation on S), $\cdot$ is relational composition, $+$ is set union, $*$ is reflexive transitive closure, $^-$ is complementation on subsets of 1_S, 1 is 1_S, and 0 is $\emptyset$. (A relational Kleene algebra with tests on S is a subalgebra of the fully relational Kleene algebra with tests on S.) Relational Kleene algebras with tests correspond to interpretations of programs in poor-test Propositional Dynamic Logic (the fragment where tests are formed using only Boolean formulas; see [8]). Another standard example of a Kleene algebra with tests is the algebra of guarded strings [18]. The class of (*-continuous, relational) Kleene algebras with tests will be denoted as KAT (KAT*, RKAT).

Let $B = \{b_n \mid n \in \omega\}$ be a set of variables disjoint from P ("Boolean variables"). The set of KAT terms, Tm_{KAT}, is defined using the following grammar invoking the sub-sort of *Boolean terms* Bm (instead of using different variables for KAT terms, we reuse the variables ranging over KA terms, with the hope that the context will disambiguate):

$$Bm \quad b, c := b_n \mid 1 \mid 0 \mid b \cdot c \mid b + c \mid \bar{b}$$
$$Tm_{KAT} \quad p, q := p_n \mid b \mid p \cdot q \mid p + q \mid p^*$$

(Hence, every Boolean term is a KAT term. Therefore, b^* is a KAT term, although it is not a Boolean term.)

A KAT equation is an ordered pair of KAT terms. A KAT model is a pair $(\mathcal{K}, v)$ where $\mathcal{K} \in$ KAT and v is a homomorphism from Tm_{KAT} to $\mathcal{K}$ such that $v(b_n) \in B$ for all n. The various notions of validity are defined similarly as in the case of Kleene algebra, and the (quasi)equational theories of classes of Kleene algebras with test are defined as expected.

Even though the classes KA, KA* and RKA are mutually distinct [13], their equational theories coincide [15]. Similarly, KAT, KAT* and RKAT are mutually distinct, but their equational theories coincide [18]. It is well known that the equational theories of KA and KAT are PSPACE-complete [2,15].

3   Kleene Algebra with Domain

In this section we recall some necessary background on Kleene algebra with domain. We assume the "internal" formulation of Kleene algebra with domain [4,5]. In this formulation, Kleene algebras with domain are expansions of Kleene algebras with a unary *antidomain* operator.[3] The properties of the antidomain operator assumed in the definition below generalize the properties of the relational antidomain operator $A : 2^{S \times S} \to 2^{S \times S}$ defined by

$$A(R) = \{(s, s) \mid \neg \exists t.(s, t) \in R\}$$

---

[3] In the original formulation [3], Kleene algebras with domain were expansions of Kleene algebras with test.

If R is seen as the input-output relation determined by a program π, then $A(R)$ is the input-output relation determined by a *test if π diverges*. It is interesting to note that the same operation is used to interpret negation in Dynamic Predicate Logic [7].

Definition 3. *A* Kleene algebra with domain *is a structure of the form*

$$\mathcal{D} = (K, \cdot, +, \,^*, \mathsf{a}, 1, 0)$$

such that $(K, \cdot, +, \,^*, 1, 0)$ *is a Kleene algebra,* $\mathsf{a} : K \to K$ *and the following are satisfied for all* $x, y, z \in K$*, assuming that* $\mathsf{d}(x) := \mathsf{a}(\mathsf{a}(x))$*:*

$$\mathsf{a}(x)x = 0 \tag{6}$$
$$\mathsf{a}(xy) \le \mathsf{a}(x\,\mathsf{d}(y)) \tag{7}$$
$$\mathsf{d}(x) + \mathsf{a}(x) = 1 \tag{8}$$

A Kleene algebra with domain *is* *-continuous iff its underlying Kleene algebra is *-continuous.*

A standard example of a (*-continuous) Kleene algebra with domain is the full relational Kleene algebra with domain over a set S that expands the full relational Kleene algebra over S with the relational antidomain operation A. Note that the relational domain operation

$$D(R) := A(A(R)) = \{(s, s) \mid \exists t.(s, t) \in R\}$$

is related to the projection operation familiar from relational databases. If R is seen as the input-output relation determined by a program π, then $D(R)$ is the input-output relation determined by a *test if π halts*. The Kleene algebra of regular languages over a finite alphabet Σ can be extended to a Kleene algebra with domain by adding $\mathsf{a} : 2^{\Sigma^*} \to 2^{\Sigma^*}$ such that

$$\mathsf{a}(L) = \begin{cases} \{\epsilon\} & \text{if } L = \emptyset \\ \emptyset & \text{otherwise.} \end{cases}$$

The quasivariety of (*-continuous, relational) Kleene algebras with domain will be denoted as KAD (KAD*, RKAD). Not every Kleene algebra expands to a Kleene algebra with domain [5], but the above example of a Kleene algebra of regular languages with domain shows that the equational theory of KAD (defined as expected[4]) expands the equational theory of KA conservatively. We will show below that the equational theories of KAD and KAD* coincide.[5] To the best of

---

[4] As before, we save letters by re-using p, q etc. as variables ranging over KAD terms, Tm_{KAD} (defined as expected), and letting the context disambiguate.

[5] McLean [20] studies the equational theory of relational Kleene algebras expanded by the relational domain operator D seen as primitive. He shows that this theory is decidable, with a 3EXPTIME upper bound. The *-free fragment of $Eq(\mathsf{RKAD})$ is not finitely based [9], in contrast to the *-free fragment of $Eq(\mathsf{KAD})$.

our knowledge, the only explicit complexity result in the literature on Kleene algebra with domain is that the equational theory of left-inductive extensional Kleene algebra with domain is in EXPTIME [21].[6]

For each p, let p' be the term that results from p by replacing, for each $n > 0$, every occurrence of p_n by an occurrence of p_{2n} (renaming of variables). It is immediate that $\mathcal{D} \models p \approx q$ iff $\mathcal{D} \models p' \approx q'$. If $q = p'$ for some p, then q is called an *even* term.

Assuming any fixed Kleene algebra with domain with universe K, let $\mathsf{d}(K) = \{\mathsf{d}(x) \mid x \in K\}$. Elements of $\mathsf{d}(K)$ are called *domain elements* of the underlying Kleene algebra with domain. The following proposition lists some useful properties of the antidomain and domain operations.

Proposition 1. *The following hold in each Kleene algebra with domain, for all* x, y, z:

1. $\mathsf{d}(x) \le 1$ *(domain elements are subidentities)*
2. $\mathsf{d}(x)\mathsf{a}(x) = 0$ *(law of noncontradiction)*
3. $\mathsf{d}(x)x = x$ *(left invariant)*
4. $\mathsf{d}(x\mathsf{d}(y)) = \mathsf{d}(xy)$ *(locality)*
5. $\mathsf{d}(x + y) = \mathsf{d}(x) + \mathsf{d}(y)$ *(additivity)*
6. $\mathsf{d}(x)\mathsf{d}(y) = \mathsf{d}(\mathsf{d}(x)\mathsf{d}(y))$ *(d-multiplication)*
7. $\mathsf{d}(x) + \mathsf{d}(y) = \mathsf{d}(\mathsf{d}(x) + \mathsf{d}(y))$ *(d-addition)*
8. $\mathsf{d}(1) = 1$ *and* $\mathsf{d}(0) = 0$ *(seriality and normality)*
9. $\mathsf{d}(x)\mathsf{d}(y) = \mathsf{d}(y)\mathsf{d}(x)$ *(domain elements are commutative)*
10. $\mathsf{d}(x)\mathsf{d}(x) = \mathsf{d}(x)$ *(domain elements are idempotent)*
11. $\mathsf{a}(x) = \mathsf{d}(\mathsf{a}(x))$ *and* $\mathsf{d}(x) = \mathsf{d}(\mathsf{d}(x))$ *(triple negation)*
12. $x \le \mathsf{d}(y)x$ *iff* $\mathsf{d}(x) \le \mathsf{d}(y)$ *(least left preserver)*

Proof. These are standard facts; see [5] for instance.   □

Lemma 1. *For each Kleene algebra with domain* $\mathcal{D}$, *the algebra*

$$\mathsf{d}(\mathcal{D}) = (\mathsf{d}(K), \cdot, +, \mathsf{a}, 1, 0)$$

is a subalgebra of $\mathcal{D}$ *which is also a Boolean algebra.*

Proof. See [5]. $\mathsf{d}(\mathcal{D})$ is a distributive lattice since $\mathsf{d}(x) \le 1$, $(\mathsf{d}x)(\mathsf{d}x) = \mathsf{d}x$, and $(\mathsf{d}x)(\mathsf{d}y) = (\mathsf{d}y)(\mathsf{d}x)$; moreover, $\mathsf{a}(x)$ is a Boolean complement of $\mathsf{d}(x)$ by definition.   □

It follows from the previous lemma that every $\mathcal{D} \in \mathsf{KAD}$ gives rise to a Kleene algebra with tests, namely, $(K, \mathsf{d}(\mathcal{D}), \cdot, +, {}^*, \mathsf{a}, 1, 0)$.[7] In fact, it can be shown that the equational theory of KAT embeds into the equational theory of KAD.

---

[6] Extensionality is closely related to the property of *separability*, well known from the literature on dynamic algebra [12, 22].

[7] Strictly speaking, we should replace a (a total operation) by its restriction to $\mathsf{d}(\mathcal{D})$, but we will not bother with this detail.

Theorem 1. *There is a function* $\eta : Tm_{KAT} \to Tm_{KAD}$ *such that* $\mathsf{KAT} \models p \approx q$ *iff* $\mathsf{KAD} \models \eta(p) \approx \eta(q)$.

Proof. We sketch the proof (the result itself seems to be folklore in the literature on KAD). Let $\eta(\mathsf{p}_n) = \mathsf{p}_{2n}$, $\eta(\mathsf{b}_n) = \mathsf{d}(\mathsf{p}_{2n+1})$, $\eta(\bar{b}) = \mathsf{a}(\eta(b))$, and then let η commute with the Kleene algebra operations (note that for each $b \in Bm$, $\mathsf{KAD} \models \eta(b) \approx \mathsf{d}(p)$ for some $p \in Tm_{KAD}$). Now, if $\mathsf{KAT} \not\models p \approx q$, then $\mathsf{RKAT} \not\models p \approx q$. Let $\mathcal{K} \in \mathsf{RKAT}$ be such that $v(p) \neq v(q)$ for some v. Take the $\mathcal{D} \in \mathsf{RKAD}$ arising from $\mathcal{K}$ by forgetting about B and replacing $^-$ by the relational antidomain operation A. Define w as the unique homomorphism $Tm_{KAD} \to \mathcal{D}$ such that $w(\mathsf{p}_{2n}) = v(\mathsf{p}_n)$ and $w(\mathsf{p}_{2n+1}) = v(\mathsf{b}_n)$. It can be proven by induction on $r \in Tm_{KAT}$ that $v(r) = w(\eta(r))$. Hence, $\mathsf{KAD} \not\models \eta(p) \approx \eta(q)$. Conversely, if $w(\eta(p)) \neq w(\eta(q))$ for some $(\mathcal{D}, w)$ where $\mathcal{D} \in \mathsf{KAD}$, then take $\mathcal{K} = (K, \mathsf{d}(\mathcal{D}), \cdot, +, ^*, \mathsf{a}, 1, 0) \in \mathsf{KAT}$ and define $v : Tm_{KAT} \to \mathcal{K}$ as the unique homomorphism such that $v(\mathsf{p}_n) = w(\mathsf{p}_{2n})$ and $v(\mathsf{b}_n) = w(\mathsf{d}(\mathsf{p}_{2n+1}))$. It can be proven by induction on $r \in Tm_{KAT}$ that $v(r) = w(\eta(r))$, and so $\mathsf{KAT} \not\models p \approx q$. $\square$

For each Kleene algebra with domain $\mathcal{D} = (K, \cdot, +, ^*, \mathsf{a}, 1, 0)$, we will refer to $\mathsf{d}(\mathcal{D})$ as the *domain algebra of* $\mathcal{D}$. While elements of $\mathcal{D}$ can be informally seen as representing (nondeterministic) *actions*, elements of the domain algebra represent a special sort of actions, namely, *tests* or *statements*. In the domain algebra, the operations $\cdot, +$ and a represent the Boolean operations of conjunction, disjunction and negation. Moreover, the domain operation d can be used to define a modal operator on the domain algebra: for $d \in \mathsf{d}(K)$, let $\langle x \rangle d := \mathsf{d}(xd)$, intuitively representing the statement that d is true after *some* execution of x. In fact, thanks to the properties of d in KAD (see Proposition 1) this modal operator shares many properties of the possibility operator from modal logic and Propositional Dynamic Logic.

Lemma 2. *The following hold in all Kleene algebras with domain, for all* $x, y \in K$ *and all* $d, e \in \mathsf{d}(K)$:

1. $\langle x \rangle 0 = 0$ *and* $\langle 1 \rangle d = d$
2. $\langle x \rangle (d + e) = \langle x \rangle d + \langle x \rangle e$
3. $\langle x + y \rangle d = \langle x \rangle d + \langle y \rangle e$
4. $\langle xy \rangle d = \langle x \rangle \langle y \rangle d$
5. $\langle d \rangle e = de$
6. $\langle x^* \rangle d = d + \langle x \rangle \langle x^* \rangle d$
7. $d + \langle x \rangle e \leq e \to \langle x^* \rangle d \leq e$

Proof. The proof uses facts stated in Proposition 1. The first half of (1) is normality and the second half follows from triple negation; (2) and (3) follow from additivity (and distribution of $\cdot$ over $+$); (4) follows from locality and (5) follows from d-multiplication. (6) is established using Kleene algebra reasoning: $x^* = 1 + xx^*$ holds in all Kleene algebras, and so $x^*d = d + xx^*d$ holds in each Kleene algebra with domain, which entails (6) using additivity, triple negation and locality. (7) is established as follows. By the least left preserver property, it is sufficient to establish that

$$d \leq ed \text{ and } xe \leq exe \text{ only if } x^*d \leq ex^*d.$$

Since $1 \leq x^*$, the first assumption entails that $d \leq ex^*$. The second assumption entails

$$xe \leq ex \qquad \qquad \text{(subidentity)}$$
$$xex^*d \leq exx^*d$$
$$xex^*d \leq ex^*d \qquad \qquad (xx^* \leq x^*)$$

Hence, $d + xex^*d \leq ex^*d$, and so $x^*d \leq ex^*d$ using (3). $\qquad\qquad\square$

(We note that the proof of (7) above, not requiring the assumption of *-continuity, was suggested by an anonymous reviewer.)

4 Relational Test Algebra

Relational test algebras are algebraic counterparts of Kripke models for Propositional Dynamic Logic. They will play a crucial role in the proof of our main result. This section gives the necessary information on relational test algebra; we follow the exposition given in [10].

Definition 4. *A relational test algebra is a structure of the form*

$$\mathcal{T} = (\mathcal{K}(S), \mathcal{B}(S), \langle \rangle, ?)$$

where

- $\mathcal{K}(S) = (\mathcal{P}(S \times S), \circ, \cup, *, 1_S, \emptyset)$ *is the full relational Kleene algebra over S;*
- $\mathcal{B}(S) = (\mathcal{P}(S), \cap, \cup, ^-, S, \emptyset)$ *is the Boolean algebra of subsets of S;*
- $\langle \rangle : \mathcal{K}(S) \times \mathcal{B}(S) \rightarrow \mathcal{B}(S)$ *such that $\langle R \rangle X = \{s \mid \exists t.(s,t) \in R \& t \in X\}$;*
- $? : \mathcal{B}(S) \rightarrow \mathcal{K}(S)$ *such that $X? = \{(s,s) \mid s \in X\}$.*

(We note that the operator $\langle \rangle$ is primitive and it is not to be confused with the modal operator on domain algebras discussed in the previous section. However, we will see that there is a close connection between the two justifying our overloading of notation.)

The class of relational test algebras is denoted as RTA. Subalgebras of relational test algebras correspond to complex algebras of (standard) Kripke frames for Propositional Dynamic Logic [6,8].

The sets of *programs Pr* and *formulas Fm* are defined by mutual induction as follows (using the sets of program variables P and Boolean variables B):

$$Pr \quad \alpha, \beta := \mathsf{p}_n \mid 1 \mid 0 \mid \alpha; \beta \mid \alpha \cup \beta \mid \alpha^* \mid \varphi?$$
$$Fm \quad \varphi, \psi := \mathsf{b}_n \mid \bot \mid \neg\varphi \mid \varphi \wedge \psi \mid \varphi \vee \psi \mid \langle \alpha \rangle \varphi.$$

Unsurprisingly, a program α is called *even* in case p_n occurs in α only if n is even. We define $[\alpha]\varphi := \neg\langle\alpha\rangle\neg\varphi$.

An RTA model is a pair $(\mathcal{T}, v)$ such that $\mathcal{T} \in$ RTA and v is a homomorphism from $Pr \cup Fm$ into $\mathcal{T}$ such that $v(p_n) \in \mathcal{K}(S)$ and $v(b_n) \in \mathcal{B}(S)$. The *program-equational theory* of RTA, denoted as Pr(RTA), is the set of equations of the form $\alpha \approx \beta$ such that $v(\alpha) = v(\beta)$ in each $(\mathcal{T}, v)$ for $\mathcal{T} \in$ RTA. The *formula-equational theory* of RTA is defined similarly. The formula-equational theory corresponds to the set of theorems of Propositional Dynamic Logic, and its finitary axiomatization is easily derived from the proof of completeness of Segerberg's axiomatization of PDL [8,17].

Improving the result of [25] where separability is assumed, Hollenberg [10] showed that the program-equational theory of RTA can be axiomatized by adding a finite number of axioms to the equational theories of Kleene algebras and Boolean algebras. Here we will present a modification of his system (mentioned also by Hollenberg) that uses Kozen's axiomatization of Kleene algebra [15].

Definition 5. *Let TC be a two-sorted quasi-equational proof system consisting of the following axioms and inference rules:*

1. *Kozen's [15] quasi-equational axiomatization of Eq(KA) (using variables in* P *and operators* $1, 0, \cdot, +$ *and* $*$ *);*
2. *your favorite equational axiomatization of the equational theory of Boolean algebras (using variables in* B *and operators* $\bot, \neg, \wedge, \vee$ *);*
3. *test algebra axioms of [25] (minus separability), de facto an equational reformulation of Segerberg's axioms for PDL:*

(T1) $\langle p\rangle\bot = \bot$	*(T6)* $\langle pq\rangle b = \langle p\rangle\langle q\rangle b$
(T2) $\langle p\rangle(b \vee c) = \langle p\rangle b \vee \langle p\rangle c$	*(T7)* $\langle p^*\rangle b = b \vee \langle p\rangle\langle p^*\rangle b$
(T3) $\langle 0\rangle b = \bot$	
(T4) $\langle 1\rangle b = b$	*(T8)* $\langle p^*\rangle b = b \vee \langle p^*\rangle(\neg b \wedge \langle p\rangle b)$
(T5) $\langle p \cup q\rangle b = \langle b\rangle p \vee \langle q\rangle p$	*(T9)* $\langle b?\rangle c = b \wedge c$

4. *additional program axioms:*

(K1) $\bot? = 0$	*(K3)* $(b \wedge c)? = b?c?$
(K2) $(b \vee c)? = b? \cup c?$	*(K4)* $(\langle p\rangle\top)?p = p$

The inference rules are (Kozen's quasi-equations for $*$ *and) the usual inference rules of equational logic and uniform (sort-respecting) substitution.*

Theorem 2. ([10]). *For all* $\alpha, \beta \in Pr$, $TC \vdash \alpha \approx \beta$ *iff* RTA $\models \alpha \approx \beta$.

We define $\phi(\alpha, \beta)$ as the formula $\langle\alpha\rangle c \leftrightarrow \langle\beta\rangle c$ where $c \in$ B is the first Boolean variable not appearing in α or β. It is easily seen that the following holds.

Theorem 3. *1.* RTA $\models \alpha \approx \beta$ *iff* PDL $\vdash \phi(\alpha, \beta)$.
2. PDL $\models \varphi$ *iff* RTA $\models (\varphi?) \approx 1$.
3. Pr(RTA) *is EXPTIME-complete.*

5 The Complexity of KAD

In this section we prove that the membership problem for the equational theory of KAD is EXPTIME-complete, and that the equational theories of KAD and KAD* coincide. First we show that $Pr(\text{RTA})$ embeds into $Eq(\text{KAD})$ via a (polynomially computable) translation function $\tau : Pr \cup Fm \to Tm$. This establishes the EXPTIME lower bound. Then we show that there is a (polynomially computable) translation function $\sigma : Tm \to Pr$ such that $\text{KAD} \models p \approx \sigma(\tau(p))$ for all even terms p. It follows that $Eq(\text{KAD})$ embeds into $Pr(\text{RTA})$, establishing the EXPTIME upper bound. The result establishing that the equational theories of KAD and KAD* coincide is a corollary of one of our embedding results. We show that for all p, q the equation $p \approx q$ is valid in KAD iff $\sigma(p) \approx \sigma(q)$ is provable in TC iff $p \approx q$ is valid in KAD*.

Definition 6. *Let τ be the following function from $Pr \cup Fm \to Tm$:*

$$\tau(\mathsf{p}_{2n}) = \mathsf{p}_{2n} \qquad\qquad \tau(\mathsf{b}_n) = \mathsf{d}(\mathsf{p}_{2n+1})$$
$$\tau(\mathsf{p}_{2n+1}) = \mathsf{p}_1 \qquad\qquad \tau(\bot) = 0$$
$$\tau(1) = 1 \qquad\qquad \tau(\neg\varphi) = \mathsf{a}\tau(\varphi)$$
$$\tau(0) = 0 \qquad\qquad \tau(\varphi \wedge \psi) = \tau(\varphi) \cdot \tau(\psi)$$
$$\tau(\alpha \cup \beta) = \tau(\alpha) + \tau(\beta) \qquad\qquad \tau(\varphi \vee \psi) = \tau(\varphi) + \tau(\psi)$$
$$\tau(\alpha; \beta) = \tau(\alpha) \cdot \tau(\beta) \qquad\qquad \tau(\langle\alpha\rangle\varphi) = \mathsf{d}(\tau(\alpha) \cdot \tau(\varphi))$$
$$\tau(\alpha^*) = \tau(\alpha)^* \qquad\qquad \tau(\varphi?) = \tau(\varphi)$$

Lemma 3. *For each $\varphi \in Fm$ there is $p \in Tm$ such that $\text{KAD} \models \tau(\varphi) \approx \mathsf{d}(p)$.*

Proof. Structural induction on φ. The base case $\varphi = \mathsf{b}$ holds by definition. The cases of the induction step are established using Proposition 1 as follows. (i) $\tau(\bot) = 0$ and $\text{KAD} \models 0 \approx \mathsf{d}(0)$. (ii) $\tau(\neg\varphi) = \mathsf{a}\tau(\varphi)$, and so $\text{KAD} \models \tau(\neg\varphi) \approx \mathsf{ad}(p)$ for some p by the induction hypothesis; however, $\text{KAD} \models \mathsf{ad}(p) \approx \mathsf{da}(p)$. (iii) $\text{KAD} \models \tau(\varphi) \cdot \tau(\psi) \approx \mathsf{d}(p) \cdot \mathsf{d}(q)$ for some p, q by the induction hypothesis, and $\text{KAD} \models \mathsf{d}(p) \cdot \mathsf{d}(q) \approx \mathsf{d}(\mathsf{d}(p) \cdot \mathsf{d}(q))$. (iv) $\text{KAD} \models \tau(\varphi) + \tau(\psi) \approx \mathsf{d}(p) + \mathsf{d}(q)$ for some p, q by the induction hypothesis and $\text{KAD} \models \mathsf{d}(p) + \mathsf{d}(q) \approx \mathsf{d}(\mathsf{d}(p) + \mathsf{d}(q))$. (v) $\tau(\langle\alpha\rangle\varphi)$ is equivalent to a term of the form $\mathsf{d}(p)$ by definition. $\qquad\square$

Theorem 4. *For all even $\alpha, \beta \in Pr$:*

1. $TC \vdash \alpha \approx \beta$ iff $\text{KAD} \models \tau(\alpha) \approx \tau(\beta)$;
2. $TC \vdash \alpha \approx \beta$ iff $\text{KAD}^* \models \tau(\alpha) \approx \tau(\beta)$.

Proof. We prove the first claim and we will point out that the proof establishes the second claim as well.

Right to left. If $TC \nvdash \alpha \approx \beta$, then there is a test algebra model $(\mathcal{T}, v)$ based on a set S such that $v(\alpha) \neq v(\beta)$. Now consider the full relational Kleene algebra $\mathcal{D}$ based on S and define a KAD-model $(\mathcal{D}, w)$ where w is the unique homomorphism such that

$$w(\mathsf{p}_{2n}) = v(\mathsf{p}_{2n}) \qquad w(\mathsf{p}_{2n+1}) = v(\mathsf{b}_n?).$$

Claim. Let α be an arbitrary even program and let φ be an arbitrary formula. Then

$$v(\alpha) = w(\tau(\alpha)) \tag{9}$$
$$v(\varphi?) = w(\tau(\varphi)) \tag{10}$$

The proof is by simultaneous structural induction on α and φ. The base case holds by definition, and the fact that $w(\mathsf{p}_{2n+1}) = \mathsf{d}(w(\mathsf{p}_{2n+1}))$. The induction step for Kleene algebra operations and Boolean connectives is trivial. For example, take $\cup$ and $\vee$. First, $v(\alpha \cup \beta) = v(p) \cup v(\beta) = w(\tau(\alpha)) \cup w(\tau(\beta))$ by the induction hypothesis; and $w(\tau(\alpha)) \cup w(\tau(\beta)) = w(\tau(\alpha) + \tau(\beta)) = w(\tau(\alpha \cup \beta))$ by definition. Second, $v((\varphi \vee \psi)?) = v(\varphi? \cup \psi?)$ by Theorem 2, $v(\varphi? \cup \psi?) = v(\varphi?) \cup v(\psi?) = w(\tau(\varphi)) \cup w(\tau(\psi))$ by the induction hypothesis, and $w(\tau(\varphi)) \cup w(\tau(\psi)) = w(\tau(\varphi) + \tau(\psi)) = w(\tau(\varphi \vee \psi)) = w(\tau((\varphi \vee \psi)?))$ by definition.

The induction step for negation is established as follows: $v((\neg\varphi)?) = \{(s,s) \mid s \not\models \varphi\} = \{(s,s) \mid \neg\exists t.(s,t) \in v(\varphi?)\} = A(v(\varphi?)) = A(w(\tau(\varphi)))$ by the induction hypothesis; but clearly $A(w(\tau(\varphi))) = w(\mathsf{a}(\tau(\varphi))) = w(\tau(\neg\varphi)) = w(\tau((\neg\varphi)?))$.

The induction step for ? is established as follows: $v(\varphi?) = w(\tau(\varphi))$ by the induction hypothesis (in proving (9) for $\varphi?$, we assume that (10) holds for φ, which has lower complexity than $\varphi?$; this kind of simultaneous inductive proof is standard in Propositional Dynamic Logic); but $w(\tau(\varphi)) = w(\tau(\varphi?))$ by definition.

Finally, the induction step for the diamond operator $\langle\rangle$ is established as follows:

$$
\begin{aligned}
v((\langle\alpha\rangle\varphi)?) &= \{(s,s) \mid s \in v(\langle\alpha\rangle\varphi)\} \\
&= \{(s,s) \mid \exists t.(s,t) \in v(\alpha) \& (t,t) \in v(\varphi?)\} \\
&= \{(s,s) \mid \exists t.(s,t) \in w(\tau(\alpha)) \& (t,t) \in w(\tau(\varphi))\} \quad \text{(ind. hyp.)} \\
&= \{(s,s) \mid \exists t.(s,t) \in w(\tau(\alpha) \cdot \tau(\varphi))\} \qquad\qquad w(\tau(\varphi)) \subseteq 1_S \\
&= D(w(\tau(\alpha) \cdot \tau(\varphi))) = w(\tau(\langle\alpha\rangle\varphi)) \\
&= w(\tau((\langle\alpha\rangle\varphi)?))
\end{aligned}
$$

This concludes the proof of the claim, and so we have shown that, for even α and β, $TC \not\vdash \alpha \approx \beta$ entails $\mathsf{KAD} \not\models \tau(\alpha) \approx \tau(\beta)$. Since the $\mathcal{D}$ constructed in the proof is a *-continuous Kleene algebra with domain, we have also established the right-to-left implication of the second part of the theorem.

Left to right. We prove by induction on the length of proofs in TC that $TC \vdash \alpha \approx \beta$ implies $\mathsf{KAD} \models \tau(\alpha) \approx \tau(\beta)$, and so $\mathsf{KAD}^* \models \tau(\alpha) \approx \tau(\beta)$, establishing the left-to-right implication of the second part of the theorem as well. Translations of the Kleene algebra axioms are clearly valid (resp. preserve validity) since τ commutes with Kleene algebra operators. Translations of the Boolean algebra axioms are valid thanks to Lemma 1. Validity of translations of the test algebra axioms is established as follows.

Note that, thanks to Lemma 3, for each $\varphi \in Fm$ and each Kleene algebra with domain model $(\mathcal{D}, v)$, $v(\tau(\varphi))$ is a domain element. Hence, we may use Lemma 2 to show that translations of the test algebra axioms are valid. In fact, translations of all axioms except T8 correspond straightforwardly to cases listed in Lemma 2. For instance, $\tau(\langle p \rangle \bot) = \mathsf{d}(\tau(p)0)$, but $\mathsf{KAD} \models \langle p \rangle 0 \approx 0$ by Lemma 2, where $\langle p \rangle 0$ is defined as $\mathsf{d}(p0)$; and $\tau(\langle pq \rangle b) = \mathsf{d}(\tau(p)\tau(q)\tau(b))$, but $\mathsf{KAD} \models \langle pq \rangle d \approx \langle p \rangle \langle q \rangle d$ by Lemma 2 (d is an arbitrary term of the form $\mathsf{d}(r)$ for some term r).

Consider the translation of T8:

$$\mathsf{d}\big(\tau(\mathsf{p})^* \cdot \tau(\mathsf{b})\big) = \tau(\mathsf{b}) + \mathsf{d}\big(\tau(\mathsf{p})^* \cdot \mathsf{a}(\tau(\mathsf{b})) \cdot \mathsf{d}(\tau(\mathsf{p}) \cdot \tau(\mathsf{b}))\big)$$

To prove that the translation of T8 is valid, it is sufficient to show that

$$\mathsf{d}(x^*e) = e + \mathsf{d}(x^*\mathsf{a}(e)\mathsf{d}(xe)) \tag{11}$$

holds in all Kleene algebras with domain for all $x \in K$ and $e \in \mathsf{d}(K)$. The inequality from right to left is established as follows: $e \leq x^*e$, and so $e = \mathsf{d}(e) \leq \mathsf{d}(x^*e)$; moreover, $\mathsf{d}(x^*\mathsf{a}(e)\mathsf{d}(xe)) \leq \mathsf{d}(x^*\mathsf{d}(xe))$ since $\mathsf{a}(e) \leq 1$, and $\mathsf{d}(x^*\mathsf{d}(xe)) = \mathsf{d}(x^*xe) \leq \mathsf{d}(x^*e)$.

The inequality in (11) from left to right is established as follows. Since

$$\underbrace{e + \mathsf{d}(x^*\mathsf{a}(e)\mathsf{d}(xe))}_{d} \in \mathsf{d}(K) \quad \text{and} \quad \underbrace{\mathsf{a}(e)\mathsf{d}(xe)}_{f} \in \mathsf{d}(K)$$

by Proposition 1, it is sufficient to show that

$$e + \mathsf{d}(xd) \leq d \tag{12}$$

and $\mathsf{d}(x^*e) \leq d$ will follow using the last item of Lemma 2. We reason as follows:

$$
\begin{aligned}
e + \mathsf{d}(xd) &= e + \mathsf{d}(x(e + x^*f)) \\
&= e + \mathsf{d}(xe + xx^*f) \\
&\leq e + \mathsf{d}(xe + x^*f) \\
&= e + \mathsf{d}(xe) + \mathsf{d}(x^*f) \\
&= e + (e + \mathsf{a}(e))\mathsf{d}(xe) + \mathsf{d}(x^*f) \\
&= e + e\mathsf{d}(xe) + f + \mathsf{d}(x^*f) \\
&= e(1 + \mathsf{d}(xe)) + f + \mathsf{d}(x^*f) \\
&\leq e + f + \mathsf{d}(x^*f) \\
&\leq e + \mathsf{d}(x^*f) + \mathsf{d}(x^*f) = d
\end{aligned}
$$

The last inequality holds since $f \leq \mathsf{d}(x^*f)$: $1 \leq x^*$ implies $f \leq x^*f$, which implies $f = \mathsf{d}(f) \leq \mathsf{d}(x^*f)$.

Validity of the additional program equations is established as follows ($p \equiv q$ means that $\mathsf{KAD} \models p \approx q$):

- K1. $\tau(\bot?) = \tau(\bot) = 0 = \tau(0)$.
- K2. $\tau((\varphi \lor \psi)?) = \tau(\varphi) + \tau(\psi) = \tau(\varphi?) + \tau(\psi?) = \tau(\varphi? \cup \psi?)$.
- K3. $\tau((\varphi \land \psi)?) = \tau(\varphi) \cdot \tau(\psi) = \tau(\varphi?) \cdot \tau(\psi?) = \tau(\varphi?\psi?)$.
- K4. $\tau((\langle \alpha \rangle \top)?\alpha) = \tau(\langle \alpha \rangle \top) \cdot \tau(\alpha) = \mathsf{d}(\tau(\alpha) \cdot \tau(\top)) \cdot \tau(\alpha) \equiv \mathsf{d}\tau(\alpha)\tau(\alpha) \equiv \tau(\alpha)$.

Finally, it is trivial to show that translations of the inference rules preserve membership in $Eq(\mathsf{KAD})$. □

Definition 7. *Let $\sigma : Tm \to Pr$ be defined as follows:*

$$\sigma(\mathsf{p}_n) = \mathsf{p}_n$$
$$\sigma(1) = \top?$$
$$\sigma(0) = \bot?$$
$$\sigma(pq) = \sigma(p); \sigma(q)$$
$$\sigma(p + q) = \sigma(p) + \sigma(q)$$
$$\sigma(p^*) = \sigma(p)^*$$
$$\sigma(\mathsf{a}(p)) = (\lceil \sigma(p) \rceil \bot)?$$

Lemma 4. *For each even term p, $\mathsf{KAD} \models p \approx \tau\sigma(p)$.*

Proof. Induction on the complexity of p. The base case: $\tau\sigma(\mathsf{p}_{2n}) = \tau(\mathsf{p}_{2n}) = \tau(\mathsf{p}_{2n}) = \mathsf{p}_{2n}$. The induction step:

- $\tau\sigma(1) = \tau(\top?) = \tau(\top) \equiv 1$;
- $\tau\sigma(0) = \tau(\bot?) = \tau(\bot) = 0$;
- $\tau\sigma(pq) = \tau(\sigma(p); \sigma(q)) = \tau\sigma(p) \cdot \tau\sigma(q) \equiv pq$;
- $\tau\sigma(p + q) = \tau(\sigma(p) \cup \sigma(q)) = \tau\sigma(p) + \tau\sigma(q) \equiv p + q$;
- $\tau\sigma(p^*) = \tau(\sigma(p)^*) = (\tau\sigma(p))^* \equiv p^*$;
- $\tau\sigma(\mathsf{a}p) = \tau(\lceil \sigma(p) \rceil \bot) \equiv \mathsf{a}(\tau\sigma(p) \cdot \mathsf{a}\tau(\bot)) \equiv \mathsf{a}(\tau\sigma(p)) \equiv \mathsf{a}p$. □

Note that Lemma 4 entails that for each even term p, $\mathsf{KAD}^* \models p \approx \tau\sigma(p)$ since $Eq(\mathsf{KAD}) \subseteq Eq(\mathsf{KAD}^*)$.

Theorem 5. *For all even $p, q \in Tm$:*

1. $\mathsf{KAD} \models p \approx q$ *iff* $TC \vdash \sigma(p) \approx \sigma(q)$;
2. $\mathsf{KAD}^* \models p \approx q$ *iff* $TC \vdash \sigma(p) \approx \sigma(q)$.

Hence, $Eq(\mathsf{KAD}) = Eq(\mathsf{KAD}^*)$.

Proof. Note that if r is even, then so is $\sigma(r)$. Hence, we may reason as follows. First part: $TC \vdash \sigma(p) \approx \sigma(q)$ iff $\mathsf{KAD} \models \tau\sigma(p) \approx \tau\sigma(q)$ (Theorem 4, first part) iff $\mathsf{KAD} \models p \approx q$ (Lemma 4). Second part: $TC \vdash \sigma(p) \approx \sigma(q)$ iff $\mathsf{KAD}^* \models \tau\sigma(p) \approx \tau\sigma(q)$ (Theorem 4, second part) iff $\mathsf{KAD}^* \models p \approx q$ (Lemma 4). □

Theorem 6. *The membership problem for Eq*(KAD) *is EXPTIME-complete.*

Proof. The lower bound (EXPTIME-hardness) follows from Theorem 4: RTA $\models$ $\alpha \approx \beta$ iff RTA $\models \alpha' \approx \beta'$ iff $TC \vdash \alpha' \approx \beta'$ iff KAD $\models \tau(\alpha') \approx \tau(\beta')$. Pr(RTA) is EXPTIME-hard by Theorem 3.

The upper bound (membership in EXPTIME) follows from Theorem 5: KAD $\models p \approx q$ iff KAD $\models p' \approx q'$ iff $TC \vdash \sigma(p') \approx \sigma(q')$ iff RTA $\models \sigma(p') \approx \sigma(q')$. Pr(RTA) is in EXPTIME by Theorem 3. □

6 Conclusion

Our main result is that the membership problem for the equational theory of Kleene algebra with domain is EXPTIME-complete. An interesting result we obtained on the way (with the help of an anonymous reviewer) is that the equational theory of KAD coincides with the equational theory of KAD*. Our embedding results we used to obtain the main result show that, essentially, KAD has the same expressive power as PDL or RTA.

A natural open problem is the question whether there are variants of KAD that are closer to have the same complexity as KAT. More precisely: Are there classes K of algebras of the KAD type such that: (i) KAD $\subseteq$ K, (ii) Eq(KAT) embeds into Eq(K), and (ii) Eq(K) is PSPACE-complete?

A natural candidate for such a class of algebras is "weak Kleene algebra with domain" (our name) briefly discussed in [1]. A weak Kleene algebra with domain is an expansion of a Kleene algebra with a $: K \rightarrow K$ satisfying (d $:=$ a^2)

$$d(1) = 1$$
$$d(d(x)d(y)) = d(y)d(x)$$
$$a(x)d(x) = 0$$
$$a(x) + a(y) = a(d(x)d(y))$$

Many laws valid in KAD fail in weak KAD, most importantly locality, both directions of additivity, and the least left preserver property. However, in weak Kleene algebra with domain, the set of domain elements is still closed under $\cdot, +$ and a, it contains 1 and 0 and, moreover, domain elements are subidentities, and they are commutative and idempotent. Hence, domain algebras in weak Kleene algebras with domain are Boolean algebras.

Another interesting problem is the problem of identifying free algebras in KAD and weak KAD (McLean [20] identifies free relational Kleene algebras with domain and Mbacke [19] identifies free *-continuous Kleene algebras with domain.)

Acknowledgement. The author is grateful to three anonymous reviewers for valuable comments, and to Johann J. Wannenburg and Adam Přenosil for helpful discussions. Work on this paper was supported by the long term strategic development financing of the Institute of Computer Science (RVO:67985807).

References

1. Armstrong, A., Gomes, V.B.F., Struth, G.: Building program construction and verification tools from algebraic principles. Form. Aspects Comput. **28**(2), 265–293 (2016)
2. Cohen, E., Kozen, D., Smith, F.: The complexity of Kleene algebra with tests. Technical report TR96-1598, Computer Science Department, Cornell University, July 1996
3. Desharnais, J., Möller, B., Struth, G.: Kleene algebra with domain. ACM Trans. Comput. Log. **7**(4), 798–833 (2006)
4. Desharnais, J., Struth, G.: Modal semirings revisited. In: Audebaud, P., Paulin-Mohring, C. (eds.) MPC 2008. LNCS, vol. 5133, pp. 360–387. Springer, Heidelberg (2008). https://doi.org/10.1007/978-3-540-70594-9_19
5. Desharnais, J., Struth, G.: Internal axioms for domain semirings. Sci. Comput. Program. **76**(3), 181–203 (2011). Special issue on the Mathematics of Program Construction (MPC 2008)
6. Fischer, M.J., Ladner, R.E.: Propositional dynamic logic of regular programs. J. Comput. Syst. Sci. **18**, 194–211 (1979)
7. Groenendijk, J., Stokhof, M.: Dynamic predicate logic. Linguist. Philos. **14**(1), 39–100 (1991)
8. Harel, D., Kozen, D., Tiuryn, J.: Dynamic Logic. MIT Press, Cambridge (2000)
9. Hirsch, R., Mikulás, S.: Axiomatizability of representable domain algebras. J. Log. Algebr. Program. **80**(2), 75–91 (2011)
10. Hollenberg, M.: Equational axioms of test algebra. In: Nielsen, M., Thomas, W. (eds.) CSL 1997. LNCS, vol. 1414, pp. 295–310. Springer, Heidelberg (1998). https://doi.org/10.1007/BFb0028021
11. Hopcroft, J.E., Ullman, J.D.: Introduction to Automata Theory, Languages and Computation. Addison-Wesley Publishing Company, Boston (1979)
12. Kozen, D.: A representation theorem for models of *-free PDL. In: de Bakker, J., van Leeuwen, J. (eds.) ICALP 1980. LNCS, vol. 85, pp. 351–362. Springer, Heidelberg (1980). https://doi.org/10.1007/3-540-10003-2_83
13. Kozen, D.: On Kleene algebras and closed semirings. In: Rovan, B. (ed.) MFCS 1990. LNCS, vol. 452, pp. 26–47. Springer, Heidelberg (1990). https://doi.org/10.1007/BFb0029594
14. Kozen, D.: The Design and Analysis of Algorithms. Springer, New York (1992). https://doi.org/10.1007/978-1-4614-1701-9
15. Kozen, D.: A completeness theorem for Kleene algebras and the algebra of regular events. Inf. Comput. **110**(2), 366–390 (1994)
16. Kozen, D.: Kleene algebra with tests. ACM Trans. Program. Lang. Syst. **19**(3), 427–443 (1997)
17. Kozen, D., Parikh, R.: An elementary proof of the completeness of PDL. Theor. Comput. Sci. **14**, 113–118 (1981)
18. Kozen, D., Smith, F.: Kleene algebra with tests: completeness and decidability. In: van Dalen, D., Bezem, M. (eds.) CSL 1996. LNCS, vol. 1258, pp. 244–259. Springer, Heidelberg (1997). https://doi.org/10.1007/3-540-63172-0_43
19. Mbacke, S.D.: Completeness for Domain Semirings and Star-continuous Kleene Algebras with Domain. mathesis, Université Laval (2018)
20. McLean, B.: Free Kleene algebras with domain. J. Log. Algebr. Methods Program. **117**, 100606 (2020)

21. Möller, B., Struth, G.: Algebras of modal operators and partial correctness. Theor. Comput. Sci. **351**(2):221–239 (2006). Algebraic Methodology and Software Technology

22. Pratt, V.: Dynamic algebras: examples, constructions, applications. Stud. Log. **50**(3), 571–605 (1991)

23. Pratt, V.R.: Models of program logics. In: Proceedings of the 20th Annual Symposium on Foundations of Computer Science, FOCS 1979, pp. 115–122. IEEE Computer Society, USA (1979)

24. Struth, G.: On the expressive power of Kleene algebra with domain. Inf. Process. Lett. **116**(4), 284–288 (2016)

25. Trnková, V., Reiterman, J.: Dynamic algebras with test. J. Comput. Syst. Sci. **35**(2), 229–242 (1987)

Enumerating, Cataloguing and Classifying All Quantales on up to Nine Elements

Arman Shamsgovara[✉]

Department of Computing Science, Umeå University, Umeå, Sweden
ens12asa@cs.umu.se

Abstract. Using computer software, every quantale on up to nine elements has been enumerated up to isomorphism, catalogued and classified with respect to various properties. In order to achieve this the enumeration was branched by partitioning the search space based on various isomorphic invariants of quantales.

Keywords: Quantales · Enumeration · Classification

1 Introduction

Quantales are algebraic structures that on the one hand are semigroups, and on the other hand are complete lattices. They encode a notion of order as well as a composition that distributes over joins. This makes quantales useful as model structures in many-valued and fuzzy logic, as the elements of a quantale can be interpreted as truth values of different degrees, and the semigroup operator as a logical operation on these truth values. Certain types such as Girard quantales are also related to linear logic [21].

The aim of this paper is to report the compilation of a catalogue of every quantale on up to nine elements, as well as their properties, up to isomorphism. A catalogue of small-order quantales and their properties can be very useful in both theory and practice. On the one hand, a rich collection of examples, numbers and patterns surrounding quantales can stimulate theoretical developments. On the other hand, there are recent ideas of using quantales in practice, opening up concrete use cases for the produced catalogue. Hypothetical applications of quantales include modelling diagnosis systems in healthcare [7,10], or modelling many-valued logic circuits [8]. Understanding quantales as a design space will surely be helpful in guiding practical applications of them.

To put the work developed here into context, the quantales on up to 3 elements were enumerated prior to the book [9], the author then enumerated all quantales on up to 6 elements using SAT solvers [20], and up to 9 elements using the software Mace4 [17]. The quantales up to 6 elements have been made available as a mobile application for easier browsing [19]. Beyond the author's own work on enumerating quantales, a prior paper used a backtracking approach to enumerate residuated finite lattices up to order 12, which are essentially

R. Glück et al. (Eds.): RAMiCS 2023, LNCS 13896, pp. 224–240, 2023.
https://doi.org/10.1007/978-3-031-28083-2_14

quantales that are restricted to be integral and commutative [1]. Furthermore, semigroups [4,5] have been enumerated using both constraint- and SAT-solvers in much more involved ways than what is presented in this document, and been released as a GAP package called SmallSemi. Lastly, lattices [14] have been enumerated using a variety of different algorithms, consult the sources in the cited OEIS entry for more information.

The organization of this paper is as follows. In Sect. 2 we define quantales and their properties, then discuss how they were enumerated in Sect. 3. In Sect. 4 we present a summary of some results obtained, and we conclude with a discussion on how to proceed with further research in Sect. 5.

2   Quantales

We will classify quantales with respect to quite a few properties, hence we state their definitions in this section for the reader's perusal. A few of these are novel, and pointed out as such. The rest can be found in textbooks such as [9,16]. We will use $\perp$ and $\top$ to denote the bottom and top elements, respectively.

Definition 1. *A (finite)* quantale *is an algebraic structure consisting of a set X and two binary operators $*$ and $\vee$ such that X is a semigroup under $*$, a complete suplattice under $\vee$, $\perp$ is a zero element under $*$ and $*$ distributes over $\vee$:*

$$\perp * x = x * \perp = \perp, \tag{1}$$

$$x * (y \vee z) = (x * y) \vee (x * z), \tag{2}$$

$$(x \vee y) * z = (x * z) \vee (y * z). \tag{3}$$

For infinite quantales one additionally requires that the semigroup operation distributes over arbitrary joins ($\vee$), but we only consider finite quantales below.

We consider in Table 1 a rather broad collection of quantale properties considered in the literature, some of which rely on concepts such as prime or cyclic elements. We thus state some definitions for these concepts next.

Definition 2. *An element $p \neq \top$ of a quantale is* prime *if*

$$x * y \leq p \Rightarrow x * \top \leq p \text{ or } \top * y \leq p \tag{4}$$

and strong prime *if*

$$x * y \leq p \Rightarrow (x * \top) \vee x \leq p \text{ or } (\top * y) \vee y \leq p. \tag{5}$$

Definition 3. *Given a quantale, we define two additional binary operations $\leftarrow$ and $\rightarrow$ on it via the defining equations*

$$x * y \leq z \Leftrightarrow x \leq z \leftarrow y, \tag{6}$$

$$x * y \leq z \Leftrightarrow y \leq x \rightarrow z. \tag{7}$$

We call $\leftarrow$ left residuation/implication, and $\rightarrow$ right residuation/implication.

Semantically, the residuations can be interpreted as the maximal solutions for the inequality $x * y \leq z$, where the left one solves for x given y and z, and the right one solves for y given x and z. They can be viewed as many-valued analogues of the two-valued implication from Boolean logic. Given a quantale, both residuations exist and are uniquely determined.

Definition 4. *An element d of a quantale is* dualizing *if for all elements x*

$$d \leftarrow (x \rightarrow d) = (d \leftarrow x) \rightarrow d. \tag{8}$$

Definition 5. *An element d of a quantale is* cyclic *if for all elements x*

$$x \rightarrow d = d \leftarrow x. \tag{9}$$

Definition 6. *Consult Table 1 for a condensed presentation of several properties of quantales along with their definitions.*

Table 1. List of various properties that quantales can have, with their definitions. We have classified the enumerated quantales w.r.t. these properties, i.e. determined for each quantale whether or not it satisfied these properties. An asterisk after the property name indicates that this, to the author's knowledge, is the first time these definitions appear in print in relation to quantales.

Property	Definition
Semi-unital	For all x: $x \leq x * \top$ and $x \leq \top * x$
Unital	There exists a unit w.r.t. $*$
Left-sided	For all x: $\top * x \leq x$
Right-sided	For all x: $x * \top \leq x$
Strictly left-sided	For all x: $\top * x = x$
Strictly right-sided	For all x: $x * \top = x$
Two-sided	Left-sided and Right-sided
Integral	The top element $\top$ is a unit
Spatial	Every element except $\top$ is a meet of primes
Strongly spatial (*)	Every element except $\top$ is a meet of strong primes
Balanced	$\top * \top = \top$
Idempotent	For all x: $x * x = x$
Semi-integral	For all elements x, y: $x * \top * y \leq x * y$
Commutative	For all elements x, y: $x * y = y * x$
Bisymmetric	For all x, y, z, w: $(x * y) * (z * w) = (x * z) * (y * w)$
Factor	Only $\bot$ and $\top$ are two-sided
Prime (*)	Every element except $\top$ is prime
Strong prime (*)	Every element except $\top$ is a strong prime
Inf-distributive	$*$ distributes over $\wedge$
Completely distributive	The underlying lattice is completely distributive
Frobenius	There is a dualizing element in the quantale
Girard	There is a cyclic and dualizing element in the quantale

Example 1. An example of a quantale is shown in Fig. 1. The left and right residuations are also presented. The quantale is semi-unital but not unital, right-but not left-sided, as well as spatial and balanced. In fact, it is prime, since both 0, 1 and 2 are prime elements. The only two-sided elements are verified to be 0 and 3, so the quantale is factor. There is no dualizing element, but 0 and 3 are cyclic.

*	0 1 2 3		←	0 1 2 3		→	0 1 2 3
0	0 0 0 0		0	3 3 3 3		0	3 3 3 3
1	0 1 1 1		1	0 1 2 3		1	0 3 0 3
2	0 2 2 2		2	0 1 2 3		2	0 0 3 3
3	0 3 3 3		3	0 1 2 3		3	0 0 0 3

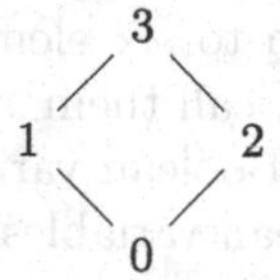

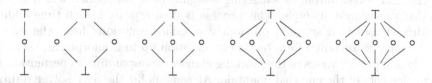

Fig. 1. A quantale on four elements, left and right implications included. Due to the choice of notation, the table for left implication should be read in a transposed fashion, e.g. $2 \leftarrow 1 = 2$. This is intentional, and is because using an arrow symbol that points in the correct (left) direction of the implication makes more notational sense than trying to represent an implication "to the left" with an expression that is read left to right, as advocated in e.g. [9].

Lastly, we will present numbers for the quantales on two special types of lattices as well, so we will define them here for completeness. Consult Fig. 2 for illustrating Hasse diagrams.

Definition 7. *A complete lattice is a chain if every pair of elements is comparable with each other.*

Definition 8. *An N-element complete lattice is a diamond if it consists of a top element, a bottom element, and N − 2 elements with pairwise meets ⊥ and pairwise joins ⊤.*

Fig. 2. The first few diamond lattices. The leftmost one is also a chain.

3   Method

To enumerate quantales, we initially used SAT solvers to enumerate quantales on up to 6 elements with a relatively naive and straight-forward approach. We then switched over to the model enumeration software Mace4 for orders 7 through 9 because it was faster than what the author was capable of accomplishing with their SAT models. Some ad-hoc approaches were necessary to push the boundary for what Mace4 could do and enumerate the quantales without crashing from memory issues. Obtained quantales were then classified using custom scripts.

3.1 SAT Solvers

We will briefly mention the SAT-based approach used in [20] to enumerate the quantales on up to six elements. The idea is to fix the number n of elements of our quantale, call them x_1 through x_n and encode the semigroup operator $*$ by introducing Boolean variables $*_{ijk}$ that encode the statements $x_i * x_j = x_k$. Similarly, Boolean variables $\vee_{ijk}$ are introduced to encode the operator $\vee$. The quantale axioms can now be encoded as Boolean formulae. As an example, left distributivity of $*$ over $\vee$ can be encoded viz.

$$\forall x, y, z : x * \underbrace{\overbrace{(y \vee z)}^{d}}_{a} = \underbrace{\overbrace{(x * y)}^{d}}_{b} \vee \underbrace{(x * z)}_{c} \xrightarrow{\text{encoding}}$$

$$\bigwedge_{x,y,z,a,b,c,d} \left(\vee_{yza} \wedge *_{xyb} \wedge *_{xzc} \rightarrow (*_{xad} \leftrightarrow \vee_{bcd}) \right) \quad (10)$$

where the universal quantifier has been replaced by a conjunction running over all the domain elements, and some auxiliary indices have been introduced to keep track of intermediate values (visualized above using braces for the readers convenience). The other quantale axioms are encoded in the same fashion, fixing one of the index values to represent $\bot$. Additional formulas are also created to ensure that for every choice of x and y, the expressions $x * y$ and $x \vee y$ attain exactly one value each. These formulas are then converted to conjunctive normal form using the standard algorithm (see any textbook on logic in computer science, e.g. [2]), and the model is fed to a solver such as Minisat [6] or the more modern PLingeling [3].

The SAT solver outputs a satisfying assignment to the model, which can be decoded to obtain a quantale. This process is then repeated, each time adding blocking clauses that exclude previously obtained solutions from the search space. Since we are only interested in enumeration up to isomorphism, we additionally add permutations of the blocking clauses corresponding to permutations of the elements of the encoded quantale. At some point the SAT solver returns no more satisfying assignments, instead giving an unsatisfiability certificate.

Although this method is sound and does work, the blocking clauses make the method unfeasible for bigger orders, as they make the model size quickly grow unfeasibly large. Another weakness is that most SAT-solvers available only find one solution at a time and then terminate, meaning that one will have to rerun the solver from scratch once for every structure up to isomorphism, a great potential time sink. One final weakness of the particular approach as described here is that the described model does not incorporate any particular tricks to break symmetries or otherwise constrain the search space. It can certainly be done to great effect, as shown by prior work in enumerating semigroups [4,5], but it requires quite a bit more expertise and the author chose to use the model finder Mace4 for the sake of convenience.

3.2   Mace4

Mace4 is a piece of software specializing in enumerating algebraic structures with unary and binary operators [11]. It uses a recursive backtracking approach to, given a list of input algebraic axioms, generate models that satisfy the axioms. It has a convenient syntax for specifying algebraic axioms and is, in our experience, a faster and less error prone alternative than the naive SAT-based approach described previously. This is in part since Mace4 has built-in heuristics to prune away some isomorphic search branches at runtime, and in part since it can find every satisfying assignment in a single run of the program. Furthermore, as the syntax is very human-friendly, it is much easier to debug and automatically or manually generate input files. A basic Mace4 file is presented in Fig. 3 for the reader's convenience.

```
assign(max_models, -1).
assign(domain_size, 4).
formulas(assumptions).
(x*y)*z = x*(y*z).
(x v y) v z = x v (y v z).
x v y = y v x.
x v x = x.
x*(y v z) = (x*y) v (x*z).
(x v y)*z = (x*z) v (y*z).
0*x = 0.
x*0 = 0.
0 v x = x.
3 v x = 3.
end_of_list.
formulas(goals).
end_of_list.
```

Fig. 3. A straight-forward Mace4 input file that enumerates all quantales on 4 elements, hard-coding element 0 as the bottom and element 3 as the top. By modifying the domain_size argument this file, together with the program isofilter that accompanies Mace4, is sufficient to relatively instantly enumerate all quantales on 6 elements or less. One can also set the max_models property to any positive number if only a fixed maximum number of structures is desired. This file is also presented in the author's OEIS entry A354493 [18]. For more information on how to use a file like this with Mace4, we refer to the user manual [11].

3.3 Branching Scheme

Since Mace4 encounters memory-related issues if it finds too many structures in a single run, a branching scheme is necessary in order to partition the search space and find a manageable number of quantales per run. This method requires some thought to prevent the possibility of finding isomorphic quantales in different search branches. To this end, we branch on isomorphically invariant properties, i.e. properties that must provably be preserved by isomorphisms.

During our enumerations, the following scheme was used, proceeding to the next branching only if the program crashed. It was created ad-hoc as needed to achieve progress. Each branching was implemented by means of generating new input Mace4 files with added constraints to express the branching criteria.

1. Lattice table, expressed by hard-coding what values the join would take on for every pair of elements.
2. The number of idempotent elements,
3. The number of leftsided elements and the number of rightsided elements,
4. The number of elements x such that $x * x = \top$,
5. Tailor-made branchings based on the underlying lattice.

Taking criterion 1 as an example, it was implemented by generating one Mace4 input file for every lattice, each file containing the axioms for a quantale as well as statements hard-coding the lattice table, that is, adding one line each of the form a v b = c. for every pair of numbers a,b $\in \{0,\dots,(N-1)\}$, c being the value of the join in the lattice. Similarly, adding criterion 2 generated an additional factor 10 new input files, one for each value $n = 0, 1, \dots, 9$ of idempotent elements. Step 3 created one file for every tuple (l, r) of integers corresponding to l leftsided elements and r rightsided ones, and so on for step 4.

As for implementing branchings 2–4, each of them are based on adding statements of the form "exactly n elements satisfy P", P being a property such as "is idempotent", to the existing Mace4 model, and step 3 in particular is a conjunction of two such statements. These statements were respectively expressed in Mace4 syntax using single expressions that run over all subsets $S \subseteq \{0, 1, \dots, (N-1)\}$, consisting of n elements, N denoting the number of elements considered, and form a disjunction over all statements of the form "these n elements satisfy P, and the others do not". In mathematical notation, the encodings may be stated viz.

$$\bigvee_{\substack{S \subseteq \{0,\dots,(N-1)\} \\ |S|=n}} \left(\bigwedge_{x \in S} P(x) \wedge \bigwedge_{x \notin S} \neg P(x) \right) \tag{11}$$

although in the input files the same statements are written by explicitly rolling out all the conjunctions and disjunctions.

The key insight used in step 5 was that once the lattice table is hard-coded in step 1, the set of valid isomorphisms is greatly constrained by the lattice table. In the case of the chain lattice, notorious in our searches for having vast numbers of quantales compared to other lattices, once the lattice table is fixed the only possible quantale isomorphism that preserves the lattice table is easily seen to be the identity. This makes it possible to, for example, pick an arbitrary cell of the semigroup table and branch by letting that cell attain each different value of the structure. Consider Fig. 4 for the specific branching rules used per lattice.

In practice, step 1 was enough to enumerate the quantales of the vast majority of lattices: on 8 elements only the chain needed step 2, and on 9 elements only 149 of the 1078 lattices had to enter step 2 of the scheme. Out of these, 99 entered step 3, and out of these 30 needed further branching. The chain lattice in particular was handled directly in step 5, skipping step 4, but the remaining 29 lattices entered step 4. Only 6 lattices (see Fig. 4) needed to be handled separately in step 5. For the chain in particular we did not perform step 4 as it was deemed not necessary with the right tailor-made branching (namely, branching on all 10 possible values of $\top * \top$).

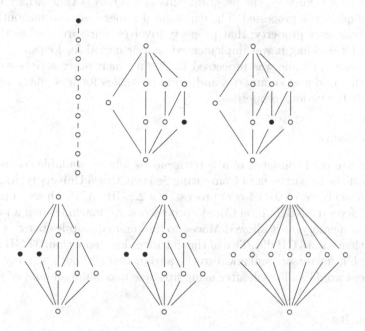

Fig. 4. The 6 lattices where tailor-made branchings were made. For the lattices in the top row, branching was made on the 10 possible values of $x * x$, where x is the element represented by a filled in circle. The two non-diamond lattices in the bottom row were handled by branching on whether $x * x = \top$ holds for one, both or none of the two filled in circles. For the diamond, branching was done on how many of the middle elements x satisfy $x * y = \top$ for all y.

3.4   Post-processing the Obtained Quantales

Once the quantales were enumerated and stored on disk, they were classified with respect to each of the listed properties in Sect. 2 using a custom program. In essence, the quantales were streamed one at a time from the Mace4 output, the meet, left and right residuations were computed from their definitions, each property considered was checked to see if the quantale satisfied it, special elements such as the unit, left-sided elements, et cetera were identified as needed by the definitions, and finally a summary of the properties and special elements was printed to a dedicated output file.

From a complexity point of view, the classification algorithm uses $\mathcal{O}(N^2)$ space, N being the number of elements in a quantale. This is since only one quantale at a time is processed and stored in memory, and since the only other data computed by the program to check for properties are a constant number of N-element Boolean arrays encoding whether each element is left-sided, idempotent, et cetera. Further, the program runs in $\mathcal{O}(Q \cdot N^4)$ time, where Q is the number of quantales processed. The exponent 4 comes from the subroutine that checks bisymmetry property; that property involves four variables, so the naive algorithm for deciding it was implemented as four nested for-loops.

Once every quantale was processed in this manner, other scripts were used to parse the produced summaries and compute tallies for how many quantales that satisfy the various properties.

3.5   Hardware

The hardware used consisted of a heterogeneous mix of available computer lab machines at the Department of Computing Science, Umeå University, including a 3.4 GHz AMD Epyc 7702P 64-core processor, a 2.2 GHz AMD Opteron processor 6272, and several 3.5 GHz Intel Core i5 processors. All machines had a minimum of 32 GB of memory, as it allowed Mace4 to find more quantales per run without crashing than e.g. 16 GB did. Two of the machines had more than 32 GB memory available, but for reasons unknown to the author that did not yield any benefits since Mace4 would still crash after using up some maximum amount of memory.

4   Results

Aside from tallies of the number of quantales up to isomorphism with respect to various properties, we present a few other findings such as notable examples of quantales and insights reflected in the distribution of quantales. See Table 2 for the number of quantales in relation to the number of semigroups and lattices. The number of quantales of order up to 6 have been verified by comparing the number of quantales obtained using Mace4 and those using the SAT method.

On a practical note, over a hundred thousand Mace4 input files were generated to perform all of the branchings and the full dataset of quantales, storing each quantale as two tables for $*$ and $\vee$, takes up around 400 GB of storage,

Table 2. Number of semigroups [15], lattices [14] and quantales [18] on N elements. For lattices the terms beyond $N = 9$ have been omitted.

Structure	OEIS	1	2	3	4	5	6	7	8	9
Semigroups	A027851	1	5	24	188	1915	28634	1627672	3684030417	105978177936292
Quantales	A354493	1	2	12	129	1852	33391	729629	19174600	658343783
Lattices	A006966	1	1	1	2	5	15	53	222	1078

with about 380 GB being taken by the 9 element quantales, and 17 GB by the quantales on the 9 element chain alone. These numbers are halved by storing only the semigroup tables and relegating the identical lattice tables to a separate data source. The full set of quantales are available in a GitLab repository hosted by Umeå University[1].

Anecdotally, it took a couple of months total (wall-clock) time to enumerate all the quantales on up to 9 elements, using around 50 parallel instances of Mace4 processing search branches at any given time. The execution time of the classification algorithm, in comparison, was on the scale of a few hours to process the entire dataset.

4.1 The Number of Quantales that Have Various Properties

For each of the properties defined in Sect. 2, we have counts for how many quantales satisfy those properties. We present these in Table 3. We also present the corresponding counts for quantales on the chain lattices in Table 4, and for the diamond lattices (compare with Fig. 2) in Table 5. There are many more numbers that could be presented, but we will omit them for the sake of space.

4.2 Lattices with the Most and Least Quantales

If one compares the different lattices with respect to how many quantales up to isomorphism they admit, it turns out that as a rule of thumb, taller lattices generally have more quantales than shorter ones. That is, on the one hand we have lattices like the diamonds, that have only a few tens of thousands of quantales total even on nine elements, but on the other we have the chains that on nine elements admit tens of millions. Other lattices fall in between these two extremes, having more quantales the more chain-like they are. It must be emphasised that this is only a vague guideline, and in fact there is (exactly) one lattice with even fewer quantales than the diamond on nine elements, see Fig. 5.

4.3 Examples of Quantales with New Properties

Since we did present a few novel definitions in Sect. 2, it makes sense to present some examples of such quantales. Consider Fig. 6 for a quantale that is strong

---

[1] https://git.cs.umu.se/ens12asa/quantales_up_to_9_elements.

Table 3. Number of quantales on N elements, satisfying various properties.

Property	1	2	3	4	5	6	7	8	9
Quantale	1	2	12	129	1852	33391	729629	19174600	658343783
Semi-unital	1	1	6	64	939	17578	403060	11327795	440735463
Unital	1	1	3	20	149	1488	18554	295292	6105814
Left/Right-sided	1	2	9	60	497	4968	58507	897338	13341730
Strictly Left/Right-sided	1	1	4	23	164	1482	15838	197262	2830649
Two-sided	1	2	8	47	354	3277	36506	490983	8301353
Integral	1	1	2	9	49	364	3335	37026	496241
Spatial	1	2	10	71	570	5147	51248	557143	6557759
Strongly Spatial	1	1	4	21	121	818	6236	53077	498046
Balanced	1	1	9	106	1597	29720	663897	17747907	620659554
Idempotent	1	1	4	24	169	1404	13104	134464	1492598
Semi-integral	1	2	11	96	1041	13669	211561	3780964	77057208
Commutative	1	2	8	57	550	6639	96264	1639905	32781241
Bisymmetric	1	2	12	125	1691	28249	565046	13553879	448314086
Factor	1	2	4	38	519	9442	219222	6538004	296594240
Prime	1	2	10	70	559	4989	49154	529433	6181882
Strong Prime	1	1	4	20	115	764	5749	48413	450342
Inf-distributive	1	2	12	108	1124	13256	172535	2452680	38098425
Completely distributive	1	2	12	129	1437	19047	269739	4207822	132177828
Frobenius	1	1	2	8	19	91	267	1388	4881
Girard	1	1	2	8	19	91	262	1359	4710

Table 4. Number of quantales on the N-element chain, satisfying various properties.

Property	1	2	3	4	5	6	7	8	9
Quantale	1	2	12	101	1003	11329	142094	1957183	29634185
Semi-unital	1	1	6	45	414	4324	49997	631949	8681521
Unital	1	1	3	15	84	575	4687	45223	516882
Left/Right-sided	1	2	9	55	413	3728	39627	492535	7308241
Strictly Left/Right-sided	1	1	4	20	133	1087	10512	118112	1527872
Two-sided	1	2	8	44	308	2641	27120	332507	5035455
Integral	1	1	2	8	44	308	2641	27120	332507
Spatial	1	2	10	55	293	1536	8007	41663	216626
Strongly Spatial	1	1	4	14	48	164	560	1912	6528
Balanced	1	1	9	82	846	9774	124258	1720426	25819824
Idempotent	1	1	4	17	82	422	2274	12665	72326
Semi-Integral	1	2	11	79	661	6487	73605	954581	14220741
Commutative	1	2	8	41	241	1553	10704	77811	591441
Bisymmetric	1	2	12	97	877	8677	92268	1047921	12933247
Factor	1	2	4	24	187	1737	18423	218026	2846283
Prime	1	2	10	55	293	1536	8007	41663	216626
Strong Prime	1	1	4	14	48	164	560	1912	6528
Inf-distributive	1	2	12	101	1003	11329	142094	1957183	29634185
Completely distributive	1	2	12	101	1003	11329	142094	1957183	29634185
Frobenius	1	1	2	4	8	17	38	91	222
Girard	1	1	2	4	8	17	38	91	222

Table 5. Number of quantales on the diamond lattices, satisfying various properties. Column labels indicate the total number of elements in the lattice, i.e. 3 stands for the chain on 3 elements, 4 for the diamond with two middle elements, et cetera.

Property	3	4	5	6	7	8	9
Quantale	12	28	78	262	1036	5129	48299
Semi-unital	6	19	67	249	1021	5112	48280
Unital	3	5	8	17	42	176	1421
Left/Right-sided	9	5	3	3	3	3	3
Strictly Left/Right-sided	4	3	2	2	2	2	2
Two-sided	8	3	1	1	1	1	1
Integral	2	1	0	0	0	0	0
Spatial	10	16	19	23	27	31	35
Strongly Spatial	4	7	8	10	12	14	16
Balanced	9	24	74	258	1032	5125	48295
Idempotent	4	7	9	11	13	15	17
Semi-Integral	11	17	21	25	29	33	37
Commutative	8	16	35	89	240	696	2244
Bisymmetric	12	28	77	243	869	3966	40351
Factor	4	14	49	187	772	4053	42192
Prime	10	15	19	23	27	31	35
Strong Prime	4	6	8	10	12	14	16
Inf-distributive	12	7	1	1	1	1	1
Completely distributive	12	28	0	0	0	0	0
Frobenius	2	4	5	10	14	30	45
Girard	2	4	5	10	13	25	34

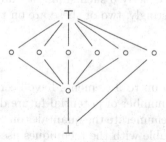

Fig. 5. The lattice on 9 elements with the fewest (19447) quantales up to isomorphism. Notably, it is not a diamond lattice, whereas the lattices on up to 8 elements with the fewest quantales are exclusively diamonds.

prime, hence also prime and strongly spatial. In Fig. 7 we present a quantale on the four-element diamond that is strongly spatial, hence spatial, but not prime. This quantale explains why there are 16 (strongly) spatial quantales on that lattice, but only 15 (strong) prime ones, as seen in Table 5.

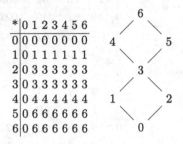

$$
\begin{array}{c|ccccccc}
* & 0 & 1 & 2 & 3 & 4 & 5 & 6 \\
\hline
0 & 0 & 0 & 0 & 0 & 0 & 0 & 0 \\
1 & 0 & 1 & 1 & 1 & 1 & 1 & 1 \\
2 & 0 & 3 & 3 & 3 & 3 & 3 & 3 \\
3 & 0 & 3 & 3 & 3 & 3 & 3 & 3 \\
4 & 0 & 4 & 4 & 4 & 4 & 4 & 4 \\
5 & 0 & 6 & 6 & 6 & 6 & 6 & 6 \\
6 & 0 & 6 & 6 & 6 & 6 & 6 & 6 \\
\end{array}
$$

Fig. 6. A quantale where every element but ⊤ is a strong prime. It is strongly spatial.

$$
\begin{array}{c|cccc}
* & 0 & 1 & 2 & 3 \\
\hline
0 & 0 & 0 & 0 & 0 \\
1 & 0 & 1 & 0 & 1 \\
2 & 0 & 0 & 2 & 2 \\
3 & 0 & 1 & 2 & 3 \\
\end{array}
$$

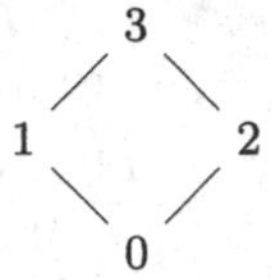

Fig. 7. A quantale on four elements that is strongly spatial, yet not prime. Elements 1 and 2 are (strong) primes, but not 0.

4.4 Minimal Quantales that are Frobenius and not Girard

As seen in Table 3, the smallest quantales that are Frobenius but not Girard have seven elements. There are exactly 5 such smallest examples up to isomorphism, presented in Fig. 8. Interestingly, two of these are on the same lattice.

5   Discussion

Now that all quantales on up to 9 elements have been enumerated, catalogued and classified, there are a number of potential future directions of research. The most obvious would be to enumerate the quantales on 10 elements. At present we anticipate it should be doable with the techniques used here, although it would require not only a nontrivial factor of additional computation time, but also additional branchings on invariant properties or tailor-made symmetry breaking on a per-lattice basis. Extrapolating from what we know about quantales up to 9 elements, storing them would require terabytes of storage if further analysis is desired, but that is mainly a matter of acquiring the hardware to store them.

One could also study what was done to enumerate semigroups in [4,5] in more detail and try to mimic their methods to enumerate quantales, as they

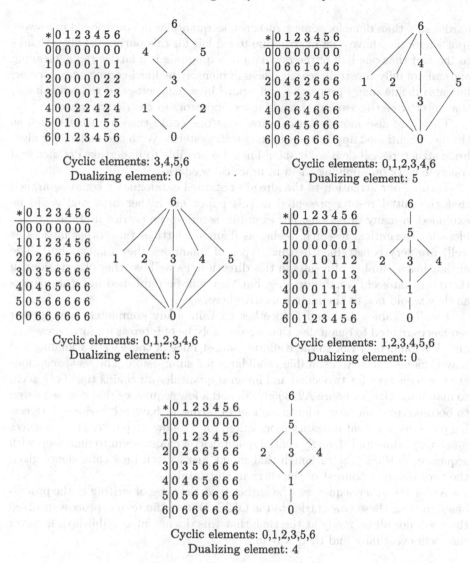

Fig. 8. The five smallest quantales that are Frobenius, but not Girard.

(albeit with a lot more effort) managed to enumerate orders of magnitude more structures than we have; quantales are semigroups over lattices after all. Our partitioning approach with Mace4 certainly has a lot of room for improvement as well, especially when it comes to breaking lattice symmetries and exploiting the lack of such where applicable. Perhaps by studying how quantales distribute w.r.t. various properties one can also devise even more effective branching rules.

One could also attempt enumeration of other algebraic structures using our techniques, especially the branchings on isomorphic invariants. This is a bit harder said than done for some structures, as quantales are positioned in a sweet spot where they have a lot of structure to exploit for this purpose. This is thanks to the fact that one half of the raw data of a quantale is a lattice table, meaning one can not only constrain search spaces immensely by hard-coding one operator, but also define many properties and branchings (e.g. left- and right-sidedness) that intertwine the semigroup and lattice operators in restrictive ways.

There are also more practical directions one could take from here, such as the aforementioned applications to diagnosis systems. With the underlying algebraic structures catalogued, it should now be possible to undertake detailed and concrete work on applying them to practical scenarios.

Turning our attention to the already obtained catalogue, it could be argued that the initial results presented in this paper are rather brief and could be extended in many directions. For example, some of the entries in Table 5 evoke ideas for theoretical research, seeing as diamond lattices (and other lattices as well) are very constrained in what types of quantales they might admit. The author has a number of results in this direction, as well as other observations in the data that seem worth pursuing, but they will be published in future papers as they would not fit the main narrative here.

Finally, Tables 3, 4 and 5 together contain many completely new integer sequences related to quantales, that would likely be of interest for future research and development. Prior to these efforts, almost none of these were available, and may thus be considered suitable candidates for submission. The few exceptions are the sequences for two-sided and integral quantales on chains that both seem to match up with sequence A253950 [13], and a few sequences that can be proven to become constant after a finite number of terms, or have otherwise uninteresting patterns, but that is a story more suited for a follow-up paper. The sequences for strong prime and strongly spatial quantales on chains seem to match up with sequence A204089 [12], although that might just as well be a coincidence given the very different context of the entry in the database.

As for the other sequences, the author is at the time of writing in the process of submitting these new entries to the OEIS. Due to the review process involved they will not all be ready at the time that this document is published, however they will eventually find their way in.

Acknowledgement. This work has been supported by the project "A digital twin to support sustainable and available production as a service" (DT-SAPS), funded by Produktion2030, the Strategic innovation programme for sustainable production in Sweden, and the project "A general digital twin driving mining innovation through statistical and logical modelling" (DT-MINN), funded by the Swedish Mining Innovation 2022. The author would also like to thank professor Patrik Eklund for many discussions surrounding quantales and other matters, some of which inspired this project to happen. The anonymous reviewers also deserve kudos for having provided thorough and useful feedback on this manuscript.

References

1. Belohlavek, R., Vychodil, V.: Residuated lattices of size ≤ 12. Order **27**(2), 147–161 (2010)
2. Ben-Ari, M.: Mathematical Logic for Computer Science. Springer, Heidelberg (2012)
3. Biere, A., et al.: Lingeling, Plingeling and Treengeling entering the SAT competition 2013. Proc. SAT Compet. **2013**, 1 (2013)
4. Distler, A., Jefferson, C., Kelsey, T., Kotthoff, L.: The semigroups of order 10. In: Milano, M. (ed.) CP 2012. LNCS, pp. 883–899. Springer, Heidelberg (2012). https://doi.org/10.1007/978-3-642-33558-7_63
5. Distler, A., Kelsey, T.: The semigroups of order 9 and their automorphism groups. Semigroup Forum **88**(1), 93–112 (2013). https://doi.org/10.1007/s00233-013-9504-9
6. Eén, N., Sörensson, N.: An extensible SAT-solver. In: Giunchiglia, E., Tacchella, A. (eds.) SAT 2003. LNCS, vol. 2919, pp. 502–518. Springer, Heidelberg (2004). https://doi.org/10.1007/978-3-540-24605-3_37
7. Eklund, P.: Information and process in health. In: Integrated Care and Fall Prevention in Active and Healthy Aging, pp. 263–279. IGI Global (2021)
8. Eklund, P.: Quantales in circuit design. In: 2021 IEEE 51st International Symposium on Multiple-Valued Logic (ISMVL), pp. 39–42. IEEE (2021)
9. Eklund, P., García, J.G., Höhle, U., Kortelainen, J.: Semigroups in complete lattices. Quant. Modul. Related Topics **54** (2018)
10. Eklund, P., Höhle, U., Kortelainen, J.: Modules in health classifications. In: 2017 IEEE International Conference on Fuzzy Systems (FUZZ-IEEE), pp. 1–6. IEEE (2017)
11. McCune, W.: Prover9 and Mace4 (2005–2010). https://www.cs.unm.edu/~mccune/prover9
12. OEIS Foundation Inc.: The number of 1 by n haunted mirror maze puzzles with a unique solution ending with a mirror, where mirror orientation is fixed., entry A204089 in the On-Line Encyclopedia of Integer Sequences (2022). https://oeis.org/204089
13. OEIS Foundation Inc.: Number of finite, negative, totally ordered monoids of size n (semi-groups with a neutral element that is also the top element), entry A253950 in the On-Line Encyclopedia of Integer Sequences (2022). https://oeis.org/253950
14. OEIS Foundation Inc.: Number of lattices on n unlabeled nodes, entry A006966 in the On-Line Encyclopedia of Integer Sequences (2022). https://oeis.org/A006966
15. OEIS Foundation Inc.: Number of nonisomorphic semigroups of order n, entry A027851 in the On-Line Encyclopedia of Integer Sequences (2022). https://oeis.org/A027851
16. Rosenthal, K.I.: Quantales and their applications. Longman Scientific and Technical (1990)
17. Shamsgovara, A.: A catalogue of every quantale of order up to 9. In: 39th Linz Seminar on Fuzzy Set Theory, Linz, Austria. LINZ2022 (2022)
18. Shamsgovara, A.: Number of quantales on n elements, up to isomorphism, entry A354493 in the On-Line Encyclopedia of Integer Sequences (2022). https://oeis.org/A354493

19. Shamsgovara, A.: Quantales (2022). https://apps.apple.com/se/app/quantales/id1627990198
20. Shamsgovara, A., Eklund, P., Winter, M.: A catalogue of finite quantales. GLIOC Notes (2019)
21. Yetter, D.N.: Quantales and (noncommutative) linear logic. J. Symb. Logic **55**(1), 41–64 (1990)

Duoidally Enriched Freyd Categories

Chris Heunen[iD] and Jesse Sigal[✉][iD]

School of Informatics, University of Edinburgh, Edinburgh, UK
{chris.heunen,jesse.sigal}@ed.ac.uk

Abstract. Freyd categories provide a semantics for first-order effectful programming languages by capturing the two different orders of evaluation for products. We enrich Freyd categories in a duoidal category, which provides a new, third choice of parallel composition. Duoidal categories have two monoidal structures which account for the sequential and parallel compositions. The traditional setting is recovered as a full coreflective subcategory for a judicious choice of duoidal category. We give several worked examples of this uniform framework, including the parameterised state monad, basic separation semantics for resources, and interesting cases of change of enrichment.

Keywords: Freyd category · Duoidal category · Kleisli category · Lawvere theory · Monad

1 Introduction

Computational effects encapsulate interactions of a computer program with its environment in a modular way, and are a staple of modern programming languages [17]. Originally captured by strong monads [15], they have been extended to Arrows to deal with input as well as output [12], to Lawvere theories to better combine effects algebraically [20], to PROs and PROPs to deal with non-cartesian settings [13], and to Freyd categories for non-higher-order effects [14].

Freyd categories let one compose effectful computations both in sequence and, to some extent, in parallel, and reason about such compositions rigorously. For an effectful computation $f: a \to b$, we may embed it, the domain, and the codomain into a larger context by extending with $- \otimes c$ for any object c and monoidal-like operation $\otimes$, which we write as $f \otimes \mathrm{id}: a \otimes c \to b \otimes c$. Intuitively, $f \otimes \mathrm{id}$ does not interact with c. Effectful computations need not commute as they may alter the environment: $(f \otimes \mathrm{id}).(\mathrm{id} \otimes g) \neq (\mathrm{id} \otimes g).(f \otimes \mathrm{id})$ in general.

But what if we want to track more data about computations than just types and effects? For example, suppose we want to annotate every computation with its resource needs: there could $e.g.$ be a set R of resources, and every computation f requires a certain subset $P \subseteq R$ of resources for it to execute. Sequencing two computations needs all resources to execute both, so if $f: a \to b$ and $g: b \to c$

J. Sigal—Partly funded by Huawei.

require resources P and Q respectively, then $g.f$ requires $P \cup Q$. The same is true for parallel composition: if $f_1 : a_1 \to b_1$ and $f_2 : a_2 \to b_2$ require P_1 and P_2 respectively, then $f_1 \otimes f_2 : a_1 \otimes a_2 \to b_1 \otimes b_2$ requires $P_1 \cup P_2$. However, it is often desirable to restrict P_1 and P_2 by requiring $P_1 \cap P_2 = \emptyset$ so that morphisms composed in parallel use different resources. If we have an identity map $\mathrm{id} : a \to a$ for all a which requires $\emptyset \subseteq R$, then we can always form $f \otimes \mathrm{id}$ for any f, but what of the general case?

This article proposes a solution that achieves just this: enrich Freyd categories in duoidal categories. Duoidal categories carry two interacting monoidal structures that will account for the sequential and parallel composition of both the effectful computations and the extra data we want to track, such as the resources above. We provide a concrete example for resources in Sect. 3.1.

Section 2 introduces duoidally enriched Freyd categories. Section 3 shows the breadth of such categories by treating disparate examples: separation semantics for resources as above, indexed state monads, and Kleisli categories of Lawvere theories. Section 4 shows that a judicious choice of duoidal enriching category recovers traditional Freyd categories as a full coreflective subcategory, and Sect. 5 gives an abstract characterisation of duoidally enriched Freyd categories in purely algebraic terms. Section 6 considers changing the enriching duoidal category, accounting for *e.g.* changing the underlying permission model in the example above. Section 7 concludes and suggests directions for future work. Due to space constraints, please see the extended preprint on arXiv for appendices.

Related Work. Morrison and Penneys define a $\mathbf{V}$-monoidal category [16] for braided monoidal $\mathbf{V}$ as a $\mathbf{V}$-category with parallel composition that interacts well with the braid. In the case $\mathbf{V}$ is braided (and thus duoidal), our definition of a $\mathbf{V}$-Freyd category is similar. However, we also require bifunctorality of the hom objects, an important difference for some of our constructions.

The abstract characterisation in Sect. 5 is inspired by Fujii's characterisation of PROs and PROPs [7] as monoids in $\mathbf{MonCat}_{\mathrm{lax}}(\mathbf{N}^{\mathrm{op}} \times \mathbf{N}, \mathbf{Set})$ and $\mathbf{MonCat}_{\mathrm{lax}}(\mathbf{P}^{\mathrm{op}} \times \mathbf{P}, \mathbf{Set})$ respectively, where $\mathbf{N}$ and $\mathbf{P}$ have natural numbers as objects and equalities respectively bijections as morphisms.

Garner and López Franco describe a general framework for commutativity using categories enriched in the sequential product of a duoidal category [8]. Their framework requires the duoidal category to be *normal*, meaning that the two units are isomorphic. Only with this requirement and others do they define a monoidal structure on their category of enriched categories, and do not define a monoidal enriched category. We do not require normality.

Finally, Forcey [6], and Batanin and Markl [4] enrich over duoidal categories, but using the parallel product instead. We choose to enrich over the sequential product in order to define examples in which this is the appropriate choice.

2   Duoidally Enriched Freyd Categories

This section introduces duoidally enriched Freyd categories (in Sect. 2.3), but first we discuss Freyd categories (in Sect. 2.1) and duoidal categories (in Sect. 2.2).

2.1 Freyd Categories

Freyd categories provide semantics for first-order call-by-value programming languages with effects [20]. We will generalise the definition of a Freyd category slightly so that the effect free fragment need not have products, beginning with the following preliminary definitions [14,18].

Definition 1. *A category* $\mathbf{C}$ *is* binoidal *when it comes with endofunctors* $(-) \ltimes x$ *and* $x \rtimes (-)$ *for each object* x *such that* $x \ltimes y = x \rtimes y$ *for all* y; *write* $x \otimes y$ *for this object. A morphism* $f: x \to y$ *is* central *if for any morphism* $g: x' \to y'$ *the two maps* $(y \rtimes g).(f \ltimes x')$ *and* $(f \ltimes y').(x \rtimes g)$ *of type* $x \otimes x' \to y \otimes y'$ *are equal, as are the two maps* $(y' \rtimes f).(g \ltimes x)$ *and* $(g \ltimes y).(x' \rtimes f)$ *of type* $x' \otimes x \to y' \otimes y$. *Central morphisms form a wide subcategory* $Z(\mathbf{C})$ *called the* centre.

Definition 2. *A* binoidal *category* $\mathbf{C}$ *is* premonoidal *when equipped with an object* e *and families of central isomorphisms* $\alpha: (x \otimes y) \otimes z \to x \otimes (y \otimes z)$, $\lambda: e \otimes x \to x$, *and* $\rho: x \otimes e \to x$ *that are natural in each component and satisfy triangle and pentagon equations.*

Definition 3. *A functor* $F: \mathbf{C} \to \mathbf{D}$ *between premonoidal categories is a* premonoidal functor *when equipped with central morphisms* $\eta: e_{\mathbf{D}} \to F(e_{\mathbf{C}})$ *and* $\mu: F(x) \otimes_{\mathbf{D}} F(y) \to F(x \otimes_{\mathbf{C}} y)$ *such that* μ *is natural in each component, and the following diagrams commute:*

$$(F(x) \otimes_{\mathbf{D}} F(y)) \otimes_{\mathbf{D}} F(z) \xrightarrow{\alpha_{\mathbf{D}}} F(x) \otimes_{\mathbf{D}} (F(y) \otimes_{\mathbf{D}} F(z))$$

$$\mu \otimes \mathrm{id} \downarrow \qquad\qquad\qquad \downarrow \mathrm{id} \otimes \mu$$

$$F(x \otimes_{\mathbf{C}} y) \otimes_{\mathbf{D}} F(z) \qquad\qquad F(x) \otimes_{\mathbf{D}} F(y \otimes_{\mathbf{C}} z)$$

$$\mu \downarrow \qquad\qquad\qquad \downarrow \mu$$

$$F((x \otimes_{\mathbf{C}} y) \otimes_{\mathbf{C}} z) \xrightarrow{F\alpha_{\mathbf{C}}} F(x \otimes_{\mathbf{C}} (y \otimes_{\mathbf{C}} z))$$

$$e_{\mathbf{D}} \otimes_{\mathbf{D}} F(x) \xrightarrow{\eta \otimes \mathrm{id}} F(e_{\mathbf{C}}) \otimes_{\mathbf{D}} F(x)$$
$$\lambda_{\mathbf{D}} \downarrow \qquad\qquad \downarrow \mu$$
$$F(x) \xleftarrow{F\lambda_{\mathbf{C}}} F(e_{\mathbf{C}} \otimes_{\mathbf{C}} x)$$

$$F(x) \otimes_{\mathbf{D}} e_{\mathbf{D}} \xrightarrow{\mathrm{id} \otimes \eta} F(x) \otimes_{\mathbf{D}} F(e_{\mathbf{C}})$$
$$\rho_{\mathbf{D}} \downarrow \qquad\qquad \downarrow \mu$$
$$F(x) \xleftarrow{F\rho_{\mathbf{C}}} F(x \otimes_{\mathbf{C}} e_{\mathbf{C}})$$

A premonoidal functor is strong (strict) *when* η *and* μ *are isomorphisms (identities).*

Note that a strict premonoidal functor F preserves associators and unitors on the nose. Recall that a functor $F: \mathbf{C} \to \mathbf{D}$ between monoidal categories is *lax monoidal* when it comes with a morphism $\eta: I \to F(I)$ and a natural transformation $\mu: F(X) \otimes F(Y) \to F(X \otimes Y)$ satisfying coherence conditions. It is *strong monoidal* when η and μ are invertible. Lax/strong monoidal functors are closed under composition. Here now is our definition of a Freyd category.

Definition 4. *A* Freyd category *consists of a monoidal category* $\mathbf{M}$ *and a premonoidal category* $\mathbf{C}$ *with the same objects, and an identity-on-objects strict premonoidal functor* $J: \mathbf{M} \to \mathbf{C}$ *whose image lies in* $Z(\mathbf{C})$. *A morphism* $J \to J'$ *of Freyd categories consists of a strong monoidal functor* $F_0: \mathbf{M} \to \mathbf{M}'$ *and a strong premonoidal functor* $F_1: \mathbf{C} \to \mathbf{C}'$ *such that* $F_1 J = J' F_0$. *Freyd categories and their morphisms form a category* **Freyd**.

2.2 Duoidal Categories

A duoidal category carries two interacting monoidal structures, that one may intuitively think of as sequential and parallel composition, but let us give the definition [2, Definition 6.1] before examples.

Definition 5. *A category* $\mathbf{V}$ *is* duoidal *when it comes with two monoidal structures* $(\mathbf{V}, *, J)$ *and* $(\mathbf{V}, \circ, I)$, *a natural transformation* $\zeta_{A,B,C,D} \colon (A \circ B) * (C \circ D) \to (A * C) \circ (B * D)$, *and three morphisms* $\Delta \colon J \to J \circ J$, $\nabla \colon I * I \to I$, *and* $\epsilon \colon J \to I$ *such that* (I, ∇, ϵ) *is a monoid in* $(\mathbf{V}, *, J)$ *and* (J, Δ, ϵ) *is a comonoid in* $(\mathbf{V}, \circ, I)$, *and the following diagrams commute:*

$$((A \circ B) * (C \circ D)) * (E \circ F) \xrightarrow{\alpha} (A \circ B) * ((C \circ D) * (E \circ F))$$

$$\Big\downarrow \zeta * \mathrm{id} \qquad\qquad\qquad\qquad \Big\downarrow \mathrm{id} * \zeta$$

$$((A * C) \circ (B * D)) * (E \circ F) \qquad (A \circ B) * ((C * E) \circ (D * F))$$

$$\Big\downarrow \zeta \qquad\qquad\qquad\qquad\qquad \Big\downarrow \zeta$$

$$((A * C) * E) \circ ((B * D) * F) \xrightarrow{\alpha \circ \alpha} (A * (C * E)) \circ (B * (D * F))$$

$$J * (A \circ B) \xrightarrow{\Delta * \mathrm{id}} (J \circ J) * (A \circ B)$$
$$\Big\downarrow \lambda \qquad\qquad\qquad \Big\downarrow \zeta$$
$$A \circ B \xleftarrow{\lambda \circ \lambda} (J * A) \circ (J * B)$$

$$(A \circ B) * J \xrightarrow{\mathrm{id} * \Delta} (A \circ B) * (J \circ J)$$
$$\Big\downarrow \rho \qquad\qquad\qquad \Big\downarrow \zeta$$
$$A \circ B \xleftarrow{\rho \circ \rho} (A * J) \circ (B * J)$$

$$((A \circ B) \circ C) * ((D \circ E) \circ F) \xrightarrow{\alpha * \alpha} (A \circ (B \circ C)) * (D \circ (E \circ F))$$

$$\Big\downarrow \zeta \qquad\qquad\qquad\qquad\qquad \Big\downarrow \zeta$$

$$((A \circ B) * (D \circ E)) \circ (C * F) \qquad (A * D) \circ ((B \circ C) * (E \circ F))$$

$$\Big\downarrow \zeta \circ \mathrm{id} \qquad\qquad\qquad\qquad \Big\downarrow \mathrm{id} \circ \zeta$$

$$((A * D) \circ (B * E)) \circ (C * F) \xrightarrow{\alpha} (A * D) \circ ((B * E) \circ (C * F))$$

$$I \circ (A * B) \xleftarrow{\nabla \circ \mathrm{id}} (I * I) \circ (A * B)$$
$$\Big\downarrow \lambda \qquad\qquad\qquad \Big\uparrow \zeta$$
$$A * B \xleftarrow{\lambda * \lambda} (I \circ A) * (I \circ B)$$

$$(A * B) \circ I \xleftarrow{\mathrm{id} \circ \nabla} (A * B) \circ (I * I)$$
$$\Big\downarrow \rho \qquad\qquad\qquad \Big\uparrow \zeta$$
$$A * B \xleftarrow{\rho * \rho} (A \circ I) * (B \circ I)$$

We may write $(\mathbf{V}, *, J, \circ, I)$ *or* $(\mathbf{V}, *, \circ)$ *to be explicit about the role of each monoidal structure.*

Example 1. Any braided monoidal category becomes duoidal by letting both monoidal structures coincide and ζ be the middle-four interchange $x \otimes y \otimes z \otimes w \to x \otimes z \otimes y \otimes w$ up to associativity. In particular, any symmetric or cartesian monoidal category is duoidal [2, Proposition 6.10, Example 6.19].

Example 2. If $(\mathbf{V}, *, J, \circ, I)$ is duoidal, so is $(\mathbf{V}^{\mathrm{op}}, \circ, I, *, J)$, with opposite structure maps [2, Section 6.1.2].

Example 3. If $(\mathbf{V}, \otimes, I)$ is a monoidal category with products, $(\mathbf{V}, \otimes, I, \times, 1)$ is duoidal with $\zeta = \langle \pi_1 \otimes \pi_1, \pi_2 \otimes \pi_2 \rangle$, $\Delta = \langle \mathrm{id}, \mathrm{id} \rangle$, and ∇ and ϵ terminal maps. Similarly, if a monoidal category $\mathbf{V}$ has coproducts, $(\mathbf{V}, +, 0, \otimes, I)$ is duoidal [2, Example 6.19].

Example 4. If $(\mathbf{V}, *, J, \circ, I)$ is small and duoidal, straightforward calculation shows Day convolution [5] of each monoidal structure makes the category of presheaves $([\mathbf{V}^{\mathrm{op}}, \mathbf{Set}], *_{\mathrm{Day}}, \mathbf{V}(-, J), \circ_{\mathrm{Day}}, \mathbf{V}(-, I))$ again duoidal where

$$(F *_{\mathrm{Day}} G)(A) = \int^{B,C} \mathbf{V}(A, B * C) \times F(B) \times G(C)$$

and likewise for $\circ_{\mathrm{Day}}$. An analogous construction holds for $[\mathbf{V}, \mathbf{Set}]$ by starting with $\mathbf{V}^{\mathrm{op}}$.

Example 5. An endofunctor on **Set** is finitary when it preserves filtered colimits and is therefore determined on finite sets. Finitary endofunctors are closed under functor composition, $\circ$, with unit Id; closed under Day convolution with products, $\times_{\mathrm{Day}}$, with unit $\mathbf{Set}(1, -) \cong \mathrm{Id}$; making $([\mathbf{Set}, \mathbf{Set}]_f, \times_{\mathrm{Day}}, \mathrm{Id}, \circ, \mathrm{Id})$ a duoidal category. [8]

Example 6. For a small monoidal category $(\mathbf{M}, \oplus, e)$, the category of **Set**-valued endoprofunctors $\mathbf{Prof}(\mathbf{M}) := [\mathbf{M}^{\mathrm{op}} \times \mathbf{M}, \mathbf{Set}]$ is duoidal $(\mathbf{Prof}(\mathbf{M}), \oplus_{\mathrm{Day}}, \diamond)$ with profunctor composition $(P \diamond Q)(a, c) := \int^b P(a, b) \times Q(b, c)$ (having unit $\mathbf{M}(-, -)$) and Day convolution of $\oplus$ on both sides $(P \oplus_{\mathrm{Day}} Q)(a, b) := \int^{a_1, a_2, b_2, b_2} \mathbf{M}(a, a_1 \oplus a_2) \times \mathbf{M}(b_1 \oplus b_2, b) \times P(a_1, b_1) \times Q(a_2, b_2)$ (having unit $\mathbf{M}(-, e) \times \mathbf{M}(e, -)$). [8]

Example 7. An important example for us is the category **Subset** of *distinguished subsets*. Objects are pairs of sets (X, A) such that $X \subseteq A$ and morphisms $f \colon (X, A) \to (Y, B)$ are functions $f \colon A \to B$ with $f(X) \subseteq Y$. We call X the *distinguished subset*. Composition and identities are as in **Set**. We may suppress the distinguished subset X by writing $a \notin A$ when $a \in X$. Next, we give two monoidal structures on **Subset**.

The first is the cartesian product: $(X, A) \times (Y, B) := (X \times Y, A \times B)$ on objects, and $f \times g$ as in **Set** on morphisms, with unit $(1, 1)$. Associators and unitors are as in **Set**. This is also a categorical product.

The second is the *disjunctive product*: on objects $(X, A) \otimes (Y, B)$ is defined as $(X \times Y, (A \times Y) \cup (X \times B))$ with unit $(1, 1)$. We again have $f \times g$ on morphisms, which is well-defined. Finally, the coherence maps are restricted versions of those for the cartesian product.

Now $(\mathbf{Subset}, \otimes, (1, 1), \times, (1, 1))$ is duoidal by Example 3: Δ and ∇ are unitors, ϵ is the identity, and $\zeta \colon ((X, A) \times (Y, B)) \otimes ((Z, C) \times (W, D)) \to ((X, A) \otimes (Z, C)) \times ((Y, B) \otimes (W, D))$ is the restricted middle-four interchange; all axioms are inherited from $(\mathbf{Set}, \times, 1)$ via Example 1.

The important difference between $(\mathbf{Subset}, \otimes, \times)$ and $(\mathbf{Set}, \times, \times)$ is that ζ is not invertible in the former (as it is not surjective as a **Set** map). This allows Freyd categories enriched in **Subset** a premonoidal-like structure.

2.3   Concrete Definition

We are now ready for the titular notion of this paper. We first give a concrete definition, leaving an abstract characterisation to Sect. 5.

Definition 6. *Let* $(\mathbf{V}, *, J, \circ, I)$ *be a duoidal category and* $(\mathbf{M}, \oplus, e)$ *a monoidal category. A* $\mathbf{V}$*-Freyd category over* $\mathbf{M}$ *consists of*

- *a bifunctor* $\mathbf{C} \colon \mathbf{M}^{\mathrm{op}} \times \mathbf{M} \to \mathbf{V}$
- *an extranatural family* $\mathsf{idt} \colon I \to \mathbf{C}(a,a)$, *meaning* $\mathbf{C}(\mathrm{id}, f).\mathsf{idt} = \mathbf{C}(f, \mathrm{id}).\mathsf{idt}$
- *an extranatural family* $\mathsf{seq} \colon \mathbf{C}(a,b) \circ \mathbf{C}(b,c) \to \mathbf{C}(a,c)$, *meaning* seq *is natural in* a *and* c, *and* $\mathsf{seq}.(\mathrm{id} \circ \mathbf{C}(f, \mathrm{id})) = \mathsf{seq}.(\mathbf{C}(\mathrm{id}, f) \circ \mathrm{id})$
- *a morphism* $\mathsf{zero} \colon J \to \mathbf{C}(e,e)$
- *a natural family* $\mathsf{par} \colon \mathbf{C}(a_1, b_1) * \mathbf{C}(a_2, b_2) \to \mathbf{C}(a_1 \oplus a_2, b_1 \oplus b_2)$

satisfying the following axioms:

- *(i)* idt *is the identity for* seq, *that is,* $\mathsf{seq}.(\mathsf{idt} \circ \mathrm{id}) = \lambda$ *and symmetrically;*
- *(ii)* seq *is associative, that is,* $\mathsf{seq}.(\mathsf{seq} \circ \mathrm{id}) = \mathsf{seq}.(\mathrm{id} \circ \mathsf{seq}).\alpha;$
- *(iii)* zero *is the identity for* par, *that is,* $\mathbf{C}(\lambda^{-1}, \lambda).\mathsf{par}.(\mathsf{zero} * \mathrm{id}) = \lambda$ *and symmetrically;*
- *(iv)* par *is associative, that is,* $\mathbf{C}(\alpha^{-1}, \alpha).\mathsf{par}.(\mathsf{par} * \mathrm{id}) = \mathsf{par}.(\mathrm{id} * \mathsf{par}).\alpha;$
- *(v)* idt *respects* zero *via* $\mathsf{idt}.\epsilon = \mathsf{zero};$
- *(vi)* idt *respects* par *via* $\mathsf{idt}.\nabla = \mathsf{par}.(\mathsf{idt} * \mathsf{idt});$
- *(vii)* seq *respects* zero *via* $\mathsf{seq}.(\mathsf{zero} \circ \mathsf{zero}).\Delta = \mathsf{zero};$
- *(viii)* seq *respects* par *via* $\mathsf{seq}.(\mathsf{par} \circ \mathsf{par}).\zeta = \mathsf{par}.(\mathsf{seq} * \mathsf{seq}).$

See Appendix A of the extended preprint for diagrams expressing the axioms.

Definition 7. *A morphism of* $\mathbf{V}$*-Freyd categories consists of a strong monoidal functor* $F_0 \colon \mathbf{M} \to \mathbf{M}'$ *and a natural transformation* $F_1 \colon \mathbf{C}(a,b) \to \mathbf{C}'(F_0 a, F_0 b)$ *satisfying:*

- $F_1.\mathsf{idt} = \mathsf{idt}';$
- $F_1.\mathsf{seq} = \mathsf{seq}'.(F_1 \circ F_1);$
- $\mathbf{C}'(\mathrm{id}, \mu).\mathsf{par}'.(F_1 * F_1) = \mathbf{C}'(\mu, \mathrm{id}).F_1.\mathsf{par}.$

$\mathbf{V}$*-Freyd categories and morphisms between them form a category* $\mathbf{V}$**-Freyd***.*

Our definition differs from the duoidally enriched categories of Batanin and Markl [4] in a few important ways. They use $*$ for sequencing and $\circ$ for parallel composition. Their analogues to axioms v to viii are $\mathsf{idt} = \mathsf{zero}.\epsilon$, $\mathsf{idt} = \mathsf{par}.(\mathsf{idt} \circ \mathsf{idt}).\Delta$, $\mathsf{seq}.(\mathsf{zero} * \mathsf{zero}) = \mathsf{zero}.\nabla$, and $\mathsf{seq}.(\mathsf{par} * \mathsf{par}) = \mathsf{par}.(\mathsf{seq} \circ \mathsf{seq}).\zeta$. Additionally, their monoidal structure is more enriched while we inherit ours from a **Set**-category, namely $\mathbf{M}$. Thus, we believe both notions are not inter-expressible.

3   Examples

This section works out three applications of duoidally enriched Freyd categories: resource management (in Sect. 3.1), indexed state (in Sect. 3.2), and Kleisli categories of Lawvere theories (in Sect. 3.3).

3.1   Stateful Functions and Separated Monoids

To deal with resources abstractly, we first introduce the novel notion of a separated monoid.

Definition 8. *A monoid $(M, \bullet, e)$ is separated when it comes with a binary relation $\|$ such that: $e\|m$ and $m\|e$; and $mm'\|n$ iff $m\|n$ and $m'\|n$; and $m\|nn'$ iff $m\|n$ and $m\|n'$.*

Examples include $(\mathbb{N}, +, 0)$ with $x\|y$ iff $x = 0$ or $y = 0$; finite subsets $(\mathcal{P}_f(R), \cup, \emptyset)$ of a fixed set R, with $P\|Q$ iff $P \cap Q = \emptyset$; and products of separated monoids under pointwise separation. Separated monoids parametrise duoidal categories of resources as follows.

Definition 9. *Let $(M, \|)$ be a separated monoid. The category $\mathbf{Label}_M$ of M-labelled sets has as objects functions $\ell\colon A \to M$ and as morphisms functions $f\colon A \to A'$ with $\ell' f = \ell$. This category has a monoidal structure $\bullet$ as follows: on objects, $\ell \bullet \ell'\colon A \times A' \to M$ sends (a, a') to $\ell(a) \bullet \ell'(a')$; on morphisms, $f \bullet f' = f \times f'$; the unit $\mathrm{cst}_e\colon 1 \to M$ picks out $e \in M$. There is a second monoidal structure $\|$ as follows: on objects, $\ell\|\ell'$ is the restriction of $\ell \bullet \ell'$ to $\{(a, a') \mid \ell(a)\|\ell'(a')\}$; on morphisms, $f\|f' = f \times f'$. The category $(\mathbf{Label}_M, \|, \mathrm{cst}_e, \bullet, \mathrm{cst}_e)$ is duoidal with $\zeta\colon (\ell_1 \bullet \ell'_1) \| (\ell_2 \bullet \ell'_2) \to (\ell_1\|\ell_2) \bullet (\ell'_1\|\ell'_2)$ the restricted version of the ζ for $(\mathbf{Set}, \times, 1, \times, 1)$.*

Think of objects in $\mathbf{Label}_M$ as sets of elements labelled with their resource needs. The multiplication of M combines resources, and the separation $\|$ relates non-conflicting resources. We will now describe an enriched Freyd category where morphisms are labelled by resources as in the introduction.

Fix a countable family $R = \{x, y, z, \ldots\}$ of sets which we think of as resources. The set $\mathcal{P}_f(R)$ of finite subsets of R is a monoid under union, and becomes a separated monoid under disjointness. For set of resources $Q \in \mathcal{P}_f(R)$, fix a product of sets $\Pi_{x \in Q} x =: \Pi_Q$ which thus combines the resources in Q. Write $\pi_{Q'}\colon \Pi_Q \to \Pi_{Q'}$ for the projection if $Q' \subseteq Q$, and given a map $f\colon a \times \Pi_{Q'} \to b \times \Pi_{Q'}$ for sets a and b, write $f_{Q'}^Q$ for the map $a \times \Pi_Q \to b \times \Pi_Q$ induced by f when $Q' \subseteq Q$ which leaves the extra resources $Q \setminus Q'$ unchanged.

We will define a $\mathbf{Label}_{\mathcal{P}_f(R)}$-Freyd category over $\mathbf{Set}$ of state-transforming functions. Let $\mathbf{C}(a, b)$ be the function from the disjoint union of $\mathbf{Set}(a \times \Pi_Q, b \times \Pi_Q)$ over $Q \in \mathcal{P}_f(R)$ to $\mathcal{P}_f(R)$, that sends $f\colon a \times \Pi_Q \to b \times \Pi_Q$ to Q. Thus, a map $f \in \mathbf{C}(a, b)$ with label Q is an effectful computation from a to b which can effect only resources in Q. This becomes a bifunctor under pre- and post-composition. Writing $\cup$ for $\bullet$ and $\cap$ for $\|$ for the sake of concreteness, the structure maps are:

$$\mathrm{idt}\colon \mathrm{cst}_\emptyset \to \mathbf{C}(a, a) \qquad\qquad \mathrm{zero}\colon \mathrm{cst}_\emptyset \to \mathbf{C}(1, 1)$$
$$\star \mapsto (\emptyset, \mathrm{id}_{a \times 1}) \qquad\qquad\qquad \star \mapsto (\emptyset, \mathrm{id})$$

$$\mathrm{seq}\colon \mathbf{C}(a, b) \cup \mathbf{C}(b, c) \to \mathbf{C}(a, c)$$
$$((P, f), (Q, g)) \mapsto \left(P \cup Q, g_Q^{P \cup Q} . f_P^{P \cup Q}\right)$$

par: $\mathbf{C}(a,b) \cap \mathbf{C}(a',b') \to \mathbf{C}(a \times a', b \times b')$

$\quad ((Q,f),(Q',f')) \mapsto$

$$\left(Q \cup Q', \left(\mathrm{id} \times \langle \pi_Q, \pi_{Q'} \rangle^{-1} \right) m^{-1}.(f \times f').m.\left(\mathrm{id} \times \langle \pi_Q, \pi_{Q'} \rangle \right) \right)$$

where $\langle \pi_Q, \pi_{Q'} \rangle : \Pi_{Q \cup Q'} \to \Pi_Q \times \Pi_{Q'}$ is invertible because $Q \cap Q' = \emptyset$ and m is middle-four interchange. So par places maps in parallel up to rearranging state.

3.2   Indexed State

An important computational effect is global state. However, it is often inflexible as the type of storage remains constant over time. In this example the type can vary. We use the duoidal category of finitary endofunctors on **Set** of Example 5 to give a $[\mathbf{Set}, \mathbf{Set}]_f$-Freyd category over **Set** based on the state monad $(s \times (-))^s$, extending Atkey's example [3]. Define $\mathbf{C}(a,b) = (b \times (-))^a$, which is a bifunctor via pre- and post-composition. The natural structure maps are:

$$\mathrm{idt}_X : X \to (a \times X)^a \qquad\qquad \mathrm{zero}_X : X \to (1 \times X)^1$$
$$x \mapsto \lambda a.(x,a) \qquad\qquad\qquad x \mapsto \lambda \star .(x, \star)$$

$$\mathrm{seq}_X : \left(b \times \left((c \times X)^b \right) \right)^a \to (c \times X)^a$$
$$f \mapsto \mathrm{eval}.f$$

$$\mathrm{par}_X : \int^{Y,Z} X^{Y \times Z} \times (b \times Y)^a \times (c \times Z)^{a'} \to ((b \times c) \times X)^{a \times a'}$$
$$(k,f,g) \mapsto (\mathrm{id} \times k).m.(f \times g)$$

where $\mathrm{eval} : b \times (c \times X)^b \to c \times X$ is the evaluation map and m is the middle-four interchange. idt and seq are the unit and multiplication of a state monad but with varying types of state.

3.3   Kleisli Categories of Lawvere Theories

Lawvere theories model effectful computations. Functional programmers might be more familiar with Kleisli categories of monads, to which they are closely related. Here we describe an indexed version, which models independent effects in parallel. Let **Law** be the category of Lawvere theories. Its initial object is the theory $\mathcal{S}$ of sets, the unit for the *tensor product* $\otimes$ of Lawvere theories [10]. This makes **Law** a symmetric monoidal category, with the special property that there exist inclusion maps $\phi_i : \mathcal{L}_i \to \mathcal{L}_1 \otimes \mathcal{L}_2$. Thus the functor category $[\mathbf{Law}, \mathbf{Set}]$ is monoidal under Day convolution with unit the constant functor $\mathbf{Law}(\mathcal{S}, -) \simeq \mathbb{1}$. As this category also has products, Example 3 makes it duoidal.

Now, **Law** is equivalent to the category of finitary monads [1, Chapter 3]: any Lawvere theory $\mathcal{L}$ induces a monad $T(\mathcal{L})$, and any map θ of Lawvere theories induces a monad morphism $T(\theta)$. Every monad T on **Set** is canonically bistrong: there are maps $\mathrm{st}_T : a \times Tb \to T(a \times b)$ and $\mathrm{st}'_T : Ta \times b \to T(a \times b)$ making the two induced maps $(a \times Tb) \times c \to T((a \times b) \times c)$ equal. Each monad morphism $T(\theta)$ preserves strength: $T(\theta)_{a \times b}.\mathrm{st}_{T(\mathcal{L})} = \mathrm{st}_{T(\mathcal{L}')}.(\mathrm{id} \times T(\theta)_b)$.

We now show a $[\mathbf{Law}, \mathbf{Set}]$-Freyd category over $\mathbf{Set}$ given by the Kleisli construction on Lawvere theories. Define on objects $\mathbf{C}(a,b) = T(-)(b)^a$, and on morphisms $\mathbf{C}(f,g)\colon \mathbf{C}(a,b) \Rightarrow \mathbf{C}(a',b')$ by $\mathbf{C}(f,g)_{\mathcal{L}}(k) = T(\mathcal{L})(g).k.f$, finally:

$$\mathrm{idt}_{\mathcal{L}}\colon 1 \to T(\mathcal{L})(a)^a \qquad\qquad \mathrm{zero}_{\mathcal{L}}\colon 1 \to T(\mathcal{L})(1)^1$$
$$\star \mapsto \eta \qquad\qquad\qquad\qquad\qquad \star \mapsto \eta$$

$$\mathrm{seq}_{\mathcal{L}}\colon T(\mathcal{L})(b)^a \times T(\mathcal{L})(c)^b \to T(\mathcal{L})(c)^a$$
$$(f,g) \mapsto \mu.T(\mathcal{L})g.f$$

$$\mathrm{par}_{\mathcal{L}}\colon \int^{\mathcal{L}_1,\mathcal{L}_2} \mathbf{Law}(\mathcal{L}_1 \otimes \mathcal{L}_2, \mathcal{L}) \times T(\mathcal{L}_1)(b_1)^{a_1} \times T(\mathcal{L}_2)(b_2)^{a_2} \to T(\mathcal{L})(b_1 \times b_2)^{a_1 \times a_2}$$
$$(\theta, f_1, f_2) \mapsto T(\theta).\mu.T(\mathcal{L}_1 \otimes \mathcal{L}_2)(\mathrm{st}').\mathrm{st}.\, (T(\phi_1) \times T(\phi_2)) \cdot (f_1 \times f_2)$$

Intuitively, par lets us put Kleisli maps in parallel as long as their effects are forced to commute (by $\otimes$). So $\mathrm{idt}_{\mathcal{L}}$ and $\mathrm{seq}_{\mathcal{L}}$ are the identity and composition for the Kleisli category of $T(\mathcal{L})$. The definition of $\mathrm{par}_{\mathcal{L}}$ seems noncanonical because of the use of $T(\mathcal{L}_1 \otimes \mathcal{L}_2)(\mathrm{st}').\mathrm{st}$, but it is not: $\mu.T(\mathcal{L}_1 \otimes \mathcal{L}_2)(\mathrm{st}').\mathrm{st}.\, (T(\phi_1) \times T(\phi_2))$ and $\mu.T(\mathcal{L}_1 \otimes \mathcal{L}_2)(\mathrm{st}).\mathrm{st}'.\, (T(\phi_1) \times T(\phi_2))$ are equal by definition of $\mathcal{L}_1 \otimes \mathcal{L}_2$.

4   Adjunction Between Subset-Freyd and Freyd

Now let us explain how $\mathbf{V}$-Freyd categories generalise Freyd categories. Our approach is similar to Power's [19] in that we work with $\mathbf{Subset}$-enriched categories. Take $\mathbf{V} = \mathbf{Subset}$ and consider a $\mathbf{Subset}$-Freyd category $\mathbf{C}\colon \mathbf{M}^{\mathrm{op}} \times \mathbf{M} \to \mathbf{Subset}$; it comes equipped with a premonoidal-like structure via par and idt. We call a morphism $f \in \mathbf{C}(a,b)$ which is a member of the distinguished subset a *distinguished morphism*. We will show they are central in the premonoidal sense.

First observe that $\mathrm{idt}\colon (1,1) \to \mathbf{C}(a,a)$ is a $\mathbf{Subset}$ morphism, so $\mathrm{idt}(\star) \in \mathbf{C}(a,a)$ is distinguished. Thus, for $g \in \mathbf{C}(a',b')$ we find $(\mathrm{idt}(\star),g) \in \mathbf{C}(a,a) \otimes \mathbf{C}(a',b')$ by definition of $\otimes$. Hence the pair is in the domain of par, giving $\mathrm{par}(\mathrm{idt}(\star),g) \in \mathbf{C}(a \oplus a', a \oplus b')$ which we denote by $a \ltimes_{\mathrm{par}} g$. Similarly, for any $f \in \mathbf{C}(a,b)$ we have $f \ltimes_{\mathrm{par}} b' \in \mathbf{C}(a \oplus b', b \oplus b')$. We may also construct $f \ltimes_{\mathrm{par}} a'$ and $b \ltimes_{\mathrm{par}} g$. Hence it makes sense to ask if $\mathrm{seq}(a \ltimes_{\mathrm{par}} g, f \ltimes_{\mathrm{par}} b') = \mathrm{seq}(f \ltimes_{\mathrm{par}} a', b \ltimes_{\mathrm{par}} g)$, and if this equation (and its mirrored version by placing g on the left) holds for all f, we call g *central* in analogy to the binoidal case from Definition 1.

Next we claim that distinguished morphisms $g \in \mathbf{C}(a',b')$ are central. Note that $(\mathrm{idt}(\star),g) \in \mathbf{C}(a',a') \times \mathbf{C}(a',b')$ and $(g,\mathrm{idt}(\star)) \in \mathbf{C}(a',b') \times \mathbf{C}(b',b')$ are distinguished and in the domain of seq. For any $f \in \mathbf{C}(a,b)$, we have $((\mathrm{idt}(\star),f),(g,\mathrm{idt}(\star))) \in (\mathbf{C}(a,a) \times \mathbf{C}(a,b)) \otimes (\mathbf{C}(a',b') \times \mathbf{C}(b',b'))$ and similarly $((f,\mathrm{idt}(\star)),(\mathrm{idt}(\star)),g) \in (\mathbf{C}(a,b) \times \mathbf{C}(b,b)) \otimes (\mathbf{C}(a',a') \times \mathbf{C}(a',b'))$ by definition of $\otimes$ and are thus in the domain of $\mathrm{seq} \otimes \mathrm{seq}$. We now apply $\mathrm{par}.(\mathrm{seq} \otimes \mathrm{seq})$ to each pair and find they equal $\mathrm{par}(f,g)$. Axiom viii states $\mathrm{par}.(\mathrm{seq} \otimes \mathrm{seq}) = \mathrm{seq}.(\mathrm{par} \times \mathrm{par}).\zeta$ and therefore $\mathrm{seq}(a \ltimes_{\mathrm{par}} g, f \ltimes_{\mathrm{par}} b') = \mathrm{par}(f,g) = \mathrm{seq}(f \ltimes_{\mathrm{par}} a', b \ltimes_{\mathrm{par}} g)$ (and the mirrored equation analogously), so g is central.

Distinguished morphisms have their centrality preserved by **Subset**-Freyd maps as they are mapped to distinguished morphisms, but central morphisms need not be distinguished. Thus, Definition 7 ensures that membership in the distinguished subset is preserved by **Subset**-Freyd maps, so centrality of distinguished morphisms of $\mathbf{C}$ is preserved by all maps. Furthermore, bifunctorality of $\mathbf{C}$ ensures that for all $f \in \mathbf{M}(a, b)$, $\mathbf{C}(\mathrm{id}, f)(\mathrm{idt}(\star)) \in \mathbf{C}(a, b)$, and so the image of $\mathbf{M}$ is central and this centrality is preserved. The same is true for a Freyd category $J \colon \mathbf{M} \to \mathbf{C}$, the image of $\mathbf{M}$ under J is central and this centrality is preserved by all morphisms of Freyd categories. This preservation requirement is the difference between Freyd categories and **Subset**-Freyd categories: the latter can require more central morphisms than the image of $\mathbf{M}$ to have centrality preserved. The rest of this subsection proves that there is an adjunction between **Freyd** and **Subset**-Freyd. The left adjoint $\mathfrak{F} \colon$ **Freyd** $\to$ **Subset**-Freyd is a free functor that only requires the image of $\mathbf{M}$ to be preserved. The right adjoint $\mathfrak{U} \colon$ **Subset**-Freyd $\to$ **Freyd** forgets the extra distinguished central morphisms.

Proposition 1. *There is a functor* $\mathfrak{F} \colon$ **Freyd** $\to$ **Subset**-Freyd *defined on objects as* $\mathfrak{F}(\mathbf{C})(a, b) = \bigl(J(\mathbf{M}(a, b)), \mathbf{C}(a, b)\bigr)$ *and* $\mathfrak{F}(\mathbf{C})(f, g) = \mathbf{C}(Jf, Jg)$.

Proof (Proof sketch). $\mathfrak{F}(\mathbf{C})$ is well-defined on morphisms because J is identity-on-objects, and it is bifunctorial by bifunctorality of hom and functorality of J. The structure maps are:

- $\mathsf{idt} \colon (1, 1) \to \mathfrak{F}(\mathbf{C})(a, a)$ is $\star \mapsto \mathrm{id}$;
- $\mathsf{seq} \colon \mathfrak{F}(\mathbf{C})(a, b) \times \mathfrak{F}(\mathbf{C})(b, c) \to \mathfrak{F}(\mathbf{C})(a, c)$ is $(f, g) \mapsto g.f$;
- $\mathsf{zero} \colon (1, 1) \to \mathfrak{F}(\mathbf{C})(e, e)$ is $\star \mapsto \mathrm{id}$;
- $\mathsf{par} \colon \mathfrak{F}(\mathbf{C})(a_1, b_1) \otimes \mathfrak{F}(\mathbf{C})(a_2, b_2) \to \mathfrak{F}(\mathbf{C})(a_1 \oplus a_2, b_1 \oplus b_2)$ is $(f_1, f_2) \mapsto f_1 \otimes f_2$; this is well-defined whether (f_1, f_2) is in $J(\mathbf{M}(a_1, b_1)) \times \mathbf{C}(a_2, b_2)$ or is in $\mathbf{C}(a_1, b_1) \times J(\mathbf{M}(a_2, b_2))$ as J preserves centrality of $\mathbf{M} = Z(\mathbf{M})$.

The (extra)naturality of the structure maps comes from the extranaturality of composition, functorality of $\mathbf{M}$'s monoidal product, and J being a strict premonoidal functor preserving centrality. Axioms i and ii are true by $\mathbf{C}$'s composition, axioms iii and iv follow from the strict premonoidality of J and the naturality of unitors and associators, and axioms v to vii are trivial. Finally, axioms vi to viii follow from $\mathbf{C}$'s premonoidal structure.

Finally, it is easy to check that $\mathfrak{F}(F) = F$ is well-defined and functorial.

Proposition 2. *There is a functor* $\mathfrak{U} \colon$ **Subset**-Freyd $\to$ **Freyd** *that sends an object* $\mathbf{C} \colon \mathbf{M}^{\mathrm{op}} \times \mathbf{M} \to$ **Subset** *to the functor* $J \colon \mathbf{M} \to \mathfrak{U}(\mathbf{C})$ *defined as follows:*

- *the category* $\mathfrak{U}(\mathbf{C})$ *has the same objects as* $\mathbf{M}$ *but homsets* $\mathfrak{U}(\mathbf{C})(a, b) = A$ *where* $(X, A) := \mathbf{C}(a, b)$, *with composition* $g.f = \mathsf{seq}(f, g)$, *and identity* $\mathrm{id}_a = \mathsf{idt}(\star)$;
- *the functor* J *is the identity on objects and* $J(f) = \mathbf{C}(\mathrm{id}_a, f)(\mathsf{idt}(\star))$ *on morphisms;*
- *the binoidal structure on* $\mathfrak{U}(\mathbf{C})$ *is* $a \ltimes b = a \ltimes b = a \oplus_{\mathbf{M}} b$ *on objects and* $a \ltimes f = \mathsf{par}(\mathsf{idt}(\star), f)$ *and* $f \ltimes b = \mathsf{par}(f, \mathsf{idt}(\star))$ *on morphisms.*

Proof (Proof sketch). It is mechanical to check that $\mathfrak{U}(\mathbf{C})$ is a well-defined Freyd category. Given a morphism $F = (F_0, F_1)$ from $\mathbf{C} \colon \mathbf{M}^{\mathrm{op}} \times \mathbf{M} \to \mathbf{Subset}$ to $\mathbf{C}' \colon \mathbf{M}'^{\mathrm{op}} \times \mathbf{M}' \to \mathbf{Subset}$, we must define a morphism $\mathfrak{U}(F) \colon J_{\mathfrak{U}(\mathbf{C})} \to J_{\mathfrak{U}(\mathbf{C}')}$. We define $\mathfrak{U}(F)_0$ to be the strong monoidal functor F_0, and define $\mathfrak{U}(F)_1$ as F_0 on objects and as F_1 on homsets. This is a well-defined morphism of Freyd categories. It is straightforward to verify that $\mathfrak{U}$ is functorial.

Theorem 1. *The functors of Propositions 1 and 2 form an adjunction $\mathfrak{F} \dashv \mathfrak{U}$.*

Proof (Proof sketch). For the unit η of the adjunction we may take the identity as a short calculation shows that $\mathfrak{U}\mathfrak{F} = \mathrm{Id}_{\mathbf{Freyd}}$. A second calculation shows that for a $\mathbf{Subset}$-Freyd category $\mathbf{C} \colon \mathbf{M}^{\mathrm{op}} \times \mathbf{M} \to \mathbf{Subset}$, we have $\mathfrak{F}\mathfrak{U}(\mathbf{C})(a,b) = (\mathbf{C}(\mathrm{id}, \mathbf{M}(a,b))(\mathrm{idt}(\star)), \mathbf{C}(a,b))$, and so each component $\epsilon_{\mathbf{C}} \colon \mathfrak{F}\mathfrak{U}(\mathbf{C}) \to \mathbf{C}$ of the counit can be defined as $\epsilon_{\mathbf{C}0} = \mathrm{Id}_{\mathbf{M}}$ and $\epsilon_{\mathbf{C}1} = \mathrm{id}_{\mathbf{C}(a,b)} \colon \mathfrak{F}\mathfrak{U}(\mathbf{C})(a,b) \to \mathbf{C}(a,b)$. Note that the underlying $\mathbf{Set}$ map for $\epsilon_{\mathbf{C}1}$ is the identity map, but this is not an identity in $\mathbf{Subset}$. This counit is natural, and this unit and counit satisfy the zig-zag identities for an adjunction.

Recall that an adjunction $F \dashv G$ with unit $\eta \colon \mathrm{Id} \to GF$ and counit $\epsilon \colon FG \to \mathrm{Id}$ is *idempotent* if any of $F\eta$, ϵF, ηG, or $G\epsilon$ are invertible [9, Section 3.8]. In the case of the previous theorem, clearly $\mathfrak{F}\eta$ is invertible as η is the identity, so this adjunction is idempotent. This leads to the following theorem detailing just how $\mathbf{Subset}$-$\mathbf{Freyd}$ generalises $\mathbf{Freyd}$.

Theorem 2. *The full coreflective subcategory of $\mathbf{Subset}$-$\mathbf{Freyd}$ consisting of objects $\mathbf{C} \colon \mathbf{M}^{\mathrm{op}} \times \mathbf{M} \to \mathbf{Subset}$ for which $\mathbf{C}(a,b)$ has the distinguished subset $\mathbf{C}(\mathrm{id}, \mathbf{M}(a,b))(\mathrm{idt}(\star))$ is equivalent to $\mathbf{Freyd}$.*

Proof (Proof sketch). The following is a general fact about idempotent adjunctions [9, Section 3.8]: if $F \dashv G$ is an idempotent adjunction with associated monad $T = GF$ and comonad $S = FG \colon \mathbf{A} \to \mathbf{A}$, then the category of algebras of T is equivalent to the category of coalgebras of S, and the category of coalgebras of S is a full coreflective subcategory of $\mathbf{A}$ given by the objects of $\mathbf{A}$ for which $\epsilon \colon SA \to A$ is invertible.

The category of algebras for the monad $\mathfrak{U}\mathfrak{F} = \mathrm{Id}$ is equivalent to $\mathbf{Freyd}$, which is therefore a full coreflective subcategory of $\mathbf{Subset}$-$\mathbf{Freyd}$. Furthermore, we can characterize the objects of this subcategory as $\mathbf{Subset}$-Freyd categories $\mathbf{C}$ for which to $\epsilon \colon \mathfrak{F}\mathfrak{U}(\mathbf{C}) \to \mathbf{C}$ is invertible. Concretely, this means $\epsilon_{\mathbf{C}1}$ must be invertible in $\mathbf{Subset}$. But the underlying $\mathbf{Set}$ map is the identity, establishing the claim.

5 Abstract Characterisation

Definition 6 is a very concrete way to specify a $\mathbf{V}$-Freyd category, involving a nontrivial amount of data and axioms. Yet it fits together, as we show in this subsection by giving a characterisation in the style of [12]. Recall that a natural transformation between lax monoidal functors is *monoidal* when it respects the coherence maps μ and η. Write $\mathbf{MonCat}_{lax}(\mathbf{C}, \mathbf{D})$ for the category of lax monoidal functors from $\mathbf{C}$ to $\mathbf{D}$ and monoidal natural transformations between them. If $\mathbf{A}$ and $\mathbf{B}$ are monoidal categories, so are $\mathbf{A}^{op}$ and $\mathbf{A} \times \mathbf{B}$, with componentwise structure. Thus we may consider $\mathbf{MonCat}_{lax}(\mathbf{M}^{op} \times \mathbf{M}, \mathbf{V})$ for the monoidal category $(\mathbf{V}, *, J)$. We will lift the other monoidal structure $(\mathbf{V}, \circ, I)$ to $\mathbf{MonCat}_{lax}(\mathbf{M}^{op} \times \mathbf{M}, \mathbf{V})$ and prove that a $\mathbf{V}$-Freyd category is exactly a monoid with respect to this monoidal structure, under additional assumptions on $\mathbf{V}$. Most proofs are deferred to Appendix B of the extended preprint.

Definition 10. *A duoidal category* $\mathbf{V}$ *is a* cocomplete duoidal category *if* $\mathbf{V}$ *is cocomplete and* $*$ *and* $\circ$ *are cocontinuous in each argument. In a cocomplete duoidal category, the following diagrams and their symmetric versions commute:*

$$J * colim(D) \xleftarrow{\cong} colim(J * D) \qquad I \circ colim(D) \xleftarrow{\cong} colim(I \circ D)$$
$$\nwarrow_{\cong} \quad \nearrow_{\cong} \qquad\qquad \nwarrow_{\cong} \quad \nearrow_{\cong}$$
$$colim(D) \qquad\qquad\qquad colim(D)$$

where the top isomorphism is colimit preservation and the others are induced by unitors.

The rest of this subsection assumes that $\mathbf{V}$ is a cocomplete duoidal category; importantly, this is satisfied for presheaf categories. This restriction will be mitigated in Sect. 6.2 for small $\mathbf{V}$. We also assume that $\mathbf{M}$ is small. All laxness is with respect to $(\mathbf{V}, *, J)$. We now lift $(\mathbf{V}, \circ, I)$; first the unit, then composition.

Proposition 3. *There is a lax monoidal functor* $\underline{hom}_{\mathbf{M}} : \mathbf{M}^{op} \times \mathbf{M} \to \mathbf{V}$ *defined on objects as* $\underline{hom}_{\mathbf{M}}(a, b) = \coprod_{\sigma \in hom_{\mathbf{M}}(a,b)} I$.

Proposition 4. *If* $S, T : \mathbf{M}^{op} \times \mathbf{M} \to \mathbf{V}$ *are lax monoidal functors, the functor* $S \hat{\circ} T : \mathbf{M}^{op} \times \mathbf{M} \to \mathbf{V}$ *defined using coends as* $(S \hat{\circ} T)(a, c) = \int^b T(a, b) \circ S(b, c)$ *is lax monoidal.*

Proposition 5. $\left(\mathbf{MonCat}_{lax}(\mathbf{M}^{op} \times \mathbf{M}, \mathbf{V}), \hat{\circ}, \underline{hom}_{\mathbf{M}}\right)$ *is a monoidal category.*

Proof. Lemmas 5 to 7 in Appendix B of the extended preprint show that the $\circ$-composition is functorial, associative, and has $\underline{hom}_{\mathbf{M}}$ as left and right unit. That leaves only the triangle and pentagon identities, which follow from cocontinuity and the equivalent identities for $\circ$.

With these preparations we can characterise $\mathbf{V}$-Freyd categories abstractly.

Theorem 3. *Let* $\mathbf{V}$ *be a cocomplete duoidal category. Then a* $\mathbf{V}$-*Freyd category* $\mathbf{C}: \mathbf{M} \times \mathbf{M}^{op} \to \mathbf{V}$ *is exactly a monoid in* $\mathbf{MonCat}_{lax}(\mathbf{M}^{op} \times \mathbf{M}, \mathbf{V})$.

Proof (Proof sketch). A monoid $\mathbf{C}$ in $\mathbf{MonCat}_{lax}(\mathbf{M}^{op} \times \mathbf{M}, \mathbf{V})$ consists of two maps $e: \underline{hom}_{\mathbf{M}} \to \mathbf{C}$ and $m: \mathbf{C} \hat{\circ} \mathbf{C} \to \mathbf{C}$, inducing idt and seq satisfying unit and associativity conditions. The lax monoidal structure of $\mathbf{C}$ gives zero and par respectively, so identity and associativity conditions follow. Finally, the components of e and m are monoidal natural transformations, ensuring that idt and seq respect zero and par.

We note that by Fujii's observations [7], PROs and PROPs are equivalent to **Set**-Freyd categories over $\mathbf{N}$ and $\mathbf{P}$ respectively because $(\mathbf{Set}, \times, \times)$ is a cocomplete duoidal category.

6 Change of Enrichment

After defining enriched categories, a natural next step is to consider a change of enrichment. Any monoidal functor $\mathbf{V} \to \mathbf{W}$ induces a functor $\mathbf{V}$-**Cat** $\to \mathbf{W}$-**Cat**. We will show that the same holds for the appropriate type of functors between duoidal categories and enriched Freyd categories (in Sect. 6.1). We will then use that to alleviate the restriction of duoidal cocompleteness on the abstract characterisation of Sect. 5 (in Sect. 6.2) at the cost of losing a direction of the correspondence. Finally, changing enrichment along a forgetful functor gives an underlying (unenriched) Freyd category $J: \mathbf{M} \to \mathbf{C}$ with $\mathbf{C}$ monoidal, which we show recovers the pure computations in the examples of Sect. 3 (in Sect. 6.3).

6.1 Lifting Duoidal Functors

To talk about change of enrichment, we first need to define the appropriate type of functor between the enriching categories along which to change.

Definition 11. *[2, Definition 6.54] Take duoidal categories* $(\mathbf{V}, *_\mathbf{V}, J_\mathbf{V}, \circ_\mathbf{V}, I_\mathbf{V})$ *and* $(\mathbf{W}, *_\mathbf{W}, J_\mathbf{W}, \circ_\mathbf{W}, I_\mathbf{W})$. *A functor* $F: \mathbf{V} \to \mathbf{W}$ *is a* double lax monoidal func-*tor* when equipped with η_*, μ_*, $\eta_\circ$, and $\mu_\circ$ *such that* (F, η_*, μ_*) *is lax monoidal for* $*_\mathbf{V}$ *and* $*_\mathbf{W}$, $(F, \eta_\circ, \mu_\circ)$ *is lax monoidal for* $\circ_\mathbf{V}$ *and* $\circ_\mathbf{W}$, *and the following diagrams commute:*

$$(F(A) \circ_\mathbf{W} F(B)) *_\mathbf{W} (F(C) \circ_\mathbf{W} F(D)) \xrightarrow{\zeta} (F(A) *_\mathbf{W} F(C)) \circ_\mathbf{W} (F(B) *_\mathbf{W} F(D))$$

$$\begin{array}{ccc}
\mu_\circ * \mu_\circ \downarrow & & \downarrow \mu_* \circ \mu_* \\
F(A \circ_\mathbf{V} B) *_\mathbf{W} F(C \circ_\mathbf{V} D) & & F(A *_\mathbf{V} C) \circ_\mathbf{W} F(B *_\mathbf{V} D) \\
\mu_* \downarrow & & \downarrow \mu_\circ \\
F((A \circ_\mathbf{V} B) *_\mathbf{V} (C \circ_\mathbf{V} D)) \xrightarrow{\quad F\zeta \quad} F((A *_\mathbf{V} C) \circ_\mathbf{V} (B *_\mathbf{V} D))
\end{array}$$

$$\begin{array}{cc}
F(J_\mathbf{V}) \xrightarrow{F\epsilon} F(I_\mathbf{V}) \\
\eta_* \uparrow \qquad \uparrow \eta_\circ \\
J_\mathbf{W} \xrightarrow{\epsilon} I_\mathbf{W}
\end{array}$$

$$\begin{array}{ccc}
J_\mathbf{W} \xrightarrow{\eta_*} F(J_\mathbf{V}) \xrightarrow{F\Delta} F(J_\mathbf{V} \circ_\mathbf{V} J_\mathbf{V}) & & I_\mathbf{W} \xrightarrow{\eta_\circ} F(I_\mathbf{V}) \xleftarrow{F\nabla} F(I_\mathbf{V} *_\mathbf{V} I_\mathbf{V}) \\
\Delta \downarrow \qquad\qquad \uparrow \mu_\circ & & \nabla \uparrow \qquad\qquad \uparrow \mu_* \\
J_\mathbf{W} \circ_\mathbf{W} J_\mathbf{W} \xrightarrow{\eta_* \circ \eta_*} F(J_\mathbf{V}) \circ_\mathbf{W} F(J_\mathbf{V}) \quad I_\mathbf{W} *_\mathbf{W} I_\mathbf{W} \xrightarrow{\eta_\circ * \eta_\circ} F(I_\mathbf{V}) *_\mathbf{W} F(I_\mathbf{V})
\end{array}$$

Here now is the change-of-enrichment theorem for duoidally enriched Freyd categories.

Theorem 4. *Let* $F: \mathbf{V} \to \mathbf{W}$ *be a double lax monoidal functor. For a* $\mathbf{V}$-*Freyd category* $\mathbf{C}: \mathbf{M}^{\mathrm{op}} \times \mathbf{M} \to \mathbf{V}$, *define* $\overline{F}(\mathbf{C})(a, b) := F(\mathbf{C}(a, b))$ *with structure maps* $\mathrm{idt}_F := F\mathrm{idt}.\eta_\circ$, $\mathrm{seq}_F := F\mathrm{seq}.\mu_\circ$, $\mathrm{zero}_F := F\mathrm{zero}.\eta_*$, *and* $\mathrm{par}_F := F\mathrm{par}.\mu_*$. *For a map* $G = (G_0, G_1): \mathbf{C} \to \mathbf{C}'$, *define* $\overline{F}(G) := (G_0, FG_1)$. *This* $\overline{F}$ *is a functor* $\mathbf{V}$-**Freyd** $\to \mathbf{W}$-**Freyd**.

Proof. See Appendix C of the extended preprint.

Example 8. Let M and N be separated monoids and $\phi\colon M \to N$ a homomorphism such that $\phi(m) \parallel \phi(m')$ implies $m \parallel m'$. Then ϕ induces a double lax monoidal functor $\phi_*\colon \mathbf{Label}_M \to \mathbf{Label}_N$ given by $\ell \mapsto \phi.\ell$ on objects and $f \mapsto f$ on morphisms. The maps η_*, μ_*, and $\eta_\circ$ are all identities, while $\mu_\circ\colon \{(a, a') \mid \phi.\ell(a) \Vert \phi.\ell'(a')\} \to \{(a, a') \mid \ell(a) \Vert \ell'(a')\}$ is the inclusion, and so ϕ_* is clearly double lax monoidal. Apply Theorem 4 to the example from Sect. 3.1 along the map $\mathcal{P}_f(!)\colon \mathcal{P}_f(R) \to \mathcal{P}_f(1)$, which is a homomorphism such that $\mathcal{P}_f(!)(P) \cap \mathcal{P}_f(!)(Q) = \emptyset$ implies $P \cap Q = \emptyset$. We get $\mathcal{P}_f(!)_*(\mathbf{C})(a, b) = \sum_{Q \in \mathcal{P}_f(R)} (\mathbf{Set}(a \times \Pi_Q, b \times \Pi_Q)) \to \mathcal{P}_f(1)$, $(Q, f) \mapsto \emptyset$ if $Q = \emptyset$, else 1. This change of enrichment alters the example to only allowing maps to be put in parallel if at least one of them requires no resources.

Example 9. We can use change of enrichment for the indexed state example of Sect. 3.2. Consider Example 6 for $(\mathbf{Set}, \times, 1)$ (using universes for this example to avoid size issues). There, the definition of Day convolution $\times_{\text{Day}}$ simplifies to $(P \times_{\text{Day}} Q)(a, b) = \int^{b_2, b_2} \mathbf{Set}(b_1 \times b_2, b) \times P(a, b_1) \times Q(a, b_2)$ and its unit becomes $k(a, b) = b$. The Kleisli construction turns a finitary endofunctor on $\mathbf{Set}$ into a profunctor as follows. Define $\mathrm{Kl}\colon [\mathbf{Set}, \mathbf{Set}]_f \to \mathbf{Prof}(\mathbf{Set})$ by $\mathrm{Kl}(F)(a, b) = \mathbf{Set}(a, Fb)$, and coherence maps:

$$\eta_*\colon k \to \mathrm{Kl}(\mathrm{Id}) \qquad \mu_*\colon \mathrm{Kl}(F_1) \times_{\text{Day}} \mathrm{Kl}(F_2) \to \mathrm{Kl}(F_1 \times_{\text{Day}} F_2)$$
$$b \mapsto \mathrm{cst}_b \qquad\qquad (k, f_1, f_2) \mapsto \lambda a.(k, f_1(a), f_2(a))$$

$$\eta_\circ\colon \hom \to \mathrm{Kl}(\mathrm{Id}) \qquad \mu_\circ\colon \mathrm{Kl}(F) \diamond \mathrm{Kl}(G) \to \mathrm{Kl}(F \circ G)$$
$$f \mapsto f \qquad\qquad (f, g) \mapsto Fg.f$$

This makes Kl a double lax monoidal functor. Theorem 4 then gives a $\mathbf{Prof}(\mathbf{Set})$-Freyd category defined by $\mathrm{Kl}(\mathbf{C})(a, b)(x, y) := \mathbf{Set}(x, (b \times y)^a)$.

6.2   Yoneda Embedding

The Yoneda embedding of a small monoidal category is a strong monoidal functor with respect to Day convolution. This extends to small duoidal categories.

Proposition 6. *The Yoneda embedding* $\mathbf{V} \to [\mathbf{V}^{\mathrm{op}}, \mathbf{Set}]$ *is a double lax monoidal functor from small* $(\mathbf{V}, *, J, \circ, I)$ *to* $([\mathbf{V}^{\mathrm{op}}, \mathbf{Set}], *_{\text{Day}}, \mathbf{V}(-, J), \circ_{\text{Day}}, \mathbf{V}(-, I))$.

Proof. See [11] for the fact that it is lax monoidal for each monoidal structure separately. The diagrams of Definition 11 are verified straightforwardly.

It follows from Theorem 4 that every $\mathbf{V}$-Freyd category for small $\mathbf{V}$ induces a $[\mathbf{V}^{\mathrm{op}}, \mathbf{Set}]$-Freyd category. But $[\mathbf{V}^{\mathrm{op}}, \mathbf{Set}]$ is duoidally cocomplete, so the setting in which the abstract characterisation of Theorem 3 applies. We conclude that the characterisation extends beyond the duoidally cocomplete setting in the sense that every $\mathbf{V}$-Freyd category for small $\mathbf{V}$ induces a monoid in $\mathbf{MonCat}_{\text{lax}}(\mathbf{M}^{\mathrm{op}} \times \mathbf{M}, [\mathbf{V}^{\mathrm{op}}, \mathbf{Set}])$.

6.3   Forgetful Functors

Any category enriched in a monoidal category $\mathbf{V}$ has an underlying (unenriched) category, got by changing the enrichment along the 'forgetful' monoidal functor $\mathbf{V}(I, -)\colon \mathbf{V} \to \mathbf{Set}$. A similar process plays out for duoidal categories.

Proposition 7. *Let $(\mathbf{V}, *, J, \circ, I)$ be a duoidal category and write $\phi\colon J \to J * J$ for the inverse of the unitors. Then $\mathbf{V}(J, -)\colon \mathbf{V} \to \mathbf{Set}$ is a double lax monoidal functor with coherence maps:*

$$\eta_*\colon 1 \to \mathbf{V}(J, J) \qquad\qquad \mu_*\colon \mathbf{V}(J, A_1) \times \mathbf{V}(J, A_2) \to \mathbf{V}(J, A_1 * A_2)$$
$$\star \mapsto \mathrm{id} \qquad\qquad\qquad\qquad (f_1, f_2) \mapsto (f_1 * f_2).\phi$$

$$\eta_\circ\colon 1 \to \mathbf{V}(J, I) \qquad\qquad \mu_\circ\colon \mathbf{V}(J, A_1) \times \mathbf{V}(J, A_2) \to \mathbf{V}(J, A_1 \circ A_2)$$
$$\star \mapsto \epsilon \qquad\qquad\qquad\qquad (f_1, f_2) \mapsto (f_1 \circ f_2).\Delta$$

Applying Theorem 4 along the forgetful functor of the previous proposition in the case of the examples of Sect. 3 will show that this recovers the underlying pure computations. Note that a $\mathbf{Set}$-Freyd category $\mathbf{C}$ has a trivial instance of the exchange axiom, axiom viii, and so $\mathbf{C}$ is a monoidal category with identity-on-objects monoidal functor $J\colon \mathbf{M} \to \mathbf{C}$.

Example 10. Applying the forgetful functor to the stateful function example of Sect. 3.1 results in the (unenriched) category with $\mathbf{Label}_{\mathcal{P}_f(R)}(\mathrm{cst}_\emptyset, \mathbf{C}(a, b))$ as the homsets. Because labels are preserved, the morphisms in this (unenriched) category are exactly the elements of $\mathbf{C}(a, b)$ which have label $\emptyset$, *i.e.* maps $a \times 1 \to b \times 1$ which are pure functions.

Example 11. Changing the enrichment of the indexed state example from Sect. 3.2 along the forgetful functor gives the (unenriched) category with homsets $[\mathbf{Set}, \mathbf{Set}]_f (\mathrm{Id}, \mathbf{C}(a, b))$. If $\phi\colon \mathrm{Id} \to (b \times (-))^a$ is such a natural transformation, then the function $\phi_1\colon 1 \to (b \times 1)^a$, which is equivalent to choosing a function $f\colon a \to b$, completely determines ϕ, because for any set X and $x \in X$ by naturality $1 \xrightarrow{x} X \xrightarrow{\phi X} (b \times X)^a = 1 \xrightarrow{\phi_1} (b \times 1)^a \xrightarrow{(\mathrm{id} \times x).-} (b \times X)^a$, whence $\phi_X(x)(a) = (f(a), x)$. Therefore the morphisms in this (unenriched) category are all functions $a \to b$.

Example 12. Changing the enrichment of the Kleisli categories of Lawvere theories example from Sect. 3.3 along the forgetful functor gives the (unenriched) category with homsets $[\mathbf{Law}, \mathbf{Set}](\mathbb{1}, \mathbf{C}(a, b))$. Consider such a natural transformation $\phi\colon \mathbb{1} \to T(-)(b)^a$. It is completely determined by its component at $\mathcal{S}$. For any $\mathcal{L}$ let $\iota\colon \mathcal{S} \to \mathcal{L}$ be the unique map, then naturality implies $\phi_\mathcal{L} = T(\iota)\phi_\mathcal{S}$. Furthermore, $\phi_\mathcal{S}(\star) \in T(\mathcal{S})(b)^a = b^a$. So the morphisms in this (unenriched) category again are all functions $a \to b$.

7   Conclusion

We have defined a version of Freyd categories enriched over any duoidal category $\mathbf{V}$, and morphisms between them. We used various duoidal categories to give examples based on separation of resources, parameterised monads, and the Kleisli construction for Lawvere theories. By enriching with **Subset**, we have proven that the category of Freyd categories **Freyd** is a full coreflective subcategory of **Subset-Freyd**, thus establishing that $\mathbf{V}$-Freyd categories indeed generalise Freyd categories. Additionally, we proved an abstract characterisation of $\mathbf{V}$-Freyd categories over small $\mathbf{M}$ for duoidally cocomplete $\mathbf{V}$, they are monoids in $\mathbf{MonCat_{lax}}(\mathbf{M}^{op} \times \mathbf{M}, \mathbf{V})$. Finally, we provided change of enrichment and examples thereof.

Future Work. There are several directions for further investigation:

- The abstract characterisation of Sect. 5 may be part of a larger structure, namely a *bicategory with proarrow equipment*, whose objects are monoidal categories, arrows are strong monoidal functors, proarrows are lax monoidal profunctors, and cells are lax monoidal natural transformations. In this setting, a $\mathbf{V}$-Freyd category would be a monad and the vertical monad morphisms would be a $\mathbf{V}$-Freyd morphism. This would enable applying general constructions for monads in a bicategory.
- Relatedly, an *fc-multicategory* structure on $\mathbf{MonCat_{lax}}(\mathbf{M}^{op} \times \mathbf{M}, \mathbf{V})$ may bypass cocompleteness in characterising $\mathbf{V}$-Freyd categories as monoids.
- The abstract characterisation of Sect. 5 also uses the free $\mathbf{V}$-category on $\mathbf{M}$. It may be fruitful to change the definition of a $\mathbf{V}$-Freyd category to be a $\mathbf{V}$-functor $J\colon \mathbf{M} \to \mathbf{C}$ where we extend $\mathbf{V}$-categories in a way similar to Morrison and Penneys [16].
- Freyd categories can have the property of being *closed*. In this case they induce a strong monad. A similar definition may be possible for $\mathbf{V}$-Freyd categories. This could determine a higher-order semantics for effectful programs based on duoidal categories. A nontrivial definition of closure may require a $\mathbf{V}$-category $\mathbf{M}$ that is not free.
- Our original motivation stemmed from the desire for semantics combining differentiable and probabilistic programming, in particular, the possibility of having a linear structure for the probabilistic fragment and a cartesian one for differentiable terms. **Prof**-Freyd categories may provide a useful separation to aid the desired distinction between linear and cartesian properties.

Acknowledgments. We would like to thank Robin Kaarsgaard, Ohad Kammar, and Matthew Di Meglio for their input and encouragement, as well as the reviewers of all versions of this work.

References

1. Adamek, J., Rosicky, J.: Locally Presentable and Accessible Categories. Cambridge University Press, Cambridge (1994). https://doi.org/10.1017/CBO9780511600579

2. Aguiar, M., Mahajan, S.: Monoidal Functors, Species and Hopf Algebras. American Mathematical Society, Providence (2010). https://doi.org/10.1090/crmm/029
3. Atkey, R.: Algebras for parameterised monads. In: Kurz, A., Lenisa, M., Tarlecki, A. (eds.) CALCO 2009. LNCS, vol. 5728, pp. 3–17. Springer, Heidelberg (2009). https://doi.org/10.1007/978-3-642-03741-2_2
4. Batanin, M., Markl, M.: Centers and homotopy centers in enriched monoidal categories. Adv. Math. **230**, 1811–1858 (2012). https://doi.org/10.1016/j.aim.2012.04.011
5. Day, B.: On closed categories of functors. In: Midwest Category Seminar. Lecture Notes in Mathematics, vol. 137, pp. 1–38. Springer, Berlin (1970). https://doi.org/10.1007/BFb0060438
6. Forcey, S.: Enrichment over iterated monoidal categories. Algebraic Geometric Topology **4**, 95–119 (2004). https://doi.org/10.2140/agt.2004.4.95
7. Fujii, S.: A unified framework for notions of algebraic theory. Theory Appl. Categories **34**(40), 1246–1316 (2019)
8. Garner, R., López Franco, I.: Commutativity. J. Pure Appl. Algebra **220**(5), 1707–1751 (2016). https://doi.org/10.1016/j.jpaa.2015.09.003
9. Grandis, M.: Category Theory and Applications: A Textbook for Beginners, 2nd edn. World Scientific, Singapore (2021). https://doi.org/10.1142/12253
10. Hyland, M., Plotkin, G., Power, J.: Combining effects: sum and tensor. Theor. Comput. Sci. **357**(1–3), 70–99 (2006). https://doi.org/10.1016/j.tcs.2006.03.013
11. Im, G.B., Kelly, G.M.: A universal property of the convolution monoidal structure. J. Pure Appl. Algebra **43**, 75–88 (1986). https://doi.org/10.1016/0022-4049(86)90005-8
12. Jacobs, B., Heunen, C., Hasuo, I.: Categorical semantics for arrows. J. Funct. Program. **19**(3–4), 403–438 (2009). https://doi.org/10.1017/S0956796809007308
13. Lack, S.: Composing PROPs. Theory Appl. Categories **13**(9), 147–163 (2004)
14. Levy, P.B., Power, J., Thielecke, H.: Modelling environments in call-by-value programming languages. Inf. Comput. **185**(2), 182–210 (2003). https://doi.org/10.1016/S0890-5401(03)00088-9
15. Moggi, E.: Notions of computation and monads. Inf. Comput. **93**, 55–92 (1991). https://doi.org/10.1016/0890-5401(91)90052-4
16. Morrison, S., Penneys, D.: Monoidal categories enriched in braided monoidal categories. Int. Math. Res. Notes **11**, 3527–3579 (2019). https://doi.org/10.1093/imrn/rnx217
17. Plotkin, G., Power, J.: Computational effects and operations: an overview. In: Domains. Electronic Notes in Theoretical Computer Science, vol. 73, pp. 149–163 (2004). https://doi.org/10.1016/j.entcs.2004.08.008
18. Power, J., Robinson, E.: Premonoidal categories and notions of computation. Math. Struct. Comput. Sci. **7**(5), 453–468 (1997). https://doi.org/10.1017/S0960129597002375
19. Power, J.: Premonoidal categories as categories with algebraic structure. Theor. Comput. Sci. **278**(1–2), 303–321 (2002) https://doi.org/10.1016/S0304-3975(00)00340-6. https://www.sciencedirect.com/science/article/pii/S0304397500003406
20. Staton, S.: Freyd categories are enriched Lawvere theories. In: Proceedings of the Workshop on Algebra, Coalgebra and Topology. Electronic Notes in Theoretical Computer Science, vol. 303, pp. 197–206 (2014). https://doi.org/10.1016/j.entcs.2014.02.010

Towards a Theory of Conversion Relations for Prefixed Units of Measure

Baltasar Trancón y Widemann[1,2(✉)] and Markus Lepper[2]

[1] Nordakademie, Elmshorn, Germany
baltasar@trancon.de
[2] semantics gGmbH, Berlin, Germany

Abstract. Units of measure with prefixes and conversion rules are given
a formal semantic model in terms of categorial group theory. Basic structures and both natural and contingent semantic operations are defined.
Conversion rules are represented as a class of ternary relations with
both group-like and category-like properties. A hierarchy of subclasses
is explored, each with better algebraic behavior than the preceding, culminating in a direct efficient conversion-by-rewriting algorithm.

1 Introduction

In the mathematics of science, *dimensions* and *units of measure* are used as static
metadata for understanding and checking *quantity* relations [Wal22]. Quantities
are variables that may assume a value expressing a fact about a model, in terms
of a formal multiplication of a *numerical magnitude* with a unit of measure. [1]

Clearly, neither half of the pair is sufficient for interpretation: The magnitude
carries the actual data, and the unit of measure the context of reference. Each
quantity is assigned a dimension, such that only quantities of identical dimension
are *commensurable*, that is, may be added or compared (as apples with apples),
and may furthermore be associated with preferred units of measurement. [2]

In scientific programming, traditional practice expresses only the magnitude
part of quantity values, whereas units of measure, and their consistency among
related quantities, are left implicit. The evident potential for disastrous software errors inherent in that practice has been demonstrated, for instance, by
the famous crash of the Mars Climate Orbiter probe, where *pounds of force*
were confounded with *newtons* at a subsystem interface [Ste+99]. Practical tool
solutions abound [MBBS20]. Theoretical foundations are rare, but essential for
program correctness, specification and verification.

In the seminal work of Kennedy [Ken96], the checking of stated or implied
units of measure is cast formally as a *type inference* problem. The approach has
been implemented successfully, for example in F# [Ken10] or Haskell [Gun15].

---

[1] For example, a *distance* of 42195 m is a quantity value. Quantities themselves have
a complicated and controversial ontology, which shall not be discussed here.
[2] For example, the quantities of *molecular bond length* and *planetary perihelion* are
both of dimension *length* and thus formally commensurable, but traditionally use
quite different units of measure.

R. Glück et al. (Eds.): RAMiCS 2023, LNCS 13896, pp. 258–273, 2023.
https://doi.org/10.1007/978-3-031-28083-2_16

However, the underlying concept of units is strongly simplified, and falls far short of the traditional scientific practice, embodied in the International Systems of Units and Quantities (SI, ISQ [ISO09], respectively).

A particular challenge is the concept of *convertible* units, where the scaling factors for equivalent transformation of magnitudes relative to some other unit are fixed and known, and hence could be applied implicitly.[3] Furthermore, as a special case of convertibility of such practical importance that it comes with its own notation, there are *prefixes* that can be applied to any unit, and modify the scaling factor in a uniform way.[4]

Traditional concepts of units of measure also have their own curious, historically grown idiosyncrasies and restrictions. Particularly worrying from the algebraic viewpoint is the contradiction between *reductionist* definitions and *non-compositional* notation: On the one hand, units are defined in terms of some expression over other, more basic units. On the other hand, the definiens often cannot be substituted for the definiendum without violating syntactic rules.[5]

1.1 Contributions

The present work[6] lays the foundation for a formal model of dimensions and units of measure that extends and complements the findings of [Ken96]. We refine the mathematical structures outlined there, in order to accomodate unit conversions and prefixes. In addition, the proposed model rectifies irrelevant, historically accrued restrictions of traditional notation systems. A clear line is drawn between *necessary* and *contingent* properties of the model; respectively, those that follow from the mathematical structure without regard to the meaning of particular symbols, and those that represent the actual semantic conventions of science but could be extended or modified without breaking the logic.

In particular, the main contributions of the present work are as follows:

- A formal mathematical model of dimensions and units with optional prefixes, separated into the generic structures shared by all unit systems (Sect. 3) and the properties to be interpreted contingently by each unit system (Sect. 4), together with a hierarchy of equivalence relations.
- A formal mathematical model of unit conversion rules as a class of ternary relations with a closure operator, that function both as groups and as categories (Sect. 5), together with a hierarchy of six subclasses, each with better algebraic behavior than the preceding (Sect. 6), and in particular an efficient rewriting procedure for calculating conversion factors.

---

[3] In the Mars Climate Orbiter case, implicit multiplication by 4.4482216152605 would have avoided the observed failure, and thus potentially saved the mission.

[4] For example, multiplication by 1000 serves to convert from seconds to milliseconds, and exactly in the same way from meters to millimeters.

[5] For example, the *area* unit *hectare* (ha) is defined as the prefix *hecto-*, implying a multiplicator of 100, being applied to the French Revolutionary unit *are* (a), which in turn is defined as the square of a *decameter* (dam^2); however, the expanded expression $\mathrm{h(dam}^2)$ is syntactically invalid.

[6] See [TL22] for an extended version with proofs and additional examples.

2   Prerequisites

We assume that the reader is familiar with basic concepts of abstract algebra and category theory. In this section, we shall recall some established facts, without individual reference or proof, and define notation to be used in the following sections. For encyclopedic reference, we recommend [AHS06, §3–6,19–20].

2.1   Abelian Groups

The following formal model is based on abelian groups and their homomorphisms. In the usual sense of a concrete category, we distinguish an abelian group from its carrier set. We write $\mathcal{U}(G)$ for the carrier set of a group G. A homomorphism $f : G \to H$ is a map $f : \mathcal{U}(G) \to \mathcal{U}(H)$ that commutes with the group operation. The carrier-set operator $\mathcal{U}$, together with the identity operation on homomorphism maps, $\mathcal{U}(f) = f$, is the forgetful functor from the category **Ab** of abelian groups to the category **Set** of sets. Wherever a generic group variable G is mentioned, we write $\diamond$ for the group operation, e for the neutral element, and † for inversion. Thus, we could work with formal tuples $G = (\mathcal{U}(G), \diamond, e, ^{\dagger})$, but the need does not arise.

All actual numbers in the model are from the set $\mathbb{Q}_{+}$ of positive rationals. For arithmetics only the multiplicative group structure Qm on that set is required. We refer to these numbers as *ratios*.

Kennedy [Ken96] observed that the algebras of dimensions and units are essentially *free abelian groups*.

2.2   Free Abelian Groups

We write $\mathcal{A}(X)$ for the free abelian group over a set X of generators. Whereas the concept is specified only up to isomorphism in category theory, we conceive of a particular construction of the carrier set $\mathcal{U}(\mathcal{A}(X))$, namely the space $\mathrm{Zf}(X)$ of finitely supported integer-valued functions on X:

$$\mathrm{Zf}(X) = \{f : X \to \mathbb{Z} \mid \mathrm{supp}(f) \text{ finite}\} \qquad \mathrm{supp}(f) = \{x \in X \mid f(x) \neq 0\}$$

If X is finite, then the group $\mathcal{A}(X)$ is called *finitely generated*. Note that the finite-support constraint is irrelevant in this case.

For denoting a particular element of a free abelian group directly, it suffices to refer to the support. The finite partial maps of type $f : X \to \mathbb{Z} \setminus \{0\}$ are in one-to-one correspondence with group elements $f_{/0} \in \mathrm{Zf}(X)$:

$$f_{/0}(x) = \begin{cases} f(x) & \text{if defined} \\ 0 & \text{otherwise} \end{cases}$$

The operation that turns the set $\mathrm{Zf}(X)$ into the abelian group $\mathcal{A}(X)$ is realized by pointwise addition. However, since the standard interpretation of function values is *power exponents*, we shall use multiplicative notation, such that $(fg)(x) = f(x) + g(x)$, with neutral element $\varnothing_{/0}$.

For the following definitions, it is useful to introduce the *Iverson bracket*: For any proposition p, $[p] = 1$ if p holds, otherwise $[p] = 0$.

Zf is a functor, and a monad with *unit*[7] δ and *multiplication* λ:

$$Zf(f)(g)(y) = \sum_{x \in X} [f(x) = y] \cdot g(x)$$

$$\delta : 1 \Rightarrow Zf \qquad \delta_X(x)(y) = [x = y] \text{ i.e. } \delta(x) = \{x \mapsto 1\}_{/0}$$

$$\lambda : Zf^2 \Rightarrow Zf \qquad \lambda_X(f)(x) = \sum_{g \in Zf(X)} f(g) \cdot g(x)$$

Monads and their operations are often best understood as abstract syntax: The elements of $Zf(X)$ represent group words. The unit δ embeds the generators into the words as *literals*. The multiplication λ *flattens* two layers of *nested* words.

Lemma 1. *With the usual interpretation of integer powers of elements of a multiplicative group, any group element can be written, uniquely up to permutation of factors, as a finite product of nonzeroth powers:*

$$\{x_1 \mapsto z_1, \ldots, x_n \mapsto z_n\}_{/0} = \delta(x_1)^{z_1} \cdots \delta(x_n)^{z_n}$$

The monad Zf arises from the composition of functors $\mathcal{U} \circ \mathcal{A}$, where $\mathcal{A}$ is *left adjoint* to $\mathcal{U}$. Hence there is also a *counit* ε, defined recursively by:

$$\varepsilon : \mathcal{A}\mathcal{U} \Rightarrow 1 \qquad \begin{array}{ll} \varepsilon_G(\varnothing_{/0}) = e & \varepsilon_G(gh) = \varepsilon_G(g) \diamond \varepsilon_G(h) \\ \varepsilon_G(\delta_{\mathcal{U}(G)}(x)) = x & \varepsilon_G(g^{-1}) = \varepsilon_G(g)^\dagger \end{array}$$

The counit embodies the concept of interpreting words over a particular target group by "multiplying things out". It follows that $\lambda = \mathcal{U}\varepsilon\mathcal{A}$.

2.3 Pairing with an Abelian Group

Consider the functor that pairs elements of arbitrary sets X with elements of (the carrier of) a fixed abelian group G:

$$\mathcal{C}_G(X) = \mathcal{U}(G) \times X \qquad \mathcal{C}_G(f) = \text{id}_{\mathcal{U}(G)} \times f$$

From the group structure of G (or actually any monoid), we obtain a simple monad, with unit η_G and multiplication μ_G:

$$\eta_G : 1 \Rightarrow \mathcal{C}_G \qquad \eta_{G,X}(x) = (e, x)$$

$$\mu_G : \mathcal{C}_G^2 \Rightarrow \mathcal{C}_G \qquad \mu_{G,X}(a, (b, x)) = (a \diamond b, x)$$

When the second component is restricted to abelian groups as well, a corresponding functor on **Ab** is obtained, which creates the *direct sum* of abelian groups instead of the Cartesian product of carrier sets:

$$\mathcal{D}_G(H) = G \times H \qquad \mathcal{D}_G(f) = \text{id}_G \times f$$

The two are related by commuting with the forgetful functor: $\mathcal{C}_G \circ \mathcal{U} = \mathcal{U} \circ \mathcal{D}_G$.

---

[7] Monad units are unrelated to units of measure.

2.4   Monad Composition

There is a natural transformation in **Ab** that exchanges pairing and free abelian group construction:

$$\beta_G : \mathcal{A}\mathcal{C}_G \Rightarrow \mathcal{D}_G\,\mathcal{A} \qquad\qquad \beta_{G,X} = \langle \varepsilon_G \circ \mathcal{A}(\pi_1), \mathcal{A}(\pi_2) \rangle$$

Back down in **Set**, this is a natural transformation $\mathcal{U}(\beta_G) : \mathrm{Zf}\,\mathcal{C}_G \Rightarrow \mathcal{C}_G\,\mathrm{Zf}$ which can be shown to be a distributive law between the monads $\mathcal{C}_G$ and Zf. This turns the composite functor $\mathcal{C}_G \circ \mathrm{Zf}$ into a monad, with unit θ_G and multiplication ξ_G:

$$\theta_G : 1 \Rightarrow \mathcal{C}_G\,\mathrm{Zf} \qquad\qquad \theta_G = \eta_G\delta$$
$$\xi_G : (\mathcal{C}_G\,\mathrm{Zf})^2 \Rightarrow \mathcal{C}_G\,\mathrm{Zf} \qquad\qquad \xi_G = \mu_G\lambda \circ \mathcal{C}_G\beta_G\mathrm{Zf}$$

3   Generic Structures

This section defines the structures of a formal model of dimensions and units. These are *generic*: no interpretation of base symbols is presupposed. This is achieved by δ, λ, ε, π_1, π_2, η, μ, β, θ, ξ all being *natural transformations*: They are parametric families of maps that transform data between *shapes* specified by functors, in a way that is logically independent of the elementary *content* specified by the type argument.

Definition 1 (Dimension). *The* dimensions Dm *are the free abelian group over* $\mathrm{Dm_b}$, *a given finite set of* base dimensions:

$$\mathrm{Dm} = \mathcal{A}(\mathrm{Dm_b})$$

Example 1 (SI Dimensions). The SI/ISQ recognizes seven physical base dimensions, $\mathrm{Dm_b^{SI}} = \{\mathsf{L}, \mathsf{T}, \mathsf{M}, \mathsf{I}, \Theta, \mathsf{N}, \mathsf{J}\}$. From these, compound dimensions can be formed; for example, quantities of *thermodynamical entropy* are associated with $d_H = \{\mathsf{L} \mapsto 2, \mathsf{M} \mapsto 1, \mathsf{T} \mapsto -2, \Theta \mapsto -1\}_{/0}$. However, the actual physical interpretation of these symbols need not be considered at all for formal analysis.

Remark 1. We do not use the product-of-powers notation of Lemma 1 for concrete elements.[8]

Definition 2 (Root Unit). *The* root units $\mathrm{Ut_r}$ *are the free abelian group over* $\mathrm{Ut_b}$, *a given finite set of* base units:

$$\mathrm{Ut_r} = \mathcal{A}(\mathrm{Ut_b})$$

---

[8] Neither do we recommend that use for formal analysis, because it is known to cause subtle misunderstandings, in particular the highly overloaded expression for the empty case, $1 = \varnothing_{/0}$. For example, ISO 80000 [ISO09, §5] states, without satisfactory justification, that "1 is *not* a dimension". The statements in [ISO09, §6.5.3] about "1 as a derived unit" [sic!] are even less clear.

Example 2 (SI Root Units). The SI/ISQ recognizes one canonical base unit per base dimension, $\mathrm{Ut_b^{SI_0}} = \{\mathrm{m, s, g, A, K, mol, cd}\}$. From these, compound units can be formed, for example, quantities of *velocity* are associated with $u_v = \{\mathrm{m} \mapsto 1, \mathrm{s} \mapsto -1\}_{/0}$ which by tradition is notated multiplicatively as m/s or $\mathrm{m \cdot s^{-1}}$.[9]

Definition 3 (Prefix). *The* prefixes Px *are the free abelian group over* $\mathrm{Px_b}$, *a given finite set of* base prefixes:

$$\mathrm{Px} = \mathcal{A}(\mathrm{Px_b})$$

Example 3 (SI Prefixes). Three families of prefixes are recognized in combination with the SI units. The symbols come with associated numerical values, discussed in detail in Sect. 4 below. For now it suffices to simply name them:[10]

$$\mathrm{Px_b^{SI}} = \left\{ \begin{array}{l} \mathrm{d, c, m, m, n, p, f, a, z, y, r, q,} \\ \mathrm{da, h, k, M, G, T, P, E, Z, Y, R, Q,} \\ \mathrm{ki, Mi, Gi, Ti, Pi, Ei, Zi, Yi} \end{array} \right\}$$

Remark 2. Composite prefixes are forbidden in modern scientific notation. However, the empty prefix $\varnothing_{/0}$ is always allowed, some traditional usage of double prefixes is known, the inner logic of each SI prefix family is a geometric sequence, and composite prefixes arise virtually in the algebra of composite units. Thus, the generalization to the free abelian group of prefixes is a theoretical simplification and unification.

Definition 4 (Preunit). *The* preunits $\mathrm{Ut_p}$ *are the Cartesian pairs of a prefix, forgetting the group structure, and a base unit:*

$$\mathrm{Ut_p} = \mathcal{C}_{\mathrm{Px}}(\mathrm{Ut_b})$$

Example 4. The SI unit *kilogram*, already discussed above, is represented formally as a preunit, $\mathrm{kg} = (\delta_{\mathrm{Px_b}}(\mathrm{k}), \mathrm{g})$.

Definition 5 (Unit). *The* units Ut *are the free abelian group over the preunits:*

$$\mathrm{Ut} = \mathcal{A}(\mathrm{Ut_p})$$

Remark 3. Unlike in the preceding constructions, the generator set $\mathrm{Ut_p}$ is generally infinite. Fortunately, this causes no problems for the remainder of the present work; see Theorem 13 below.

---

[9] For historical reasons, the SI unit *gram* (g) is not considered the canonical unit of quantities of *mass*. Rather, this distinguished role is played by the unit *kilogram* (kg). However, the latter unit is best understood not as a base unit, but as a compound of a prefix *kilo-* and a base unit *gram*.

[10] The symbol m occurs as both a base prefix and a base unit in the SI vocabulary. However, this causes ambiguity issues only for parsing traditional notations, where prefix and unit symbols are simply concatenated; the formal semantics discussed here are not affected. See Example 6 below for unambiguous usage.

Since the structure Ut arises from the composition of two monadic functors, the composition of monad unit maps is bound to occur often in its use. We write just the bracket $\lfloor \_ \rfloor : \mathrm{Ut_b} \to \mathrm{Ut}$ for the precise but verbose map $(\delta\eta_{\mathrm{Px}})_{\mathrm{Ut_b}} = \delta_{\mathrm{Ut_p}} \circ \eta_{\mathrm{Px},\mathrm{Ut_b}} = \mathrm{Zf}(\eta_{\mathrm{Px},\mathrm{Ut_b}}) \circ \delta_{\mathrm{Ut_b}}$. In other occurrences, the subscripts of δ and η are omitted in applications, and can be inferred from the context.

Example 5. There are additional so-called *derived* units in the SI, which have a base unit symbol of their own, but whose semantics are defined by reduction to other (composite) units, for example the *newton*, $\lfloor \mathrm{N} \rfloor \equiv \{\mathrm{kg} \mapsto 1, \eta(\mathrm{m}) \mapsto 1, \eta(\mathrm{s}) \mapsto -2\}_{/0}$, where $\equiv$ is some semantic equivalence relation yet to be specified; see Definitions 10, 18, 22 and 27 below.

Definition 6 (Prefix/Root of a Unit). *Any unit can be decomposed into a prefix and a root unit, and the root units embedded back into the units, by means of natural group homomorphisms:*

$$
\begin{aligned}
\mathrm{pref} &: \mathrm{Ut} \to \mathrm{Px} & \mathrm{pref} &= \pi_1 \circ \beta_{\mathrm{Px},\mathrm{Ut_b}} = \varepsilon_{\mathrm{Px}} \circ \mathcal{A}(\pi_1) \\
\mathrm{root} &: \mathrm{Ut} \to \mathrm{Ut_r} & \mathrm{root} &= \pi_2 \circ \beta_{\mathrm{Px},\mathrm{Ut_b}} = \mathcal{A}(\pi_2) \\
\mathrm{unroot} &: \mathrm{Ut_r} \to \mathrm{Ut} & \mathrm{unroot} &= \mathcal{A}(\eta_{\mathrm{Px},\mathrm{Ut_b}}) \\
\mathrm{strip} &: \mathrm{Ut} \to \mathrm{Ut} & \mathrm{strip} &= \mathrm{unroot} \circ \mathrm{root}
\end{aligned}
$$

Definition 7 (Root Equivalence). *Two units are called* root equivalent, *written $\simeq_{\mathrm{r}}$, iff their roots coincide:*

$$
u \simeq_{\mathrm{r}} v \iff \mathrm{root}(u) = \mathrm{root}(v)
$$

The units as defined above are a faithful semantic model of traditional notations, confer [ISO09]. However, an algebraically more well-behaved structure can be derived by transposing prefixes and free abelian group construction:

Definition 8 (Normalized Unit). *The* normalized units $\mathrm{Ut_n}$ *are the direct sum of a prefix and a root unit:*

$$
\mathrm{Ut_n} = \mathcal{D}_{\mathrm{Px}}(\mathrm{Ut_r})
$$

Definition 9 (Normalization). *The normalized units are derived naturally from the units proper, in terms of the natural group homomorphism β:*

$$
\mathrm{norm} : \mathrm{Ut} \to \mathrm{Ut_n} \qquad \mathrm{norm} = \beta_{\mathrm{Px},\mathrm{Ut_b}} = \langle \mathrm{pref}, \mathrm{root} \rangle
$$

The composite monad structure of $\mathcal{U}(\mathrm{Ut_n}) = \mathcal{C}_{\mathrm{Px}}\mathrm{Zf}(\mathrm{Ut_b})$ comes with a multiplication $\xi_{\mathrm{Px},\mathrm{Ut_b}}$, and thus provides exactly the compositionality lacking in the traditional model Ut: In a normalized unit, other normalized units can be substituted for base units and just "multiplied out". In Ut, this is formally impossible:

Theorem 1. *The group $\mathrm{Ut} = \mathrm{Zf}\mathcal{C}_{\mathrm{Px}}(\mathrm{Ut_b})$ cannot be given a composite monadic structure in the same way; except for degenerate cases, there is no distributive law $\mathcal{C}_G \mathrm{Zf} \Rightarrow \mathrm{Zf}\,\mathcal{C}_G$ that is also a group homomorphism.*

Remark 4. Normalization is generally not injective, since the actual distribution of partial prefixes is also "multiplied out", and forgotten. For example, *micrometer-per-microsecond* cancels to *meter-per-second*: $\text{norm}(\{(\delta(m), m) \mapsto 1, (\delta(m), s) \mapsto -1\}_{/0}) = \text{norm}(\{\eta(m) \mapsto 1, \eta(s) \mapsto -1\}_{/0}) = \theta(m)\,\theta(s)^{-1}$.

Definition 10 (Normal Equivalence). *Two units are called* normally equivalent, *written* $\simeq_n$, *iff their normalizations coincide:*

$$u \simeq_n v \iff \text{norm}(u) = \text{norm}(v)$$

Example 6 (**Precipitation**). Normalization is not part of the tradition notation of units. However, it has the beneficial property that entangled, redundant prefixes and base units are cancelled out orthogonally:

Consider a meteorological unit for *amount of precipitation, litres-per-square-meter* (L/m^2), where a *litre* (L) is defined as the third power of a *decimeter* (dm), which parses as a simple preunit analogous to kg; that is formally $p = \{(\delta(d), m) \mapsto 3, \eta(m) \mapsto -2\}_{/0}$. By normalization, a root unit factor of $\delta(m)^2$ is cancelled out: $\text{norm}(p) = (\delta(d)^3, \delta(m))$ Thus we find that p is normally equivalent to a *deci-deci-deci-meter*, $p \simeq_n \delta((\delta(d)^3, m))$, but not to a *millimeter*; $p \not\simeq_n \delta((\delta(m), m))$.

In the semantic structures presented so far, the base symbols are free; they stand only for themselves, operated on exclusively by natural transformations, and do not carry any attributes for comparison. As the last example has shown, the resulting notions of semantic equivalence may be narrower than intended. The following section introduces one such attribute each for prefixes and units, and explores the semantic consequences. Note that any actual assignment of attribute values is *contingent*; it could well be different in another possible world, i.e., system of units, whereas all of the preceding reasoning is logically *necessary*.

4 Specific Attributes

Definition 11 (Base Prefix Value). *Every base prefix shall be assigned a ratio as its numerical value.*

$$\text{val}_b : Px_b \to \mathbb{Q}_+$$

Example 7 (SI Prefix Values). The three families of SI prefixes are defined numerically as negative and positive (mostly triple) powers of ten, and positive (tenfold) powers of two, respectively: [11]

$$\text{val}_b^{SI} = \begin{cases} d \mapsto 10^{-1}, c \mapsto 10^{-2}, m \mapsto 10^{-3}, m \mapsto 10^{-6}, n \mapsto 10^{-9}, \dots \\ da \mapsto 10^{+1}, h \mapsto 10^{+2}, k \mapsto 10^{+3}, M \mapsto 10^{+6}, G \mapsto 10^{+9}, \dots \\ ki \mapsto 2^{+10}, Mi \mapsto 2^{+20}, Gi \mapsto 2^{+30}, \dots \end{cases}$$

---

[11] It is scientific standard to assign context-free numerical values to prefixes; some traditional notations do not follow the practice. For example, consider the popular (ab)use of *kilobyte*, which has given rise to the binary family for proper distinction.

Definition 12 (Base Unit Dimension). *Every base unit shall be assigned a (possibly compound) dimension.*

$$\dim_b : Ut_b \to \mathcal{U}(Dm)$$

Example 8 (SI Unit Dimensions). The SI base units correspond to the SI base dimensions in the respective order presented above, that is

$$\dim_b^{SI}(m) = \delta(L) \qquad\qquad \dim_b^{SI}(s) = \delta(T)$$

etc., whereas derived SI units may have more complex dimensions:

$$\dim_b^{SI}(N) = \{L \mapsto 1, T \mapsto -2, M \mapsto 1\}_{/0}$$

Definition 13 (Prefix Value). *Prefix value assignment lifts naturally to all concerned structures:*

$$
\begin{aligned}
val &: Px \to Qm & val &= \varepsilon_{Qm} \circ \mathcal{A}(val_b)\\
pval_p &: Ut_p \to \mathbb{Q}_+ & pval_p &= val \circ \pi_1\\
pval &: Ut \to Qm & pval &= \varepsilon_{Qm} \circ \mathcal{A}(pval_p)\\
pval_n &: Ut_n \to Qm & pval_n &= val \circ \pi_1
\end{aligned}
$$

Definition 14 (Unit Dimension). *Dimension assignment lifts naturally to all concerned structures:*

$$
\begin{aligned}
\dim_r &: Ut_r \to Dm & \dim_r &= \varepsilon_{Dm} \circ \mathcal{A}(\dim_b)\\
\dim_p &: Ut_p \to \mathcal{U}(Dm) & \dim_p &= \dim_b \circ \pi_2\\
\dim &: Ut \to Dm & \dim &= \varepsilon_{Dm} \circ \mathcal{A}(\dim_p)\\
\dim_n &: Ut_n \to Dm & \dim_n &= \dim_r \circ \pi_2
\end{aligned}
$$

Example 9 (Density). Consider a unit of *mass density*, kg/cm^3, that is formally $q = \{(\delta(k), g) \mapsto 1, (\delta(c), m) \mapsto -3\}_{/0}$. In the SI interpretation, it follows that $pval(q) = 10^3 \cdot (10^{-2})^{-3} = 10^9$ and $\dim(q) = \{L \mapsto -3, M \mapsto 1\}_{/0}$.

Definition 15 (Codimensionality). *Two (base, pre-, ...) units are called codimensional, written $\sim_\square$, with the appropriate initial substituted for $\square$, iff their assigned dimensions coincide:*

$$u \sim_\square v \iff \dim_\square(u) = \dim_\square(v)$$

Definition 16 (Evaluated Unit). *The evaluated units Ut_e are the direct sum of a ratio and a root unit:*

$$Ut_e = D_{Qm}(Ut_r)$$

Definition 17 (Evaluation). *Evaluation separates prefix value and root unit:*

$$eval : Ut \to Ut_e \qquad eval = \langle pval, root \rangle = (val \times id_{Ut_r}) \circ norm$$

Definition 18 (Numerical Equivalence). *Two units are called* numerically *equivalent, written $\simeq_e$, iff their evaluations coincide:*

$$u \simeq_e v \iff \mathrm{eval}(u) = \mathrm{eval}(v)$$

Theorem 2. *Normalization equivalence entails numerical equivalence, which entails root equivalence, which entails codimensionality:*

$$u \simeq_n v \implies u \simeq_e v \implies u \simeq_r \implies u \sim v$$

Numerical equivalence is coarser than normalization equivalence, because it uses an additional source of information; whereas the latter depends only on the universal group structures of prefixes and units, the former takes contingent value assignments (val_b) into account.

Example 10 (Precipitation revisited). Continuing Example 6, we find that, given the SI interpretation of prefix values, the precipitation unit p is indeed numerically equivalent to the *millimeter*:

$$\mathrm{eval}^{\mathrm{SI}}(p) = \left(10^{-3}, \delta(\mathrm{m})\right) = \mathrm{eval}^{\mathrm{SI}}\left(\delta((\delta(\mathrm{m}), \mathrm{m}))\right) \implies p \simeq_e \delta((\delta(\mathrm{m}), \mathrm{m}))$$

5 Unit Conversion Relations

In the following sections, we deal with ternary relations of a particular type, namely the carrier of a direct sum of three groups:

$$\mathrm{Conv} = \mathrm{Ut} \times \mathrm{Qm} \times \mathrm{Ut} \qquad \mathcal{U}(\mathrm{Conv}) = \mathrm{Zf}(\mathrm{Ut_p}) \times \mathbb{Q}_+ \times \mathrm{Zf}(\mathrm{Ut_p})$$

The middle component is called a *conversion ratio*. We shall write $u \xrightarrow{r}_R v$ for $(u, r, v) \in R$, a notation that alludes to categorial diagrams; see Theorem 4 below.

Definition 19 (Unit Conversion). *A relation $C \subseteq \mathcal{U}(\mathrm{Conv})$ is called a* (unit) *conversion, iff it is dimensionally consistent and functional in its arguments of unit type:*

$$u \xrightarrow{r}_C v \implies u \sim v \tag{1a}$$

$$u \xrightarrow{r}_C v \wedge u \xrightarrow{r'}_C v \implies r = r' \tag{1b}$$

Remark 5. A judgement $u \xrightarrow{r}_C v$ means "One u is r vs," and hence introduces an algebraic rewriting rule for quantity values that converts (values measured in) u to v, since multiplication with the conversion factor r is taken as associative:

$$x\,u = x\,(r\,v) = (xr)\,v$$

Example 11 (UK Units). Standards in the United Kingdom define the regional customary units *pound* (lb) and *pint* (pt) with the following conversion rules:

$$\lfloor \mathrm{lb} \rfloor \xrightarrow{453.59237} \lfloor \mathrm{g} \rfloor \qquad\qquad \lfloor \mathrm{pt} \rfloor \xrightarrow{568.26125} \delta((\delta(\mathrm{c}), \mathrm{m}))^3$$

Definition 20 (Conversion Closure). *Let* $C \subseteq \mathcal{U}(\mathrm{Conv})$ *be a unit conversion. Its* (conversion) *closure is the smallest relation* C^* *with* $C \subseteq C^* \subseteq \mathcal{U}(\mathrm{Conv})$ *such that the following axioms hold:*

$$u \xrightarrow{r}_{C^*} v \wedge u' \xrightarrow{r'}_{C^*} v' \implies uu' \xrightarrow{rr'}_{C^*} vv' \tag{2a}$$

$$u \xrightarrow{r}_{C^*} v \implies u^{-1} \xrightarrow{r^{-1}}_{C^*} v^{-1} \tag{2b}$$

$$u \xrightarrow{\mathrm{pval}(u)}_{C^*} \mathrm{strip}(u) \tag{2c}$$

Remark 6. The closure of a unit conversion may fail to be a unit conversion itself, since contradictory factors can arise from closure axioms, violating (1b). For example, take both $u \xrightarrow{2}_C v$ and $u^{-1} \xrightarrow{3}_C v^{-1}$, but clearly $2 \neq 3^{-1}$.

Remark 7. Closure ensures operational *completeness* of the reasoning, namely that all potential rewriting rules for composite units implied by rules for their constituents are actually available. For example, if we know how to convert from *furlongs* to *meters*, and from *seconds* to *fortnights*, then we can deduce how to convert from *furlongs-per-fortnight* to *millimeters-per-second*.

Conversion closures have a rich algebraic structure:

Theorem 3. *The closure of a unit conversion forms a subgroup of* Conv.

Theorem 4. *The closure of a unit conversion forms an invertible category, the categorial generalization of an equivalence relation, with unit objects and conversion factor morphisms.*

Remark 8. The closure of a unit conversion is again a conversion iff it is *thin* as a category. Such relations shall take center stage in the next section.

Theorem 5. *Conversions can be decomposed and partitioned by dimension:*

$$\mathrm{Ut}_{(d)} = \{u \in \mathrm{Ut} \mid \dim(u) = d\} \subseteq \mathcal{U}(\mathrm{Ut})$$
$$C_{(d)} = C \cap (\mathrm{Ut}_{(d)} \times \mathbb{Q}_+ \times \mathrm{Ut}_{(d)}) \qquad C = \bigcup_{d \in \mathrm{Dm}} C_{(d)}$$

Remark 9. A non-converting theory of units, such as [Ken96], is characterized by being partitioned into *trivial* subgroups. This demonstrates that the theory being presented here is a complementary extension to previous work.

Remark 10. Closure can not be performed on the partitions individually, mostly because of axiom (2a) that allows for multiplication of units with orthogonal dimensions. Thus it is generally the case that:

$$C^* \neq \bigcup_{d \in \mathrm{Dm}} C_{(d)}{}^*$$

Definition 21 (Unit Convertibility). *Two units are called* convertible, *with respect to a unit conversion* C, *written* $\propto_C$, *iff they are related by some factor:*

$$u \propto_C v \iff u \xrightarrow{\exists r}_C v$$

Definition 22 (Unit Coherence). *Two units are called* coherent, *with respect to a unit conversion C, written $\cong_C$, iff they are related by the factor one:*

$$u \cong_C v \iff u \xrightarrow{1}_C v$$

Theorem 6. *Coherence entails convertibility, which entails codimensionality:*

$$u \cong_C v \implies u \propto_C v \implies u \sim v$$

Definition 23 (Conversion Coherence). *By extension, a unit conversion C is called* coherent *iff all convertible units are coherent:*

$$u \cong_C v \iff u \propto_C v$$

Example 12 (SI Unit Coherence). The conversion relation of the seven canonical base units of the SI is trivially coherent, since they are pairwise incovertible. In addition, the SI recognizes 22 derived units with coherent conversion rules. By contrast, many traditional units, such as the *hour* (h), are convertible but not coherently so: $h \xrightarrow{3600} s$.

5.1 Special Cases

Interesting subclasses of conversions arise as the closures of syntactically restricted generators.

Definition 24 (Defining Conversion). *A unit conversion C is called* defining *iff it is basic and functional in its first component:*

$$u \xrightarrow{r}_C v \implies \exists u_0 . u = \lfloor u_0 \rfloor \tag{3a}$$

$$u \xrightarrow{r}_C v \wedge u \xrightarrow{r'}_C v' \implies v = v' \tag{3b}$$

Definition 25 (Definition Expansion). *Every defining unit conversion C gives rise to a totalized expansion function:*

$$\mathrm{xpd}(C) : \mathrm{Ut_b} \to \mathcal{C}_{\mathrm{Qm}}(\mathrm{Ut}) \qquad \mathrm{xpd}(C)(u_0) = \begin{cases} (r, v) & \text{if } \lfloor u_0 \rfloor \xrightarrow{r}_C v \\ \eta(\lfloor u_0 \rfloor) & \text{if no match} \end{cases}$$

This in turn gives rise to a mapping of base to evaluated units, and ultimately to an iterable rewriting operation on evaluated units:

$$\mathrm{rwr_b}(C) : \mathrm{Ut_b} \to \mathrm{Ut_e} \qquad \mathrm{rwr_b}(C) = \mu_{\mathrm{Qm,Ut_r}} \circ \mathcal{C}_{\mathrm{Qm}}(\mathrm{eval}) \circ \mathrm{xpd}(C)$$

$$\mathrm{rwr_e}(C) : \mathrm{Ut_e} \to \mathrm{Ut_e} \qquad \mathrm{rwr_e}(C) = \xi_{\mathrm{Qm,Ut_b}} \circ \mathcal{C}_{\mathrm{Qm}} \mathrm{Zf}(\mathrm{rwr_b}(C))$$

Definition 26 (Dependency Order). *Let C be a defining unit conversion. The relation $(>_C) \subseteq \mathcal{U}(\mathrm{Ut_b})^2$ is the smallest transitive relation such that:*

$$\lfloor u_0 \rfloor \propto_C v \wedge v_0 \in \mathrm{supp}(\mathrm{root}(v)) \implies u_0 >_C v_0$$

We say that, with respect to C, u_0 depends on v_0.

Definition 27 (Well-Defining Conversion). *A defining unit conversion C is called* well-defining *iff its dependency order $>_C$ is well-founded. Since $\mathrm{Ut_b}$ is finite, this is already the case if $>_C$ is antireflexive.*

Theorem 7. *Let C be a well-defining conversion. The iteration of $\mathrm{rwr_e}(C)$ has a fixed point, which is reached after a number N_C of steps that is bounded by the number of distinct base units, and independent of the input unit:*

$$\lim_{n \to \infty} \mathrm{rwr_e}(C)^n = \mathrm{rwr_e}(C)^{N_C \leq |\mathrm{Ut_b}|}$$

Example 13 (SI Definitions). Traditional systems of units, including the SI, follow a well-defining approach: Starting from an irreducible set of base units, additional "derived" base units are added in a stratified way, by defining them as convertible to (expressions over) preexisting ones.

6   The Conversion Hierarchy

Definition 28 (Conversion Hierarchy). *A conversion is called ...*

1. consistent *iff its closure is again a conversion;*
2. closed *iff it is its own closure;*
3. finitely generated *iff it is the closure of a finite conversion;*
4. defined *iff it is the closure of a defining conversion;*
5. well-defined *iff it is the closure of a well-defining conversion;*
6. regular *iff it is the closure of an empty conversion.*

Theorem 8. *Each property in the conversion hierarchy entails the preceding.*

Theorem 9. *Consistency is non-local; namely the following three statements are equivalent in the closure of a conversion C:*

a. *contradictory factors exist for some pair of units;*
b. *contradictory factors exist for all convertible pairs of units;*
c. $\varnothing/0 \xrightarrow{\neq 1}_{C^*} \varnothing/0.$

Theorem 10. *For closed conversions, convertibility and coherence are group congruence relations.*

Theorem 11. *For closed conversions, the resulting category is strict dagger compact.*

Theorem 12. *In a closed conversion C, any two units are convertible if they are root equivalent, and coherent if they are numerically equivalent.*

$$u \simeq_r v \implies u \propto_C v \qquad\qquad u \simeq_e v \implies u \cong_C v$$

Most of the entailments are generally proper, but one is not:

Theorem 13. *All closed conversions are finitely generated.*

Remark 11. Being finitely generated as a closed conversion in this sense is entailed by being finitely generated as an abelian group in the sense of Theorem 3, but *not* vice versa; the former could be "larger" due to axiom (2c).

Theorem 14. *Every well-defining conversion C is consistent, and hence gives rise to a well-defined C^*.*

Theorem 15. *In a regular conversion C, in addition to Theorem 12, any two units are convertible if and only if they are root equivalent, and coherent if and only if they are numerically equivalent:*

$$u \simeq_r v \iff u \propto_C v \qquad\qquad u \simeq_e v \iff u \cong_C v$$

Remark 12. The convertibility problem for closed conversions can encode the *word problem* for quotients of unit groups, thus there is potential danger of undecidability. We conjecture that, by Theorem 13, closed conversions are *residually finite*, such that a known decision algorithm [Rob96] could be used in principle.

For well-defined conversions, a more direct and efficient algorithm exists.

Definition 29 (Exhaustive Rewriting). *Given a well-defining conversion C, units can be evaluated and exhaustively rewritten:*

$$\mathrm{rwr}^*(C) : \mathrm{U}\iota \to \mathrm{U}\mathrm{t}_e \qquad\qquad \mathrm{rwr}^*(C) = \mathrm{rwr}_e(C)^{N_C} \circ \mathrm{eval}$$

Each step requires a linearly bounded number of group operations. The kernel of this map can be upgraded to a ternary relation $C^\sharp \subseteq \mathcal{U}(\mathrm{Conv})$:

$$C^\sharp = \left\{ (u, rs^{-1}, v) \in \mathcal{U}(\mathrm{Conv}) \,\middle|\, \begin{matrix} \mathrm{rwr}^*(C)(u) = (r, u') \\ \mathrm{rwr}^*(C)(v) = (s, v') \end{matrix} \wedge u' = v' \right\}$$

Theorem 16. *Exhaustive rewriting solves the well-defined conversion problem; namely, let C be a well-defining conversion, then:*

$$C^* = C^\sharp$$

Example 14 (Mars Climate Orbiter). The subsystems of the Orbiter attempted to communicate using different units of *linear momentum*, namely *pound-force-seconds* $u = \lfloor \mathrm{lbf} \rfloor \lfloor \mathrm{s} \rfloor$ vs. *newton-seconds* $v = \lfloor \mathrm{N} \rfloor \lfloor \mathrm{s} \rfloor$. The relevant base units are well-defined w.r.t the irreducible SI units as $\lfloor \mathrm{lbf} \rfloor \xrightarrow{1} \lfloor \mathrm{lb} \rfloor \lfloor g_n \rfloor$ and $\lfloor \mathrm{N} \rfloor \xrightarrow{1} \delta(\mathrm{kg}) \lfloor \mathrm{m} \rfloor \lfloor \mathrm{s} \rfloor^{-2}$, with the auxiliary units *pound* $\lfloor \mathrm{lb} \rfloor \xrightarrow{a} \lfloor \mathrm{g} \rfloor$ (see Example 11) and *norm gravity* $\lfloor g_n \rfloor \xrightarrow{b} \lfloor \mathrm{m} \rfloor \lfloor \mathrm{s} \rfloor^{-2}$, and the conversion factors $a = 453.59237$ and $b = 9.80665$. Exhaustive rewriting yields $\mathrm{rwr}^*(C)(u) = (ab, \lfloor \mathrm{g} \rfloor \lfloor \mathrm{m} \rfloor \lfloor \mathrm{s} \rfloor^{-1})$ and $\mathrm{rwr}^*(C)(v) = \left(1000, \lfloor \mathrm{g} \rfloor \lfloor \mathrm{m} \rfloor \lfloor \mathrm{s} \rfloor^{-1}\right)$, and hence $u \xrightarrow{ab/1000} v$.[12]

---

[12] Which reconstructs the factor given in the introduction.

7   Conclusion

We have presented a group-theoretic formal model of units of measure, to our knowledge the first one that supports prefixes and arbitrary conversion factors. The model is based on two composeable monadic structures, and hence has the compositionality required for reasoning with substitution and for denotational semantics. The model is epistemologically stratified, and distinguishes cleanly between necessary properties and natural operations on the one hand, and contingent properties and definable interpretations on the other, leading to a hierarchy of semantic equivalences.

We have characterized unit conversion rules by a class of ternary relations equipped with a closure operator, that function both as a group congruence and as a category, together with a six-tiered hierarchy of subclasses, each with better algebraic behavior than the preceding. This model reconstructs and justifies the reductionistic approach of traditional scientific unit systems.

7.1   Related Work

[Ken94] has found existing related work then to be generally lacking in both formal rigor and universality. With regard to the research programme outlined by [Ken96], matters have improved only partially; see [MBBS20] for a recent critical survey. Formally rigorous and expressive type systems have been proposed for various contexts, such as C [HCR12], UML/OCL [MWV16], generic (but instantiated for Java) [XLD20]. However, in that line of research, neither flexibly convertible units in general nor prefixes in particular are supported.

An object-oriented solution with convertible units is described by [All+04]. Their approach, like Kennedy's, is syntactic; abelian groups are added as an ad-hoc language extension with "abelian classes" and a normalization procedure. Prefixes are not recognized. Unit conversion is defined always in relation to a fixed reference unit per dimension. This implies that all codimensional units are convertible, which is manifestly unsound. [13] Nevertheless, the approach has been reiterated in later work such as [CM07]. By contrast, [KL78] had already proposed a more flexible relational approach to unit conversion, with methods of matrix calculus for checking consistency.

References

[AHS06] Adamek, J., Herrlich, H., Strecker, G.E.: Abstract and Concrete Categories. Free reprint (2006). http://www.tac.mta.ca/tac/reprints/articles/17/tr17.pdf

[All+04] Allen, E., et al.: Object-oriented units of measurement. In: SIGPLAN Not. 39.10, pp. 384–403 (2004). https://doi.org/10.1145/1035292.1029008

---

[13] Consider the codimensional units of *energy* and *torque*, *joule* and *newton-meter*, or of *absorbed dose* and *equivalent dose of ionizing ration*, *gray* and *sievert*, resp.

[CM07] Cooper, J., McKeever, S.: A model-driven approach to automatic conversion of physical units. Softw. Pract. Exp. **34**(4), 337–359 (2007). https://doi.org/10.1002/spe.828

[Gun15] Gundry, A.: A typechecker plugin for units of measure. In: Proceedings of Haskell Workshop. ACM (2015). https://doi.org/10.1145/2804302.2804305

[HCR12] Hills, M., Chen, F., Roşu, G.: A rewriting logic approach to static checking of units of measurement in C. ENTCS **290**, 51–67 (2012). https://doi.org/10.1016/j.entcs.2012.11.011

[KL78] Karr, M., Loveman III, D.B.: Incorporation of units into programming languages. Comm. ACM **21**(5), 385–391 (1978)

[Ken94] Kennedy, A.: Dimension types. In: Sannella, D. (ed.) ESOP 1994. LNCS, vol. 788, pp. 348–362. Springer, Heidelberg (1994). https://doi.org/10.1007/3-540-57880-3_23

[Ken96] Kennedy, A.: Programming Languages and Dimensions. PhD Diss. University of Cambridge (1996). https://www.cl.cam.ac.uk/techreports/UCAM-CL-TR-391.pdf

[Ken10] Kennedy, A.: Types for units-of-measure: theory and practice. In: Horváth, Z., Plasmeijer, R., Zsók, V. (eds.) CEFP 2009. LNCS, vol. 6299, pp. 268–305. Springer, Heidelberg (2010). https://doi.org/10.1007/978-3-642-17685-2_8

[MWV16] Mayerhofer, T., Wimmer, M., Vallecillo, A.: Adding uncertainty and units to quantity types in software models. In: Proceedings of SLE 2016, pp. 118–131. ACM (2016). https://doi.org/10.1145/2997364.2997376

[MBBS20] McKeever, S., Bennich-Björkman, O., Salah, O.-A.: Unit of measurement libraries, their popularity and suitability. Softw. Pract. Exp. **51**, 711–734 (2020). https://doi.org/10.1002/spe.2926

[ISO09] Quantities and units – Part 1: General. Standard ISO/IEC 80000-1:2009. International Organization for Standardization, 2009. https://www.iso.org/obp/ui/#iso:std:iso:80000:-1:ed-1:v1:en

[Rob96] Robinson, D.: A Course in the Theory of Groups. Springer, Heidelberg (1996). isbn: 978-1-4419-8594-1

[Ste+99] Stephenson, A.G., et al.: Mars Climate Orbiter Mishap Investigation Board Phase I Report. NASA (1999). https://llis.nasa.gov/llis_lib/pdf/1009464main1_0641-mr.pdf

[TL22] Trancón y Widemann, B., Lepper, M.: Towards a theory of con- version relations for prefixed units of measure. Extended report (2022). arXiv: 2212.11580

[Wal22] Wallot, J.: Die physikalischen und technischen Einheiten. Elek- trotechnische Zeitschrift **43**, 1329–1333 (1922)

[XLD20] Xiang, T., Luo, J.Y., Dietl, W.: Precise inference of expressive units of measurement types. In: Proceedings of ACM Programming Language (OOPSLA 2020), vol. 4, no. 142, pp. 1–28 (2020). https://doi.org/10.1145/3428210

Relational Algebraic Approach to the Real Numbers the Additive Group

Michael Winter[✉]

Department of Computer Science, Brock University, St. Catharines, ON, Canada
mwinter@brocku.ca

Abstract. In this paper we start the investigation of an object representing the real numbers in categories of relations. Our axiomatization uses the construction of a relation power, i.e., an abstract version of power sets within the category. This allows us to utilize a relation algebraic version of Tarski's axioms of the real numbers as a first-order definition of a real number object. The current paper focuses on the addition operation of the real number object. It is shown that addition forms a densely and linearly ordered abelian group.

1 Introduction

Relation algebraic methods have been used to specify, implement, and verify programs. In fact, this is a major thread within the RAMiCS community. If the approach utilizes categories of relations, objects usually represent types, and relations usually represent programs of the programming language. Therefore, constructions on objects model typical type constructions in programming languages. For example, the relational sum is used to model a disjoint union of types, and a relational product is used to model pairing. Similarly, certain objects such as unit or such as a natural number object can be used to model basic types in programming languages. This paper starts the investigation of a real number object, i.e., an object that models the real numbers.

Our framework will be based on Heyting categories with relational powers. The relational power is an abstract version of the power set within these categories. It is worth mentioning that this framework is a purely equational theory that allows formulating properties that are usually formulated using second-order formulas as equations as well. Therefore, our axiomatization of a real number object utilizes Tarski's axioms of the real numbers [7] translated into the equational theory of Heyting categories with relational powers.

The theory presented in this paper is complement free, i.e., all theorems follow without the use of Boolean complements. We will maintain this approach in future work on this topic. Our main motivation for this is that the results transfer immediately to so-called *L*-fuzzy relations, i.e., to relations that use a Heyting algebra *L* as truth values instead of the Boolean truth values **true** and

The author gratefully acknowledges support from the Natural Sciences and Engineering Research Council of Canada (283267).

R. Glück et al. (Eds.): RAMiCS 2023, LNCS 13896, pp. 274–292, 2023.
https://doi.org/10.1007/978-3-031-28083-2_17

`false`. This will make it possible to investigate the real numbers also in the fuzzy case. Please note that Axiom 4 (see Definition 4) is the only axiom that involve the power object. In future work we will be interested in replacing this construction in the context of L-fuzzy relation with the fuzzy power object, i.e., the set of all fuzzy subsets, in order to obtain a fuzzy version of the real numbers.

In this paper we will focus on the addition operation of the real number object. It is shown that addition forms a densely and linearly ordered abelian group.

2 Mathematical Preliminaries

In this chapter we want to introduce the mathematical notions used in this paper. We start with the theory of relations that we will be using.

2.1 Heyting Categories

In this section we want to recall some basic notions from categories and allegories [1,9,10]. First, we are going to introduce Heyting categories as an extension of division allegories defined in [1], i.e., a Heyting category is a division allegory where the lattice of relations between two objects is a Heyting algebra instead of just a distributive lattice. Heyting categories are also a version of Dedekind categories introduced in [2,3] without the requirement of completeness of lattice of relations between two objects.

We will write $R : A \to B$ to indicate that a morphism R of a category $\mathcal{C}$ has source A and target B. Composition and the identity morphism are denoted by ; and $\mathbb{I}_A$, respectively. Please note that composition has to be read from left to right, i.e., $Q; R$ means first Q and then R.

Definition 1. *A Heyting category $\mathcal{R}$ is a category satisfying the following:*

1. *For all objects A and B the collection $\mathcal{R}[A, B]$ is a Heyting algebra. Meet, join, relative pseudo-complement, the induced ordering, the least and the greatest element are denoted by $\sqcap, \sqcup, \to, \sqsubseteq, \bot\!\!\!\bot_{AB}, \top\!\!\!\top_{AB}$, respectively.*
2. *$Q; \bot\!\!\!\bot_{BC} = \bot\!\!\!\bot_{AC}$ for all relations $Q : A \to B$.*
3. *There is a monotone operation $\smile$ (called converse) mapping a relation $Q : A \to B$ to $Q^{\smile} : B \to A$ such that for all relations $Q : A \to B$ and $R : B \to C$ the following holds: $(Q; R)^{\smile} = R^{\smile}; Q^{\smile}$ and $(Q^{\smile})^{\smile} = Q$.*
4. *For all relations $Q : A \to B, R : B \to C$ and $S : A \to C$ the modular inclusion $(Q; R) \sqcap S \sqsubseteq Q; (R \sqcap (Q^{\smile}; S))$ holds.*
5. *For all relations $R : B \to C$ and $S : A \to C$ there is a relation $S/R : A \to B$ (called the right residual of S and R) such that for all $X : A \to B$ the following holds: $X; R \sqsubseteq S \iff X \sqsubseteq S/R$.*

If we define the left residual $Q \backslash R \cdot R \to C$ of two relations $Q : A \to B$ and $R : A \to C$ by $Q \backslash R := (R^{\smile}/Q^{\smile})^{\smile}$ we immediately obtain $X \sqsubseteq Q \backslash R$ iff $Q; X \sqsubseteq R$. Using both residuals we define the symmetric quotient as $\mathrm{syQ}(Q, R) =$

$(Q \backslash R) \sqcap (Q^{\smile}/R^{\smile})$. This construction is characterized by $X \sqsubseteq \mathrm{syQ}(Q, R)$ iff $Q; X \sqsubseteq R$ and $R; X^{\smile} \sqsubseteq Q$. Please note that if the Heyting algebra of relation is a Boolean algebra we get $Q \backslash R = \overline{Q^{\smile}; \overline{R}}$ where $\overline{R}$ is the complement of R. A similar definition for arbitrary Heyting algebra does not exist.

Throughout the paper we will use the axioms and some basics facts such as monotonicity of the operations without mentioning. The following lemma summarizes some basic properties that will be used throughout the paper. A proof can be found in [4,5,8,9].

Lemma 1. *Let* $Q : A \to B$, $R : A \to C$ *and* $S : C \to D$ *be relations, and* $i : A \to A$ *be a partial identity* $i \sqsubseteq \mathbb{I}_A$. *Then we have:*

1. $i^{\smile} = i$.
2. $i; i = i$.
3. $(Q; \mathbb{T}_{BC} \sqcap R); S = Q; \mathbb{T}_{BD} \sqcap R; S$.

A relation $Q : A \to B$ is called univalent (or partial function) iff $Q^{\smile}; Q \sqsubseteq \mathbb{I}_B$, total iff $\mathbb{I}_A \sqsubseteq Q; Q^{\smile}$, injective iff $Q^{\smile}$ is univalent, surjective iff $Q^{\smile}$ is total, a map iff Q is total and univalent. The following lemma states some basic properties of univalent relations and maps. Again, a proof can be found in [4,5,8,9].

Lemma 2. *Let* $f : A \to B$ *be a mapping,* $g : B \to A$ *univalent,* $Q : C \to A$, $R : C \to B$, $S : A \to C$ *and* $T, U : A \to D$. *Then we have:*

1. $Q; f \sqsubseteq R$ *iff* $Q \sqsubseteq R; f^{\smile}$.
2. $(Q; g^{\smile} \sqcap R); g = Q \sqcap R; g$.
3. $g; (T \sqcap U) = g; T \sqcap f; U$.

For a singleton set $\{*\}$ we obviously have $\mathbb{I}_{\{*\}} = \mathbb{T}_{\{*\}\{*\}}$. Furthermore, for any set A the relation $\mathbb{T}_{A\{*\}}$ is actually a map. The first property together with the totality in the second property also characterize singleton sets up to isomorphism. Therefore, we define a unit 1 as an abstract version of a singleton set by $\mathbb{I}_1 = \mathbb{T}_{11}$ and $\mathbb{T}_{A1}$ is total for every object A.

Considering concrete relation a map $p : 1 \to A$, i.e., a relation that maps $*$ to one element a in A, can be identified with the element a. Therefore we call a map $p : 1 \to A$ a point (of A).

Another important concept is the notion of a relational product, i.e., an abstract version of the Cartesian product of sets. The object $A \times B$ is characterized by the projection relations $\pi : A \times B \to A$ and $\rho : A \times B \to B$ satisfying

$$\pi^{\smile}; \pi \sqsubseteq \mathbb{I}_A, \quad \rho^{\smile}; \rho \sqsubseteq \mathbb{I}_B, \quad \pi; \pi^{\smile} \sqcap \rho; \rho^{\smile} = \mathbb{I}_{A \times B}, \quad \pi^{\smile}; \rho = \mathbb{T}_{AB}.$$

Given relational products we will use the following abbreviations

$$Q \otimes R = Q; \pi^{\smile} \sqcap R; \rho^{\smile},$$
$$Q \otimes S = \pi; Q \sqcap \rho; S,$$
$$Q \otimes T = \pi; Q; \pi^{\smile} \sqcap \rho; T; \rho^{\smile} = Q; \pi^{\smile} \otimes T; \rho^{\smile} = \pi; Q \otimes \rho; T,$$

and obtain the following properties [6].

Lemma 3. *If all relational products exist, then we have the following:*

1. $(Q \otimes R)^{\smile} = Q^{\smile} \otimes R^{\smile}$ *and* $(Q \ominus S)^{\smile} = Q^{\smile} \ominus S^{\smile}$.
2. *If R is total, then $(Q \otimes R); \pi = Q$ and if Q is total, then $(Q \otimes R); \rho = R$.*
3. *If S is surjective, then $\pi^{\smile}; (Q \ominus S) = Q$ and if Q is surjective, then $\rho^{\smile}; (Q \ominus S) = S$.*
4. *If f is univalent, then $f; (Q \otimes R) = f; Q \otimes f; R$ and if g is injective, then $(Q \ominus S); g = Q; g \ominus S; g$.*
5. $(Q \otimes R); (T \ominus U) = Q; T \sqcap R; U$.
6. $(Q \otimes R); (T \otimes V) = Q; T \otimes R; V$ *and* $(Q \otimes X); (T \ominus U) = Q; T \otimes X; U$.
7. $Q; \pi^{\smile} \ominus (R \ominus S) = (Q \ominus R) \ominus \rho; S$.

We also use the following two bijective mappings assoc $: A \times (B \times C) \rightarrow (A \times B) \times C$ and swap $: A \times B \rightarrow B \times A$ defined by

$$\text{assoc} = \pi; \pi^{\smile}; \pi^{\smile} \sqcap \rho; \pi; \rho^{\smile}; \pi^{\smile} \sqcap \rho; \rho; \rho^{\smile} = (\mathbb{I}_A \otimes \pi) \ominus \rho; \rho = \pi^{\smile}; \pi^{\smile} \ominus (\rho^{\smile} \otimes \mathbb{I}_C),$$
$$\text{swap} = \pi; \rho^{\smile} \sqcap \rho; \pi^{\smile} = \rho \otimes \pi = \rho^{\smile} \ominus \pi^{\smile}.$$

The following properties are easy to verify. A proof can also be found in [11].

Lemma 4. *1.* swap$^{\smile}$ = swap.
2. $(Q \ominus R); \text{swap} = R \ominus Q$ *and* swap$; (Q \ominus S) = S \ominus Q$.
3. swap$; (Q \otimes T) = (T \otimes Q); \text{swap}$.
4. $(U \ominus (Q \ominus R)); \text{assoc} = (U \ominus Q) \ominus R$ *and* assoc$; ((Q \ominus S) \ominus V) = Q \ominus (S \ominus V)$.
5. assoc$; ((Q \otimes T) \otimes X) = (Q \otimes (T \otimes X)); \text{assoc}$.

With the maps above we are now ready to define an abelian group within a Heyting category.

Definition 2. *A quadruple (A, e, f, n) in a Heyting category $\mathcal{R}$ is called an abelian group iff A is an object, $e : 1 \rightarrow A$ is a point, and $f : A \times A \rightarrow A$ and $n : A \rightarrow A$ are maps satisfying:*

1. *f is associative, i.e., $(\mathbb{I}_A \otimes f); f = \text{assoc}; (f \otimes \mathbb{I}_A); f$,*
2. *e is the neutral element of f, i.e., $(\mathbb{I}_A \ominus \mathbb{T}_{A1}; e); f = \mathbb{I}_A$,*
3. *n is the complement map, i.e., $(\mathbb{I}_A \ominus n); f = \mathbb{T}_{A1}; e$,*
4. *f is commutative. i.e., swap$; f = f$.*

A relation $C : X \rightarrow X$ is called transitive if $C; C \sqsubseteq C$, dense if $C \sqsubseteq C; C$, asymmetric if $C \sqcap C^{\smile} = \perp\!\!\!\perp_{XX}$, a strict-order if C is transitive and asymmetric, a linear strict-order if C is a strict-order and $\mathbb{I}_X \sqcup C \sqcup C^{\smile} = \mathbb{T}_{XX}$.

A linear strict-order does always have a complement. This and related properties are summarized in the next lemma. The proof is straight forward and left to the reader.

Lemma 5. *If $C : X \rightarrow X$ is a linear strict-order, then we have the following:*

1. $C \sqcap \mathbb{I}_X = \perp\!\!\!\perp_{XX}$.
2. $C \sqcap (C^{\smile} \sqcup \mathbb{I}_X) = \perp\!\!\!\perp_{XX}$.

3. $Y \sqsubseteq C$ *iff* $Y \sqcap (C^\smile \sqcup \mathbb{I}_X) = \bot\!\!\!\bot_{XX}$.
4. $Y \sqsubseteq C^\smile \sqcup \mathbb{I}_X$ *iff* $Y \sqcap C = \bot\!\!\!\bot_{XX}$.

The relation algebraic version of a power set is given by a so-called relational (or direct) power.

Definition 3. *An object* $\mathcal{P}(A)$ *together with a relation* $\varepsilon : A \to \mathcal{P}(A)$ *is called a relational (or direct) power of* A *iff*

$$\text{syQ}(\varepsilon, \varepsilon) = \mathbb{I}_{\mathcal{P}(A)} \quad and \quad \text{syQ}(Q, \varepsilon) \text{ is total for every } Q : A \to B.$$

Please note that $\text{syQ}(R^\smile, \varepsilon)$ is a map for every relation $R : B \to A$. In fact, this construction is the existential image of R, i.e., x is mapped by $\text{syQ}(R^\smile, \varepsilon)$ to the set $\{y \mid (x, y) \in R\}$ for concrete relations.

3   Real Number Object

We will use a relation algebraic version of Tarski's axioms for the real number for defining a real number object in a Heyting category with relational powers. First, we recall Tarski's axioms as they were stated in [7]. His axioms are based on the language $\mathbb{R}, <, +, 1$:

Axiom 1: If $x \neq y$, then $x < y$ or $y < x$.
Axiom 2: If $x < y$, then $y \not< x$.
Axiom 3: If $x < z$, then there is a y such that $x < y$ and $y < z$.
Axiom 4: If $X \subseteq \mathbb{R}$ and $Y \subseteq \mathbb{R}$ so that for every $x \in X$ and every $y \in Y$ we have
$x < y$, then there is a z so that for all $x \in X$ and $y \in Y$ we have $x \leqslant z$ and
$z \leqslant y$ ($x \leqslant y$ shorthand for $x < y$ or $x = y$).
Axiom 5: $x + (y + z) = (x + z) + y$.
Axiom 6: For every x and y there is a z such that $x = y + z$.
Axiom 7: If $x + z < y + t$, then $x < y$ or $z < t$.
Axiom 8: $1 \in \mathbb{R}$.
Axiom 9: $1 < 1 + 1$.

A suitable translation of the axioms above into the language of relations leads to the following definition. Please note that we added Axiom 0 that states that add is a map explicitly since we are dealing with relations rather than functions. In addition, it is worth mentioning that the axioms below do not use the notion of a complement and, therefore, are suitable in the framework of Heyting categories.

Definition 4. *An object* $\mathbb{R}$ *together with three relations* $1 : 1 \to \mathbb{R}$, $C : \mathbb{R} \to \mathbb{R}$ *and* add $: \mathbb{R} \times \mathbb{R} \to \mathbb{R}$ *is called a real number object if the following holds:*

0. add *is a map.*
1. $\mathbb{I}_\mathbb{R} \sqcup C \sqcup C^\smile = \top\!\!\top_{\mathbb{R}\mathbb{R}}$.
2. $C \sqcap C^\smile = \bot\!\!\!\bot_{\mathbb{R}\mathbb{R}}$.

3. $C \sqsubseteq C; C.$

4. $\varepsilon \backslash (C/\varepsilon^{\smile}) \sqsubseteq (\varepsilon \backslash (C \sqcup \mathbb{I}_\mathbb{R})); (\varepsilon \backslash (C \sqcup \mathbb{I}_\mathbb{R})^{\smile})^{\smile}.$

5. $(\mathbb{I}_\mathbb{R} \otimes \mathrm{add}); \mathrm{add} = (\mathbb{I}_\mathbb{R} \otimes \mathrm{swap}); \mathrm{assoc}; (\mathrm{add} \otimes \mathbb{I}_\mathbb{R}); \mathrm{add}.$

6. $\pi^{\smile}; \mathrm{add} = \top_{\mathbb{R}\mathbb{R}}.$

7. $\mathrm{add}; C; \mathrm{add}^{\smile} \sqsubseteq \pi; C; \pi^{\smile} \sqcup \rho; C; \rho^{\smile}.$

8. $\mathbb{1}$ is a map, i.e., a point.

9. $\mathbb{1} \sqsubseteq \mathbb{1}; (\mathbb{I}_\mathbb{R} \ominus \mathbb{I}_\mathbb{R}); \mathrm{add}; C^{\smile}.$

First we define abstract versions of the number 0 and of the negation operation on the real numbers by $Z = (\mathrm{add}^{\smile} \sqcap \pi^{\smile}); \rho$ and $\mathrm{neg} = \pi^{\smile}; (\mathrm{add}; Z^{\smile} \sqcap \rho).$

Lemma 6. 1. $\mathrm{swap}; \mathrm{add} = \mathrm{add}.$

2. $(\mathbb{I}_\mathbb{R} \otimes \mathrm{add}); \mathrm{add} = \mathrm{assoc}; (\mathrm{add} \otimes \mathbb{I}_\mathbb{R}); \mathrm{add}.$

3. $\top_{\mathbb{R}\mathbb{R}}; Z = Z.$

4. $(\mathbb{I}_\mathbb{R} \ominus Z); \mathrm{add} = \mathbb{I}_\mathbb{R}.$

5. $\mathbb{I}_\mathbb{R} \ominus Z \sqsubseteq \mathrm{add}^{\smile}.$

6. $\mathrm{neg}^{\smile} = \mathrm{neg}.$

Proof. 1. In order to show the assertion we first compute several properties:

(a) $(\pi^{\smile} \ominus (\pi \ominus \mathrm{add}))^{\smile}; (\pi^{\smile} \ominus (\pi \ominus \mathrm{add})) = \mathbb{I}_{\mathbb{R} \times \mathbb{R}}$:

$$(\pi^{\smile} \ominus (\pi \ominus \mathrm{add}))^{\smile}; (\pi^{\smile} \ominus (\pi \ominus \mathrm{add}))$$

$$= (\pi \ominus (\pi \ominus \mathrm{add})^{\smile}); (\pi^{\smile} \ominus (\pi \ominus \mathrm{add})) \qquad \text{Lemma 3(1)}$$

$$= \pi; \pi^{\smile} \sqcap (\pi \ominus \mathrm{add})^{\smile}; (\pi \ominus \mathrm{add}) \qquad \text{Lemma 3(5)}$$

$$= \pi; \pi^{\smile} \sqcap (\pi \ominus \mathrm{add})^{\smile}; (\pi; \pi^{\smile} \sqcap \mathrm{add}; \rho^{\smile})$$

$$= \pi; \pi^{\smile} \sqcap (\pi \ominus \mathrm{add})^{\smile}; ((\pi \ominus \mathrm{add}); \pi; \pi^{\smile} \sqcap \mathrm{add}; \rho^{\smile}) \qquad \text{Lemma 3(2)}$$

$$= \pi; \pi^{\smile} \sqcap (\pi; \pi^{\smile} \sqcap (\pi \ominus \mathrm{add})^{\smile}; \mathrm{add}; \rho^{\smile}) \qquad \text{Lemma 2(2)}$$

$$= \pi; \pi^{\smile} \sqcap (\pi; \pi^{\smile} \sqcap \rho; \mathrm{add}^{\smile}); \mathrm{add}; \rho^{\smile}$$

$$= \pi; \pi^{\smile} \sqcap (\pi; \pi^{\smile}; \mathrm{add} \sqcap \rho); \rho^{\smile} \qquad \text{Lemma 2(2)}$$

$$= \pi; \pi^{\smile} \sqcap \rho; \rho^{\smile} \qquad \text{Axiom 6., } \pi \text{ total}$$

$$= \mathbb{I}_{\mathbb{R} \times \mathbb{R}}.$$

(b) $\pi; \pi^{\smile}; \pi^{\smile} \sqcap \mathrm{swap}; (\rho^{\smile} \otimes \mathbb{I}_\mathbb{R}) = (\pi^{\smile}; \pi^{\smile} \sqcap \rho^{\smile}) \ominus \rho^{\smile}; \pi^{\smile}$:

$$\pi; \pi^{\smile}; \pi^{\smile} \sqcap \mathrm{swap}; (\rho^{\smile} \otimes \mathbb{I}_\mathbb{R})$$

$$= \pi; \pi^{\smile}; \pi^{\smile} \sqcap (\rho \ominus \pi); (\rho^{\smile} \otimes \mathbb{I}_\mathbb{R})$$

$$= \pi; \pi^{\smile}; \pi^{\smile} \sqcap (\rho; \rho^{\smile} \ominus \pi) \qquad \text{Lemma 3(6)}$$

$$= \pi; \pi^{\smile}; \pi^{\smile} \sqcap \rho; \rho^{\smile}; \pi^{\smile} \sqcap \pi; \rho^{\smile}$$

$$= (\pi^{\smile}; \pi^{\smile} \sqcap \rho^{\smile}) \ominus \rho^{\smile}; \pi^{\smile}. \qquad \text{Lemma 2(3)}$$

(c) $(\pi^\smile \ominus (\pi \ominus \text{add})); \text{add} = (\pi^\smile; \pi^\smile \ominus ((\pi^\smile; \pi^\smile \sqcap \rho^\smile) \ominus \rho^\smile; \pi^\smile)); (\text{add} \otimes \mathbb{I}_\mathbb{R}); \text{add}:$

$$(\pi^\smile \ominus (\pi \ominus \text{add})); \text{add}$$
$$= ((\mathbb{I}_\mathbb{R} \ominus \pi) \ominus \rho; \text{add}); \text{add} \qquad\qquad \text{Lemma 3(7)}$$
$$= ((\mathbb{I}_\mathbb{R} \ominus \pi) \ominus \rho); (\mathbb{I}_\mathbb{R} \otimes \text{add}); \text{add} \qquad\qquad \text{Lemma 3(6)}$$
$$= ((\mathbb{I}_\mathbb{R} \ominus \pi) \ominus \rho); (\mathbb{I}_\mathbb{R} \otimes \text{swap}); \text{assoc}; (\text{add} \otimes \mathbb{I}_\mathbb{R}); \text{add} \qquad \text{Axiom 5.}$$
$$= ((\mathbb{I}_\mathbb{R} \ominus \pi) \ominus \rho; \text{swap}); \text{assoc}; (\text{add} \otimes \mathbb{I}_\mathbb{R}); \text{add} \qquad\qquad \text{Lemma 3(6)}$$
$$= ((\mathbb{I}_\mathbb{R} \ominus \pi) \ominus \rho; \text{swap}); (\pi^\smile; \pi^\smile \ominus (\rho^\smile \otimes \mathbb{I}_\mathbb{R})); (\text{add} \otimes \mathbb{I}_\mathbb{R}); \text{add}$$
$$= ((\mathbb{I}_\mathbb{R} \ominus \pi); \pi^\smile; \pi^\smile \sqcap \rho; \text{swap}; (\rho^\smile \otimes \mathbb{I}_\mathbb{R})); (\text{add} \otimes \mathbb{I}_\mathbb{R}); \text{add} \qquad \text{Lemma 3(5)}$$
$$= (\pi; \pi^\smile; \pi^\smile \sqcap \rho; (\pi; \pi^\smile; \pi^\smile \sqcap \text{swap}; (\rho^\smile \otimes \mathbb{I}_\mathbb{R}))); (\text{add} \otimes \mathbb{I}_\mathbb{R}); \text{add} \qquad \text{Lemma 2(3)}$$
$$= (\pi^\smile; \pi^\smile \ominus (\pi; \pi^\smile; \pi^\smile \sqcap \text{swap}; (\rho^\smile \otimes \mathbb{I}_\mathbb{R}))); (\text{add} \otimes \mathbb{I}_\mathbb{R}); \text{add}$$
$$= (\pi^\smile; \pi^\smile \ominus ((\pi^\smile; \pi^\smile \sqcap \rho^\smile) \ominus \rho^\smile; \pi^\smile)); (\text{add} \otimes \mathbb{I}_\mathbb{R}); \text{add}. \qquad \text{by (b)}$$

(d) $(\pi \ominus \text{add})^\smile; ((\pi^\smile; \pi^\smile \sqcap \rho^\smile) \ominus \rho^\smile; \pi^\smile) = \pi; \pi^\smile; \pi^\smile \sqcap \pi; \rho^\smile \sqcap (\pi \ominus \text{add})^\smile; \rho; \rho^\smile; \pi^\smile:$

$$(\pi \ominus \text{add})^\smile; ((\pi^\smile; \pi^\smile \sqcap \rho^\smile) \ominus \rho^\smile; \pi^\smile)$$
$$= (\pi \ominus \text{add})^\smile; (\pi; (\pi^\smile; \pi^\smile \sqcap \rho^\smile) \sqcap \rho; \rho^\smile; \pi^\smile)$$
$$= (\pi \ominus \text{add})^\smile; ((\pi \ominus \text{add}); \pi; (\pi^\smile; \pi^\smile \sqcap \rho^\smile) \sqcap \rho; \rho^\smile; \pi^\smile) \qquad \text{Lemma 3(2)}$$
$$= \pi; (\pi^\smile; \pi^\smile \sqcap \rho^\smile) \sqcap (\pi \ominus \text{add})^\smile; \rho; \rho^\smile; \pi^\smile \qquad \text{Lemma 2(2)}$$
$$= \pi; \pi^\smile; \pi^\smile \sqcap \pi; \rho^\smile \sqcap (\pi \ominus \text{add})^\smile; \rho; \rho^\smile; \pi^\smile. \qquad \text{Lemma 2(3)}$$

(e) $(\pi^\smile \ominus (\pi \ominus \text{add}))^\smile; (\pi^\smile; \pi^\smile \ominus ((\pi^\smile; \pi^\smile \sqcap \rho^\smile) \ominus \rho^\smile; \pi^\smile)) = (\pi \ominus \text{add})^\smile \ominus \pi:$

$$(\pi^\smile \ominus (\pi \ominus \text{add}))^\smile; (\pi^\smile; \pi^\smile \ominus ((\pi^\smile; \pi^\smile \sqcap \rho^\smile) \ominus \rho^\smile; \pi^\smile))$$
$$= (\pi \ominus (\pi \ominus \text{add})^\smile); (\pi^\smile; \pi^\smile \ominus ((\pi^\smile; \pi^\smile \sqcap \rho^\smile) \ominus \rho^\smile; \pi^\smile)) \qquad \text{Lemma 3(1)}$$
$$= \pi; \pi^\smile; \pi^\smile \sqcap (\pi \ominus \text{add})^\smile; ((\pi^\smile; \pi^\smile \sqcap \rho^\smile) \ominus \rho^\smile; \pi^\smile) \qquad \text{Lemma 3(5)}$$
$$= \pi; \pi^\smile; \pi^\smile \sqcap \pi; \pi^\smile; \pi^\smile \sqcap \pi; \rho^\smile \sqcap (\pi \ominus \text{add})^\smile; \rho; \rho^\smile; \pi^\smile \qquad \text{by (d)}$$
$$= \pi; \pi^\smile; \pi^\smile \sqcap \pi; \rho^\smile \sqcap (\pi \ominus \text{add})^\smile; \rho; \rho^\smile; \pi^\smile$$
$$= (\pi \ominus \text{add})^\smile; ((\pi \ominus \text{add}); \pi; \pi^\smile; \pi^\smile \sqcap \rho; \rho^\smile; \pi^\smile) \sqcap \pi; \rho^\smile \qquad \text{Lemma 2(2)}$$
$$= (\pi \ominus \text{add})^\smile; (\pi; \pi^\smile; \pi^\smile \sqcap \rho; \rho^\smile; \pi^\smile) \sqcap \pi; \rho^\smile \qquad \text{Lemma 3(2)}$$
$$= (\pi \ominus \text{add})^\smile; (\pi; \pi^\smile \sqcap \rho; \rho^\smile); \pi^\smile \sqcap \pi; \rho^\smile \qquad \text{Lemma 2(3)}$$
$$= (\pi \ominus \text{add})^\smile; \pi^\smile \sqcap \pi; \rho^\smile$$
$$= (\pi \ominus \text{add})^\smile \ominus \pi.$$

(f) $((\pi \ominus add)^{\smile} \ominus \pi); (add \otimes \mathbb{I}_\mathbb{R}) = swap$:

$$((\pi \ominus add)^{\smile} \ominus \pi); (add \otimes \mathbb{I}_\mathbb{R})$$
$$= (\pi \ominus add)^{\smile}; add \ominus \pi \qquad\qquad . \qquad \text{Lemma 3(6)}$$
$$= (\pi; \pi^{\smile} \sqcap \rho; add^{\smile}); add \ominus \pi$$
$$= (\pi; \pi^{\smile}; add \sqcap \rho) \ominus \pi \qquad\qquad \text{Lemma 2(2)}$$
$$= \rho \ominus \pi \qquad\qquad\qquad\qquad\quad \text{Axiom 6. and } \pi \text{ total}$$
$$= swap.$$

Finally, we conclude

add
$$= (\pi^{\smile} \ominus (\pi \ominus add))^{\smile}; (\pi^{\smile} \ominus (\pi \ominus add)); add \qquad\qquad\qquad\qquad\qquad \text{by (a)}$$
$$= (\pi^{\smile} \ominus (\pi \ominus add))^{\smile}; (\pi^{\smile}; \pi^{\smile} \ominus ((\pi^{\smile}; \pi^{\smile} \sqcap \rho^{\smile}) \ominus \rho^{\smile}; \pi^{\smile})); (add \otimes \mathbb{I}_\mathbb{R}); add \quad \text{by (c)}$$
$$= ((\pi \ominus add)^{\smile} \ominus \pi); (add \otimes \mathbb{I}_\mathbb{R}); add \qquad\qquad\qquad\qquad\qquad\qquad\quad \text{by (e)}$$
$$= swap; add. \qquad\qquad\qquad\qquad\qquad\qquad\qquad\qquad\qquad\qquad\qquad\qquad\quad \text{by (f)}$$

2. This follows immediately from (1) and Axiom 5 by

$$(\mathbb{I}_\mathbb{R} \otimes add); add$$
$$= (\mathbb{I}_\mathbb{R} \otimes swap; add); add \qquad\qquad\qquad\qquad \text{by (1)}$$
$$= (\mathbb{I}_\mathbb{R} \otimes swap); (\mathbb{I}_\mathbb{R} \otimes add); add \qquad\qquad \text{Lemma 3(6)}$$
$$= (\mathbb{I}_\mathbb{R} \otimes swap); (\mathbb{I}_\mathbb{R} \otimes swap); assoc; (add \otimes \mathbb{I}_\mathbb{R}); add \qquad \text{Axiom 5.}$$

$$= (\mathbb{I}_\mathbb{R} \otimes swap; swap); assoc; (add \otimes \mathbb{I}_\mathbb{R}); add \quad \text{Lemma 3(6)}$$
$$= assoc; (add \otimes \mathbb{I}_\mathbb{R}); add. \qquad\qquad\qquad\qquad\quad \text{swap bij. and Lemma 4(1)}$$

3. We only have to show the inclusion $\sqsubseteq$. For this we use the abbreviation $X = (\mathbb{I}_\mathbb{R} \otimes (add^{\smile} \sqcap \pi^{\smile})); assoc; (add \otimes \mathbb{I}_\mathbb{R})$. Then we have

$$add^{\smile}; X; \rho = add^{\smile}; (\mathbb{I}_\mathbb{R} \otimes (add^{\smile} \sqcap \pi^{\smile})); assoc; (add \otimes \mathbb{I}_\mathbb{R}); \rho$$
$$= add^{\smile}; (\mathbb{I}_\mathbb{R} \otimes (add^{\smile} \sqcap \pi^{\smile})); assoc; \rho \qquad\qquad \text{Lemma 3(2)}$$
$$= add^{\smile}; (\mathbb{I}_\mathbb{R} \otimes (add^{\smile} \sqcap \pi^{\smile})); \rho; \rho \qquad\qquad\quad \text{Lemma 3(2)}$$
$$= add^{\smile}; \rho; (add^{\smile} \sqcap \pi^{\smile}); \rho \qquad\qquad\qquad\qquad \text{Lemma 3(2)}$$
$$= add^{\smile}; swap^{\smile}; \rho; Z \qquad\qquad\qquad\qquad\qquad \text{by (1)}$$
$$= add^{\smile}; \pi; Z \qquad\qquad\qquad\qquad\qquad\qquad\quad \text{Lemma 3(2)}$$
$$= \mathbb{T}_{\mathbb{R}\mathbb{R}}; Z. \qquad\qquad\qquad\qquad\qquad\qquad\qquad\quad \text{Axiom 6.}$$

Furthermore, we obtain the inclusion

$$
\begin{aligned}
X &= (\mathbb{I}_\mathbb{R} \otimes (\mathrm{add}^\smile \sqcap \pi^\smile)); \mathrm{assoc}; (\mathrm{add} \otimes \mathbb{I}_\mathbb{R}) \\
&= (\mathbb{I}_\mathbb{R} \otimes (\mathrm{add}^\smile \sqcap \pi^\smile)); (\pi^\smile; \pi^\smile \ominus (\rho^\smile \otimes \mathbb{I}_\mathbb{R})); (\mathrm{add} \otimes \mathbb{I}_\mathbb{R}) \\
&= (\pi; \pi^\smile; \pi^\smile \sqcap \rho; (\mathrm{add}^\smile \sqcap \pi^\smile); (\rho^\smile \otimes \mathbb{I}_\mathbb{R})); (\mathrm{add} \otimes \mathbb{I}_\mathbb{R}) && \text{Lemma 3(5)} \\
&\sqsubseteq (\pi; \pi^\smile; \pi^\smile \sqcap \rho; (\mathrm{add}^\smile \sqcap \pi^\smile); (\rho^\smile \otimes \mathbb{I}_\mathbb{R})); \pi; \mathrm{add}; \pi^\smile \\
&= (\pi; \pi^\smile \sqcap \rho; (\mathrm{add}^\smile \sqcap \pi^\smile); (\rho^\smile \otimes \mathbb{I}_\mathbb{R}); \pi); \mathrm{add}; \pi^\smile && \text{Lemma 2(2)} \\
&= (\pi; \pi^\smile \sqcap \rho; (\mathrm{add}^\smile \sqcap \pi^\smile); \pi; \rho^\smile); \mathrm{add}; \pi^\smile && \text{Lemma 3(2)} \\
&= (\pi; \pi^\smile \sqcap \rho; (\mathrm{add}^\smile; \pi \sqcap \mathbb{I}_\mathbb{R}); \rho^\smile); \mathrm{add}; \pi^\smile && \text{Lemma 2(2)} \\
&= (\pi; \pi^\smile \sqcap \rho; \rho^\smile); \mathrm{add}; \pi^\smile && \text{Axiom 6.} \\
&= \mathrm{add}; \pi^\smile
\end{aligned}
$$

and the inclusion

$$
\begin{aligned}
\mathrm{add}^\smile; X &= \mathrm{add}^\smile; (\mathbb{I}_\mathbb{R} \otimes (\mathrm{add}^\smile \sqcap \pi^\smile)); \mathrm{assoc}; (\mathrm{add} \otimes \mathbb{I}_\mathbb{R}) \\
&\sqsubseteq \mathrm{add}^\smile; (\mathbb{I}_\mathbb{R} \otimes \mathrm{add}^\smile); \mathrm{assoc}; (\mathrm{add} \otimes \mathbb{I}_\mathbb{R}) \\
&= \mathrm{add}^\smile; (\mathrm{add}^\smile \otimes \mathbb{I}_\mathbb{R}); \mathrm{assoc}^\smile; \mathrm{assoc}; (\mathrm{add} \otimes \mathbb{I}_\mathbb{R}) && \text{by (2)} \\
&= \mathrm{add}^\smile; (\mathrm{add}^\smile \otimes \mathbb{I}_\mathbb{R}); (\mathrm{add} \otimes \mathbb{I}_\mathbb{R}) && \text{assoc bij.} \\
&= \mathrm{add}^\smile; (\mathrm{add}^\smile; \mathrm{add} \otimes \mathbb{I}_\mathbb{R}) && \text{Lemma 3(6)} \\
&\sqsubseteq \mathrm{add}^\smile.
\end{aligned}
$$

Together we get

$$
\begin{aligned}
\mathbb{T}_{\mathbb{R}\mathbb{R}}; Z &= \mathrm{add}^\smile; X; \rho && \text{see above} \\
&= (\mathrm{add}^\smile; X \sqcap \mathrm{add}^\smile; X); \rho \\
&\sqsubseteq (\mathrm{add}^\smile \sqcap \mathrm{add}^\smile; \mathrm{add}; \pi^\smile); \rho && \text{see above} \\
&\sqsubseteq Z. && \text{add univalent}
\end{aligned}
$$

4. First of all, we have

$$
\begin{aligned}
&(\pi^\smile \sqcap (\mathrm{add}^\smile \sqcap \pi^\smile); \rho; \rho^\smile); \mathrm{add} \\
&= (\mathrm{add}^\smile \sqcap \pi^\smile); ((\mathrm{add} \sqcap \pi); \pi^\smile \sqcap \rho; \rho^\smile); \mathrm{add} && \text{Lemma 2(2)} \\
&= (\mathrm{add}^\smile \sqcap \pi^\smile); (\mathrm{add}; \pi^\smile \sqcap \pi; \pi^\smile \sqcap \rho; \rho^\smile); \mathrm{add} && \text{Lemma 2(3)} \\
&\sqsubseteq (\mathrm{add}^\smile \sqcap \pi^\smile); (\pi; \pi^\smile \sqcap \rho; \rho^\smile); \mathrm{add} \\
&= (\mathrm{add}^\smile \sqcap \pi^\smile); \mathrm{add}
\end{aligned}
$$

and $(\text{add}^{\smile} \sqcap \pi^{\smile}); \text{add} \sqsubseteq (\pi^{\smile} \sqcap (\text{add}^{\smile} \sqcap \pi^{\smile}); \rho; \rho^{\smile}); \text{add}$ because ρ is total. We conclude

$$
\begin{aligned}
(\mathbb{I}_\mathbb{R} \oslash Z); \text{add} &= (\pi^{\smile} \sqcap (\text{add}^{\smile} \sqcap \pi^{\smile}); \rho; \rho^{\smile}); \text{add} \\
&= (\text{add}^{\smile} \sqcap \pi^{\smile}); \text{add} && \text{see above} \\
&= \mathbb{I}_\mathbb{R} \sqcap \pi^{\smile}; \text{add} && \text{Lemma 2(2)} \\
&= \mathbb{I}_\mathbb{R}. && \text{Axiom 6.}
\end{aligned}
$$

5. This follows from (4) and Lemma 2(1).
6. From the following computation

$$
\begin{aligned}
\text{neg}^{\smile} &= (Z; \text{add}^{\smile} \sqcap \rho^{\smile}); \pi \\
&= (\mathbb{T}_{\mathbb{R}\mathbb{R}}; Z; \text{add}^{\smile} \sqcap \rho^{\smile}); \pi && \text{by (3)} \\
&= \rho^{\smile}; (\mathbb{T}_{\mathbb{R}\otimes\mathbb{R}\mathbb{R}}; Z; \text{add}^{\smile} \sqcap \mathbb{I}_{\mathbb{R}\otimes\mathbb{R}}); \pi && \text{Lemma 1(3)} \\
&= \rho^{\smile}; (\text{add}; Z^{\smile}; \mathbb{T}_{\mathbb{R}\mathbb{R}\otimes\mathbb{R}} \sqcap \mathbb{I}_{\mathbb{R}\otimes\mathbb{R}}); \pi && \text{Lemma 1(1)} \\
&= \rho^{\smile}; (\text{add}; Z^{\smile}; \mathbb{T}_{\mathbb{R}\mathbb{R}} \sqcap \pi) && \text{Lemma 1(3)} \\
&= \rho^{\smile}; (\text{add}; Z^{\smile} \sqcap \pi) && \text{by (3)} \\
&= \rho^{\smile}; (\text{swap}; \text{add}; Z^{\smile} \sqcap \pi) && \text{by (1)} \\
&= \rho^{\smile}; \text{swap}; (\text{add}; Z^{\smile} \sqcap \text{swap}; \pi) && \text{Lemma 2(2)} \\
&= \pi^{\smile}; (\text{add}; Z^{\smile} \sqcap \rho) && \text{Lemma 3(2)} \\
&= \text{neg}.
\end{aligned}
$$

we conclude the assertion. $\qquad\qquad\qquad\qquad\qquad\qquad\qquad\qquad\qquad\qquad\square$

After these preparations we are able to show our first theorem.

Theorem 1. *The quadruple* $(\mathbb{R}, 0, \text{add}, \text{neg})$ *with* $0 = \mathbb{T}_{1\mathbb{R}}; Z$ *is an abelian group.*

Proof. First of all, by Lemma 6(1) and (2) add is commutative and associative. Next, we want to show that 0 is a point. Therefore, we obtain

$$
\begin{aligned}
0^{\smile}; 0 &= Z^{\smile}; \mathbb{T}_{\mathbb{R}1}; \mathbb{T}_{1\mathbb{R}}; Z \\
&\sqsubseteq Z^{\smile}; \mathbb{T}_{\mathbb{R}\mathbb{R}}; Z \\
&= Z^{\smile}; \mathbb{T}_{\mathbb{R}\mathbb{R}} \sqcap \mathbb{T}_{\mathbb{R}\mathbb{R}} Z && \text{Lemma 1(3)} \\
&= Z^{\smile} \sqcap Z && \text{Lemma 6(3)} \\
&= (\mathbb{I}_\mathbb{R} \oslash Z); (Z^{\smile} \oslash \mathbb{I}_\mathbb{R}) && \text{Lemma 3(5)} \\
&= (\mathbb{I}_\mathbb{R} \oslash Z); \text{swap}; (\mathbb{I}_\mathbb{R} \oslash Z^{\smile}) && \text{Lemma 4(2)} \\
&= (\mathbb{I}_\mathbb{R} \oslash Z); \text{swap}; (\mathbb{I}_\mathbb{R} \oslash Z)^{\smile} && \text{Lemma 3(1)}
\end{aligned}
$$

$$\sqsubseteq \mathrm{add}^\smile; \mathrm{swap}; \mathrm{add} \qquad\qquad \text{Lemma 6(5)}$$
$$= \mathrm{add}^\smile; \mathrm{add} \qquad\qquad\qquad \text{Lemma 6(1)}$$
$$\sqsubseteq \mathbb{I}_\mathbb{R},$$

i.e., 0 is univalent. Using the following computation

$$
\begin{aligned}
\mathbb{I}_\mathbb{R} &= \mathbb{I}_\mathbb{R} \sqcap \mathrm{add}^\smile; \pi \sqcap \pi^\smile; \mathrm{add} \sqcap \mathbb{I}_\mathbb{R} && \text{Axiom 6.}\\
&= \mathrm{add}^\smile; (\mathrm{add} \sqcap \pi) \sqcap \pi^\smile; (\mathrm{add} \sqcap \pi) && \text{Lemma 2(2)}\\
&= (\mathrm{add}^\smile \sqcap \pi^\smile; (\mathrm{add} \sqcap \pi); (\mathrm{add}^\smile \sqcap \pi^\smile)); (\mathrm{add} \sqcap \pi) && \text{Lemma 2(2)}\\
&\sqsubseteq (\mathrm{add}^\smile \sqcap \pi^\smile; \pi; \pi^\smile); (\mathrm{add} \sqcap \pi)\\
&\sqsubseteq (\mathrm{add}^\smile \sqcap \pi^\smile); (\mathrm{add} \sqcap \pi) && \pi \text{ univalent}\\
&\sqsubseteq (\mathrm{add}^\smile \sqcap \pi^\smile); \rho; \rho^\smile; (\mathrm{add} \sqcap \pi) && \rho \text{ total}\\
&= Z; Z^\smile
\end{aligned}
$$

we conclude that Z is total. Since $\pi_{1\mathbb{R}}$ is also total we conclude that 0 is a map, and, hence, a point. 0 is right neutral with respect to add because of

$$
\begin{aligned}
(\mathbb{I}_\mathbb{R} \otimes \pi_{\mathbb{R}1}; 0); \mathrm{add} &= (\mathbb{I}_\mathbb{R} \otimes \pi_{\mathbb{R}\mathbb{R}}; Z); \mathrm{add} && \text{1 unit}\\
&= (\mathbb{I}_\mathbb{R} \otimes Z); \mathrm{add} && \text{Lemma 6(3)}\\
&= \mathbb{I}_\mathbb{R}. && \text{Lemma 6(4)}
\end{aligned}
$$

Before we show that neg is a mapping we want to verify that neg satisfies the right inverse law. We compute

$$
\begin{aligned}
(\mathbb{I}_\mathbb{R} \otimes \mathrm{neg}); \mathrm{add} &= (\pi^\smile \sqcap \pi^\smile; (\mathrm{add}; 0^\smile; \pi_{1\mathbb{R}} \sqcap \rho); \rho^\smile); \mathrm{add} && \text{1 unit}\\
&= \pi^\smile; (\pi; \pi^\smile \sqcap (\mathrm{add}; 0^\smile; \pi_{1\mathbb{R}} \sqcap \rho); \rho^\smile); \mathrm{add} && \text{Lemma 2(2)}\\
&= \pi^\smile; (\pi; \pi^\smile \sqcap \mathrm{add}; 0^\smile; \pi_{1\mathbb{R} \otimes \mathbb{R}} \sqcap \rho; \rho^\smile); \mathrm{add} && \text{Lemma 1(3)}\\
&= \pi^\smile; (\mathbb{I}_{\mathbb{R} \otimes \mathbb{R}} \sqcap \mathrm{add}; 0^\smile; \pi_{1\mathbb{R} \otimes \mathbb{R}}); \mathrm{add}\\
&= \pi^\smile; (\mathbb{I}_{\mathbb{R} \otimes \mathbb{R}} \sqcap \pi_{\mathbb{R} \otimes \mathbb{R}1}; 0; \mathrm{add}^\smile); \mathrm{add} && \text{Lemma 1(1)}\\
&= \pi^\smile; (\mathrm{add} \sqcap \pi_{\mathbb{R} \otimes \mathbb{R}1}; 0) && \text{Lemma 2(2)}\\
&= \pi^\smile; (\mathrm{add} \sqcap \pi; \pi_{\mathbb{R}1}; 0) && \pi \text{ total}\\
&= \pi^\smile; \mathrm{add} \sqcap \pi_{\mathbb{R}1}; 0 && \text{Lemma 2(2)}\\
&= \pi_{\mathbb{R}1}; 0 && \text{Axiom 6.}
\end{aligned}
$$

In order to show that neg is univalent we first need several additional properties:

(a) $\mathbb{I}_\mathbb{R} \otimes \mathrm{neg} \sqsubseteq Z; \mathrm{add}^\smile$:

$$
\begin{aligned}
\mathbb{I}_\mathbb{R} &\otimes \mathrm{neg}\\
&\sqsubseteq \pi_{\mathbb{R}1}; 0; \mathrm{add}^\smile && \text{see above and Lemma 2(1)}\\
&= \pi_{\mathbb{R}\mathbb{R}}; Z; \mathrm{add}^\smile && \text{1 unit}\\
&= Z; \mathrm{add}^\smile. && \text{Lemma 6(3)}
\end{aligned}
$$

(b) $(\mathbb{I}_\mathbb{R} \otimes (\pi \oslash (\mathrm{neg} \oslash \mathbb{I}_\mathbb{R}))); \rho; (\mathrm{neg} \oslash \mathbb{I}_\mathbb{R}) = \rho; (\mathrm{neg} \oslash \mathbb{I}_\mathbb{R})$:

$$(\mathbb{I}_\mathbb{R} \otimes (\pi \oslash (\mathrm{neg} \oslash \mathbb{I}_\mathbb{R}))); \rho; (\mathrm{neg} \oslash \mathbb{I}_\mathbb{R})$$

$$= \rho; (\pi \oslash (\mathrm{neg} \oslash \mathbb{I}_\mathbb{R})); (\mathrm{neg} \oslash \mathbb{I}_\mathbb{R}) \qquad \text{Lemma 3(2)}$$

$$= \rho; (\pi; \mathrm{neg} \sqcap (\mathrm{neg} \oslash \mathbb{I}_\mathbb{R})) \qquad \text{Lemma 3(5)}$$

$$= \rho; (\mathrm{neg} \oslash \mathbb{I}_\mathbb{R}), \qquad \mathrm{neg} \oslash \mathbb{I}_\mathbb{R} \sqsubseteq \pi; \mathrm{neg}$$

(c) $(\mathbb{I}_\mathbb{R} \oslash \mathrm{neg}; \pi^\smile); (\mathbb{I}_\mathbb{R} \otimes (\pi \oslash (\mathrm{neg} \oslash \mathbb{I}_\mathbb{R}))) = \mathbb{I}_\mathbb{R} \oslash \mathrm{neg}; (\mathbb{I}_\mathbb{R} \oslash \mathrm{neg})$:

$$(\mathbb{I}_\mathbb{R} \oslash \mathrm{neg}; \pi^\smile); (\mathbb{I}_\mathbb{R} \otimes (\pi \oslash (\mathrm{neg} \oslash \mathbb{I}_\mathbb{R})))$$

$$= \mathbb{I}_\mathbb{R} \oslash \mathrm{neg}; \pi^\smile; (\pi \oslash (\mathrm{neg} \oslash \mathbb{I}_\mathbb{R})) \qquad \text{Lemma 3(6)}$$

$$= \mathbb{I}_\mathbb{R} \oslash \mathrm{neg}; \pi^\smile; ((\mathbb{I}_\mathbb{R} \oslash \mathrm{neg}) \oslash \rho^\smile) \qquad \text{Lemma 3(1)\&(7)}$$

$$= \mathbb{I}_\mathbb{R} \oslash \mathrm{neg}; (\mathbb{I}_\mathbb{R} \oslash \mathrm{neg}), \qquad \text{Lemma 3(3)}$$

(d) $(\mathbb{I}_\mathbb{R} \oslash \mathrm{neg}; (\mathbb{I}_\mathbb{R} \oslash \mathrm{neg})); (\pi \oslash (\mathrm{neg}; \pi^\smile \oslash \mathbb{I}_{\mathbb{R}\times\mathbb{R}})) = \mathbb{I}_\mathbb{R} \oslash \mathrm{neg}; (\mathbb{I}_\mathbb{R} \oslash \mathrm{neg})$:

$$(\mathbb{I}_\mathbb{R} \oslash \mathrm{neg}; (\mathbb{I}_\mathbb{R} \oslash \mathrm{neg})); (\pi \oslash (\mathrm{neg}; \pi^\smile \oslash \mathbb{I}_{\mathbb{R}\times\mathbb{R}}))$$

$$= (\mathbb{I}_\mathbb{R} \oslash \mathrm{neg}; (\mathbb{I}_\mathbb{R} \oslash \mathrm{neg})); ((\mathbb{I}_\mathbb{R} \oslash \mathrm{neg}; \pi^\smile) \oslash \rho^\smile) \quad \text{Lemma 3(1)\&(7)}$$

$$= (\mathbb{I}_\mathbb{R} \oslash \mathrm{neg}; \pi^\smile) \sqcap \mathrm{neg}; (\mathbb{I}_\mathbb{R} \oslash \mathrm{neg}); \rho^\smile \qquad \text{Lemma 3(5)}$$

$$= \pi^\smile \sqcap \mathrm{neg}; \pi^\smile; \rho^\smile \sqcap \mathrm{neg}; (\mathbb{I}_\mathbb{R} \oslash \mathrm{neg}); \rho^\smile$$

$$= \pi^\smile \sqcap \mathrm{neg}; (\mathbb{I}_\mathbb{R} \oslash \mathrm{neg}); \rho^\smile \qquad \mathrm{neg}; (\mathbb{I}_\mathbb{R} \oslash \mathrm{neg}); \rho^\smile \sqsubseteq \mathrm{neg}; \pi^\smile; \rho^\smile$$

$$= \mathbb{I}_\mathbb{R} \oslash \mathrm{neg}; (\mathbb{I}_\mathbb{R} \oslash \mathrm{neg}),$$

(e) $(\pi \oslash (\mathrm{neg}; \pi^\smile \oslash \mathbb{I}_{\mathbb{R}\times\mathbb{R}})); \rho; (\mathrm{neg} \oslash \mathbb{I}_\mathbb{R}) = \mathrm{assoc}; ((\mathrm{neg} \oslash \mathbb{I}_\mathbb{R}); \mathrm{neg} \oslash \mathbb{I}_\mathbb{R})$:

$$(\pi \oslash (\mathrm{neg}; \pi^\smile \oslash \mathbb{I}_{\mathbb{R}\times\mathbb{R}})); \rho; (\mathrm{neg} \oslash \mathbb{I}_\mathbb{R})$$

$$= (\mathrm{neg}; \pi^\smile \oslash \mathbb{I}_{\mathbb{R}\times\mathbb{R}}); (\mathrm{neg} \oslash \mathbb{I}_\mathbb{R}) \qquad \text{Lemma 3(2)}$$

$$= (\mathrm{neg}; \pi^\smile \oslash (\pi^\smile \oslash \rho^\smile)); (\mathrm{neg} \oslash \mathbb{I}_\mathbb{R})$$

$$= \mathrm{assoc}; ((\mathrm{neg}; \pi^\smile \oslash \pi^\smile) \oslash \rho^\smile); (\mathrm{neg} \oslash \mathbb{I}_\mathbb{R}) \qquad \text{Lemma 4(4)}$$

$$= \mathrm{assoc}; ((\mathrm{neg} \oslash \mathbb{I}_\mathbb{R}); \pi^\smile \oslash \rho^\smile); (\mathrm{neg} \oslash \mathbb{I}_\mathbb{R}) \qquad \text{Lemma 2(3)}$$

$$= \mathrm{assoc}; ((\mathrm{neg} \oslash \mathbb{I}_\mathbb{R}) \otimes \mathbb{I}_\mathbb{R}); (\mathrm{neg} \oslash \mathbb{I}_\mathbb{R})$$

$$= \mathrm{assoc}; ((\mathrm{neg} \oslash \mathbb{I}_\mathbb{R}); \mathrm{neg} \oslash \mathbb{I}_\mathbb{R}), \qquad \text{Lemma 3(6)}$$

(f) $\mathbb{I}_\mathbb{R} \oslash \mathrm{neg}; (\mathbb{I}_\mathbb{R} \oslash \mathrm{neg}) \sqsubseteq \mathrm{add}^\smile; (\mathbb{I}_\mathbb{R} \otimes \mathrm{add}^\smile)$:

$$\mathbb{I}_\mathbb{R} \oslash \mathrm{neg}; (\mathbb{I}_\mathbb{R} \oslash \mathrm{neg})$$

$$\sqsubseteq \mathbb{I}_\mathbb{R} \oslash \mathrm{neg}; Z; \mathrm{add}^\smile \qquad \text{by (a)}$$

$$= (\mathbb{I}_\mathbb{R} \oslash \mathrm{neg}; Z); (\mathbb{I}_\mathbb{R} \otimes \mathrm{add}^\smile) \qquad \text{Lemma 3(6)}$$

$$\sqsubseteq (\mathbb{I}_\mathbb{R} \oslash \top\!\top_{\mathbb{R}\mathbb{R}}; Z); (\mathbb{I}_\mathbb{R} \otimes \mathrm{add}^\smile)$$

$$= (\mathbb{I}_\mathbb{R} \oslash Z); (\mathbb{I}_\mathbb{R} \otimes \mathrm{add}^\smile) \qquad \text{Lemma 6(3)}$$

$$\sqsubseteq \mathrm{add}^\smile; (\mathbb{I}_\mathbb{R} \otimes \mathrm{add}^\smile) \qquad \text{Lemma 6(5)}$$

(g) $(\text{neg} \ominus \mathbb{I}_\mathbb{R}); \text{neg} \ominus \mathbb{I}_\mathbb{R} \sqsubseteq (\text{add} \otimes \mathbb{I}_\mathbb{R}); \text{add}$:

$$(\text{neg} \ominus \mathbb{I}_\mathbb{R}); \text{neg} \ominus \mathbb{I}_\mathbb{R}$$

$\quad = (\text{neg}^\smile; (\text{neg}^\smile \ominus \mathbb{I}_\mathbb{R}) \ominus \mathbb{I}_\mathbb{R})^\smile$ Lemma 3(1)

$\quad = (\text{neg}; (\text{neg} \ominus \mathbb{I}_\mathbb{R}) \ominus \mathbb{I}_\mathbb{R})^\smile$ Lemma 6(6)

$\quad = ((\mathbb{I}_\mathbb{R} \ominus \text{neg}; (\mathbb{I}_\mathbb{R} \ominus \text{neg})); (\mathbb{I}_\mathbb{R} \otimes \text{swap}); \text{swap})^\smile$ Lemma 3(6)&4(2)

$\quad \sqsubseteq (\text{add}^\smile; (\mathbb{I}_\mathbb{R} \otimes \text{add}^\smile); (\mathbb{I}_\mathbb{R} \otimes \text{swap}); \text{swap})^\smile$ by (f)

$\quad = (\text{add}^\smile; (\mathbb{I}_\mathbb{R} \otimes \text{add}^\smile; \text{swap}); \text{swap})^\smile$ Lemma 3(6)

$\quad = (\text{add}^\smile; (\mathbb{I}_\mathbb{R} \otimes \text{add}^\smile); \text{swap})^\smile$ Lemma 6(1)

$\quad = (\text{add}^\smile; (\text{add}^\smile \otimes \mathbb{I}_\mathbb{R}))^\smile$ Lemma 4(2)

$\quad = (\text{add} \otimes \mathbb{I}_\mathbb{R}); \text{add}.$ Lemma 3(1)

Now we compute

$\text{neg}^\smile; \text{neg}$

$\quad = \text{neg}; \text{neg}$ Lemma 6(6)

$\quad = \text{neg}; \pi^\smile; (\text{neg} \ominus \mathbb{I}_\mathbb{R})$ Lemma 3(3)

$\quad = (\mathbb{I}_\mathbb{R} \ominus \text{neg}; \pi^\smile); \rho; (\text{neg} \ominus \mathbb{I}_\mathbb{R})$ Lemma 3(2)

$\quad = (\mathbb{I}_\mathbb{R} \ominus \text{neg}; \pi^\smile); (\mathbb{I}_\mathbb{R} \otimes (\pi \ominus (\text{neg} \ominus \mathbb{I}_\mathbb{R}))); \rho; (\text{neg} \ominus \mathbb{I}_\mathbb{R})$ by (b)

$\quad = (\mathbb{I}_\mathbb{R} \ominus \text{neg}; (\mathbb{I}_\mathbb{R} \ominus \text{neg})); \rho; (\text{neg} \ominus \mathbb{I}_\mathbb{R})$ by (c)

$\quad = (\mathbb{I}_\mathbb{R} \ominus \text{neg}; (\mathbb{I}_\mathbb{R} \ominus \text{neg})); (\pi \ominus (\text{neg}; \pi^\smile \ominus \mathbb{I}_{\mathbb{R} \times \mathbb{R}})); \rho; (\text{neg} \ominus \mathbb{I}_\mathbb{R})$ by (d)

$\quad = (\mathbb{I}_\mathbb{R} \ominus \text{neg}; (\mathbb{I}_\mathbb{R} \ominus \text{neg})); \text{assoc}; ((\text{neg} \ominus \mathbb{I}_\mathbb{R}); \text{neg} \ominus \mathbb{I}_\mathbb{R})$ by (e)

$\quad \sqsubseteq \text{add}^\smile; (\mathbb{I}_\mathbb{R} \otimes \text{add}^\smile); \text{assoc}; (\text{add} \otimes \mathbb{I}_\mathbb{R}); \text{add}$ by (f), (g)

$\quad = \text{add}^\smile; (\mathbb{I}_\mathbb{R} \otimes \text{add}^\smile); (\mathbb{I}_\mathbb{R} \otimes \text{swap}); (\mathbb{I}_\mathbb{R} \otimes \text{swap}); \text{assoc}; (\text{add} \otimes \mathbb{I}_\mathbb{R}); \text{add}$ Lemma 4(1)

$\quad = \text{add}^\smile; (\mathbb{I}_\mathbb{R} \otimes \text{add}^\smile); (\mathbb{I}_\mathbb{R} \otimes \text{swap}); (\mathbb{I}_\mathbb{R} \otimes \text{add}); \text{add}$ Axiom 5.

$\quad = \text{add}^\smile; (\mathbb{I}_\mathbb{R} \otimes \text{add}^\smile; \text{swap}; \text{add}); \text{add}$ Lemma 3(6)

$\quad = \text{add}^\smile; (\mathbb{I}_\mathbb{R} \otimes \text{add}^\smile; \text{add}); \text{add}$ Lemma 6(1)

$\quad \sqsubseteq \mathbb{I}_\mathbb{R}.$

Last but not least, that neg is total follows immediately from

$\mathbb{I}_\mathbb{R} = \pi^\smile; \text{add}; Z^\smile \sqcap \mathbb{I}_\mathbb{R}$ Axiom 6. and Z total

$\quad = \pi^\smile; (\text{add}; Z^\smile \sqcap \pi)$ Lemma 2(2)

$\quad = \pi^\smile; (\text{add}; Z^\smile; \top_{\mathbb{R}\mathbb{R}} \sqcap \pi)$ Lemma 6(3)

$\quad = \pi^\smile; (\text{add}; Z^\smile; \top_{\mathbb{R}\mathbb{R} \otimes \mathbb{R}} \sqcap \mathbb{I}_{\mathbb{R} \otimes \mathbb{R}}); \pi$ Lemma 1(3)

$\quad = \pi^\smile; (\text{add}; Z^\smile; \top_{\mathbb{R}\mathbb{R} \otimes \mathbb{R}} \sqcap \mathbb{I}_{\mathbb{R} \otimes \mathbb{R}}); (\top_{\mathbb{R} \otimes \mathbb{R}\mathbb{R}}; Z; \text{add}^\smile \sqcap \mathbb{I}_{\mathbb{R} \otimes \mathbb{R}}); \pi$ Lemma 1(1), (2)

$\quad \sqsubseteq \pi^\smile; (\text{add}; Z^\smile; \top_{\mathbb{R}\mathbb{R} \otimes \mathbb{R}} \sqcap \mathbb{I}_{\mathbb{R} \otimes \mathbb{R}}); \rho; \rho^\smile; (\top_{\mathbb{R} \otimes \mathbb{R}\mathbb{R}}; Z; \text{add}^\smile \sqcap \mathbb{I}_{\mathbb{R} \otimes \mathbb{R}}); \pi$ ρ total

$\quad = \pi^\smile; (\text{add}; Z^\smile; \top_{\mathbb{R}\mathbb{R}} \sqcap \rho); (\top_{\mathbb{R}\mathbb{R}}; Z; \text{add}^\smile \sqcap \rho^\smile); \pi$ Lemma 1(3)

$\quad = \text{neg}; \text{neg}^\smile.$ Lemma 6(3)

This completes the proof.   □

Notice that neg is a bijective mapping which follows from the fact that neg is a mapping and Lemma 6(6).

We will now turn our attention to the relation C and show that it is a dense linear strict-order. Please note that we only need to show that C is transitive since all other properties are already axioms. We start with the following lemma.

Lemma 7. *1.* $\mathrm{add}; \mathrm{add}^{\smile} \sqcap \rho; \rho^{\smile} = \mathbb{I}_{\mathsf{R} \otimes \mathsf{R}}$.
2. $\mathrm{add}; C; \mathrm{add}^{\smile} \sqcap \rho; \rho^{\smile} = C \otimes \mathbb{I}_{\mathsf{R}}$.

Proof. 1. The inclusion $\sqsupseteq$ follows immediately from the fact that add and ρ are both total. For the opposite inclusion we will use the abbreviation $X = (\mathbb{I}_{\mathsf{R}} \otimes (\mathbb{I}_{\mathsf{R}} \ominus \mathrm{neg})); \mathrm{assoc}; ((\mathrm{add} \ominus \rho) \otimes \mathbb{I}_{\mathsf{R}})$. From the two computations

$$(\mathbb{I}_{\mathsf{R}} \otimes (\mathbb{I}_{\mathsf{R}} \ominus \mathrm{neg})); \mathrm{assoc} = (\pi \ominus \rho; (\mathbb{I}_{\mathsf{R}} \ominus \mathrm{neg})); \mathrm{assoc}$$
$$= (\pi \ominus (\rho \ominus \rho; \mathrm{neg})); \mathrm{assoc} \qquad \text{Lemma 2(3)}$$
$$= (\pi \ominus \rho) \ominus \rho; \mathrm{neg} \qquad \text{Lemma 4(4)}$$
$$= \mathbb{I}_{\mathsf{R} \otimes \mathsf{R}} \ominus \rho; \mathrm{neg},$$

$$((\mathrm{add} \ominus \rho) \otimes \mathbb{I}_{\mathsf{R}}); ((\mathrm{add}^{\smile} \ominus \rho^{\smile}) \otimes \mathbb{I}_{\mathsf{R}})$$
$$= (\mathrm{add} \ominus \rho); (\mathrm{add}^{\smile} \ominus \rho^{\smile}) \otimes \mathbb{I}_{\mathsf{R}} \qquad \text{Lemma 3(6)}$$
$$= (\mathrm{add}; \mathrm{add}^{\smile} \sqcap \rho; \rho^{\smile}) \otimes \mathbb{I}_{\mathsf{R}} \qquad \text{Lemma 3(5)}$$

we obtain

$$X; X^{\smile}$$
$$= (\mathbb{I}_{\mathsf{R} \otimes \mathsf{R}} \ominus \rho; \mathrm{neg}); ((\mathrm{add}; \mathrm{add}^{\smile} \sqcap \rho; \rho^{\smile}) \otimes \mathbb{I}_{\mathsf{R}}); (\mathbb{I}_{\mathsf{R} \otimes \mathsf{R}} \ominus \mathrm{neg}; \rho^{\smile})$$
$$= ((\mathrm{add}; \mathrm{add}^{\smile} \sqcap \rho; \rho^{\smile}) \ominus \rho; \mathrm{neg}); (\mathbb{I}_{\mathsf{R} \otimes \mathsf{R}} \ominus \mathrm{neg}; \rho^{\smile}) \qquad \text{Lemma 3(6)}$$
$$= \mathrm{add}; \mathrm{add}^{\smile} \sqcap \rho; \rho^{\smile} \sqcap \rho; \mathrm{neg}; \mathrm{neg}; \rho^{\smile} \qquad \text{Lemma 3(5)}$$
$$= \mathrm{add}; \mathrm{add}^{\smile} \sqcap \rho; \rho^{\smile}. \qquad \text{neg bij.}$$

Furthermore, we compute

$$X; (\pi \otimes \mathbb{I}_{\mathsf{R}}); \mathrm{add}$$
$$= (\mathbb{I}_{\mathsf{R}} \otimes (\mathbb{I}_{\mathsf{R}} \ominus \mathrm{neg})); \mathrm{assoc}; ((\mathrm{add} \ominus \rho); \pi \otimes \mathbb{I}_{\mathsf{R}}); \mathrm{add} \qquad \text{Lemma 3(6)}$$
$$= (\mathbb{I}_{\mathsf{R}} \otimes (\mathbb{I}_{\mathsf{R}} \ominus \mathrm{neg})); \mathrm{assoc}; (\mathrm{add} \otimes \mathbb{I}_{\mathsf{R}}); \mathrm{add} \qquad \text{Lemma 3(2)}$$
$$= (\mathbb{I}_{\mathsf{R}} \otimes (\mathbb{I}_{\mathsf{R}} \ominus \mathrm{neg})); (\mathbb{I}_{\mathsf{R}} \otimes \mathrm{add}); \mathrm{add} \qquad \text{Lemma 6(2)}$$
$$= (\mathbb{I}_{\mathsf{R}} \otimes (\mathbb{I}_{\mathsf{R}} \ominus \mathrm{neg}); \mathrm{add}); \mathrm{add} \qquad \text{Lemma 3(6)}$$
$$= (\mathbb{I}_{\mathsf{R}} \otimes \top \top_{1\mathsf{R}}; 0); \mathrm{add} \qquad \text{Theorem 1}$$
$$= (\mathbb{I}_{\mathsf{R}} \otimes \top \top_{\mathsf{R} \mathsf{R}}; Z); \mathrm{add} \qquad \text{1 unit}$$
$$= (\pi; \pi^{\smile} \sqcap \rho; \top \top_{\mathsf{R} \mathsf{R}}; Z; \rho^{\smile}); \mathrm{add} \qquad \rho \text{ total}$$
$$= (\pi; \pi^{\smile} \sqcap \top \top_{\mathsf{R} \otimes \mathsf{R} \mathsf{R}}; Z; \rho^{\smile}); \mathrm{add} \qquad \rho \text{ total}$$
$$= (\pi; \pi^{\smile} \sqcap \pi; \top \top_{\mathsf{R} \mathsf{R}}; Z; \rho^{\smile}); \mathrm{add} \qquad \pi \text{ total}$$
$$= \pi; (\pi^{\smile} \sqcap \top \top_{\mathsf{R} \mathsf{R}}; Z; \rho^{\smile}); \mathrm{add} \qquad \text{Lemma 3(4)}$$

$$= \pi; (\pi^{\smile} \sqcap Z; \rho^{\smile}); \text{add} \qquad\qquad \text{Lemma 6(3)}$$
$$= \pi; (\mathbb{I}_{\mathbb{R}} \ominus Z); \text{add}$$
$$= \pi. \qquad\qquad\qquad\qquad\qquad \text{Lemma 6(4)}$$

Together we obtain

$$\text{add}; \text{add}^{\smile} \sqcap \rho; \rho^{\smile} = \text{add}; \text{add}^{\smile} \sqcap \rho; \rho^{\smile} \sqcap \rho; \rho^{\smile}$$
$$= X; X^{\smile} \sqcap \rho; \rho^{\smile} \qquad\qquad\qquad\qquad \text{see above}$$
$$\sqsubseteq X; (\pi \otimes \mathbb{I}_{\mathbb{R}}); \text{add}; \text{add}^{\smile}; (\pi^{\smile} \otimes \mathbb{I}_{\mathbb{R}}); X^{\smile} \sqcap \rho; \rho^{\smile} \quad \pi, \text{add total}$$
$$= \pi; \pi^{\smile} \sqcap \rho; \rho^{\smile} \qquad\qquad\qquad\qquad \text{see above}$$
$$= \mathbb{I}_{\mathbb{R} \otimes \mathbb{R}}.$$

2. The inclusion $\sqsubseteq$ follows immediately by

$$\text{add}; C; \text{add}^{\smile} \sqcap \rho; \rho^{\smile} \sqsubseteq (\pi; C; \pi^{\smile} \sqcup \rho; C; \rho^{\smile}) \sqcap \rho; \rho^{\smile} \qquad \text{Axiom 7.}$$
$$= (\pi; C; \pi^{\smile} \sqcap \rho; \rho^{\smile}) \sqcup (\rho; C; \rho^{\smile} \sqcap \rho; \rho^{\smile})$$
$$= (C \otimes \mathbb{I}_{\mathbb{R}}) \sqcup \rho; (C \sqcap \mathbb{I}_{\mathbb{R}}); \rho^{\smile} \qquad\qquad \text{Lemma 2(3)}$$
$$= C \otimes \mathbb{I}_{\mathbb{R}}. \qquad\qquad\qquad\qquad\qquad \text{Lemma 5(1)}$$

For the converse inclusion we first compute

$$(C \otimes \mathbb{I}_{\mathbb{R}}) \sqcap \text{add}; \text{add}^{\smile} \sqcap \rho; \rho^{\smile} = (C \otimes \mathbb{I}_{\mathbb{R}}) \sqcap \mathbb{I}_{\mathbb{R} \otimes \mathbb{R}} \qquad \text{by (1)}$$
$$= \pi; C; \pi^{\smile} \sqcap \pi; \pi^{\smile} \sqcap \rho; \rho^{\smile}$$
$$= \pi; (C \sqcap \mathbb{I}_{\mathbb{R}}); \pi^{\smile} \sqcap \rho; \rho^{\smile} \qquad \text{Lemma 2(3)}$$
$$= \bot\!\!\!\bot_{\mathbb{R} \otimes \mathbb{R} \mathbb{R} \otimes \mathbb{R}}, \qquad\qquad\qquad \text{Lemma 5(1)}$$

and $\;\;(C \otimes \mathbb{I}_{\mathbb{R}}) \sqcap \text{add}; C^{\smile}; \text{add}^{\smile} \sqcap \rho; \rho^{\smile} \sqsubseteq (C \otimes \mathbb{I}_{\mathbb{R}}) \sqcap (C^{\smile} \otimes \mathbb{I}_{\mathbb{R}}) \;\;$ see above
$$= (C \sqcap C^{\smile}) \otimes \mathbb{I}_{\mathbb{R}} \qquad \text{Lemma 2(3)}$$
$$= \bot\!\!\!\bot_{\mathbb{R} \otimes \mathbb{R} \mathbb{R} \otimes \mathbb{R}}. \qquad\qquad \text{Axiom 2.}$$

This implies

$$C \otimes \mathbb{I}_{\mathbb{R}} = (C \otimes \mathbb{I}_{\mathbb{R}}) \sqcap \text{add}; \mathbb{T}_{\mathbb{R}\mathbb{R}}; \text{add}^{\smile} \qquad\qquad \text{add total}$$
$$= (C \otimes \mathbb{I}_{\mathbb{R}}) \sqcap \text{add}; \mathbb{T}_{\mathbb{R}\mathbb{R}}; \text{add}^{\smile} \sqcap \rho; \rho^{\smile}$$
$$= (C \otimes \mathbb{I}_{\mathbb{R}}) \sqcap \text{add}; (C \sqcup C^{\smile} \sqcup \mathbb{I}_{\mathbb{R}}); \text{add}^{\smile} \sqcap \rho; \rho^{\smile} \quad \text{Axiom 1.}$$
$$= ((C \otimes \mathbb{I}_{\mathbb{R}}) \sqcap \text{add}; C; \text{add}^{\smile} \sqcap \rho; \rho^{\smile})$$
$$\qquad \sqcup ((C \otimes \mathbb{I}_{\mathbb{R}}) \sqcap \text{add}; C^{\smile} \text{add}^{\smile} \sqcap \rho; \rho^{\smile})$$
$$\qquad \sqcup ((C \otimes \mathbb{I}_{\mathbb{R}}) \sqcap \text{add}; \text{add}^{\smile} \sqcap \rho; \rho^{\smile}) \qquad\qquad \text{Lemma 2(3)}$$
$$= (C \otimes \mathbb{I}_{\mathbb{R}}) \sqcap \text{add}; C; \text{add}^{\smile} \sqcap \rho; \rho^{\smile}, \qquad\qquad \text{see above}$$

i.e., $C \otimes \mathbb{I}_{\mathbb{R}} \sqsubseteq \text{add}; C; \text{add}^{\smile} \sqcap \rho; \rho^{\smile}$. $\qquad\qquad\qquad\qquad\qquad$ $\square$

Now, we are ready to show the second main theorem.

Theorem 2. *The relation* $C : \mathbb{R} \to \mathbb{R}$ *is a dense strict linear order.*

Proof. As already mentioned above, it remains to show that C is transitive. We have

$$
\begin{aligned}
&(C \oslash \mathbb{I}_\mathbb{R}); (C \oslash \mathbb{I}_\mathbb{R}) \sqcap \text{add}; \text{add}^\smile \sqcap \rho; \rho^\smile \\
&= (C \oslash \mathbb{I}_\mathbb{R}); (C \oslash \mathbb{I}_\mathbb{R}) \sqcap \mathbb{I}_{\mathbb{R} \otimes \mathbb{R}} && \text{Lemma 7(1)} \\
&\sqsubseteq (C \oslash \mathbb{I}_\mathbb{R}); ((C \oslash \mathbb{I}_\mathbb{R}) \sqcap (C \oslash \mathbb{I}_\mathbb{R})^\smile) \\
&\sqsubseteq (C \oslash \mathbb{I}_\mathbb{R}); (C; \pi^\smile \sqcap C^\smile; \pi^\smile) \\
&= (C \oslash \mathbb{I}_\mathbb{R}); (C \sqcap C^\smile); \pi^\smile && \text{Lemma 2(3)} \\
&= {\perp\!\!\!\perp}_{\mathbb{R} \otimes \mathbb{R} \otimes \mathbb{R}}. && \text{Axiom 2.}
\end{aligned}
$$

From the two computations

$$
\begin{aligned}
&(C \oslash \mathbb{I}_\mathbb{R}); (C \oslash \mathbb{I}_\mathbb{R}) \sqcap \rho; C^\smile; \pi^\smile \\
&\sqsubseteq (C \oslash \mathbb{I}_\mathbb{R}); ((C \oslash \mathbb{I}_\mathbb{R}) \sqcap (C \oslash \mathbb{I}_\mathbb{R})^\smile; \rho; C^\smile; \pi^\smile) \\
&\sqsubseteq (C \oslash \mathbb{I}_\mathbb{R}); ((C \oslash \mathbb{I}_\mathbb{R}) \sqcap C^\smile; \pi^\smile) \\
&\sqsubseteq (C \oslash \mathbb{I}_\mathbb{R}); (C; \pi^\smile \sqcap C^\smile; \pi^\smile) \\
&= (C \oslash \mathbb{I}_\mathbb{R}); (C \sqcap C^\smile); \pi^\smile && \text{Lemma 2(3)} \\
&= {\perp\!\!\!\perp}_{\mathbb{R} \otimes \mathbb{R} \otimes \mathbb{R}}, && \text{Axiom 2.}
\end{aligned}
$$

and

$$
\begin{aligned}
&(C \oslash \mathbb{I}_\mathbb{R}); (C \oslash \mathbb{I}_\mathbb{R}) \sqcap \pi; C^\smile; \rho^\smile \\
&\sqsubseteq ((C \oslash \mathbb{I}_\mathbb{R}) \sqcap \pi; C^\smile; \rho^\smile; (C \oslash \mathbb{I}_\mathbb{R})^\smile); (C \oslash \mathbb{I}_\mathbb{R}) \\
&\sqsubseteq ((C \oslash \mathbb{I}_\mathbb{R}) \sqcap \pi; C^\smile); (C \oslash \mathbb{I}_\mathbb{R}) \\
&\sqsubseteq (\pi; C \sqcap \pi; C^\smile); (C \oslash \mathbb{I}_\mathbb{R}) \\
&= \pi; (C \sqcap C^\smile); (C \oslash \mathbb{I}_\mathbb{R}) && \text{Lemma 2(3)} \\
&= {\perp\!\!\!\perp}_{\mathbb{R} \otimes \mathbb{R} \otimes \mathbb{R}} && \text{Axiom 2.}
\end{aligned}
$$

we conclude that

$$
\begin{aligned}
&(C \oslash \mathbb{I}_\mathbb{R}); (C \oslash \mathbb{I}_\mathbb{R}) \sqcap \text{add}; C^\smile; \text{add}^\smile \sqcap \rho; \rho^\smile \\
&\sqsubseteq (C \oslash \mathbb{I}_\mathbb{R}); (C \oslash \mathbb{I}_\mathbb{R}) \sqcap \text{add}; C^\smile; \text{add}^\smile \\
&= (C \oslash \mathbb{I}_\mathbb{R}); (C \oslash \mathbb{I}_\mathbb{R}) \sqcap \text{swap}; \text{add}; C^\smile; \text{add}^\smile && \text{Lemma 6(1)} \\
&= (C \oslash \mathbb{I}_\mathbb{R}); (C \oslash \mathbb{I}_\mathbb{R}) \sqcap \text{swap}; (\pi; C^\smile; \pi^\smile \sqcup \rho; C^\smile; \rho^\smile) && \text{Axiom 7.} \\
&= ((C \oslash \mathbb{I}_\mathbb{R}); (C \oslash \mathbb{I}_\mathbb{R}) \sqcap \text{swap}; \pi; C^\smile; \pi^\smile) \\
&\quad \sqcup ((C \oslash \mathbb{I}_\mathbb{R}); (C \oslash \mathbb{I}_\mathbb{R}) \sqcap \text{swap}; \rho; C^\smile; \rho^\smile) \\
&= ((C \oslash \mathbb{I}_\mathbb{R}); (C \oslash \mathbb{I}_\mathbb{R}) \sqcap \rho; C^\smile; \pi^\smile) \\
&\quad \sqcup ((C \oslash \mathbb{I}_\mathbb{R}); (C \oslash \mathbb{I}_\mathbb{R}) \sqcap \pi; C^\smile; \rho^\smile) && \text{Lemma 3(2)} \\
&= {\perp\!\!\!\perp}_{\mathbb{R} \otimes \mathbb{R} \otimes \mathbb{R}}. && \text{see above}
\end{aligned}
$$

Using the first and the last property above we obtain

$(C \ominus \mathbb{I}_\mathbb{R}); (C \oslash \mathbb{I}_\mathbb{R})$

$= (C \ominus \mathbb{I}_\mathbb{R}); (C \oslash \mathbb{I}_\mathbb{R}) \sqcap \text{add}; \top_{\mathbb{R}\mathbb{R}}; \text{add}^\smile$ add total

$= (C \ominus \mathbb{I}_\mathbb{R}); (C \oslash \mathbb{I}_\mathbb{R}) \sqcap \text{add}; \top_{\mathbb{R}\mathbb{R}}; \text{add}^\smile \sqcap \rho; \rho^\smile$ $(C \ominus \mathbb{I}_\mathbb{R}); (C \oslash \mathbb{I}_\mathbb{R}) \sqsubseteq \rho; \rho^\smile$

$= (C \ominus \mathbb{I}_\mathbb{R}); (C \oslash \mathbb{I}_\mathbb{R}) \sqcap \text{add}; (C \sqcup C^\smile \sqcup \mathbb{I}_\mathbb{R}); \text{add}^\smile \sqcap \rho; \rho^\smile$ Axiom 1.

$= ((C \ominus \mathbb{I}_\mathbb{R}); (C \oslash \mathbb{I}_\mathbb{R}) \sqcap \text{add}; C; \text{add}^\smile \sqcap \rho; \rho^\smile)$

$\sqcup ((C \ominus \mathbb{I}_\mathbb{R}); (C \oslash \mathbb{I}_\mathbb{R}) \sqcap \text{add}; C^\smile \text{add}^\smile \sqcap \rho; \rho^\smile)$

$\sqcup ((C \ominus \mathbb{I}_\mathbb{R}); (C \oslash \mathbb{I}_\mathbb{R}) \sqcap \text{add}; \text{add}^\smile \sqcap \rho; \rho^\smile)$ Lemma 2(3)

$= (C \ominus \mathbb{I}_\mathbb{R}); (C \oslash \mathbb{I}_\mathbb{R}) \sqcap \text{add}; C; \text{add}^\smile \sqcap \rho; \rho^\smile,$ see above

i.e., $(C \ominus \mathbb{I}_\mathbb{R}); (C \oslash \mathbb{I}_\mathbb{R}) \sqsubseteq \text{add}; C; \text{add}^\smile \sqcap \rho; \rho^\smile$. From this we conclude

$C; C = \pi^\smile; (C \ominus \mathbb{I}_\mathbb{R}); (C \oslash \mathbb{I}_\mathbb{R}); \pi$ Lemma 3(2)

$\sqsubseteq \pi^\smile; (\text{add}; C; \text{add}^\smile \sqcap \rho; \rho^\smile); \pi$ see above

$= \pi^\smile; (C \otimes \mathbb{I}_\mathbb{R}); \pi$ Lemma 7(2)

$= \pi^\smile; \pi; C$ Lemma 3(2)

$\sqsubseteq C.$

This completes the proof.   □

Now, we define the order on a real number object by $E := \mathbb{I}_\mathbb{R} \sqcup C$. Our final theorem shows that add is monotone with respect to this order, i.e., add together with neg and 1 forms a linearly ordered group. But first we need the following lemma.

Lemma 8. Let $C_1 : X \to X$ and $C_2 : Y \to Y$ be strict-orderings, $E_1 = \mathbb{I}_X \sqcup C_1$ and $E_2 = \mathbb{I}_Y \sqcup C_2$ the induced order relations, and $f : X \times X \to Y$ a commutative map that is strictly monotone in its first parameter, i.e., $(C_1 \otimes \mathbb{I}_X); f \sqsubseteq f; C_2$. Then we have:

1. $(\mathbb{I}_X \otimes C_1); f \sqsubseteq f; C_2$, i.e., f is strictly monotone in its second parameter.
2. $(C_1 \otimes C_1); f \sqsubseteq f; C_2$, i.e., f is strictly monotone in both parameters.
3. $(E_1 \otimes E_1); f \sqsubseteq f; E_2$, i.e., f is monotone in both parameters.

Proof. 1. This follows immediately from the strict monotonicity in the first parameter and the commutativity of f.
2. Also this follows immediately from (1) and the assumption by splitting $(C_1 \otimes C_1)$ into $(\mathbb{I}_X \otimes C_1); (C1 \otimes \mathbb{I}_X)$.

3. From

$(E_1 \otimes E_1); f$

$= ((\mathbb{I}_X \sqcup C_1) \otimes (\mathbb{I}_X \sqcup C_1)); f$

$= ((\mathbb{I}_X \otimes \mathbb{I}_X) \sqcup (\mathbb{I}_X \otimes C_1) \sqcup (C_1 \otimes \mathbb{I}_X) \sqcup (C_1 \otimes C_1)); f$ Def. $\otimes$

$= f \sqcup (\mathbb{I}_X \otimes C_1); f \sqcup (C_1 \otimes \mathbb{I}_X); f \sqcup (C_1 \otimes C_1); f$

$\sqsubseteq f \sqcup f; C_2$ by (1),(2) and assump.

$= f; (\mathbb{I}_Y \sqcup C_2)$

$= f; E_2$

we obtain the assertion.   □

We are now ready to provide the final theorem.

Theorem 3. *We have the following:*

1. add *is strictly monotone in each parameter.*
2. add *is strictly monotone.*
3. add *is monotone.*

Proof. We only show that add is strictly monotone in its first parameter, i.e., that $(C \otimes \mathbb{I}_\mathbb{R}); \text{add} \sqsubseteq \text{add}; C$. Lemma 8 all other properties of this theorem will follow. From Lemma 7(2) we immediately conclude $C \otimes \mathbb{I}_\mathbb{R} \sqsubseteq \text{add}; C; \text{add}^\smile$ from which the assertion follows by Lemma 2(1).   □

4   Conclusion and Future Work

The current paper is just the beginning of the study of real number objects as defined here. We have shown that the addition of a real number objects forms a densely and linearly ordered abelian group. It remains to show that this group is also Archimedean. For showing this property one first has to define the operation of summing up n copies of an element a, i.e., a map $\mathbb{N} \times \mathbb{R} \to \mathbb{R}$. This requires either an external object of the natural numbers or to identify the natural numbers within the real number object.

Another paper will concentrate on the multiplicative group of a real number object. The definition of the multiplication operation requires the Archimedean property and shows that the multiplication of natural number has a unique extension in the real numbers.

Last but not least, we would like to study the topology induced by the order structure on a real number object using the relation algebraic approach to topological spaces [6].

References

1. Freyd, P., Scedrov, A.: Categories, Allegories. North-Holland Mathematical Library, Vol. 39, North-Holland, Amsterdam (1990)
2. Olivier J.P., Serrato D.: Catégories de Dedekind. Morphismes dans les Catégories de Schröder. C.R. Acad. Sci. Paris **290**, 939–941 (1980)
3. Olivier, J.P., Serrato, D.: Squares and Rectangles in Relational Categories - Three Cases: Semilattice, Distributive lattice and Boolean Non-unitary. Fuzzy Sets Syst. **72**(2), 167–178 (1995)
4. Schmidt, G., Ströhlein, T.: Relations and graphs: discrete mathematics for computer scientists. In: EATCS Monographs on Theoretical Computer Science, Springer, Heidelberg (1993). https://doi.org/10.1007/978-3-642-77968-8
5. Schmidt, G.: Relational mathematics. In: Encyplopedia of Mathematics and its Applications, vol. 132, Cambridge University Press, Cambridge (2011)
6. Schmidt, Gunther, Winter, Michael: Closures and their Aumann contacts. In: Relational Topology. LNM, vol. 2208, pp. 113–124. Springer, Cham (2018). https://doi.org/10.1007/978-3-319-74451-3_7
7. Tarski, A.: Introduction to Logic. Oxford University Press, Oxford (1941)
8. Winter, M.: Strukturtheorie heterogener Relationenalgebren mit Anwendung auf Nichtdetermismus in Programmiersprachen. Dissertationsverlag NG Kopierladen GmbH, München (1998)
9. Winter M.: Goguen categories - a categorical approach to L-fuzzy relations. In: Trends in Logic, vol. 25, Springer, Heidelberg (2007). https://doi.org/10.1007/978-1-4020-6164-6
10. Winter, M.: Arrow categories. Fuzzy Sets Syst. **160**, 2893–2909 (2009)
11. Winter, M.: Fixed point operators in Heyting categories. Part I - internal fixed point theorem and calculus. J. Pure Appl. Algebra (2022)

Author Index

R. Glück et al. (Eds.): RAMiCS 2023, LNCS 13896, p. 293, 2023.
https://doi.org/10.1007/978-3-031-28083-2